AF560052

OF FIBRE AND LOOM

OF FIBRE AND LOOM

The Indian Tradition

Lotika Varadarajan

Krishna Amin-Patel

NATIONAL INSTITUTE OF DESIGN
AHMEDABAD

MANOHAR
with support from
Development Commissioner (Handlooms)
Ministry of Textiles, Government of India

2008

First published 2008

Jointly published by
National Institute of Design
Paldi, Ahmedabad 380 007
Website: www.nid.edu

and

Ajay Kumar Jain *for*
Manohar Publishers & Distributors
4753/23 Ansari Road, Daryaganj, New Delhi 110002

With support from
Development Commissioner (Handlooms)
Ministry of Textiles, Government of India

ISBN: 81-7304-774-X

Designed by: Visual Vibe (www.visualvibe.net)
Processed by: Thomson Press, New Delhi
Printed by: Thomson Press, New Delhi

Contents

Foreword

THE REFINEMENTS in loom technology have been recorded along with the changes brought about by end use requirement of the products. Invaluable detailing of the costumes provided along with the historical background will prove to be a record for posterity of styles of draping and sewing, which have over the years, gone out of use. The traditional knowledge system of spinning, weaving, dyeing, and loom development are anchored in our roots as can be noted from the history of the trade in Indian textiles. Over time, Indian textiles influenced the design and development of textiles in the rest of the world. This work will be of immense interest and value to students of design, teachers, researchers, art historians, textile/art collectors, museums, ethnographers and also to the textile trade and exporters.

NID has invested a great deal of time and funding over the years to support the painstaking research and arduous work that has gone into this book. Several faculty and student projects have generated a large number of documents in this area which have been an invaluable resource to the two scholars who have authored the book. This backup from NID along with the intensive and wide ranging independent research by the authors has produced a comprehensive and pioneering work, based on rigorous field study and analysis.

The publication of this landmark work augments the continuity in the search for new interpretations in textiles and weaving.

The Office of the Development Commissioner for Handlooms has supported this project as a major initiative in the field of documentation for the sector. We are happy that the excellent work done by NID in bringing out this publication will strengthen our close relationship and see many such projects coming to fruition in the future.

B.K. Sinha
Development Commissioner (Handlooms)
Ministry of Textiles
Govt. of India

Preface

THIS PIONEERING scholarly work has evolved over a period of time through the painstaking efforts of Dr (Mrs) Lotika Varadarajan, well-known ethno-historian and scholar, and Ms Krishna Amin-Patel, a talented textile designer and educator. Both Dr (Mrs) Varadarajan and Ms Patel have been closely associated with NID's academic activities and research for several years. Dr (Mrs) Varadarajan has been a visiting faculty at the Institute. Ms Patel has had a long association with the Institute's Textile Design Department as a core faculty member with deep interest and rich knowledge in the area of traditional textile design and technology. The collaborative efforts and dedicated work of these two scholars have shaped the body of knowledge and experiences into this significant book, *Of Fibre and Loom: The Indian Tradition.*

While there are numerous books related to Indian textile design, technology and traditions, this volume will command a unique position since its essential focus is on an historical as well as contemporary development of 'appropriate technology infrastructure' for the versatile and diverse handloom sector of India. The volume will prove to be an essential reference book for a wide range of scholars as well as students. It will also provide a very detailed overview of the various developments that have taken place as deduced from various indigenous looms, accessories and equipments used by the vast and dispersed weaving and craft communities of India.

NID's Research & Publications Activity Chairperson, S.K. Khanna and earlier Chairperson of the same department, Ms Aditi Ranjan and the current Head, Publications, Ms Shilpa Das have provided excellent support over the years to the authors and co-publishers to facilitate the creation, development and presentation of this book.

I would like to place on record NID's sincere thanks to Mr Subas Pani, IAS the earlier Development Commissioner (Handlooms) and Mr B.K. Sinha, IAS the current incumbent of DC (H) Office for extending necessary support to this important publication by providing partial funding assistance and extending other support.

I would also like to thank Manohar Publishers for their painstaking efforts in designing, printing and publishing this work as co-publishers along with NID.

28 May 2007

Dr Darlie O. Koshy
Director
National Institute of Design
Ahmedabad, India

Introduction

THIS PROJECT WAS INITIALLY conceived as one in which the focus would be on attempting to identify the core of pre-Islamic weaving tradition in India. While Indian loom tradition has always been varied, it was felt that since Islamicized taste in textiles and costumes constituted a specific mix, the advent of demand created by the Islamicized élite could be taken as a cultural marker. The adaptation of the jacquard and the fly shuttle to suit Indian requirements at later dates could be considered in the same vein. It was important for us to understand the pre-Islamic mix so that we could analyse developments of the loom in later periods. Turning to fibre, it is known that much has been written on the subject of Indian cotton. It was, therefore, felt that Indian silk weaving tradition also solicited attention. This could not be done without analysing indigenous sources of silk. The wool weaving traditions of Kashmir, Kullu and Kinnaur were brought within the ambit of this research because initial interest had focused on the double interlock tapestry weave of *kani paśmīnā* and the dovetailing procedures present in the ornamental Kullu–Kinnaur woollen product. Having adopted the approach that loom, fabric and costume were to be studied as cross-cultural artefacts it became very difficult to define thresholds of liminality. The geographical confines accorded to the term Deccan has been that of the entire stretch of peninsular India south of the Narmada rather than the narrower one restricting the meaning to the plateau region between the rivers Narmada and Krishna. The field generated data was gathered essentially over the 1980s and the early 1990s.

Looms that have been selected show a progression from the simple to the complex. Complex weave structures however, can be achieved on the simplest of looms as is evident from the products of the loin loom/body tensioned loom in north-east India. Looms such as the multiple shaft and pedal loom of Gujarat have no equivalence in other culture horizons.

The Indian loom is to be read as a cross-cultural artefact. Additionally, the end usage of the product also determines loom procedures and techniques. Costume and styles in living cannot, therefore, be abstracted from the content of the study of the Indian loom. Given the multi-cultural contours of Indian society, attempts to introduce standardization could prove self-defeating. The transversal exposed end section of the sari could be called *pallav* in Gujarat and *ānchal* in Bengal. However, the term *pallu* is the most widespread in common parlance and has been retained in the text. Standardization of units of measurement has not been possible. Additionally, many different linguistic sectors have been scanned in course of investigation and it would be an arduous task to transcribe each linguistic unit in terms of the Roman script specific to it for purposes of transliteration. Moreover, since the technical analysis is essentially based on field generated data, local terminology has been generally used free of the constraint of an overall standardization. Diacriticals have been used essentially for Sanskrit and Sanskrit derived words. For Chinese terms used, concordance has, on the whole, been closer to the Wade rather than the Pin Yin system.

As material has been drawn from diverse sources cutting through barriers of chronology, there is also an absence of uniformity in nomenclature of place names used. This has also posed a problem in the preparation of the maps. One of the most difficult tasks has been that of attempting to synthesize information drawn from a variety of sources in which the element of time has played a volatile role. Given the highly developed oral tradition of India and the distinct area of non-verbal methods of transmission of skill, the ethnological present can simultaneously constitute the historical past. Such sources have been used despite inherent methodological problems. The decision not to frame a glossary has been consciously taken. The emphasis in this book is on processes and procedures and these defy simple alphabetical listing. However, since the text defines terms used, judicious usage of the index would serve to fill the gap of glossary.

The book embodies an effort to synthesize the discipline of design with that of ethno-history. This has not been an easy task. It was only after much hard work that the delineation of socio-cultural narrative could be made to cohere to the hard data of loom analysis in a contemporary context. This interdisciplinary approach has made it possible to deal with synergistic developments which may otherwise have eluded the net of enquiry. A case which can be cited in this context is that of the Banaras loom. This artefact has shown steady technological progression from *gatthuā* to *jāla* and thereafter to the adaptation of the jacquard mechanism. However, innovation has been carefully grafted and the Banaras weaver sees little justification in discarding the *paggiā* while adopting the jacquard.

The looms of India richly portray the changing context of India's past. Loom and fibre have sustained the eco-friendly Khadi movement. Given the mastery of the humble Indian spinner and weaver over natural fibres such as cotton, silk, and *paśm*, among others, Indian weaving tradition has played a significant role in the contouring of indigenous knowledge systems. It has also shaped methods of technological refinement and transmission of skill. At the same time such processes help to define the parameters within which factors such as creativity and articulation of innovation have operated. Present day modernization should not seek to break this chain of transmission. It may also serve to point the direction to be taken towards the path of sustainable development leading to a less painfully fractured future.

Lotika Varadarajan
Krishna Amin-Patel

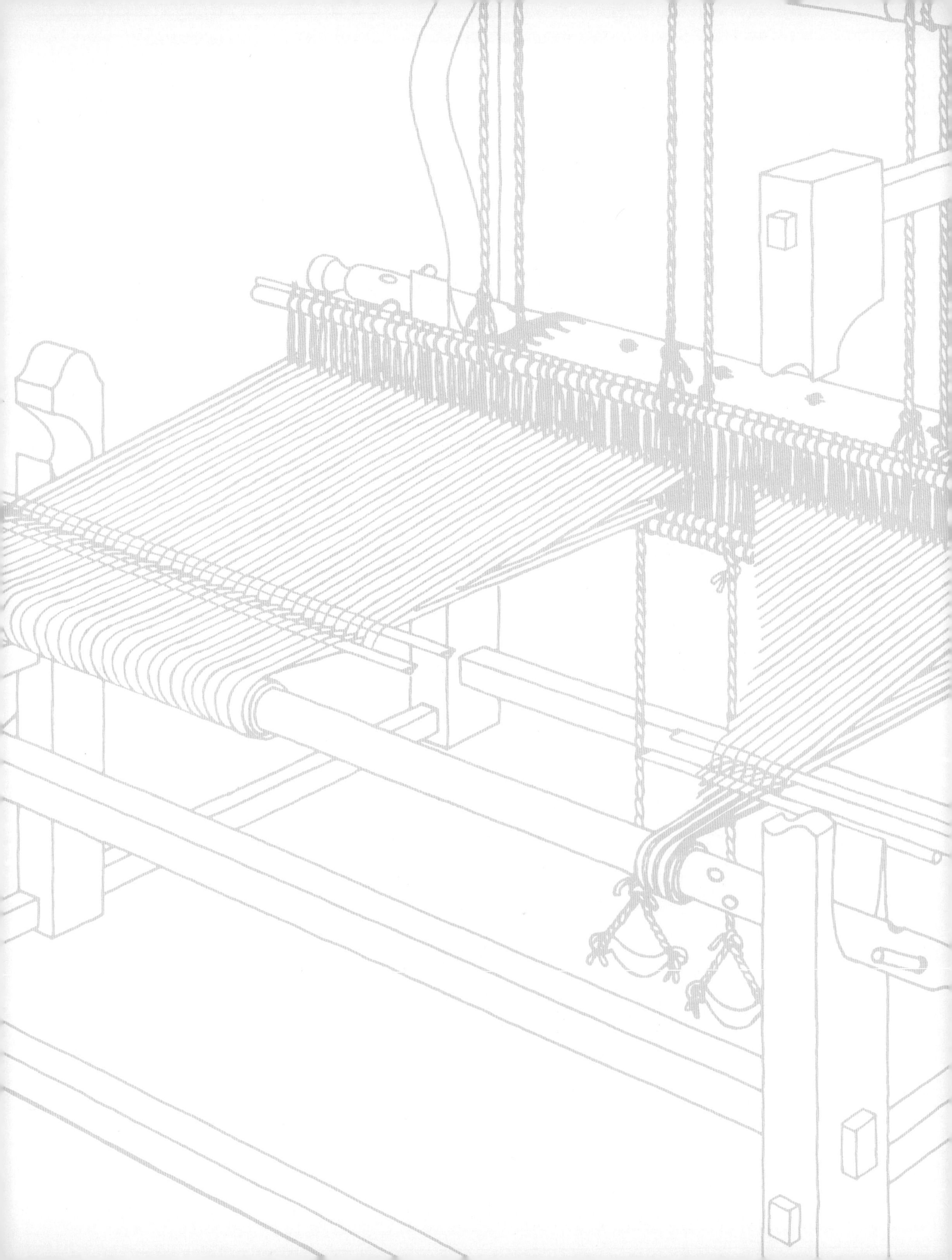

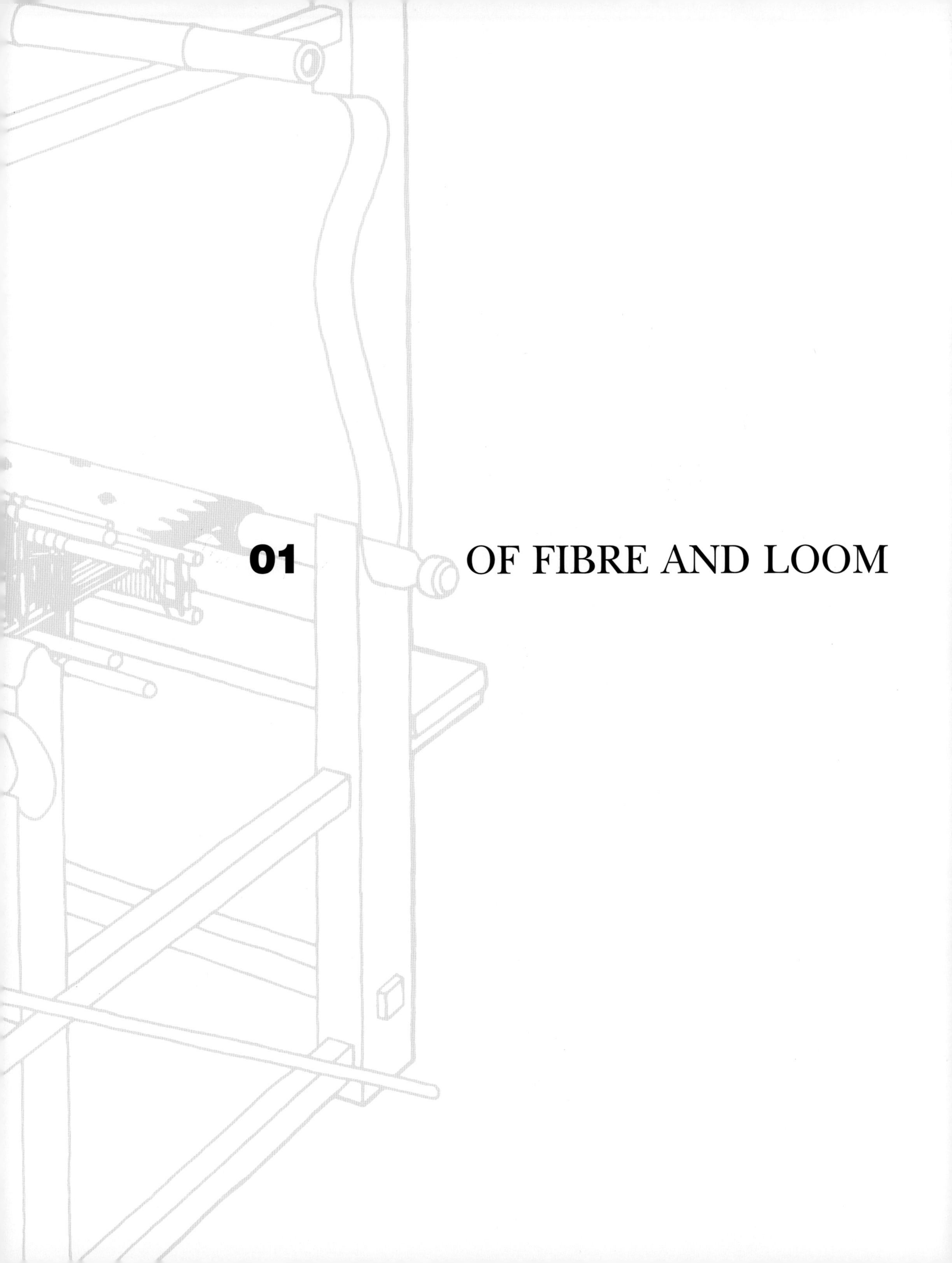

01 OF FIBRE AND LOOM

01 Of Fibre and Loom

Historical Development and Early Indications

The subject of this monograph is Indian loom typology and the fibres used for weaving with special emphasis on silk. It is curious that while weaving traditions of Africa, West Asia, East and South-East Asia have received considerable scholarly attention, that of South Asia has elicited little interest. This has provided additional incentive to work in this field.

Apart from aspects such as loom, weave and fibre, the meaning assigned to Indian tradition merits some exposition. The connotation of the term Indian extends to the subcontinent in its entirety. It does not include Myanmar (Burma) and Sri Lanka (Ceylon) subsumed within the term South Asia. The meaning assigned to tradition in the title of this work is dual in nature. At one level the cutoff date is the period when hand production was replaced by other forms of power-driven processes. At another, Islamic influences have proved very potent and one more parameter relates to the introduction of Islamic loom usages and procedures. The terms Hindu and Islamic are intended as typological parameters and are by no means value loaded. Tradition, as related to silk weaving, is taken in the sense of pre-Islamic repertoires of skill and design linked to the absence of industrial mechanization. The cutoff points are reached subsequent to the introduction of the *jāla* in the north and the evolution of the *aḍai* in the south. However, because of the aspect of continuity in loom tradition, the dobby and the jacquard have not been altogether excluded from the purview of this study. Indeed, as neither of these two mechanisms can be accommodated within the concept of non-human power-driven machines, the clash of cultures resulting from the latter form of mechanization has been absent. This has made assimilation within the traditional mix of technology feasible.

The term fibre used in the heading covers all fibres traditionally used in weaving on the subcontinent. The aspect of loom typology will be taken up first because it is at this level that the contextual subliminality of the loom in India comes into play.

The Loom

Dorothy Burnham has defined the loom as 'any device for weaving on which the warp may be arranged and openings for the passage of the weft formed through it by a shedding mechanism' (Hoffman, 1977: 7). Indian loom typology follows the contours of the subcontinent rather than present day political boundaries. It is accepted that the origin of loom woven textiles may generally be dated to the Neolithic period (Broudy, 1979: 14). This statement is valid for a number of culture horizons.

Although spindle whorls of terracotta, shell and faience have been found in archaeological sites dating back to the Harappan period, in India, looms or loom parts are noticeable by their absence (see fig. 1.1). This is perhaps not surprising since the base material must have been wood. Bronze loom parts as

Spindle and whorl. Third century BC. Excavated at Kodumanal. Courtesy: Department of Archaeology, Tamil University, Thanjavur. (Fig. 1.1)

well as depictions of foot-braced looms on a bronze cowrie container have been found in bronze–iron age burial sites in Yunan Province, China, dated between 206 BC and AD 8 (Vollmer, 1977: 78–89). Since the dominant tradition in India has been cremation of the dead there are few material remains, which could serve to indicate standards attained in the past.

While India may be lacking in material evidence she is richly endowed with a vibrant oral tradition. In those parts of India where the colonial experience has been minimal it is possible to experience, even today, the tenor and tempo of a way of life which may appear anachronistic to sophisticated dwellers of metropolitan cities. It is, therefore, possible within the Indian framework to use such ethnographic material to project some form of reconstruction of the past. The validity of such an exercise would, to a large extent, depend upon the kind of questions which are being posed. Two alternative approaches have been followed in this book. While the distilling of earlier technology has been the essential problem area in the historical overview, documentation of existing work procedures has been followed in the technical analyses. Lack of a sufficiently wide database has posed a serious constraint in this latter area. Synergistic work procedures have been included provided that the loom described can still be included in the present-day handicraft sector.

The dominant Indian loom is the horizontal counter balance pit treadle (see fig. 4.45) rather than the conventional frame loom. The ethnological distinction between pit (see fig. 4.71) frame (fig. 4.7) and body tension looms (fig. 2.4) needs to be maintained. In the north-east, earlier known as the province of Assam, the frame loom is associated with silk weaving in Sualkuchi, 32 km. from Guwahati, and cotton weaving with Nowgong, 121 km. from Guwahati. With regard to Sualkuchi, major developments are traced to *c.* AD 1620 when it was taken over by the Muslim rulers of Dacca. At this time, a Hindu zamindar from East Bengal, Satrajit, was brought to Sualkuchi, and he persuaded weavers from other parts of the state to settle there (*Selected Handicrafts of Assam*, 1961: 41). It is not clear if the Nawab of Dacca introduced any innovation or whether the frame loom was already in use by the weavers brought in from other parts. Ethnic groups such as the Garos and Khasis show a predilection for the frame loom, while it is also utilized in the weaving of items such as the *moirangphee* (see fig. 4.7) in Manipur (Mohanty, 1984: 254–5, 259). In Manipur, there is an association between the Bodos, a Tibeto-Burmese group affiliated to the Garos, and the frame loom. The term used by the Meitei, a Kuki-Chin group, for this loom is *yongkham* in contradistinction to *khwang*, the term used for the loin loom.

Austronesian and Austroasiatic Connotations

The existence of frame and body tension looms in this region is to be viewed against the broader perspective of racial movements and linguistic interchange in this part of the world. Starting approximately from 3000 BC there appears to have been a major linguistic diffusion bringing to India yet another gene source of cultural diversity. This arose from the bifurcation of Austro-Thai, having its home base in South China, into the Austroasiatic and the Austronesian language segments, both of which proved to be strongly diffusionist (Adelaar, 1989: 35). The Austroasiatic family comprises about 150 languages and is made up of two branches, Munda and Mon-Khmer. The Munda family, divided into ten languages and two branches, the northern and southern, is spoken in north-east India. The Tibeto-Burman group of languages in this region has been influenced by the Munda group, and, therefore, it is often extremely difficult to demonstrate specific characteristics in material culture. The Bodo, Naga, Kuki-Chin all fall within the Tibeto-Burman grouping (Singh, 1993: 87). The most important among the Munda family of languages are Santali and Mundarī. Mon-Khmer languages are spoken throughout much of South-East Asia, mainly in Laos, Vietnam, Kampuchea, Thailand, Myanmar, the Malay Peninsula, Meghalaya (Khasi) and the Nicobar Islands. This group is divided into three branches, Northern (which includes Khasi), Eastern (which includes Khmer) and Southern (which includes the Nicobari group) (Merrit Ruhlen, 1987: 148–9, 151, 156; Bellwood, 1985: 2–3, 162, 269; Bellwood, 1989: 6, 25–6).

In present times the Austronesian family of languages is distributed from Madagascar in the west to Easter Island in the east. All Austronesian languages are derived from one common ancestral language, Proto-austronesian. This has been located in Taiwan dating to *c.* 3000 BC. It moved into Luzon in the Philippines and Proto-austronesian split into the Formosan and Malayo-Polynesian subgroups. Around 2500 BC, there was further expansion into south Philippines, Borneo, and, through the Moluccas, into Timor. The movement into Oceania started *c.* 1500 BC. Peter Bellwood, deriving his conclusions from the linguistic work of Robert Blust, divides this dispersal into areas such as central Malayo-Polynesian Austronesian which led through the Moluccas and Lesser Sundas; the Eastern Malayo-Polynesian and Oceanic Austronesian which could be traced via Halmahera, north coastal New Guinea into Oceania, and the Western Malayo-Polynesian Austronesian. The Western Malayo-Polynesian languages expanded through the Philippines to Sulawesi, Borneo, Java and then doubled back to Sumatra, Malaya and Vietnam (Bellwood, 1985: 108 ff; Bellwood, 1987: 208; Kwang-chi Chang et al., 1964: 366; Adelaar, 1989: 35).

Apart from this linguistic manifestation, Austronesian and Austroasiatic culture spheres have also been associated with ethnological artefacts such as outrigger boats and material objects such as specific kinds of pottery, the beginnings of which have been dated by Peter Bellwood to *c.* 3000 BC.[1] The shuttle used in weaving may be another case in point (see figs. 1.2, 1.3). Bronze Dong Son drums which spread to the

Shuttles used in three shuttle weaving of *moirangphee*, Manipur, Craft Museum, New Delhi. (Fig. 1.2)

Northern Thai loom. Note form of shuttle. Courtesy: National Museum, Chiengmai, Thailand. (Fig. 1.3)

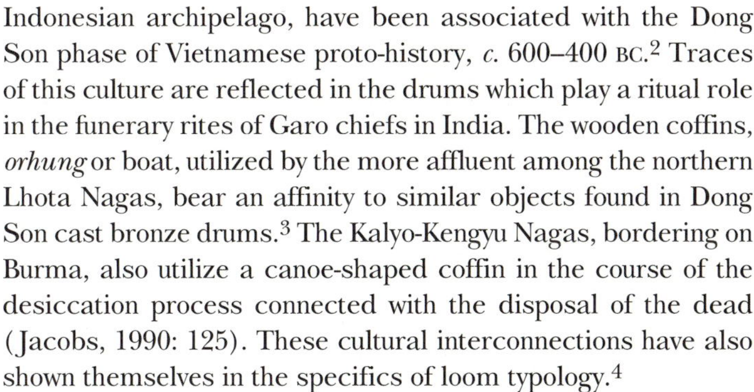

Indonesian archipelago, have been associated with the Dong Son phase of Vietnamese proto-history, *c.* 600–400 BC.[2] Traces of this culture are reflected in the drums which play a ritual role in the funerary rites of Garo chiefs in India. The wooden coffins, *orhung* or boat, utilized by the more affluent among the northern Lhota Nagas, bear an affinity to similar objects found in Dong Son cast bronze drums.[3] The Kalyo-Kengyu Nagas, bordering on Burma, also utilize a canoe-shaped coffin in the course of the desiccation process connected with the disposal of the dead (Jacobs, 1990: 125). These cultural interconnections have also shown themselves in the specifics of loom typology.[4]

Around 2000 BC, the loom, through its parts, *balija*, heddle stick (fig. 2.15) *belera*, beater, *balila*, lease rod (fig 1.4), were known in the ancestral language of the Malay, Batak and north Luzon languages. It is to be borne in mind that the illustrations relate to the typology of the loom accessory and are not intended to be taken as historically dated artefacts. The loom is assumed to be a body tension loom (Blust, 1976: 34, 36, 37). Peter Bellwood describes the language as Proto-Western-Malayo-Polynesian (Bellwood, 1985: 108, 152). The north-eastern region in India is the classic home of the body tension loom and this is also the area with the greatest ecological and cultural affinity with South-East Asia.

Lease/pattern rods. Nuapatna loom. Craft Museum, New Delhi. (Fig. 1.4)

Eastern Looms

The eastern looms with a bearing on Indian prototypes are partly historical and partly ethnological. The Meo/Hmong of northern Thailand, whose language falls within the domain of the Sino-Tibetan grouping, use a body tension one pedal loom which combines within itself some features of the frame loom (Fraser Lu, 1988: 35, fig. 33). In China, the loom known under its Japanese appellation, *izaribata,* has some characteristics of the Hmong loom. The Han stone relief depiction of a loom dated *c.* 206 BC–AD 220, shows a two treadle frame loom (Broudy, 1979: 111–16, figs. 6.17, 6.18). John Becker postulates that the most acceptable reconstruction of the Han loom on the basis of Han period tomb relief carving is one with a solid frame, reedless, with two beams, a slanting warp, one shed stick, one heddle shaft and two treadles (Becker, 1981: 22–3). This invites a comparison with the *paṭolā* loom. The *paṭolā* loom is without a frame and treadles but has a heddle rod and two beams. The loom would have a shed and counter shed. Such a loom would probably have only been used for weaving tabby (Becker, 1981: 22–3).

For figured silks presumably, a larger loom was used. It would have had a horizontal warp, two shafts and two treadles (see figs. 1.5, 1.6). Sufficient space would have been provided for the placement of heddle rods. Loom usage would reflect convention in ornamentation whereby half the pattern unit would be reflected in an extended mirror image, the axis being horizontal in nature.[5] The description of a loom equipped to deal with silk fibre given by Albers, conforms to this projection by John Becker.[6] Extant loom usages in South-East Asia may provide evidence as to how the Han frame loom could be linked to the *izaribata* loom.

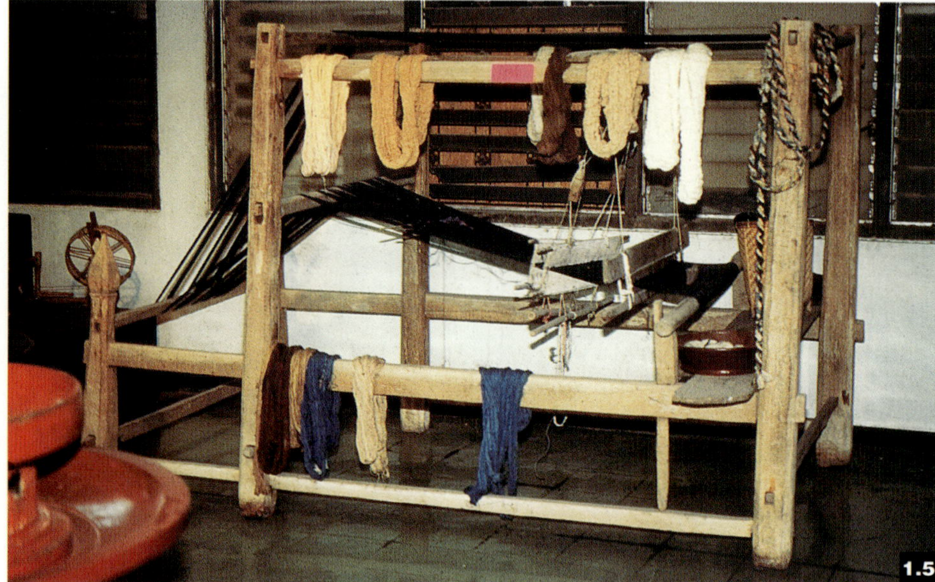

Northern Thai loom. Courtesy: Teachers' Training Centre, Chiengmai. (Fig. 1.5)

Treadles. Northern Thai loom. Courtesy: National Museum, Chiengmai, Thailand. (Fig. 1.6)

Northern Thai loom. Detail. Heddles for border weaving. Courtesy: National Museum, Chiengmai, Thailand. (Fig. 1.7)

Sonket loom. Courtesy: National Museum, Nakorn Si Thammarat, Thailand. (Fig. 1.8)

The loom used by the Tibeto-Burman Akha in north-west Thailand is a frame loom but of a rudimentary order possessing a cloth bar at one end, a stick at the other around which the warp length is wound, two heddle rods and two pedals. The weaver stands on one side of the loom and manipulates the treadles by foot. The heddle bars can be moved as desired along its supporting bars above the loom. The average width of cloth woven is 20 cm. (Fraser-Lu, 1988: 37; Gittinger: 1977, 58). Other frame looms which appear to be of high antiquity are the Thai looms (cf. figs. 1.3, 1.5, 1.7) and the *sonket* looms of Kuala Trengganu and Kota Bahru in Malaysia (figs. 1.8, 1.9, 1.10), described by D. de Jonghe and M. Tavernier.[7]

These looms show a progression in the range of patterns which could be achieved without necessitating any fundamental change in technology. The Thai loom in which heddle rods for patterning are introduced into extended loops of half heddles (fig. 1.11), documented in Bangkok, is postulated as bearing an affinity to the Han monochrome pattern loom. During weaving of the pattern, here also, the order in the pattern rods is reversed after half the motif has been woven. It is of interest that this loom mechanism is suitable only for designs with a horizontal axis in symmetry (de Jonghe, Tavernier: 1979, 39, 41, 49).

The simplest Thai loom has two heddle shafts and a reed beater. For complex weaves the number of heddle shafts can go up to sixteen. For twill weaves there can be between three to eight heddle shafts. In north-east Thailand, an area with considerable ethnological affinity with the Shan region of Burma and the states of north-east India, extra weft patterns sometimes involve the use of two shafts for plain weave binding. In addition, the warp is threaded through an extra set of heddles which, when lifted in sequential order, creates an extra weft pattern (cf. figs. 1.3 and 1.7).

The oldest method in this region was the manual insertion of the supplementary weft (fig. 4.5). The method was accelerated over time by the insertion of shed sticks (cf. fig. 1.4), or supplementary string heddles, into the warp in sequential order. This developed into a more complex loom (see fig. 1.11) found in some villages of north and north-east Thailand. Here the sticks are interwoven in a string harness which is set at right angles to the warp tensioned between the upper and lower extremities of the loom frame. The intersection of the ground weave is controlled by two shafts which the weaver activates by treadles. Pattern selection and the creation of the shed for the supplementary weft is operated by one or two assistants positioned at the far end of the loom. However, if the loom

Sonket loom. Courtesy: National Museum, Nakorn Si Thammarat, Thailand. Detail. (Fig. 1.9)

Kota bahru loom for *sonket* weaving. Side view. Courtesy: National Museum, Nakorn Si Thammarat, Thailand. (Fig. 1.10)

Extra weft weaving with lashes. Photographed at an exhibition in Bangkok. (Fig. 1.11)

frame is small and the harness within reach of the weaver, she may undertake the whole process on her own.

To set up the loom, the pattern to be woven can be first picked up row by row into a vertical string harness using bamboo shed sticks. The strings of the harness are connected to individual warp threads by string loops. As the pattern commences the weaver's assistant moves the first shed stick down towards the warp. The sword is inserted into the shed created by the lowered stick and the supplementary weft yarn. While the shed stick remains in the harness above the warp, another stick is inserted into the harness below the warp in the same shed. The upper shed stick can then be removed. One pick of the plain ground weave is then accomplished. When only one stick is found remaining in the upper part of harness, the process is reversed. It is in this way that the mirror image is woven. Such harnesses can be stored if desired by the weaver (de Jonghe, Tavernier, 1979: 39, 41, 49).

The Malay looms show a further developmental progression over the Thai looms as pattern heddle rods are replaced by lashes, which are loops tied to a cross thread. In Kuala Trengganu the lashes are tied directly on the warp ends. In the Kota Bahru loom, which suggests parallels with the south Indian *aḍai* loom, we find loops of lashes drawn in bunches along cords aligned longitudinally (see figs. 1.8, 1.10, 1.12). Such an arrangement imposes no constraints of horizontal or vertical symmetry in design.

The authors quoted do not regard the Thai and Malay *sonket* looms as being mutually exclusive. Rather, they are viewed in terms of progressive adaptation to advancement in technology and the Kota Bahru type could well have lent itself to the manufacture of Han polychrome patterned silks.[8] The existence of body tension and frame looms in north-eastern India can be accommodated within a broader context of loom tradition. The link between the *aḍai* of peninsular India, a composite loom, and the classic Malay *sonket* loom poses greater problems. However, since the south-eastern coastal belt of peninsular India had early linkages with South-East Asian trade networks, it is possible that elements of the tradition represented in the Malay loom should have been absorbed by south Indian weavers (Ray, 1990: 2). Vijaya Ramaswamy has dated the first inscriptional reference to the south Indian pattern loom, the *achchutari*, to the year AD 1001 during the reign of Rājarāja I (Ramaswamy, 1980: 232–3).

Aḍai loom. Yellow twine harness with glass beads for extra warp *zarī*. Courtesy: Weavers Services Centre, Bangalore. (Fig. 1.12)

Early Texts and Technological Inferences

Before turning to African and West Asian influences on Indian loom technology, it is of interest to take cognizance of Vedic references to the loom. The term Vedic, refers to the period 1500–800 BC, a relatively late period as compared to dates for looms in Egypt. Vedic literature in itself has serious limitations as a category of historical source material, and, for this purpose, it has to be taken in conjunction with later sources and commentaries. As such, the element of interpolation cannot be ruled out particularly in an area such as that of technology. W. Rau traces words which were in existence for warp—*tata/tantu*, and weft—*ota*, the operation of setting up the warp on the loom, warp ends—*tatavaḥ*, weft pick *oṭavaḥ*, shuttle *tasarāṇī* dual movement of weft back and forth across the warp *praveāve* (see also Macdonnel, Keith, 1958: 123–4, 298–9, 302). It would also appear that warp ends were separated by some kind of heddle rod mechanism, that there was a lease stick, the *Gāyatrī* of the text, that a beating movement took place and that an implement, similar to a sword, was used to maintain the shed once it had been formed. The loom could be dismantled and set up as required (Rau, 1970: 17, 19–21). This listing provides clear evidence of loom structure but yields sparse information with regard to loom typology.

The loom tradition westwards of India will now be analysed to determine the nature of influences emanating from this cultural sphere. Two types of

West Asian/North African looms impinge on Indian tradition: the horizontal ground loom used by nomads and the pit treadle loom. The former is well adapted to nomadic patterns of life as it can be easily dismantled and reassembled. The latter can also be dismantled and reassembled but weaving is contingent upon the presence of a pit.

Horizontal Ground Loom and the *Pañja Darī*

The earliest representation of a horizontal ground loom is to be found depicted on a pottery bowl found in an archaeological context at Badari, Central Egypt, and dated *c.* 5000 BC. It is currently housed at the Petrie Museum, University College, London. According to E. Broudy this loom possessed two beams pegged above the ground, a sword beater, a heddle and shed rod (Picard-Schmitter, 1967: 13–17; Broudy, 1979: 14; Hecht, 1989: 9). The Vedic loom discussed by W. Rau may have been a similar contrivance. By the time of the First Dynasty, 3250–2800 BC, very fine quality linen was being woven in Egypt although until *c.* 2500 BC the only weave appears to have been tabby.

The horizontal fixed heddle ground loom with its characteristic beating comb is a loom usually worked by women and found distributed over a vast tract of territory stretching from Central Asia to the Atlantic edge of Africa. It was possibly developed by nomads although now it is used by settled communities. In the absence of beams, the weaver sits on the woven area, advancing his own position as also that of the heddle as weaving progresses.[9]

The Indian tradition of *pañja* loom weaving is in evidence in Achalpur in the Vidarbha district (fig. 1.13), as also in Haryana, Punjab, Mirzapur, and Fatehpur Sikri (cf. fig. 2.30, see fig. 1.14). These looms appear to bear a close relation to the horizontal fixed heddle ground loom inclusive of the positioning of the weaver on the loom. The product woven is the *pañja darī*. *Darī* is a flat weave floor covering used extensively in India. R.L. Turner (Turner, 1966: item 6194, 354), speculates that the root is *dṛ*/to tear, or *dal*/to grind or split. The adjective forms were *dārita* or *darīka*, a split cane mat. The term *dalita* was derived from the Shina language group of Dras, Kashmir. The Sanskrit term *darī*, means a piece of matting. It is possible that the original meaning of cane matting may have extended in course of time to cover the entire category of floor coverings, the Hindi term being *darī*. There is no connotation of technique involved. Since the pile carpet was introduced into India in the medieval period by the Islamic elite, the term *darī* can be taken as evidence of an earlier tradition of flat woven floor coverings. The distinction between *shatrañjī* and *darī* was that the former was woven in multicolour stripes in a type of weft faced plain weave whereas the *darī* could be woven in one colour or be patterned through slit or dovetail tapestry weaving. It is possible that the latter schema of ornamentation may have had some Central Asian orientation. It is of interest that the weavers of Achalpur, Maharashtra, are Muslim and appear to have migrated from the Agra region which suggests a linkage with the *pañja darī* weaving usages of Fatehpur Sikri.

In all the looms mentioned above, the function of beating is taken over by a forked implement, the *pañja*, rather than the sley (cf. fig. 2.37). In Haryana the warp is wound along and around the length of the frame. There are no heddles but the warp, being

Achalpur *darī*. Interlock at border design. Courtesy: Saboo Textiles, Nagpur. (Fig. 1.13)

Pañja loom, Fatehpur Sikri. Craft Museum, New Delhi. Weaver engaged in weaving. (Fig. 1.14)

Loom used in *tangalia* weaving. Workplace of Dayabhai Karsanbhai, Chitarwada village, district Kheda. (Fig. 1.15)

wound around a frame, creates a natural division between the odd and even numbers of the ends. The weft is introduced by hand. The shed and counter shed are both manually formed. In the absence of any contrivance other than the frame, the Haryana *pañja* device may be called the *pañja* frame rather than the *pañja* loom.

The *pañja darīs* of Haryana are hand woven on this type of weaving frame in tapestry weave. The *pañja* is used to beat in the weft while making these *darīs*. These are woven by women in the villages for use in their homes. Such *darīs* were traditionally used as floor rugs, *āsans*/prayer mats used in seating and as covers for *chārpāī*/cot. The *darīs* use brightly coloured geometrical forms which can be arranged in stripes or distributed as individual motifs within the main field, bound within a frame.

With regard to the distribution of the pit loom, apart from its ubiquitous appearance in India, it stretches through a belt east of Lake Chad, across the upper reaches of the Nile in Sudan to the shores of the Red Sea and the Indian Ocean encompassing Sudan, Ethiopia, Somalia, and Kenya. It also covers West as well as Central Asia.[10]

In attempting to trace the origin of the pit loom it is difficult to accept the statement made by Venice and Alastair Lamb (Lamb and Lamb, 1980: 26) that it was a timber saving device. As an adjunct to the loom, it was particularly well adapted to the exigencies of cotton weaving as it offered the possibility of maintaining a modicum of humidity which could be varied according to the circumstances required, by keeping this area wet. Indian cotton fabrics have been widely distributed and, to that extent, it would be logical to support the point of view that the pit loom had its origins in India. Work practices in India have always been geared to a minimum usage of space (figs. 1.15, 1.16), and the deployment of the simplest tools provided these proved to be sufficient for their purposes. Since the weaver operated at floor level the pit effectively served as a space saving device for accommodating the treadle without adding additional appendages to the loom. The appearance of the pit loom with a single treadle in Egypt is dated to the seventh century AD (Carroll, 1980: 38). With regard to treadles it is possible that these developed independently in China and in India.

Dismantling loom for the night. Workplace of Kanubhai Savabhai Parmar, Chitarwada village, district Kheda. (Fig. 1.16)

From the point of view of geographical spread, the looms which enjoy virtual primacy in India include the pit loom along with sparser manifestations of the horizontal ground *pañja darī* loom and reedless looms such as the *kambli* loom of Karnataka which is described on pages 38 and 39 (see fig. 1.17). The *paṭolā* loom of Patan, Gujarat (fig. 1.18), stands out in splendid isolation while in north-east India, frame (fig. 1.2) and body tension looms (fig. 2.4), constitute areas of transition into the culture regions of South-East Asia.

Natural colour wool. Reedless *Kambli* loom. Weaver, Birappa. Navalgund. (Fig. 1.17)

Paṭolā loom. Workplace of Chotulal Salvi and his family, Patan. Note slope in loom alignment. (Fig. 1.18)

Notes and References

1. Adelaar, 1989: 26; Bellwood, 1984–5: 1, 108, 111–13; Bellwood, 1985: 2, 3 and map, 122, 232–3; Bellwood, 1987, 207–8, 210. For further historical perspectives see Kwang-chi Chang et al., 1964, 359–406; Sebeok, 1971: 5.

2. For outrigger boats see Hornell, 1930: 151–2, 160–7; for pottery see Janse, 1961: 109–11; Solheim II (1965): 196–209; Bellwood, 1985: 213–23; Bellwood, 1987: 212, 215; Bellwood, Koon, 1989: 615–18; Kwang-chi Chang et al., 1964: 164, 372–84, 401–3; Fox, 1968: 14; Guy, n.d.: 25; for Dong Son drums see Bellwood, 1985: 272–5, 280–9.

3. Pirazzoli T'ser-Stevens, 1979: 129, 134; Mills, 1922: 157–8; Vidyarthi, 1986: 76–83, 96–100; Gogoi, 1971: 1–12; Nath, 1978: 1–13, 95, 113. For Austroasiatic elements in the north-east see also Choudhury, 1966, 81–3.

4. The ethnological evidence of material artifacts is very difficult to date. Apart from the linguistic diffusion of Austroasiatic and Austronesian language groupings, there were also other contacts and Peter Bellwood has tentatively dated contacts between India and South-East Asia to as early as *c.* BC 800. Ardika, Bellwood, 1991: 221–32.

5. Krishna Riboud confirms that Han silks were woven on a pattern rod loom rather than a drawloom. Riboud, 1977: 156. Also see, *Chine, 1988*: 270–1.

6. Albers, 1965: 30. Under the Yin-Shang dynasty in China, 1523–1028 BC, a simple type of shaft loom was in use. In the absence of shafts and treadles only a plain tabby or a range of tapestry weaves must have been woven. By the turn of the Christian era satin was in evidence in China inferring usage of a multi-pedalled loom. While the drawloom was developed in Syria by the third century AD it appears to have been adopted in China during the Tang period, AD 618–907. The process of adaptation was slow and was completed only during the thirteenth century. During the intervening period pattern rods continued to be used. *Chine, 1988*: 305; Gluck et al., 1977: 173; de Jonghe, 1983: 54.

7. Blinks, 1977: 76–7; de Jonghe and Tavernier, 1979; de Jonghe and Tavernier, 1981.

8. de Jonghe and Tavernier, 1981: 72–3. For a more extended discussion on South-East Asian frame looms see Fraser-Lu, 1988: 37–9, 52–5; Gittinger, 1977: 58-60. de Jonghe (1985: 15), further makes the observation that a loom, differing from the Bangkok heddle rod loom only in small detail, was in usage among a minority group, the Pai-yi, living in the deep south of Yunan, China.

9. Picard-Schmitter, 1967: 17–21; Hecht, 1989: 61–5; Lamb and Lamb, 1980: 23, 58; Weir, 1970: 16–23.

10. Lamb and Lamb, 1980: 23, 26; Ling Roth, 1950: 62; Weir, 1970: Plates 21–4; Broudy, 1979: 105–8.

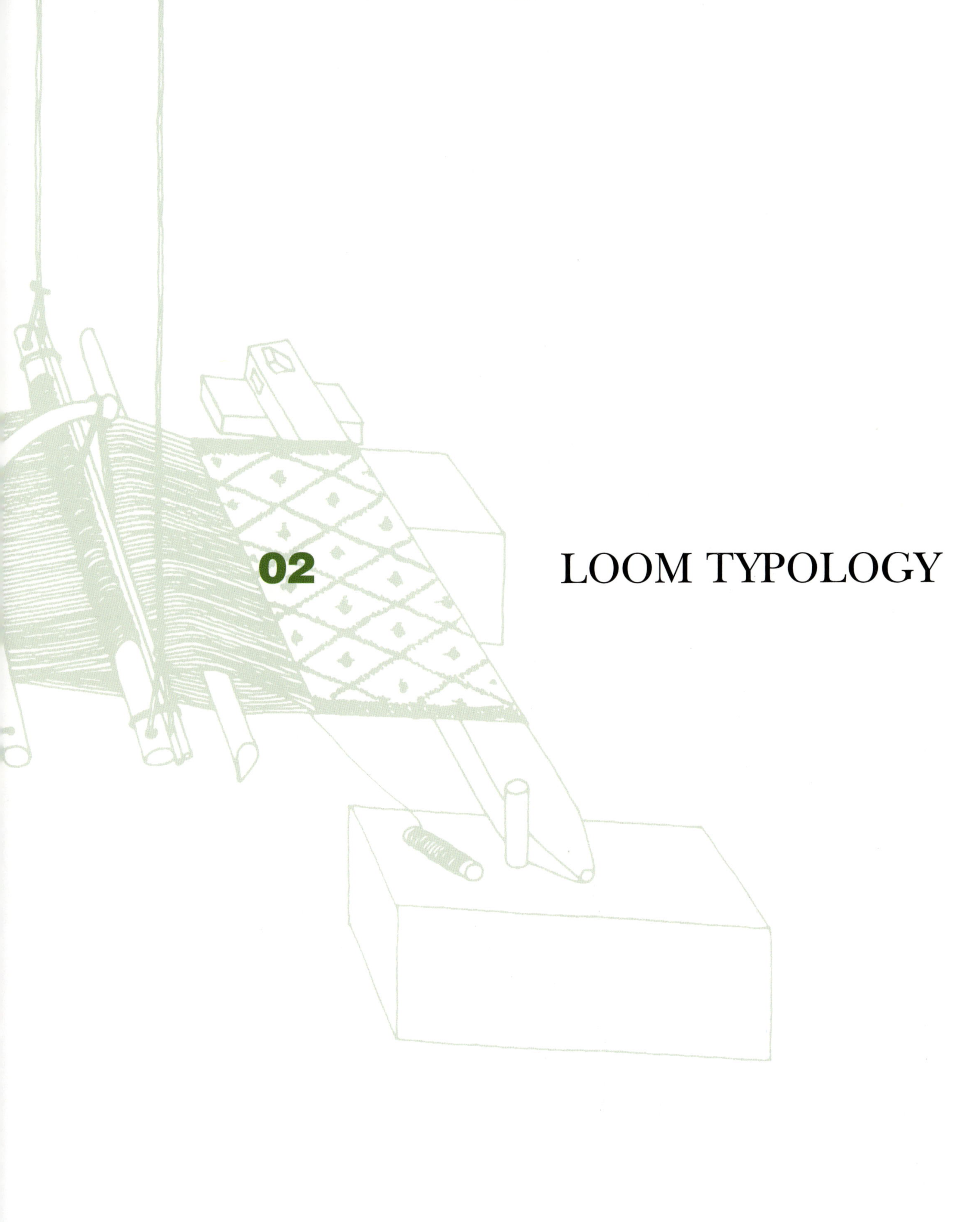

02 LOOM TYPOLOGY

02 Loom Typology

Body Tension/Loin Loom: North-East India

The loin loom or the body tension loom of the north-eastern region of India is a fairly simple device on which a variety of plain weave fabrics, plain, striped, plain with supplementary weft ornamentation, and sometimes with *soumak* ornamentation are woven.

Among the hill tribes of the north-eastern region, only the Garos and Khasis weave on the frame loom while all other hill tribes from Nagaland, Mizoram, Tripura, Arunachal Pradesh, Manipur and the Mikir Hills of Assam weave on the loin or the body tension loom.

Description of Loom

The loin loom of the north-eastern region of India (see fig. 2.1) is an indigenously made device using locally available material such as bamboo and wood.

Materials

Cotton is the main fibre used, and is grown by various tribes although nettle fibre is also known (Jacobs, 1990: 44). Cotton is de-seeded by rolling a rounded stick over it on a flat stone, or by using an indigenous ginning device. The fibre is then converted to form a sliver. This is then spun by hand using the drop spindle prior to being woven into fabric. Weaving is done by women and mainly for their own use. In Nagaland, Mizoram and Manipur, women produce shawls and sarongs for the purpose of sale (Shirali, NID, 1983: 56). In the plains of Assam and Tripura, women and men are professional weavers and weaving is also done as a commercial activity. Among the hill tribes, women produce their own requirements of fabric in their spare time between preparing meals, after returning from the fields, or after the harvest season (Shirali, NID, 1983: 56). There were situations where some tribes did not grow cotton but had good weaving skills. They were paid in kind, thus obtaining the raw material, cotton, in exchange for their labour. The Apatanis and Mishmis are known to carry out this practice.

Warp Making Process

In the north-eastern region there are two methods of making the warp. In the method used by the Nagas, the warp is made on a horizontal frame and then stretched to its full length. Then the weaver proceeds to transfer the lease on to the bamboo shed stick and to string the warp threads with yarn heddles. The other method of making the warp entails making the entire loom itself. The weaver sits on the floor and is assisted by another woman who travels back and forth with a bobbin of yarn, helping to build the entire length and width of the warp. The weaver separates the threads of the top layer from the bottom layer by placing them alternately under and over the shed stick. The threads from under the shed stick are then strung with yarn heddles. These types of heddles will be referred as half heddles through the text. In this method the warp, the shed and the yarn heddles

Loin loom and its constituent parts. (Fig. 2.1)

1. Bamboo warp beam
2. A thin bamboo lease rod or laze rod
3. A thick bamboo shed rod
4. A bamboo heedle rod
5. A narrow flat wooded sword
6. Two wooden rods which clamp the cloth in between also serving the purpose of a cloth beam
7. Back strap
8. Continuous warp
9. The woven fabric
10. Mat spread on the floor, on which the weaver sits

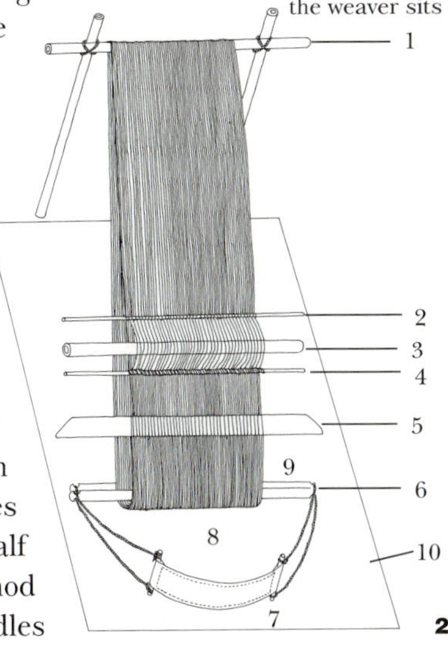

2.1

are made, thread by thread. The warps are generally narrow, not wider than 50–60 cm. in width. This loom does not have a reed and the weaver varies the density of the fabric by spacing the threads closer or further apart while making the yarn heddles (Shirali, NID, 1983: 83, 84).

Weaving Process

The definition of weaving as the working of two or more sets of elements resulting in an interlaced surface explains to us the basic requirements essential to form the structure of a woven fabric. The warp being one element and the weft the other, it would be necessary to have alternate interlacement in order to weave a simple structure. This could be achieved through two different shed formations, shed and countershed (see figs. 2.2, 2.3).

The loin loom (fig. 2.1), works on the basic principle of two different shed formations, allowing the weaver to weave a type of plain fabric. The farthest end of the warp is attached with a cord to a post or anything sturdy, whereas the other end is braced around the lower back of the weaver with the back strap. The tension of the warp is regulated by the forward and backward movement of the body of the weaver. One shed is formed when the weaver bends forward loosening the warp ends. This movement is followed by the lifting of the heddle rod along with the warp ends which have been passed through the half heddles. The warp ends that are not threaded through the half heddles remain below the lifted threads. The sword or the beater is passed through the space created by the lifting of the heddle rod, which is then turned to stand on its edge to form a shed. The weaver now swings back her body to bring the warp in tension and inserts the weft which is beaten in with the sword/beater. This is followed by moving the heddle rod closer to the bamboo shed rod, whereby the warp threads coming from over the shed rod remain above the warp ends strung by the string heddles. The sword is inserted into this gap forming the second shed through which the weft is inserted and beaten in. The process of weaving the fabric continues in this manner. From time to time, the weaver shifts the woven fabric below the cloth beam; thereby bringing the warp ends closer to her.

Fabrics and Techniques used for Weaving

The fabrics woven on the loin or body tension loom are thick as the warp is dense and covers the weft, giving a ribbed effect. This also contributes to making the fabrics predominantly warp faced (Shirali, NID, 1983: 49). The traditional loin loom fabrics of the north-eastern region are extremely bold and graphic in nature, varying from plain bold stripes in plain colours to stripe formations using geometric shapes in supplementary weft and/or supplementary warp.

Supplementary weft ornamentation is also used to form simple geometric motifs used sometimes as accents to add colour, dispersed over the main field of the fabric, or used along the two edges and ends of the fabric. The supplementary weft ornamentation is achieved in three different ways. These methods could be used individually or in combination. They are the pickup method, the pattern rod or lease rod method, and the tying of loops. Usually, the yarn for the ornamentation is thicker than the yarn used for the ground weft inserted locally.

Pick-up method: In the case of the Nagas this is achieved by picking up the desired warp ends with a bamboo stick. However, on the basis of documentation, 1939, at the Pitt River's Museum Archive, Oxford, Jacobs has illustrated the

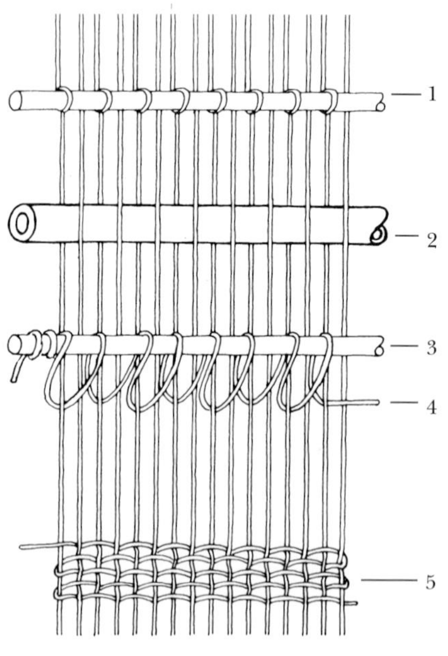

2.2

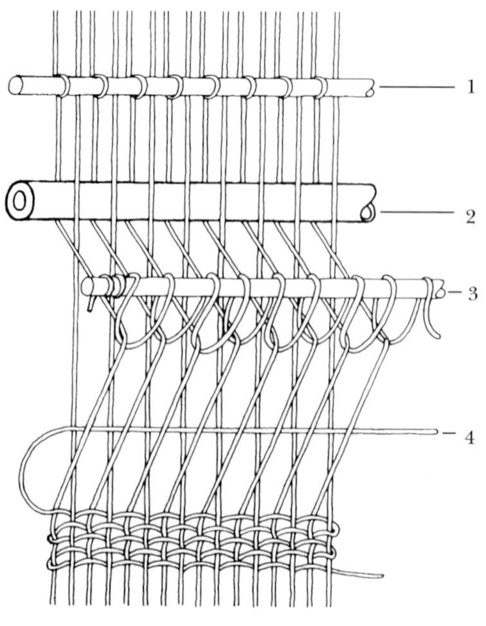

2.3

Detail drawing of lease rod and heddle rod. (Fig. 2.2)

1. Lease rod
2. Bamboo shed rod
3. Heddle rod
4. Yarn to make the heddle
5. Weft

Formation of first shed. (Fig. 2.3)

1. Lease rod
2. Bamboo shed rod
3. Lifted heddle rod
4. Weft

use of porcupine quill for placement of supplementary weft by the Naga tribe in the village of Kabui Kokodan (Jacobs, 1990: 44). Several sticks could be used, the supplementary weft yarn inserted followed by the ground weave weft (see figs. 2.4, 2.5).

Pattern rod or lease rod method: Here the desired warp threads are picked up with the pickup stick, which becomes the pattern rod when transferred to the back, and the ground weave weft is inserted. Before commencing on the next step of the motif the weaver transfers the shed created by the selected threads for that particular step of the motif, at the back of the loom on to a shed stick/pattern rod or lease stick. These shed sticks are inserted into the warp according to the requirements of the pattern. In case a mirror image of this motif is required, the shed sticks are inserted sequentially up to the central part of the motif. The order is reversed by bringing the shed sticks from the back of the loom to the front, one by one, as the weaving continues.

Tying of loops: This can be considered as an improvement on the second method where the task of picking up the warp threads was done by hand. Loops are tied around selected warp ends and are secured on sticks or left as loops. These loops are lifted to create the shed for extra weft patterning. The loops may be localized or placed across the width of the warp. This is dependent totally on the design planned by the weaver.

This kind of loin loom is used by some weavers in the Tumnoupokpi village, near Imphal, Manipur (see fig. 2.6). These types of loops with sticks are also

Foot braced loin loom, Assam. Craft Museum, New Delhi. Note continuous warp. (Fig. 2.4)

Loin loom, Sikkim. Note method of lifting warp ends for extra weft weaving. Craft Museum, New Delhi. (Fig. 2.5)

Weaving of female lower garment. Loin loom. (Fig. 2.6)

seen in the north Thailand loom (see figs. 1.3, 1.5, 1.8). As the warp ends are densely spaced and the pattern is picked up from the top half of the warp, the extra weft does not show at the back of the fabric. The method of supplementary warp ornamentation is also used by some tribes of north-east India, the Riangs, Meiteis and Karbis. The Riangs and Karbis are known for the elaborate patterns they create, using the supplementary warp techniques (Shirali, NID, 1983: 56).

Paṭolā Loom

The loom described here is the *paṭolā* loom of Patan, more particularly, that belonging to Chhotulal Salvi and his family.

One of the most visually striking aspects of this loom is that it is inclined at a slant with no rigid framework (see fig. 1.18). The slant of the loom has a definite relation to its function (see fig. 2.7). The left side of the breast/cloth beam where one of the weavers is positioned is lower than the right side, where the other weaver stands and weaves. A horizontal slant at an angle of approximately 30° is created from the left side of the cloth beam. This angle is sometimes varied to facilitate the individual requirements of the weavers but at no point does it become completely horizontal.

As the cloth beam is heavy and long it needs to be supported. It has also to be held in place while weaving in order to maintain the warp tension, and prevent the warp from swinging out. At the same time, it must be detachable to facilitate the winding and unwinding of the woven fabric. This cloth beam is supported on two cemented platforms of different heights (fig. 2.7).

Product

Before we go into the details and the purpose/function of the slant, a brief introduction to the type of fabric being woven on this loom is necessary. Patan *paṭolā* is a silk, double-*ikat* textile, woven in plain weave. Chhotulal Salvi and his family continue to weave this cloth to this day. The extraordinary technical perfection of this item, and ability to cater to an exclusive market, has ensured a steady market for the Salvi products despite its high price.

A systematic and meticulous realization at each stage is achieved through a holistic understanding of tools, materials, processes, end product and needs of the user. These processes were inherent to high quality production in the past. It is a sad commentary on present procedures that specific note has to be taken of these steps in the case of Salvi production of Patan *paṭolā*. In order to ensure technical perfection, the exercise begins with the selection of raw materials and planning of the design. It continues into the processes of resist tying and dyeing of both the warp and the weft yarns, and weaving. Without teamwork such meticulous standards could not be attained. Further refinements arising out of the needs of the immediate and ultimate user have bred development and change.

In the context of the slant of the loom, it can be stated that the ergonometric requirements of the weavers is met. When the loom is on a horizontal plane with a wide warp, a lone weaver would have a problem in throwing the shuttle with one hand and picking it up with the other at the end of the shed. While performing this function, the weavers, at the same time,

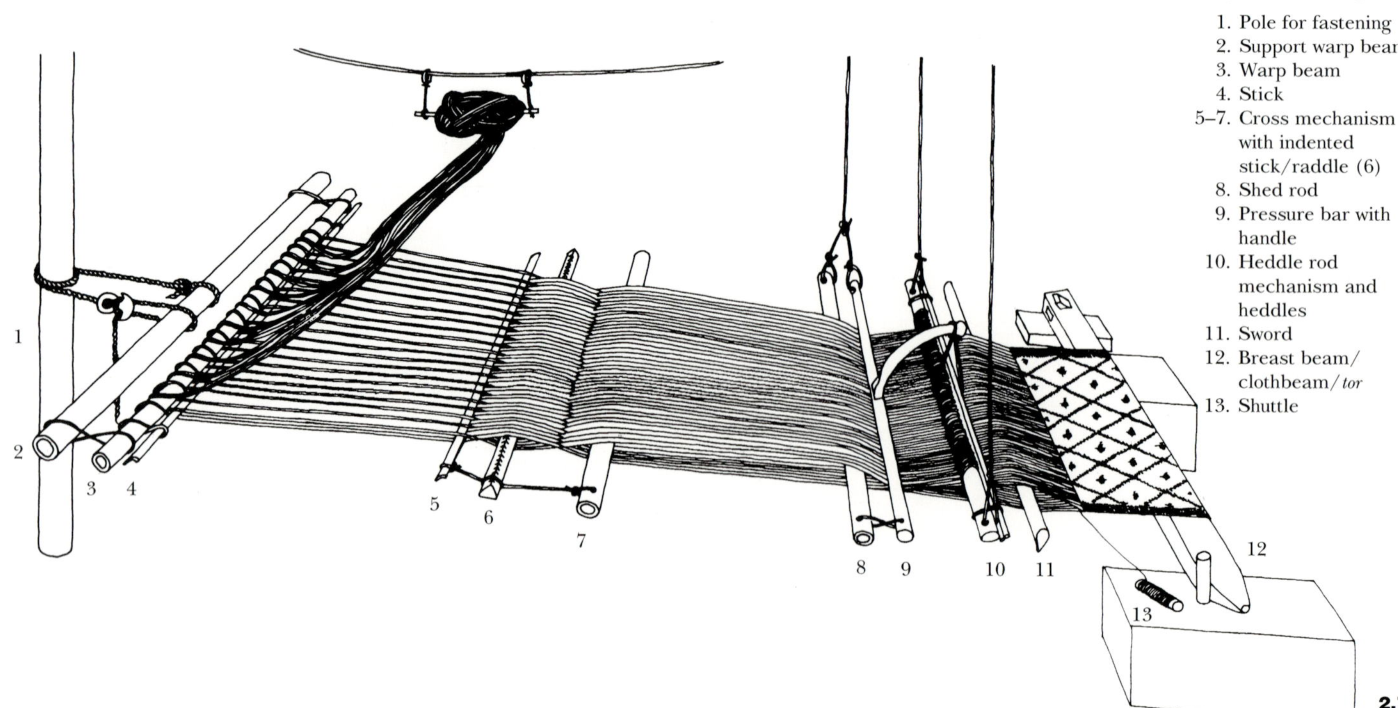

The *paṭolā* loom. (Fig. 2.7)

1. Pole for fastening
2. Support warp beam
3. Warp beam
4. Stick
5–7. Cross mechanism with indented stick/raddle (6)
8. Shed rod
9. Pressure bar with handle
10. Heddle rod mechanism and heddles
11. Sword
12. Breast beam/clothbeam/*tor*
13. Shuttle

2.7

would also need to ensure that the warp *ikat* and weft *ikat* are properly aligned, manipulate the pressure bar to create the shed, and mend broken ends. In present times it may require between six months to a year or more to weave one sari, depending on its design.

In the present situation where the loom is inclined at a horizontal slant and the span of the warp is wide, two weavers synchronize their activities organizing and distributing their tasks. As has already been pointed out one weaver is seated on the left side of the loom while the second weaver stands on the right side of the loom near the cloth beam. The person standing at the right operates the shed stick/beater, throws the shuttle from the right, and operates the pressure bar with handle/*lungī* to change the sheds.

The person sitting on the left throws the shuttle from left to right (see de Bone, 1976: 60). The perns are of different colours according to the requirements of the pattern. The weaver on the left has to ensure that each shuttle thrown has the appropriate pern. Both the weavers check the matching of the weft with the warp. Quill-like iron needles which are not very pointed at the ends are used to adjust the slight shifting that may be required to ensure perfect colour alignment.

The weaver positioned at the right side of the loom is able to see the process clearly because of the slant. Thus he can ensure that the shed is clear. The passage of the sword through it without breaking the warp ends is also facilitated. If the loom had been horizontal the weaver sitting on the left or right would have had a perspective view of the far ends of the woven area in such a manner that it would not have been possible to assess with accuracy whether the weft *ikat* was in proper alignment with the warp *ikat* in accordance with the pattern. Thus the slanted loom, in the case of weaving the *paṭolā*, a double *ikat* was developed by the weaver, the immediate user, to achieve a product of excellence for the ultimate user of the loom, the consumer. The field survey conducted by D.K. Chakraborti and V.S. Katiyar of the loom used in the workshop of Chhotulal Salvi reconfirmed the ergonomic viability of the slant in the loom. It was also learned that one organization had tried to weave *paṭolās* with a horizontal fix of the loom in place of the slant, while all other features of the loom were retained. It was not possible to achieve the finely tuned accuracy accomplished by the slant in the loom (Chakraborti, Katiyar, NID, 1993: 4).

It was observed that the weaver on the right, responsible for manipulating the pressure bar with handle/*lungī* for shed formation, manipulated the *lungī* mainly with his left hand, and, while leaning forward, simultaneously inspected the lifting of the warp ends. Due to the slant of the loom the *lungī* came within the reach of the arm of the weaver and helped him to clearly view the lifting of the warp ends. Had the loom been horizontal, this may have proved difficult. In addition, while manipulating the *lungī*, the weaver's extended arm, did not provide any visual barrier as the matching of both warp and weft *ikat* was being checked. In a horizontally aligned loom, the action may well have hampered his vision.

The loom is arranged in such a fashion that more light falls on the left side of the loom. Natural light, coming through the open terrace and the doors of the workshop, together with artificial lights provide good illumination on the tilted left weaving area by angular reflection. Artificial lights have been mounted on the ceiling, and more tube lights were installed on the left than on the right side of the working area. This facilitates the inspection of the whole shed from right to left for the weaver on the right, as well as for the weaver seated on the left. Further, as work is usually executed from left to right, the source of light on the left provides ease in general inspection while highlighting the defective points on the surface of the fabric (Chakraborti, Katiyar, NID, 1994: 4).

Paṭolās are made of various kinds of silk. Cotton is rarely used, and never for *ikat* sections in contemporary production. Although double *ikat* tends to be the characteristic patterning technique for *paṭolā*, other methods of decoration also play a part including single warp/weft *ikat* in individual areas. According to present knowledge, only metal threads or, as a substitute, white silk threads are employed to lance and brocade individual areas of the *pallu*, and, on rare occasions, the main field. In isolated cases this kind of thread is also found in the warp of the longitudinal borders (Bühler, Fischer, 1979: 4).

As the double *ikat* patterning in the *paṭolā* demands that both the warp *ikat* and the weft *ikat* are equally emphasized on both sides of the fabric, it explains both the usage of plain weave and the type of loom used in the making of these *paṭolā*, which can provide the two basic sheds, the shed and countershed necessary for interlacing of the warp and weft (see figs. 2.8 and 2.9, also see fig. 2.7). Alfred Bühler and Eberhard Fischer mention specifically in relation to the fabrics woven on looms similar to those commonly used by the Salvis in Patan, that the plain weave often has the character of a warp rib weave. An accentuated effect of this type is, however, seen only in the borders and particularly in the selvedges, in the portions left without *ikat* patterning (Bühler, Fischer, 1979: 246).

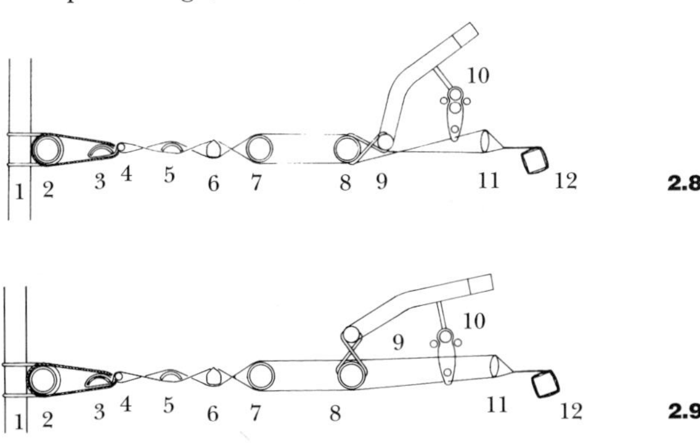

Cross section shed formation, *paṭolā* loom. (Fig. 2.8)

Cross section counter shed formation, *paṭolā* loom. (Fig. 2.9)

Fabric Structure: Loin Loom and *Paṭolā* Loom

Since there are points of similarity between the loin loom and the *paṭolā* loom it is of interest to scan the performances of the looms. As explained earlier, the majority of the fabrics woven on the loin loom or the body tension loom are predominantly warp faced as the warp is dense, covering the weft entirely and giving it a ribbed effect. Made of cotton and woven in the warp faced plain weave, the patterning is achieved through a rhythmic arrangement of coloured warp stripes and through supplementary weft/extra-weft and supplementary warp methods. However, there are a few instances, where fabrics are woven in what could be considered a type of balanced plain weave.

In both cases, the loin loom and *paṭolā* loom, the spacing of the warp ends is regulated by the heddles (see figs. 2.10, 2.11, 2.12), as no reed is used on either of these looms. In the case of the *paṭolā* loom, the weavers ensure that the heddles do not shift by securing them to the heddle rod (see fig. 2.13) as it is absolutely necessary for each unit of the warp *ikat* (warp ends) to match with each unit of the weft *ikat* (pick). Any slight variation caused by improper procedures through the process of weaving other than as originally planned, can cause distortion in the pattern.

The similarities between the loin loom of north-east India and the *paṭolā* loom of Patan are listed in the two charts which follow. The distinctive features which make the Patan *paṭolā* loom unique as also the differences in the end product achieved off both the looms are also identified.

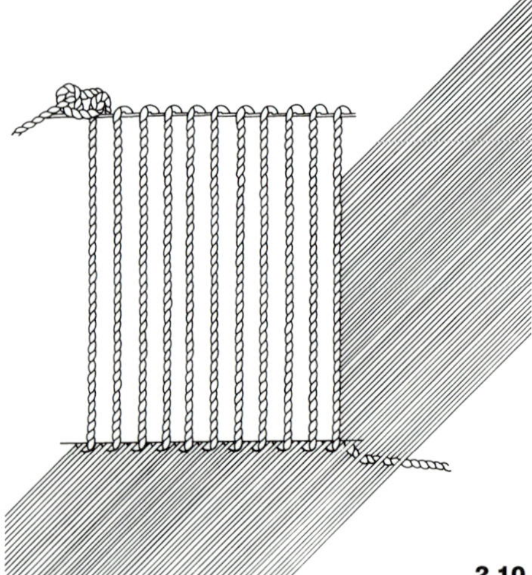

2.10

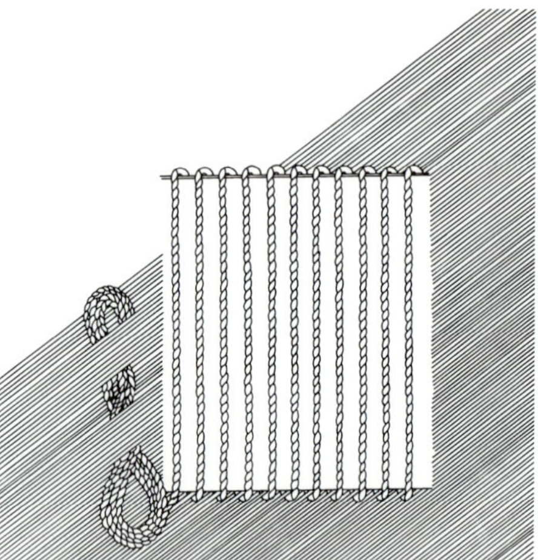

2.11

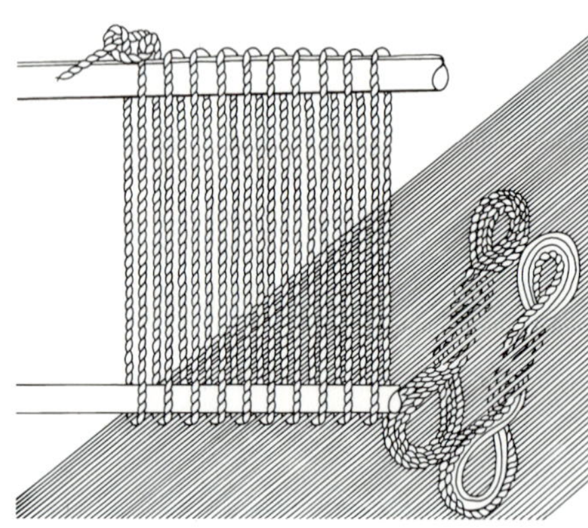

2.12

Heddle preparation. Stage I. Making of loops around *thapee*. (Fig. 2.10)

Heddle preparation. Stage II. Slipping of the loops from *thapee* to the left under the warp. (Fig. 2.11)

Heddle preparation. Stage III. Insertion of heddle bar through loops. (Fig. 2.12)

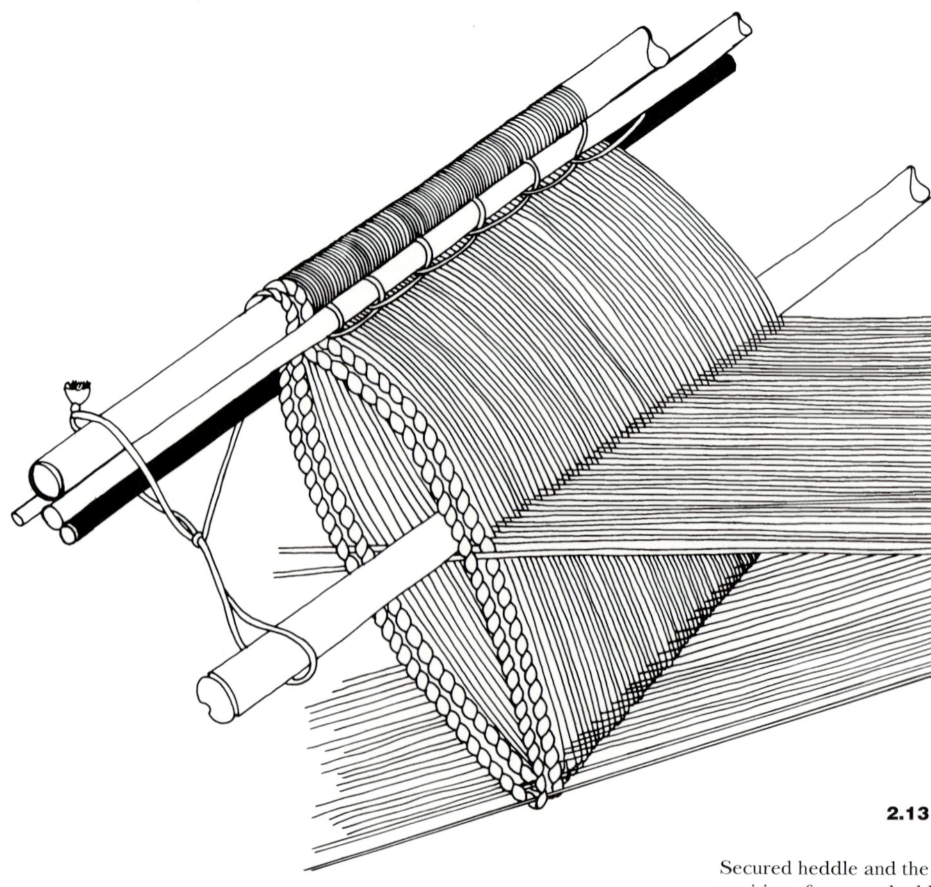

2.13

Secured heddle and the position of even and odd warp end. (Fig. 2.13)

Chart I: Functional/Operational Consonance

Loin Loom	*Paṭolā* Loom
1. A pole, tree or any object to which the warp beam can be fastened.	1. A pole for fastening the warp beam at the far end of the loom.
2. A bamboo warp beam.	2. Warp beam (double, inner one holding the warp, outer one supporting the warp beam; held together by a string. This helps inload sharing).
3. A thin bamboo lease rod.	3. Bamboo lease rod.
4. A thick bamboo shed rod.	4. Bamboo shed rod.
5. A bamboo heddle rod with half heddles to raise one set of warp threads.	5. Half heddles used; held on a wooden rod (two main and two supporting rods).
6. A narrow flat wooden sword or beater.	6. Use of a wooden sword or a beater.
7. Two wooden rods which clamp the cloth in between, serving as beam tied with back strap.	7. Cloth beam made of solid wood square beam, for cloth transversely held.
8. Raddle used by some loin loom weavers.	8. Raddle used.
9. No reed is used.	9. No reed is used.
10. No fixed framework; loom has to be set up each time a new warp is put up.	10. No fixed framework; loom has to be set up each time a new warp is put up.
11. Shuttles made of bamboo; used by some loin loom weavers of north-east India. Similar in shape to those used by *paṭolā* weavers.	11. Shuttles made of bamboo. (fig. 2.14).
12. Loom takes inclined position as warp is raised and fastened to a pole at the back, whereas the front part is braced around the hips of the seated weaver.	12. Loom takes an inclined position by its being fastened to a pole, front is lower; position is slanted horizontal (fig. 2.7).

Chart II: Functional/Operational Similarities

Loin Loom	*Paṭolā* Loom
1. Two basic sheds, shed and counter shed.	1. Only two sheds formed, shed and counter shed.
2. Shed formation by the shed rod and counter shed by manipulation of the heddle rod by body and hand.	2. Formation of shed by heddle rod and counter shed by the shed rod, both achieved by manipulation and aid of pressure bar.
3. Sword plays important role of opening shed to facilitate passage of weft and also beating it in.	3. Twofold role of sword: opening shed to facilitate passage of weft and beating it in.

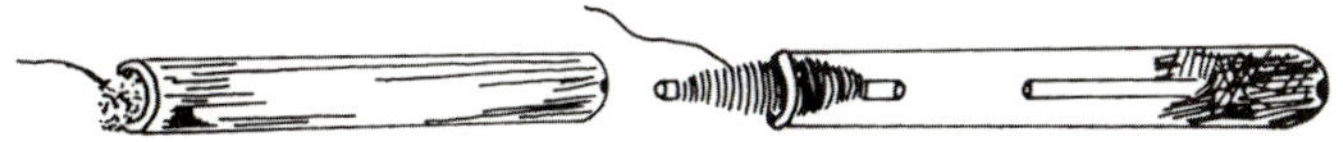

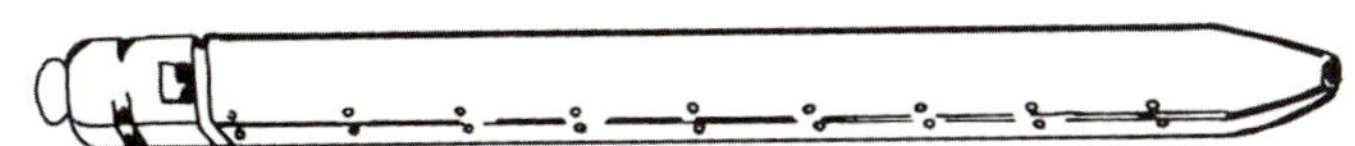

2.14

Cloth beam and bamboo shuttle. Bamboo Shuttle above. *Tor*/cloth beam below. (Fig. 2.14)

Kambli Loom of Karnataka

The *kamblis* of Banhatti, Mahalingpur and other villages in Karnataka are woven by the Kuruba community. These are made of locally available wool procured directly by the weavers from the Dhangars who rear sheep. Natural coloured wool in shades of brown and black form the colour palette of these *kamblis*. These are woven in a type of warp-faced plain weave.

The *kambli* is used by the Dhangars as a shawl or a blanket during the winter. The process of carding, spinning and warp making are all carried out by the female members of the weaver's family.

Carding

Carding is done with the help of a bow-shaped instrument. It is made locally and is called *beshi kattigiri*. A ball of woollen fibres is taken in one hand and with the help of the feet and the other hand the string on the bow is pulled. This separates each fibre and simultaneously helps remove impurities.

Spinning

Spinning is done by hand using the *charkhā*, the traditional Indian cotton spinning wheel.

Warping

Warping is done on a wooden frame with vertical pegs. The spacing of the pegs is determined by the length of the warp. One end of the woollen yarn is passed through a warping guide, a narrow hollow wooden rod about 5 in. in length. Holding this rod, each warp end is guided and wound around the vertical pegs on the frame to form the warp with a lease for the insertion of the lease sticks that help in maintaining the order of the warp threads. Once the required number of ends is wound and the lease tied, the warp is removed by holding the total group of warp threads in one hand while being held down with the other, forming a chain of links to avoid entanglement.

Sizing

The warp is then stretched on to two wooden sticks and spread evenly in accordance with the ascertained width. The sizing medium, a paste made from seeds of tamarind (*Tamarindus indica*), is spread evenly twice over the warp with the help of an indigenously made brush, after which it is left to dry.

Heddle Making

The heddles are made with twine on two wooden rods, one on top of the other. The upper one is thin and the second is thicker and hollow, the distance between the two being maintained by the experienced hand. The twine is passed from under the lower rod, brought on top and under the top rod and then over it in a figure of eight. Each alternate or top layer/odd threads, of the warp ends pass through the heddle leaving the lower part/even ends, of the warp free and without heddles. On completion of the heddles for each top layer/odd ends of the warp, the tension of the warp is adjusted and a thick hollow rod/shed rod is passed through the space created between the top and lower layer of the warp. This shed rod is secured by passing a string right through the hollow rod and around the warp to prevent it from falling.

Kambli Loom

The *kambli* loom (fig. 2.15) is a horizontal ground loom. The distinctive character of this loom is the presence of a pit in the ground. This pit enables the weaver to operate the loom efficiently while sitting comfortably. The loom does not have pedals. It marks a stage in the development of a ground loom which is fixed, unlike the Bedouin loom which is mobile. The main aspects of the loom comprise:

1. A warp beam (10) tied to a rod which is in turn tied to a pole driven into the ground (12) at the back of the loom with one end of a rope. The other end of the rope is tied to a pole on the side and behind the weaver. This helps the weaver to release the warp without having to actually get up and go to the back of the loom.
2. Lease stick (9).
3. A rod (14) placed under the warp, propped up by two wooden stumps, on either side of the loom at the level of the warp, serves the purpose of keeping the warp at the required level and also supports it, thereby preventing it from swinging while weaving.
4. Shed rod/stick (7-8).
5. Heddle rod/s (5) with a handle to facilitate lifting of the warp ends passing through the heddles.
6. Breast beam or cloth beam (3).
7. Sword (4) or a beater.

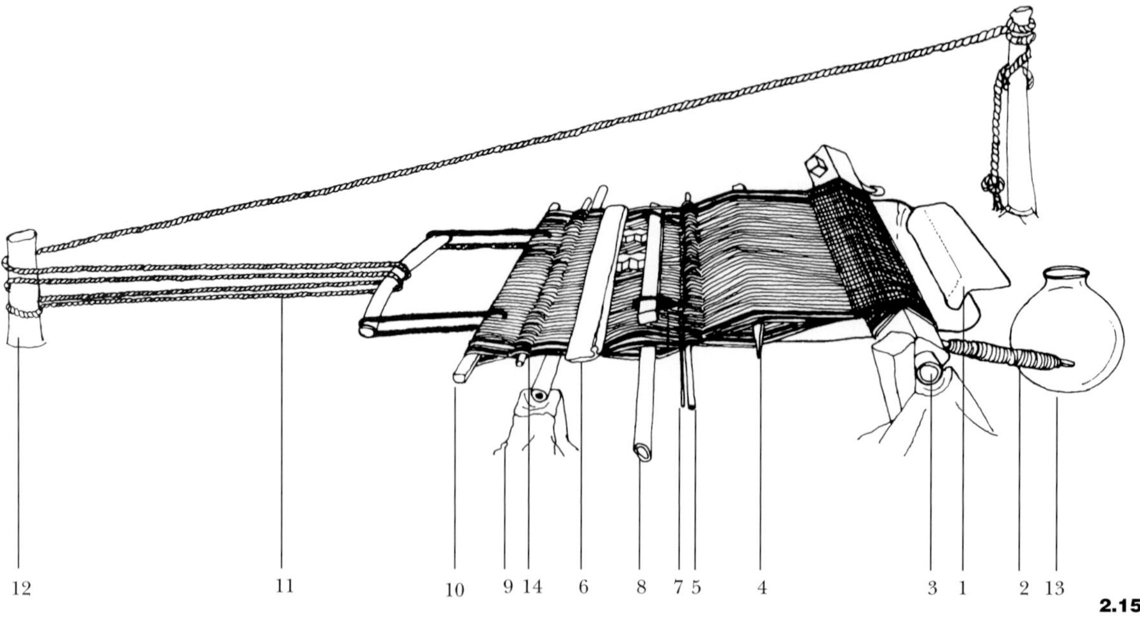

2.15

The *kambli* loom. (Fig. 2.15)

1. Pit
2. Stick shuttle
3. Cloth beam
4. Beater
5. Heddle rod
6. Heddle raiser
7. Shed rod
8. Shed rod/stick
9. Lease stick
10. Warp beam
11. Rope for tensioning
12. Peg
13. Water pot
14. Rod to prop the warp

Weaving

For the weaving process the first shed is formed by pressing the heddle rod downwards, creating a shed. A sword/ beater is inserted into this shed and propped up on its edge to enable the passage of the weft. After the insertion of the first pick, it is beaten in to put it in place and the sword is removed. To create the second shed, the heddle rod, which is attached to a handle, is raised. The shed is cleared with the sword/beater as in the first shed and the weft is then inserted. The weaver proceeds with his weaving in this manner.

The weaver applies starch water or plain water to the warp when the shed is open to enable the weft to be packed in closer. To help regulate the width of the weaving a stretcher or temple is used. This is made of wooden sticks.

Pañja Loom, The Weaving Frame of Haryana

Usually it is the women from the Jat and the Khatri communities who are involved in weaving *darīs.* They weave through the year during their free time, usually for about three to four hours a day. During the harvesting time, however, no weaving is done, as all hands are needed to share the workload in the fields. Weaving time is also a time for enjoying social contacts and interaction with other women. While one weaves another makes yarn bobbins and there is a constant undertone of conversation. Since the weaving is done entirely with the fingers, several persons can weave in unison (*Pañja Darīs,* NID, 1993: 2). As the weaving frame is portable, a group of weavers can congregate together in the open outside their homes and continue with their weaving activities. There is a constant exchange of suggestions on colour and design, comments on the appearance of the *darī* and eager enthusiasm to learn new methods and techniques.

Weaving Frame

The weaving frame used today in the villages of Haryana has seen some changes over a period of time, while maintaining the original method of weaving. Originally, four pegs were driven into the ground and two wooden rods were placed, one behind each pair of pegs demarcating the front and back, and also determining the width of the *darī.* These stayed in place due to the tension created by the warp winding around the two rods tightly to form a figure of eight.

The portability of the device was important as the women who weave could carry it out of doors, to avail of natural light while interacting with each other, and carrying it indoors whenever necessary. In traditional India there is no difference between living and work space. Such a loom had both advantages and disadvantages. The advantage was that the whole device could be rolled up and taken indoors. The disadvantage was that each time the weaving operation was recommenced, the pegs had to be fixed and the warp threads and tension reset again. This made the process time-consuming and cumbersome.

It is not known when the shift to a weaving frame occurred. However, some years ago these *darīs* could be seen being woven in Haryana on the *chārpāī* frame whereas today with the easy availability of metal pipes and the need to adapt to alternative materials due to the scarcity of wood, aluminium is used for the frame. Circular, hollow aluminium pipes, 1¾ in. in diameter, are welded together to form rectangular frames measuring 7 ft. x 4 ft. and 8 ft. x 5 ft. The frame has four legs and stands about 8 in. (20 cm.) above the ground (see fig. 2.16). A *pañja*/comb beater

Weaving frame of Haryana. (Fig. 2.16)

2.16

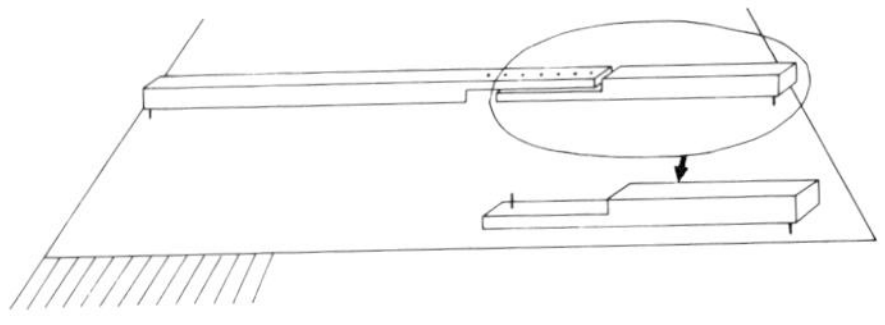

Temple/*Paṭṭī*. (Fig. 2.17) **2.17**

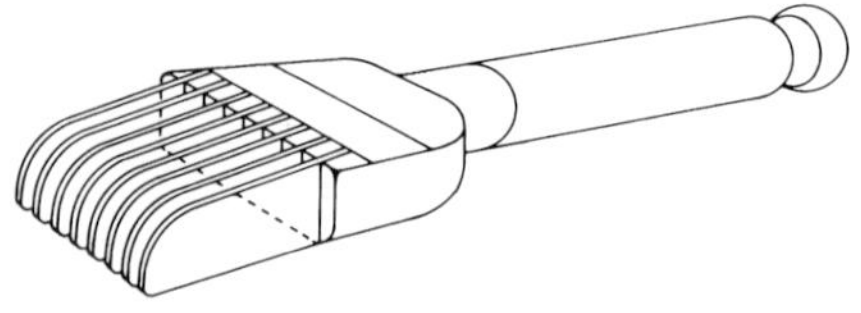

Pañja used in weaving *darī* on frame loom of Haryana. (Fig. 2.17a) **2.17a**

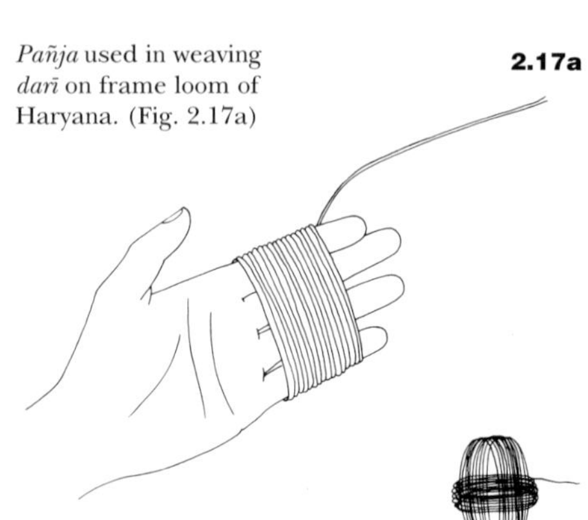

Weft yarn being made into finger hanks. (Fig. 2.18) **2.18**

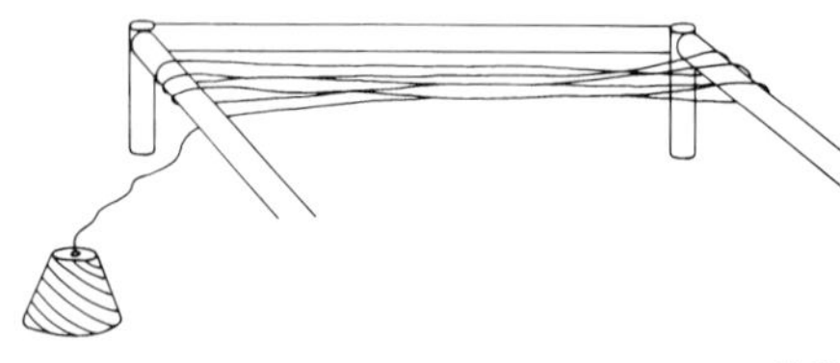

2.19 Warp making. (Fig. 2.19)

is used to beat in the weft.

Tools and Accessories

Temple/*Paṭṭī*

The temple is made of two wooden slats called *paṭṭī* with a half lap joint. The length of this *paṭṭī* can be adjusted to the width of the *darī*. One slat has a number of holes in it whereas the other has a nail which passes through the selected hole determined by the width of the *darīs*. The two farthest ends of the slats have nails fixed at right angles to the base with the pointed tips protruding downwards. The two opposite sides of this *paṭṭī* are pushed into the selvedge so that the nails pierce through it and the centre overlap is tied with a string to hold it in place (fig. 2.17). This is an extremely ingenious technical solution.

Wooden Board/*Paṭṭā*

A *paṭṭā* or flat wooden board, strong enough to hold the weight of at least two weavers, is balanced across the weaving frame. The weavers sit on this board and weave. This board is moved along as the weaving progresses.

Blade or Scissors

Ordinary shaving blades or scissors are used for cutting/trimming the yarn.

Pañja/Comb Beater/Beater-in

The *pañja* is made of iron blades inserted into a wooden handle (fig. 2.17a).

Pre-weaving Process

Today millspun white cotton yarn of count 6/6s and 4/6s is used for the warp and may be dyed to the required colour using direct dyes. For the weft loosely plied unbleached cotton handspun yarn of approximately 2s count is used. The yarn is bought from Ambala in the form of hanks and is dyed locally in the colours required by the weavers. These are then converted directly from hanks into little balls or finger hanks prior to weaving (fig. 2.18).

Dyeing/*Rangāī*

Direct dyes and common salt are used for dyeing. Prior to dyeing, the yarn is soaked in water and washed. A dye solution of the required colour is mixed in a cauldron of boiling water and the yarn turned around in it till the dye is completely absorbed. Common salt is added to the dye bath and the yarn is soaked once again. It is then removed and washed thoroughly in running water to remove the excess dye.

Warp Making

The warp is made by tying the first warp end to one of the farthest ends of the frame, the thread carried over manually to the other end and back, forming a figure of eight (fig. 2.19). The distance between each warp end is measured with the width of a fingertip, approximately 1 cm. A continuous thread is taken back and forth by the weaver each time maintaining the distance and tension across the whole warp. The selvedge yarn is double the thickness of the warp yarn which is plied and tied in a figure of eight. Both the selvedge ends are made in this manner.

Weaving

Tapestry weave is used in the making of these rugs. The thickness of the pipe around which the warp is wound automatically creates the separation between the odd and the even ends.

The weaver initiates the weaving by entering the first pick after the cross is formed by the figure of eight. This pick is beaten in with the *pañja*. In all subsequent picks, odd and even ends are manually lifted in alternate succession (*Pañja Darīs*, NID, 1993: 7).

Threads from individual coloured finger hanks are inserted in the selected shed. The weaving is not done from selvedge to selvedge. At the Khori centre in the Rewari district of Haryana, traditional rug designs using simple geometrical shapes with a triangular grid as a base are woven using bright coloured wefts. These forms are woven with individual wefts in solid colours, building up one area at a time. The motif placements are done visually and not through accurate measurements thus allowing flexibility in the size and shape of the motif.

For example, when triangular shapes are woven, the weaver starts from the base of the triangle, decreasing after every two picks by one step or one warp end, till she reaches the tip of the triangle. The negative spaces left in between are then woven by inserting a different coloured weft moving in the opposite direction. The turning points of the first two picks of the already woven motif and the one in the process of being formed face each other without overlapping or interlocking.

In the third and fourth pick, the stepping in the left hand motif is reduced by one, whereas the third and the fourth pick of the right hand motif increases by one step or one warp end. The turning point of the first and second pick of the left hand motif and the turning point of the third and fourth pick of the right hand motif are now overlapping on one common warp end. The turning points of the third and fourth pick of the right hand motif now face the turning point of the third and fourth pick of the left hand motif (cf. fig. 2.20). This type of tapestry weaving has been illustrated by Peter Collingwood (Collingwood, 1968: 152). This technique could fall under the broader category of slit tapestry. Beating in or packing in of the weft is done after every seventh or eighth pick and not after every pick.

Features of *Darī*

As the *darīs* must have a characteristic soft feel, the weft is not very tightly packed in by the weavers.

Darīs are woven today to meet the commercial demand. Starting from 2 ft. x 2 ft. to 7 ft. x 4 ft. they come in eight different sizes according to the orders received from both the internal and external market. There is a fair amount of shrinkage in width, about 2 in., which varies as the weavers sometimes intentionally distance the warp ends to increase the tension.

Slit tapestry weave. (Fig. 2.20)

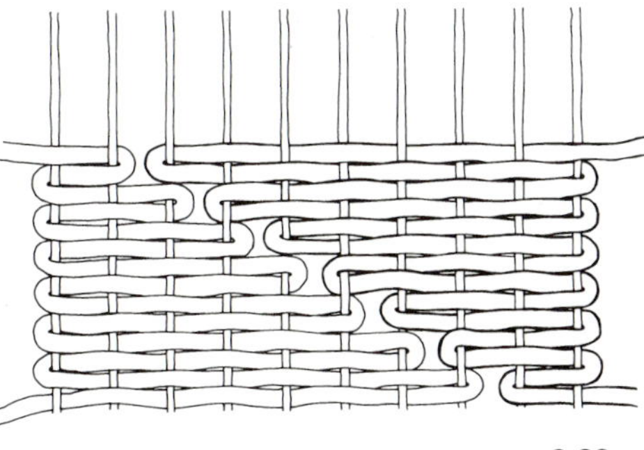

2.20

There are no set rules which define the composition of the *darīs*. Sometimes an additional row of motifs may be woven in to achieve the required length, the overall composition reflecting spontaneity. Today, traditional designs such as *Machhlī Paṭṭī*, diamond, *Laher*/wavelike and *Phūl*/flower, are combined with contemporary abstract designs to cater to the demands of the European market. Bright colours are used, in combination with dark colours. Red, blue, sky blue, leaf green, dark green, sap green, brown, and maroon were the traditional colours used.

Finishing

Trimming of protruding weft threads and tying of the warp ends to prevent the unravelling of the rug comprise the finishing process. For this a series of overhand knots are tied at both ends of the rug.

Gandha Loom of Rajasthan

Gandhas are weft faced plain woven floor coverings made by the Meghwal community of western India using cotton, camel or goat hair. The sturdiest are the ones made entirely of goat hair (*Camel Belts*, NID, 1992: 21). *Gandhas* are used in most houses as a floor covering in much the same way as a mat. Meghwals, a scheduled caste, earlier skinned dead animals and made shoes and other articles from leather. They have since given up the profession and are now leather craftsmen and weavers. The term *gandha* is self-explanatory. Due to the dirt, *gandha*, which collects on it while spread on the mud flooring of houses it is called *gandha* or dirty. Migrants of Pakistani origin in the Bikaner region, Rajasthan, however, refer to this product as *jirowi*.

Gandhas are woven to order. Due to their hardy nature they are very durable and can be used for twenty years or more without getting worn out. Therefore, the demand for them is infrequent. Among the Meghwals only a few weavers earn their livelihood from *gandha* weaving.

Material and Patterning

In *gandha* weaving, the warp is usually made from handspun goat hair, which is very strong, while softer camel hair may be used for the weft. Coloured cotton, obtained in the form of mill 'waste', is used for ornamentation as horizontal woven stripes. These horizontal stripes are usually in plain weave. However, geometric shapes may also be introduced by means of insertion of coloured yarn in the ground weave by weaving variations such as weft wrapping, slit tapestry and laid in supplementary weft. Each motif is worked with individual wefts. The picks of the weft are worked together, the left pick moving to the right while the right pick moves to the left. In this way diagonal floats which outline the motifs are created (see figs. 2.21, 2.22, 2.23). Cotton *gandhas* are also woven but these are not as hardy as those woven with camel or goat hair.

Spinning

Camel and goat hair yarns used for weaving are handspun. Spinning is done with a drop spindle called *dheria*. This is similar to the Turkish type of drop spindle (Brown, 1979: 228). The goat or camel hair is held in the left hand or under the left arm. With the right hand, the *dheria* is given a twirl in a clockwise direction, following which the fibres are drawn. Due to the momentum of the spindle the drawn fibres are twisted. The twisted yarn is wound around the *dheria* and the process of spinning is continued. When a substantial amount of yarn has been wound around the whorl, the cross bars are pulled out, leaving behind a ball of yarn.

Gandha Loom

The shaping of the outer structure of the frame loom can be observed in the *gandha* loom (see fig. 2.24). This loom is devoid of treadle and reed.

Four forked branches/poles are fixed upright to form the four corners of a rectangular space of approximately 12 ft. length and 6 ft. breadth. Two poles are set lengthwise on the forked poles connecting them and forming a basic framework. A wooden branch or rod is propped up horizontally, resting on the two longitudinal branches/poles. The heddle rod is suspended from this horizontal rod with rope tied at two ends of the heddle-rod, balancing it parallel to the overhead rod. The warp is a continuous one made of goat hair yarn or cotton yarn wound around two horizontal beams/rods which are attached, by a rope at each end to a pole. These poles are placed behind the vertical poles and are held by the tension of the warp. These may also be fixed to the vertical poles. The two warp beams are attached to another set of poles on either side with rope to facilitate the regulation of warp tension. The warp ends get tighter as weaving proceeds and the distance between two warp beams can be adjusted by releasing the rope holding the two horizontal poles together.

2.21

2.22

2.23

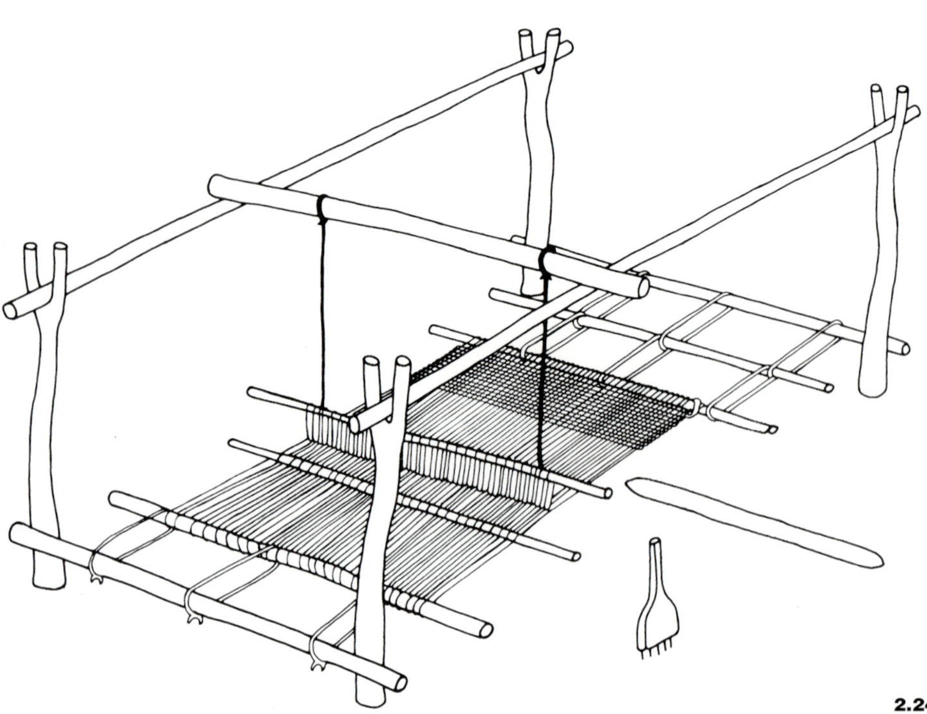
2.24

Soumak weave/weft wrapping in *gandha* floor covering. NID Coll., Ahmedabad. (Fig. 2.21)

Jāmdānī, laid in supplementary weft weave in *gandha* floor covering. NID Coll., Ahmedabad. (Fig. 2.22)

Slit tapestry weave in *gandha* floor covering. NID Coll., Ahmedabad. (Fig. 2.23)

The loom used for *gandha* weaving. (Fig. 2.24)

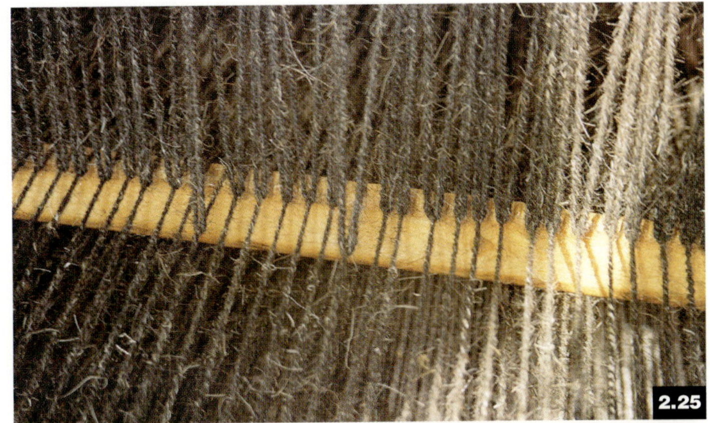

2.25

2.26

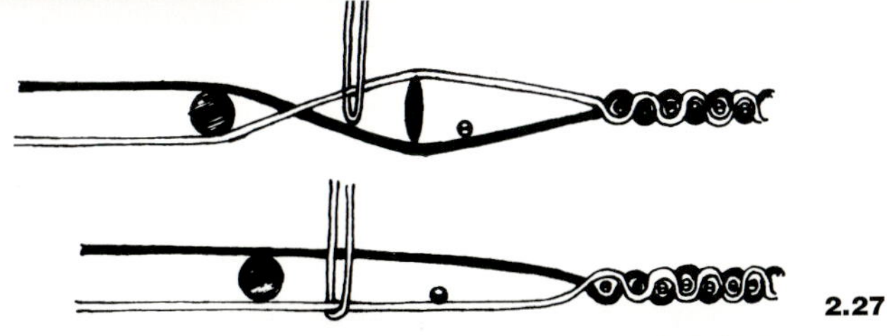

2.27

2.28

The heddle rod is made of a series of continuous long loops/half heddles, through which every alternate warp end is passed. The warp ends are spaced evenly as required by the weaver according to the knowledge gained from previous experience (see fig. 2.25).

Weaving Process

Weaving is done by two weavers sitting on either side of the warp. The first shed is created by manually lifting the heddle rod thereby also lifting the warp ends that pass through the half heddles and leaving the unthreaded ends unlifted. A space between these warp ends is created.

A shed stick is inserted into this space. The shed is cleared, and the shed stick is propped up on its edge to keep the shed open allowing the passage of the weft. After the weft is passed through, the shed stick is removed and the weft pick is beaten in place with the help of a *pañja* (fig. 2.26).

The second shed is created by bringing the lease rod close to the heddle rod, leading to the raising of the undrafted ends and lowering of the ones drafted through the heddles. A small space is created allowing the passage of the shed stick followed by the same processes as described in the previous weaving operation (see fig. 2.27).

The sliding overhead horizontal bar attached to the heddle rod is moved away from the weavers as the weaving progresses. Due to the warp ends being sparsely distributed the structure of the *gandha* is weft faced and closely woven.

Finishing

On completion, edges are finished through the usage of the ply-split braiding technique. The ply-split braiding technique is also used in the making of camel belts (Speiser, 1977: 51–2) (figs. 2.28, 2.29). This technique, like that of the *pañja darī*, appears to be an adaptation of a non-Indian prototype.

2.29

Heddle rod, *gandha* loom. Adivasi fair, Ahmedabad. (Fig. 2.25)

Beating with *pañja*, *gandha* loom. Adivasi fair, Ahmedabad. (Fig. 2.26)

First and second shed obtained by use of heddle and lease rods respectively. (Fig. 2.27)

Camel belt. NID Coll. Courtesy: Errol Pires, NID. (Fig. 2.28)

Spinning of wool, ply-split braided camel belt. Craft Museum, New Delhi. (Fig. 2.29)

Pañja Loom of Fatehpur Sikri: *Darīs* and *Farśa*

The *pañja* loom used for *darī* and *farśa* weaving of Fatehpur Sikri can be described as a horizontal ground loom with two heddle rods. This loom is devoid of reed and treadles (see fig. 2.30). It consists of three stout wooden beams. To one of these (1) is attached the warp beam at one end with the breast beam or cloth beam (2) facing it at the other end.

The distance between these two beams is determined by the length of the warp. Ordinarily, the warp is set up for two *darīs*. The size of the *darīs* can vary from 3.5 ft. x 6.5 ft. to 12 ft. x 15 ft. The two beams at the extreme ends are raised about 6 in. above the ground and are secured from either side by ropes tied to pegs driven into the ground. These pegs hold the beams in place. A third beam is attached at the far end parallel to the warp beam. In the line drawing we can consider the warp beam as (2) the stout beam attached to the warp beam as (3) and another beam attached to the farthest end of the cloth beam side as (1).

Beam numbers 2 and 3 are connected by a rope called *chikkī* in a figure of eight in five places equidistant to one another (4). A small piece of wood called *adeka* is inserted horizontally into each of the intersections formed by these ropes (5). When the warp ends have to be tightened, the *adekas* are rotated twisting the rope in the process. These are then fixed into the ground to keep them in place. A heavy stone slab is placed in the centre of the beam 2 to keep it in place and prevent it from shifting.

After the heddles are made across the width of the warp, a bamboo rod is passed through the loops above the warp ends. The cord holding the loops is tied to the two ends of the rod. When the two heddle rods are ready these are tied to the *ratch*, a wooden beam propped up on two stands placed on either side of the width of the loom perpendicular to the heddle rods. These stands are constructed by crossing and tying of two bamboo poles obliquely in two opposite directions.

The *ratch* supports two pairs of heddle raisers, *makḍā*, on both sides of the beam (fig. 2.30). The *makḍā* is a small bow-shaped piece of bamboo which rests across the *ratch*. Two grooves are carved out and the *agaḍā*/front and *pichhaḍā*/ back ropes are tied at these two points forming an inverted triangle. The *agaḍā* rope is the one closer to the weaver, while the *pichhaḍā* is the one away from him. Both these ropes are equal in length and are tied tautly at the same point to one set of heddles. This is done by tying them to the bamboo rod which supports the heddles (fig. 2.31).

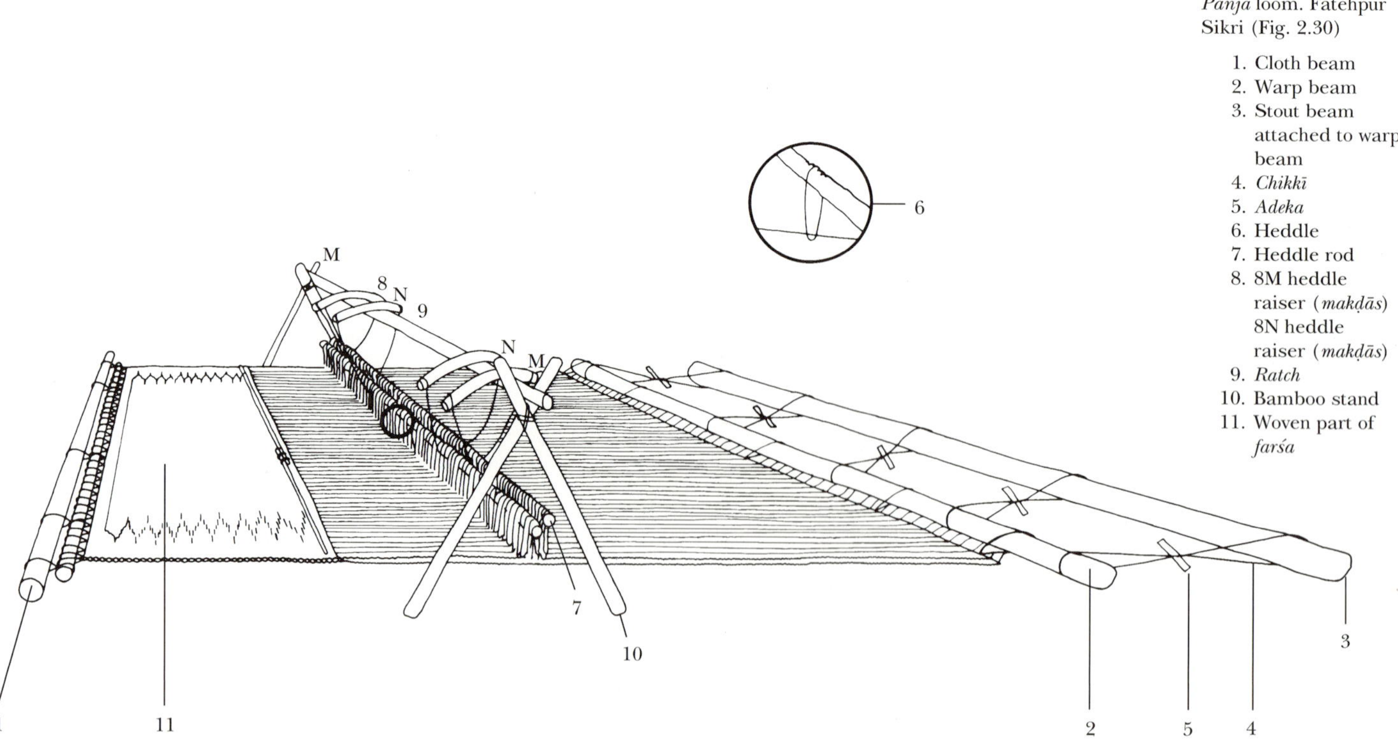

Pañja loom. Fatehpur Sikri (Fig. 2.30)

1. Cloth beam
2. Warp beam
3. Stout beam attached to warp beam
4. *Chikkī*
5. *Adeka*
6. Heddle
7. Heddle rod
8. 8M heddle raiser (*makḍās*) 8N heddle raiser (*makḍās*)
9. *Ratch*
10. Bamboo stand
11. Woven part of *farśa*

2.30

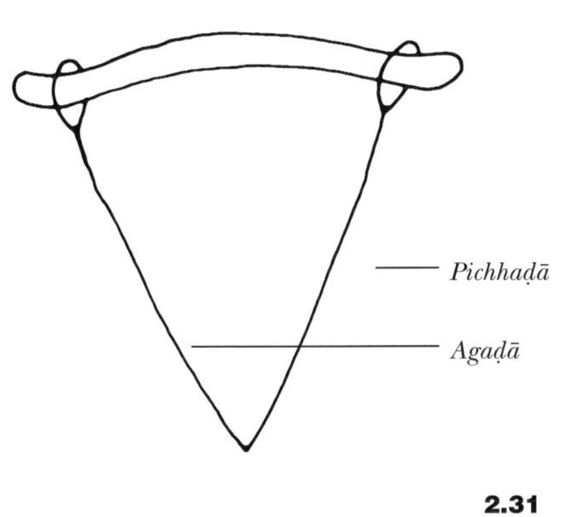

2.31

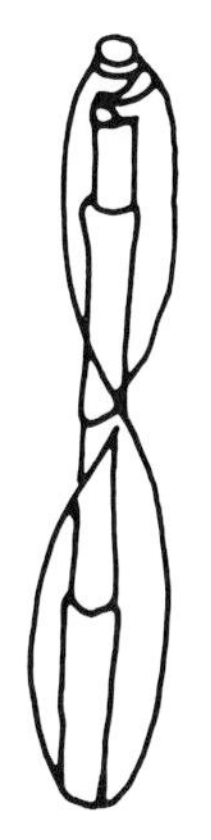
2.32

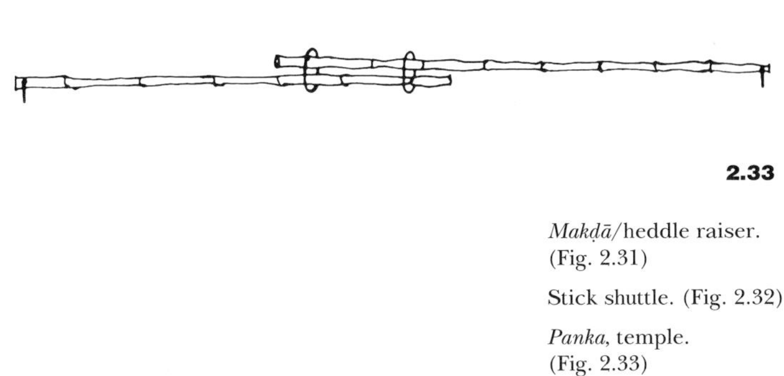
2.33

Makḍā/heddle raiser. (Fig. 2.31)

Stick shuttle. (Fig. 2.32)

Panka, temple. (Fig. 2.33)

One set each of heddle raisers on either side of the *ratch*, support one heddle rod/shaft each. Two weavers are normally required to work on the loom; each operates one set of *makḍās* coordinating simultaneously in lifting and changing of shafts.

When the weaver pulls one set of *makḍās* towards him, that particular heddle rod/shaft is lifted, forming a shed. The weft wound on stick shuttles is inserted and beaten in (fig. 2.32). The shed is changed by pushing the *makḍās* (M) in place and pulling the other set of *makḍās* (N) towards him. The weft is inserted and beaten in repeating the process and continuing the weaving. When about two feet of the *darī* is woven, the two weavers sit on a wooden board/*pāṭa*. This wooden board is placed on the woven portion to avoid direct pressure of the feet on the *darī*.

A temple or *panka* is fixed along the width of the *darī* to regulate its width. This is a simple device made of two bamboo sticks with nails attached at one end of each of the rods. These two bamboo sticks are placed along the width of the *darī*. The nail ends pierce into the selvedge, the distance being adjusted by overlapping the sticks at the centre. These are held in place by two loops (see fig. 2.33).

Warping

The warping process commences after the selvedge ends are tied on both sides. These determine the width of the *darī*. The selvedge is made of thicker yarn than that of the warp. It is made of 12 ply 2s yarn (already plied 4 times), twisted to form a rope. This helps maintain a neat selvedge and the width of the *darī*. While weaving 4 ply 6s yarn is used for making the warp. The yarn is wound into a ball, one end of which is taken by the weaver who ties it around one extreme end of the cloth beam next to the selvedge warp and then loops it around the warp beam and back to the cloth beam, walking up and down with the ball of yarn and looping it at both ends, till he completes the requisite number of ends predetermined by the width of the *darī*. The selvedge ends act as guides for the width.

The looping around the cloth beam and the warp beam is done to form a figure of eight to help maintain the order of the warp ends. At the same time as the weaver makes the warp, another weaver makes the heddles and also twines each warp end together spacing them equally and holding them in place on the warp beam. Normally the *farśa* has 12 ends and 28 picks per inch, whereas the *darī* has 12 ends and 34 picks per inch.

As the weaving progresses, the weavers move the entire weaving mechanism, i.e. the stands holding the *ratch* and the heddle rods, along the length of the warp leaving the woven areas and the warp in their original set up.

Finishing of the rugs is done by removing the knots and cutting the protruding weft yarns with the help of the *chhurrā*. The ends of the *darī* are knotted to form tassels.

Tools: for Pre-weaving and Weaving

***Kalpu* and *kani*:** *Kalpu* and *kani* are tools used for making heddles. *Kalpu* is a rectangular piece of wood about 8 in. long, 4.5 in. broad and ½ in. wide. This acts as a guide for maintaining the length of the heddles. All the edges of the *kalpu* are rounded and smooth. One end of it is shaped like a keel. This is the part through which the ready heddles are slipped off. There is a hole at the end opposite to that of the keel-shaped one. This is sufficiently large to permit passage of a long thick cord through it.

The *kani* is made of wood and is shaped like a thick needle. This needle shaped stick is about 9 in. long, an inch longer than the *kalpu*.

***Kalva* or *tairi*/stick shuttle:** A stick shuttle is used to wind the weft yarn around it in form of a figure of eight (fig. 2.32).

***Pañja*:** The *pañja* or beater-in is a toothed device. It consists of a wooden handle about 6 in. long, into the broadened end of which are driven steel or iron blades evenly spaced apart. Each blade is shaped in the form of a penknife blade with the tips all levelled and the curves flush filed. The edges, however, are not sharp. The weaver grips the handle by his hand while the tips of the blade face the woven area. He gently pushes the weft in place with the tips of the blade.

Chhurrā: A 10 in. long strip of iron $^1/_4$ in. thick is bent to form a curve. The frontal part of $1^1/_2$ in. is tapered to a point and used to pull out knots. The middle part, 5 in., is beaten flat to form a blade with a curved edge for cutting the knots and the remaining part forms the handle of the cutter or *chhurrā.*

Heddle Making Process

For heddle making, the heddle making guide, *kalpu,* is placed over the warp ends. The half-heddles are made with strong yarn along the width of the warp. The heddles for the odd and even ends are looped separately. A strong continuous yarn is passed from under an individual warp thread looped on to the *kalpu.* It is looped twice over the wooden needle placed over the *kalpu* towards the weaver who takes the end and passes it under the second warp end, looping around the *kalpu,* and twice over the wooden needle. This continues till the weaver completes one set of heddles around the warp ends from left to right. The weaver has to slip off the heddles each time the *kalpu* guide is covered with heddles. These are held in place by a cord that is threaded through the wooden needle and passed through the loops formed on the *kani.*

Techniques used in Weaving of *Darī* and *Farśa*

Darī and *farśa* are both flat pileless floor coverings woven in weft-faced plain weave. The *darīs* normally have stripes in different colours with very little or no interlocking whereas the *farśa* is a large size patterned cotton *darī* using the slit tapestry weave (fig. 2.34).

Farśa is a Persian word which literally means 'floor'. Such items are used in homes as floor coverings, for weddings and ceremonies where large spaces of the ground need to be covered. Geometric shapes are woven in the main field of the *farśa* which is surrounded by a frame in a contrasting colour complementary to that used in the main field. The most popular colours in the *farśa* are red, blue and white (*Darī Fatehpur Sikri,* NID, 1980).

India would seem to have developed considerable variation in heddle formation which appears, in general, to have allowed obviation of the rigid heddle in fibre weaving. It is in this context that the *newār* loom of Wadhwan and the *guddar* loom of Navalgund deserve special mention. The rigid heddle is, however, to be found in mat weaving looms (fig. 2.35). One of the variations in heddle formation has been noted above in the weaving of *gandha* floor coverings in Rajasthan (fig. 2.24).

Navalgund *Darī* Loom

The cotton *darīs* made by the weavers in this town are traditionally known for their excellence in quality both in terms of their construction as well as the originality in the use of colour and form, making them distinctly different from the *darīs* made in the other parts of our country.

The origin of Navalgund *darīs* can be traced back to a group of weavers who originally migrated from Bijapur to Navalgund. Navalgund town is the headquarters of the taluka of Navalgund in the Dharwad district of Karnataka (*Navalgund Durries,* NID, 1986).

There are still a few old master craftsmen left, each of who add the word *jāmkhānā*/carpet, as a suffix to his own name. Thus we have Nabu Sahib Jāmkhānā, Jandi Jāmkhānā, Makdun Jāmkhānā, and so on (Chattopadhyay, 1984: 34). These weavers used to reside and work in Jāmkhānā Galli, in Bijapur. Today only the area known as Jāmkhānā Galli exists.

During the sixteenth century, there was a tussle for supremacy between the Deccan Sultanate and the Vijayanagar Empire culminating in the battle of Talikota in AD 1565. During these troubled times *jāmkhānā* weavers migrated from Bijapur. It was in this way that these weavers formed an enclave in Navalgund where they continued with their practice of weaving.

2.34

Slit tapestry weave in *farśa* (Fig. 2.34)

Mat loom, Dokaher village, district Madhubani, Bihar. Jute warp. Weft *pater,* a vegetable fibre. (Fig. 2.35)

The exact situation and time for having made the shift from weaving of carpets to *darīs* is not clear. However, this may be attributed to the decline in the demand for carpets owing to a lack of patronage. One of the major factors could have been the cost. Not only was there a major shift in the product but also in the material. It is suggested that the weavers having come to Navalgund started weaving woollen *darīs* instead of carpets, but as the demand for cotton *darīs* increased they shifted to making these, while retaining their traditional designs and techniques.

The Navalgund Loom: Tools and Accessories

Darīs are woven all over India, and the data on *darī* loom types which emerge suggest that looms used could be classified under the broader category of horizontal looms. However, there is an exception. The Navalgund *darīs* are woven on a vertical loom. The Navalgund *darī* loom, locally called *khadav magga* means vertical or upright loom. This clearly indicates the vertical character of this loom (fig. 2.36).

Given the adaptability of the weavers in using new products and materials, it would be reasonable to assume that the weavers who were originally carpet weavers using the vertical loom, retained this form while making the shift to *darīs* — a tradition with which the weavers would not have been unfamiliar.

The *khadav magga*, made from locally available material, comprises:

1. A circular warp beam/upper beam made of wood, about 9 to 10 ft. in length with a tensioning lever.
2. A circular cloth beam/lower beam made of wood, with a tensioning lever that corresponds to the length of the warp beam.
3. Two vertical wooden slot-bearing beams at one end, measuring about 6 ft. from ground level.
4. A horizontal support bar for the heddle rods.
5. Two heddle rods.
6. Two lease sticks.
7. Heddle raiser.
8. Two bamboo poles/rods to which the warp ends are tied at the upper and lower extremities respectively called *uparkā phāsa* and *nīchekā phāsa.*
9. *Pañja* or beater-in.

The weavers have their loom permanently installed in a corner of their house usually extending along one side of the wall. Two vertical wooden beams with a bearing slot are fixed into the ground. The distance between the two is maintained by the length of the warp beam. The warp beam with a shaft at the two ends is propped horizontally on to these poles into the bearing slot. The cloth beam is also fixed to the vertical beam but is set in a pit about 6 in. below the ground. The two tensioning levers one each on the warp beam and the cloth beam, placed diagonally opposite to each other maintain the tautness or the tension of the warp.

After the warp is made and fixed on to the warp beam and the cloth beam, the two heddle rods are made by the weaver. These heddle rods are tied to the heddle raiser/*kamān*, shaped like an arch, which goes over the horizontal bamboo rod positioned 3 ft. above the ground near the wall. The weavers sit on the ground while weaving. With the dimensions of the loom given above, 8 ft. wide *darīs* can be woven. In 1986, a loom as wide as 20 ft. was seen on which 18 ft. wide *darīs* could be woven. The basic character of the pit loom are thus found combined with that of the vertical loom.

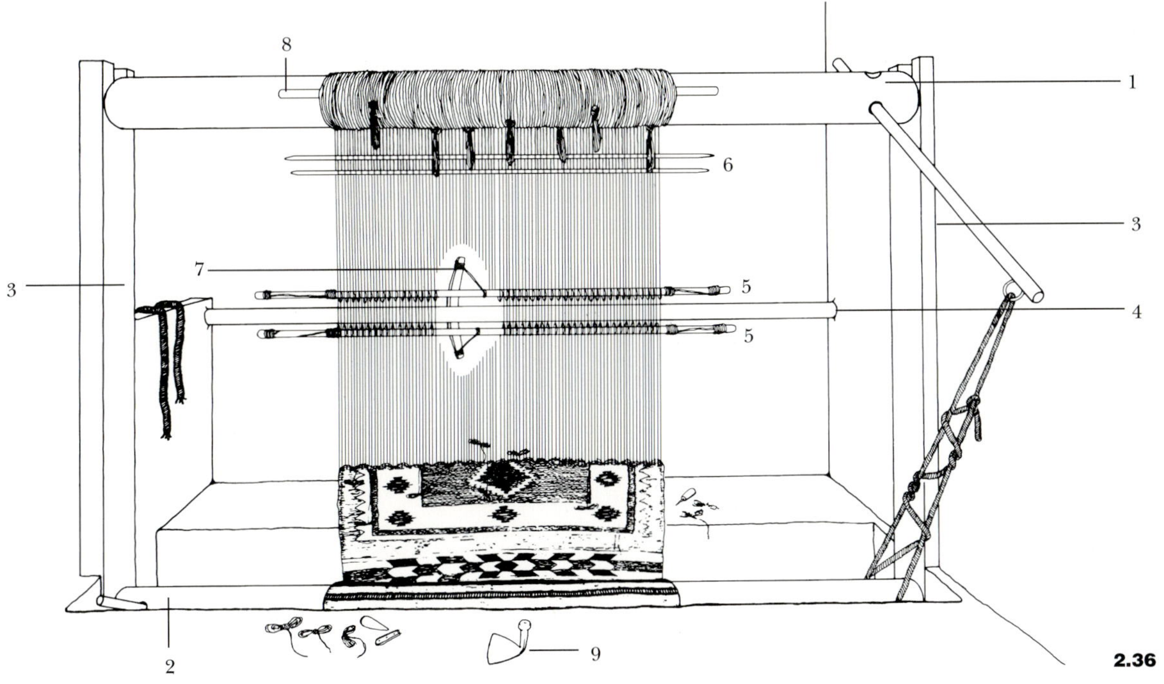

Khadav magga, the vertical loom used by weavers of Jāmkhānā. (Fig. 2.36)

1. Circular warp beam
2. Cloth beam
3. Slot bearing beam
4. Support bar
5. Heddle rod
6. Lease sticks
7. Heddle raiser
8. Bamboo pole
9. *Pañja*

2.36

Pañja/beater-in

The *pañja* is a fork-like device made of forged iron weighing from 800 gm. to 1 kg. It has a squarish handle grip with a spherically upsetted end. The solid square rod is beaten down at right angles to the handle, almost forming a sheet, diverging from the handle and fanning out into a triangular shape. The edge of the blade is fashioned like a saw with blunt parallel teeth (see fig. 2.37).

Initially, the Punjab type *pañja* was used. However, as the blades tended to break while beating in the weft, the tool was redesigned by one of the weavers. This tool is locally made by the blacksmiths.

Other Tools

Tibni: This is a wooden pointed cone-shaped tool which the weaver uses to guide down the weft after the slackness in the weft is given.

Chhurrā: A knife used for cutting the weft yarn.

Rati: Hank opener made of bamboo splits.

Pre-weaving Process

The pre-weaving process involves the following steps:

1. Warp making.
2. Heddle and heddle rod making.
3. Dyeing of weft yarn.
4. Plying of weft yarn.

Warp Making

This activity is mainly carried out by men. 3/10s unbleached cotton yarn is used for the warp. There are approximately 12 ends per inch and 28 picks per inch.

The length of the warp is dependent on the number of *darīs* to be woven, e.g. if 3 *darīs* of 6 ft. length were to be woven the weavers would make a warp of 26 ft. length.

It is commonplace in India to find the open spaces available in and around living areas being used in an extremely versatile way. One such activity seen in such spaces is warp making. It is interesting to note that a similar method of warping with pegs driven into the ground is also used by the weavers of the *kani paśmīnā* shawls in Kanihama, Kashmir.

Three wooden sticks/pegs aligned in a row are driven into the ground leaving them about 2½ ft. above the ground. The distance between the two extremes is determined by the length of warp to be made. The third stick plays the important role of guide to cross the warp ends for the purpose of maintaining their order and forming the lease.

The warp is made forming a figure of eight. After completion of the warping process, a thread is passed around the crossing of the warp ends and tied. This forms the lease to maintain the order of the ends. The warp is then rolled on to the furthest stick close to the lease, carried home, and unwound on to a rod wider than that of the required warp width.

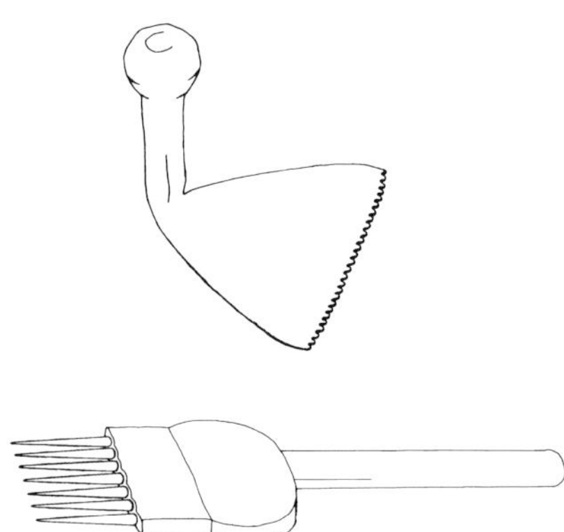

2.37

The *pañja* from Navalgund (above) and the *pañja* from Punjab (below). (Fig. 2.37)

The warp ends are spread out so that they are evenly spaced across the required warp width. The two ends of this rod are each tied with ropes which in turn are tied to the warp beam. With the help of two more people, one rotating the warp beam and the other holding the warp, even tension is maintained till the winding of the warp on to the warp beam is completed (1).

The lease thread is now replaced by lease sticks. Through the looped ends of the warp a stick is passed and tied to the cloth beam, after which the tension of the warp ends is adjusted and evened.

Heddle and Heddle Rod Making Process

Thick cotton thread is used for making the heddles. To begin with, alternate warp ends are picked up on individual lease sticks. Half heddles are made by looping the cotton thread around one set of warp ends going over the heddle rod and around the rope placed over the rod to grip and hold each heddle in place (see fig. 2.38).

After having completed one set of heddles with a heddle rod, the same process is repeated for the remaining set of alternate warp ends and the second heddle rod is also completed. These two heddle rods operate like two individual shafts. From the centre of these rods, a rope is tied to each end of the arched heddle raiser so that the heddle raiser lies tautly above the horizontal support bar for the heddle rods. The heddle raiser, when lifted upwards raises one set of warp ends to form a shed. When pulled downwards, another set of warp ends is raised to form the counter shed (7).

Heddles, detail. (Fig. 2.38)

2.38

Dyeing of Weft Yarn

Basic direct dyes are procured from Hubli and the dyeing procedure is carried out at home. The dyestuff is added to water and dissolved. Hanks of weft yarn are put into the dye bath and left to absorb the dye, after which they are taken out and dried.

Plying of Weft Yarn

6 ply 10s cotton yarn is used for the weft.

With the help of the hank opener/*rati*, the yarn is simultaneously unwound and plied six times. This plied yarn is converted to finger hanks or butterflies.

The *jāmkhānā* weaver weaves two distinctly different styles of floor coverings, the *jāmkhānā* and *jā-i-namāz*, the difference lying in their use, patterning and size.

Jāmkhānā, meaning floor covering in the local language, is mainly used during marriages and are occasionally used in the house. They are woven in three sizes: 3 ft. x 5 ft., 4 ft. x 6 ft. and 6 ft. x 9 ft. The weavers also carry out some local orders. This *darī* can be usually divided into three distinct parts; the centre of the *darī* and the two ends. The two ends of the *darī* are woven in reflected symmetry whereas the centre is woven with geometric and simplified natural forms bound within a frame, making it the focal point of the *darī*. Red is the dominant colour, used almost like a background colour, with yellow to set off the other colours placed within it. White, green and black are used as the accenting colours. There are about five variations in the design of a *jāmkhānā*. These are worked out around the broader parameters of their compositional breakdown as given above. These comprise:

1. *Sādā phūl jāmkhānā* which has a big geometrical flower placed in the centre of the *darī* with borders at two ends.
2. *Baḍa mor jāmkhānā* has a large size peacock placed in the centre of the *darī* with borders at two ends.
3. *Pagaḍī atte* has the *chaupaḍa* (traditional Indian dice game) layout in the centre of the *darī*, serving the dual purpose of offering a place to sit as well as providing the board, *chaupaḍa*, on which to play.
4. *Chār-mor-jāmkhānā* also has the *chaupaḍa* placed in the centre of the *darī* but there are four peacocks placed in the four corners of the *chaupaḍa* frame.
5. *Tottale jāmkhānā* is a small *jāmkhānā* about 1½ in. x 4 in. which is used to line the inside of a child's cradle.

A sample of a *jāmkhānā* made in 1920 from the collection of Dilip Joshi in Navalgund can be seen in (fig. 2.39).

Jā-i-namāz is a prayer mat or prayer rug used by the Muslim community. *Jā-i-namāz* is of two kinds: one, which is 2 ft. x 4 ft., is used by a single person for prayers. The main feature distinguishing it from the other *darīs* is the arch. The other is the *jā-i-namāz* used in mosques by the community. A number of people can sit in a single row on this and pray. A number of *mehrābs* or arches are repeated in this type of *jā-i-namāz* and these can be custom made to fit the size of the prayer hall of the mosque.

2.39

2.39a

Tottale jāmkhānā, 1920. Courtesy: Joshi, Navalgund. (Fig. 2.39)

Tottale jāmkhānā, 1920. Courtesy: Joshi, Navalgund. (Fig. 2.39a)

Jā-i-namāz. Note *mehrāb* motif.
(Fig. 2.40)

Vertical loom used in *jāmkhānā* weaving.
(Fig. 2.41)

Shedding mechanism of Navalgund *darī* loom
(Fig. 2.42)

1. *Khamb*/beam
2. *Tana*/warp
3. *Sabbal*/tension rod
4. *Kaichi* - crossing of warp ends
5. *Gulla*/heddle
6. *Gulleka phasa*/ heddle stick
7. *Kamanka phasa* - Bamboo holding *kaman*
8. *Kaman*
9. *Ani*/shed
10. *Rassi*/rope
11. *Tang*/tension rod

2.40

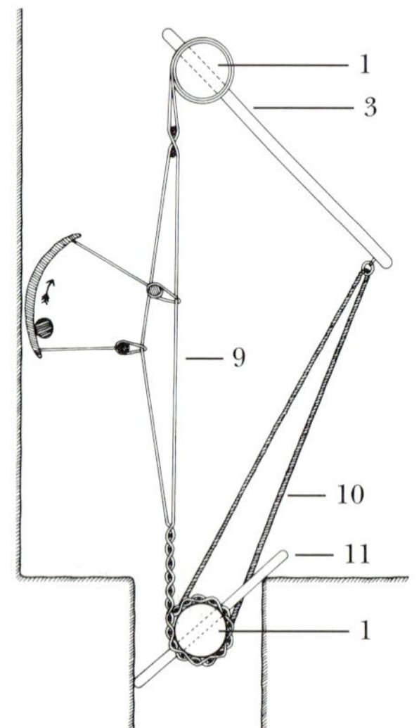

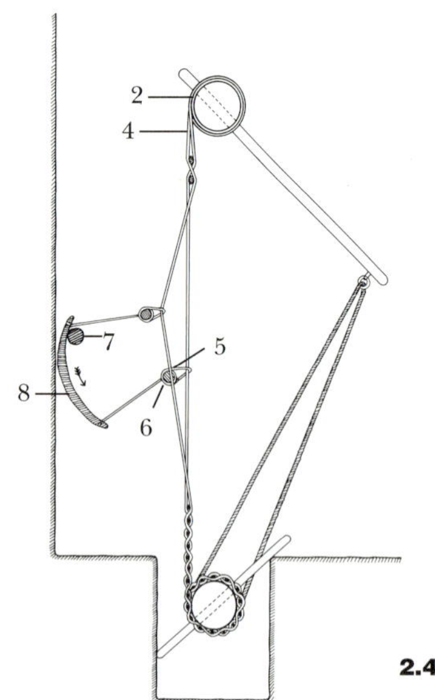

2.42

2.41

In both these types of *jā-i-namāz* the width of the *darī* is used as the length by the person who prays (fig. 2.40).

Weaving Process: Fabric, Texture and Techniques

The weaving activity is practised mainly indoors by the women of the house.

As soon as the tensioning of the warp is completed, the loom is ready for commencement of weaving. Two weavers, facing each other across the warp, work on this loom. They work in close coordination throughout the process of weaving. Their work is clearly distributed to avoid confusion (fig. 2.41).

The weaver who sits at the back of the loom operates the heddle raiser. As she manipulates it away from her, the even number of warp ends are lifted, forming a shed and when she brings the heddle raiser towards her the odd number of warp ends are lifted, forming the counter shed (fig. 2.42).

She also weaves one half of the *darī* width, whereas the weaving of the other half of the *darī* and beating in of the weft with the *pañja*, is done by the weaver sitting in front of the loom. This involves perfect co-ordination of effort.

These *darīs* are woven in a weft faced plain weave necessitating an extra length of weft to be inserted so that the warp may be totally covered. The *tibni* is used to guide the weft down to its place after having given the extra length in weaving.

When a continuous weft is required to be woven-in from end to end, one weaver inserts the pick into the shed until the central point of the warp is reached. The second weaver continues the pick to its point of termination. This is then beaten down. The main patterning in these *darīs* is achieved through the dovetailed and slit tapestry weave whereas the extra weft technique is used sparingly to form borders across the width of the *darī*.

Different coloured wefts, inserted from finger hanks are woven into one shed, all moving in one direction. These then return in the opposite direction when the next shed is formed.

The weavers do not have any reference graphs while executing the patterns and motifs. Each weaver weaves directly from memory. Also, being spontaneously woven, no two *darī* designs are woven alike.

The selvedge is rather unique. The outermost side of the selvedge warp is thick; the number of warp ends being about 10. This is gently graduated till the selvedge merges into the ground fabric leading to a gradual transition and avoiding the sudden bulge at the selvedge visible in other *darīs* (fig. 2.43). To help maintain the width of the *darī*, strong, twisted thread is inserted through both the selvedges and then tied to the vertical pillars of the loom.

The vertical lines in the *darī* are formed by the 1/1 single dovetailing technique (fig. 2.44).

The diagonal lines are achieved by the use of the slit tapestry weave (figs. 2.45, 2.46, 2.47). However the slits are too small to be visible. As Irene Emery, has explained, the term 'slit tapestry' is often used specifically to imply visible slits or long slits that have been sewn. The word 'unlocked' has been used to describe the slit-tapestry structure where the slits are too small to be visually significant, although 'unjoined' might be better for the purpose since it could not be construed as involving a distinction between interlocking and dovetailing. However, since the weave structure is not affected by the length of the slit, there should be no need for a structural term that makes a distinction based solely on the aspect of the slit. All slit tapestry could, of course, be described as unjoined (Emery, 1980: 89).

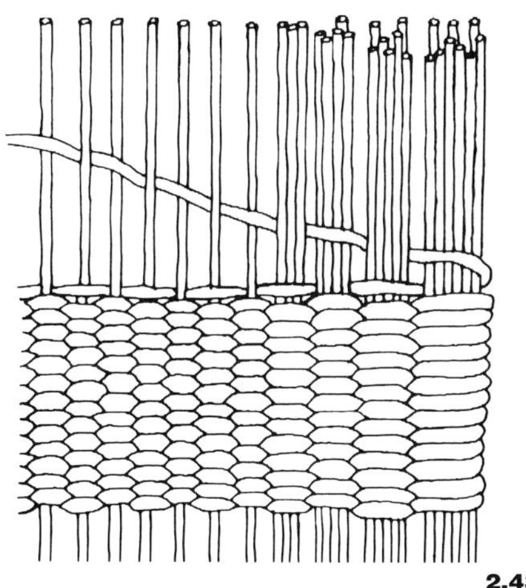

2.43

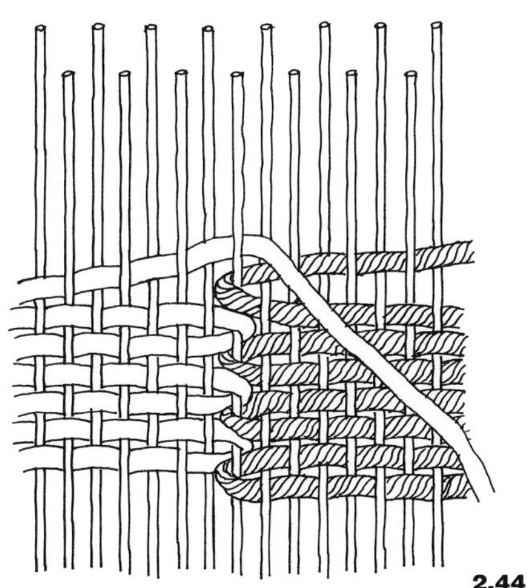

2.44

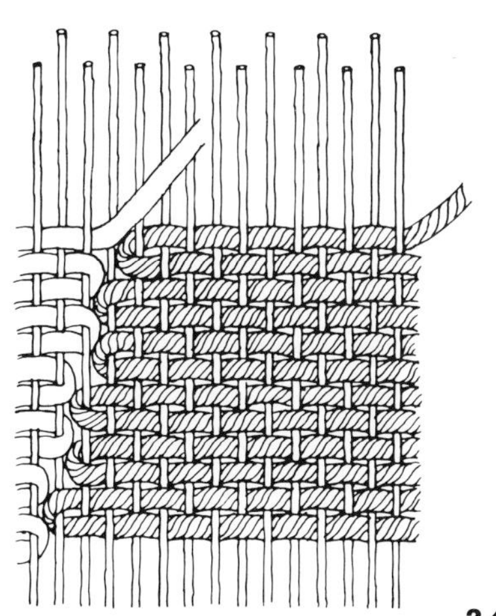

2.45

2.46

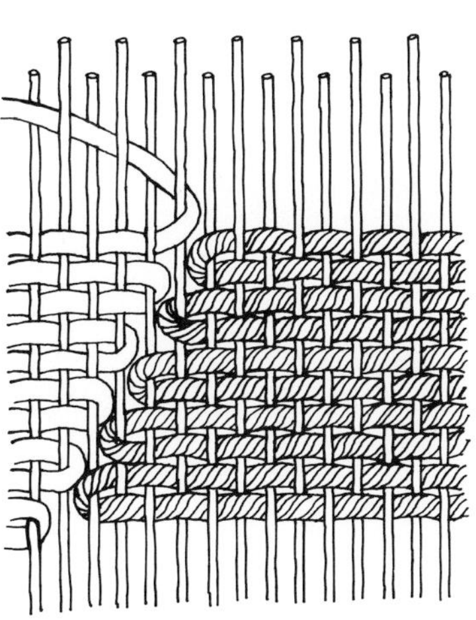

2.47

Selvedge, detail. (Fig. 2.43)

Weaving of vertical line. (Fig. 2.44)

Slit tapestry weave: a. demonstration of movement of 2 coloured wefts to achieve an angle of approximately 80° required in the patterning. (Fig. 2.45)

Slit tapestry weave: b. demonstration of movement of 2 coloured wefts to achieve an angle of approximately 45° required in the patterning. (Fig. 2.46)

Slit tapestry weave: c. demonstration of movement of 2 coloured wefts to achieve an angle of approximately 60° required in the patterning. (Fig. 2.47)

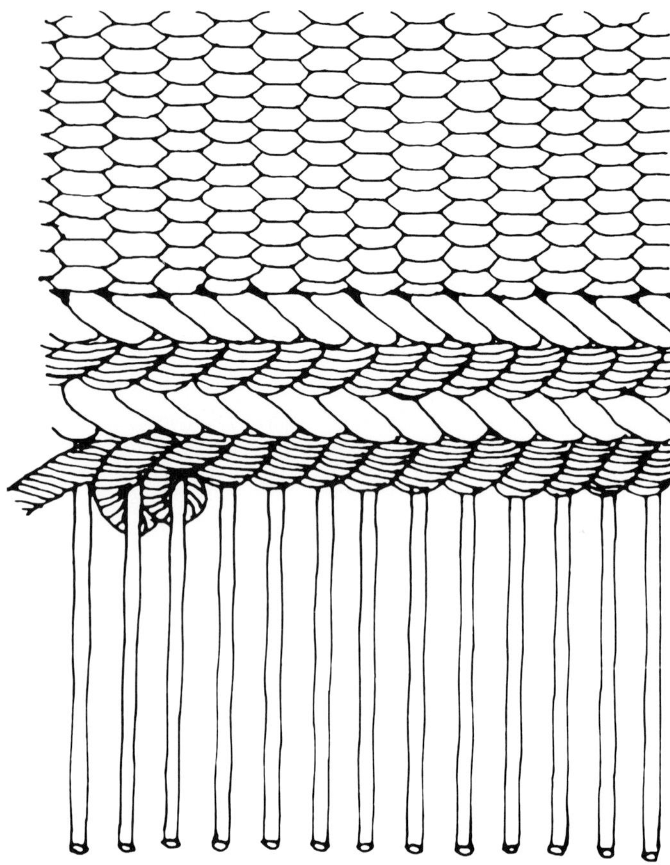

2.48

Finishing of edge. Navalgund *darī*. (Fig. 2.48)

Finishing Process

Once the *darī* is completed, it is taken off the loom and all weft threads left hanging are trimmed. The loose warp ends are grouped into the required sets and finished, using a series of overhand knots to form tassels. The rug is sometimes also finished with some rows of *soumak* or weft wrapping (see fig. 2.48).

Motifs Used in *Jāmkhānā* and *Jā-i-namāz*

The motifs and patterns used in *jāmkhānā* weaving are based on geometrical shapes. Prayer rugs are ornamented with the *mehrāb*, the Islamic niche motif. The names of the patterns are drawn from the local Kannada language, as well as from Urdu and Hindi. The terms used are:

Bel	Creeper pattern with zigzag edge
Chinda	Vertical stripes
Phūl	Geometrical flower
Kangorā	Comb like pattern
Baḍī gharī	Zigzag diamond shape with double edge
Nanhī gharī	Zigzag diamond shape with single edge
Laherī	Geometrical pattern like a wave
Dhārā	Horizontal stripes
Chunnaṭ	Twill diamonds with extra weft
Pagaḍī atte-Chaupaḍa	The board of a traditional Indian dice game
Mor/Navelu	Peacock (Kannada)

It is said that after coming to Navalgund, *jāmkhānā* weavers started using *mor*, the peacock motif, in the *jāmkhānā*.

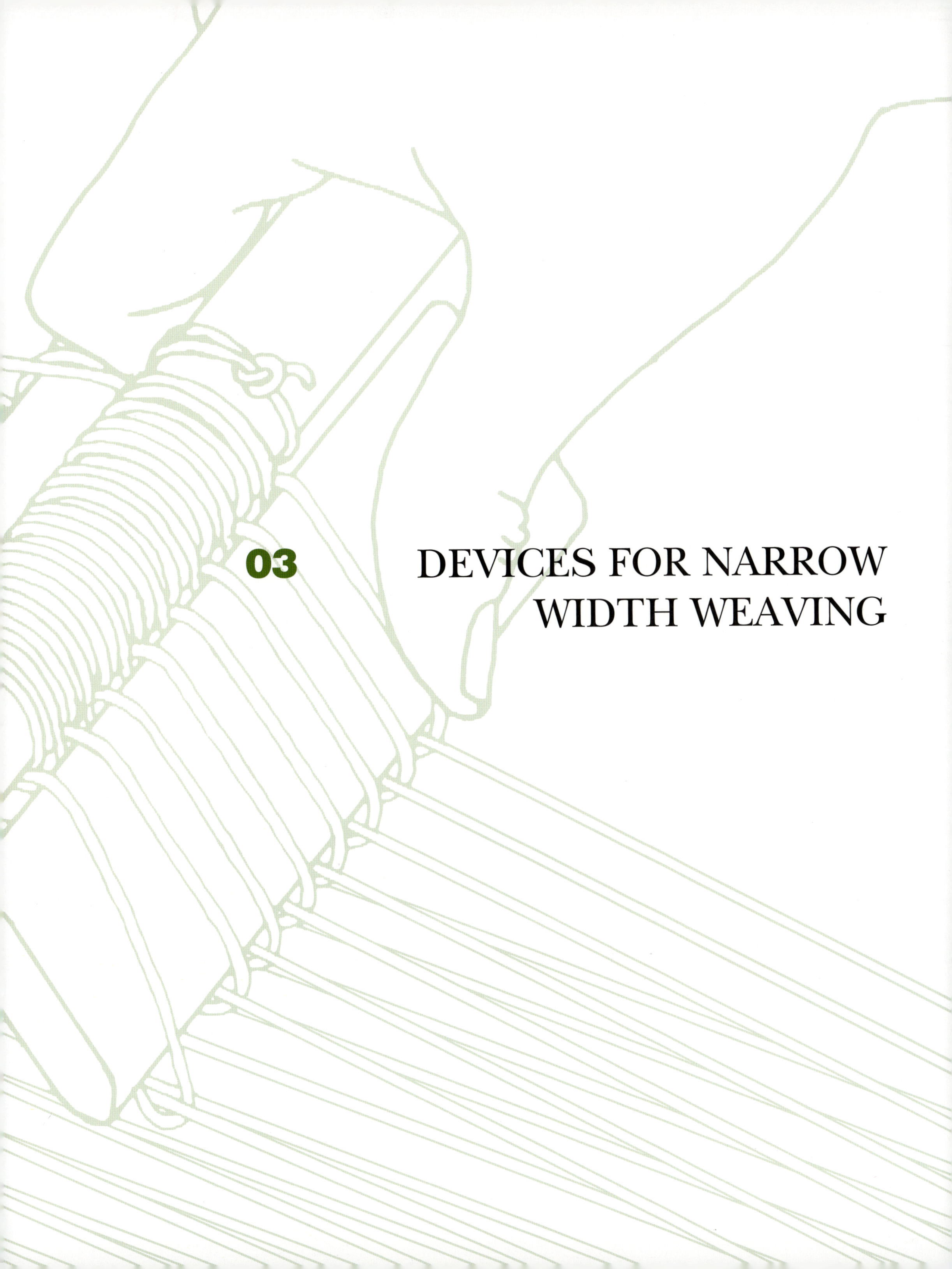

03 DEVICES FOR NARROW WIDTH WEAVING

03

Devices for Narrow Width Weaving

Apart from the looms described in the previous chapter, there are other devices through which fabric can be created. Among these the implements used in card weaving can be cited. In this context it is important to remember that even though Marie Schuette infers that India was a centre of card weaving, ethnological evidence points to the existence of a varied tradition of narrow looms in this country (Schuette, 1956: 11). This includes the loom used by Parsis to weave *kushti*, the Parsi sacred thread, worn by both men and women after the *navjyot* ceremony which is usually performed at the age of seven. Card weaving may be discerned in the tape used for decoration of the Kashmiri *phiren* worn by women (figs. 3.7, 3.9), and in the finishing of *dhablas* in Gujarat (figs. 3.36, 3.37).

There is a sample of tablet woven tape with invocatory names of Hindu deities, in the collection of the Ethnological Museum, Berlin. Wardle describes the functioning of a card contrivance at Banaras where a similar tape appears to have been woven (Wardle, 1886: 27–8). One pattern included twenty-nine names of Viṣṇu in Sanskrit and Gujarati while another had repeats of *Om namaḥ śivāya.* These sacred ribbons as he calls them, were worn during worship at temples or in homes. Today, the card weavers of Banaras are of Gujarati origin and appear to be Hindu (oral communication: R. Shingah, Varanasi). In the making of invocatory braids as many as forty to fifty cards, called *jhariā*, may be used. The warp could be silk and the weft cotton (oral communication: Sasiben, Varanasi). M.R. Majmudar, however, asserts that in western India similar tapes were used by Jains and Hindus to tie up manuscripts (Majmudar, 1968: 148, Plate LXVII). This appears to be a more plausible explanation for the usage of such tapes. According to him, such tapes were called *khorakhor*, meaning border. Laxmanbhai Bhojak, L.D. Institute, Ahmedabad, however, avers that *khor* or *khola* is to be taken in the sense of *khole* or covering case rather than that of border. In view of the usages to which tapes were put, both explanations may be correct. K.V. Sheth, L.D. Institute of Indology, Ahmedabad, has drawn attention (oral communication to Lotika Varadarajan), to the fact that such a tape called, *bandhan nī paṭṭī*, exists at the L.D. Institute, Ahmedabad (fig. 3.1). It is used to tie up the wooden covers of palm leaf manuscripts (figs. 3.2 and 3.3), demonstrating how these *paṭṭīs* could have been used in the past. Similar narrow card woven bands, *khor*, have been stitched together to make a head covering (fig. 3.3), and a prayer bag, *gaumukhī* (fig. 3.4).

Silk *bandhan nī paṭṭī*. Warp, red, yellow and green. The *Śiva panchakshara stotra* forms the pattern, the five syllabic refrain reading *namaḥ śivāya.* According to S. Andhare, L.D. Institute of Indology, Museum, Ahmedabad, the original silk cover/ *khole*, to which this cord had been attached, had disintegrated. (Fig. 3.1)

3.1

3.2

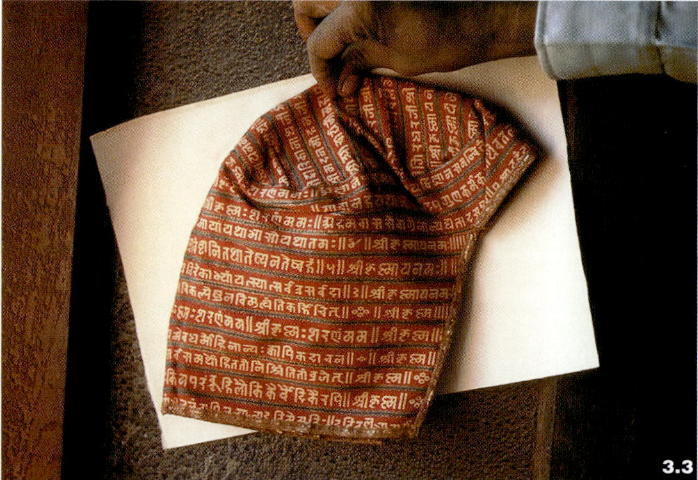

3.3

These, along with a *bandhan nī paṭṭī*, are to be found in the collection of the Baroda Museum and Picture Gallery, Vadodara. The *gaumukhī* is used by Vaiśṇavites to rotate a *mālā*/rosary with 108 beads within it while reciting appropriate invocations. It is thus obvious that card woven tapes were not only used as binding material but could be sewn together and put to other uses.

3.4

The technique of tablet weaving is widely distributed in the old world, but there is no evidence of it in Australia, the Americas and South Africa. According to A. Van Gennep, tablet weaving was invented in Egypt prior to 4000 BC and reached its full development *c.* 3000 to 2400 BC (Schuette, 1956: 17). The earliest sample of card weaving is the girdle of Ramases III, *c.* 1198–1166 BC, housed in the Liverpool Museum, England (Staudigel, 1975: 80, 97).

3.6

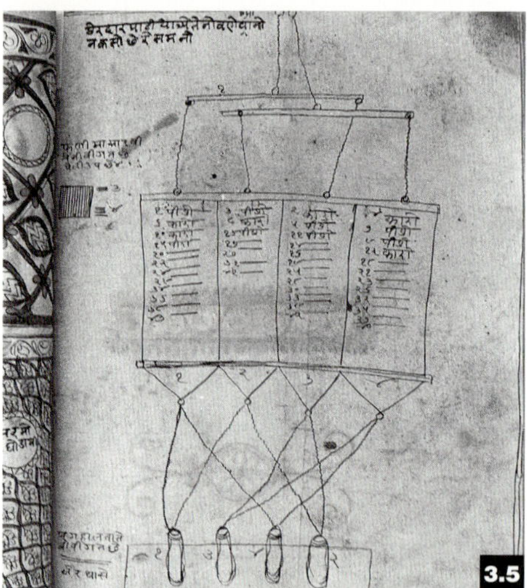

3.5

3.7

In card weaving, tablets or cards, made of leather, clay, wood, shell, ivory, and, in present times, cardboard, are used to control the shedding mechanism. The number and method of distribution of the holes on individual sets of cards is determined by the structure and form of the pattern selected. Once the cards are threaded with warp ends and anchored for tension, the stage is set for weaving. With each turn of the card a new shed is created and each full rotation introduces a twist. In order to avoid a build up of twists at the end of the warp, the weaver may periodically reverse the turning direction of the cards. The pattern and structure can change depending on how many turns are given in one direction and where reversals occur. Card weaving lends itself to double weave, double faced weave and tubular weave (Crockett, 1977: 27).

On the whole, the Indian unstitched garment is geared more towards a hot climate and card woven accessories are in evidence. The *kachchhā* form of wearing the *dhotī* leaves the calves exposed. In present times some elderly men who wear *dhotīs* wrap *paṭṭī* around their calves in lieu of socks during the winter months.

Detail of fig. 3.1. (Fig. 3.2)

Cap. Card woven. Courtesy: Baroda Museum and Picture Gallery, Vadodara. (Fig. 3.3)

Card woven panels joined to form *gaumukhī*, Courtesy: Baroda Museum and Picture Gallery, Vadodara. (Fig. 3.4)

Paṭṭī weaving loom. From Sample Book *c.* AD 1880. Courtesy: Kutch Museum, Bhuj. (Fig. 3.5)

Card weaving for *phiren*, Srinagar. Stage 1. (Fig. 3.6)

Card weaving for *phiren*, Srinagar. Stage 2. (Fig. 3.7)

In Surendranagar the weavers manufacture card woven *paṭṭī*. Those involved in finishing the ends of the *dhabla* are called *dhabla-na-cheda na vankar*, weavers who finish the ends. There are many variations in finishing methods and these include the making of fringes and tassels. Alternatively, an individual segment of the *paṭṭī* could be separately cardwoven and attached to the ends of the garment (*Rhythm of Fingers*, NID: 7–9). In Kutch another type of loom appears to have been used in the past for the same purpose (fig. 3.5).

Dorī of Kashmir

The tablet weaving device is used by the Hindu Kashmiris who weave a red silk or rayon tape which is used as a trim on the *phiren*, the traditional female garment. In the Kashmiri language the tablet weaving device is called the *dori* loom and is used for weaving *paṭṭī*. The tablets are made of walnut wood. They are square in shape with a hole in each corner. Fifteen tablets are used for *paṭṭī* weaving.

A wooden comb/*kangu* or *kangī* with twenty-two spaces between the teeth is used. The warp ends are guided through the slits between the teeth of the comb. The chief function of the comb is to spread the warp out, so that the tablets are easier to turn and a regular width is maintained (Collingwood, 1982: 50).

A wooden beater called *dajepat* is used to beat in the weft. The spool on which the weft yarn is wound is called *pull*. The approximate width of the *paṭṭī* is 5 cm. (oral communication: J.L. Bhan, Srinagar) (see figs. 3.6, 3.7, 3.8). The way in which the *paṭṭī* is used may be seen in fig. 3.9.

Newār/Nāḍā

The craft of weaving *newār*/tapes and *nāḍās*/drawstrings is a traditional occupation among the Rawal community of Wadhwan taluka, district Surendranagar. The skills are transmitted across successive generations. *Newār* is still extensively used for making beds, swings and

3.8

3.9

Card weaving for *phiren*, Srinagar. Stage 3. (Fig. 3.8)

Red card woven tape used as trim on *phiren* worn by Hindu lady, Srinagar. (Fig. 3.9)

seats. A variety of traditional lower garments in India utilize the *nāḍās* as drawstrings, for tying up and securing the garment in place. These are also used as trims on a number of products.

Newār/Nāḍā Loom

The *newār/nāḍā* loom, made of wood (fig. 3.10), is a rectangular frame of the dimensions of 1½ ft. x 2 ft. It is placed on the ground and oriented horizontally or diagonally across the room. It has fixed heddles. Alternate warp ends pass through these heddles creating one fixed shed. The second shed is created by manipulating the warp ends that do not pass through these loops.

The base or bottom slab of the wooden frame extends to a length of about 1 ft. on either side of the loom. A second slab is fixed at the centre at right angles to the bottom slab. As the loom is not fixed to the ground some weight, which could be a familiar domestic object such as a grinding stone is placed over this junction to provide stability to the loom. This remains in position at all times.

Another wooden slab approximately 8 in. from the ground is fixed at the back of the loom. This is nailed and firmly tied by means of a cord. To the base of this slab a row of nails is fixed to perform the function of the warp spacer or raddle. This also functions as a lease.

Weaving device for *newār* and *nāḍā*. (Fig. 3.10)

1. Supporting slab for frame
2. Warp spacer
3. Half heddles
4. Wooden frame

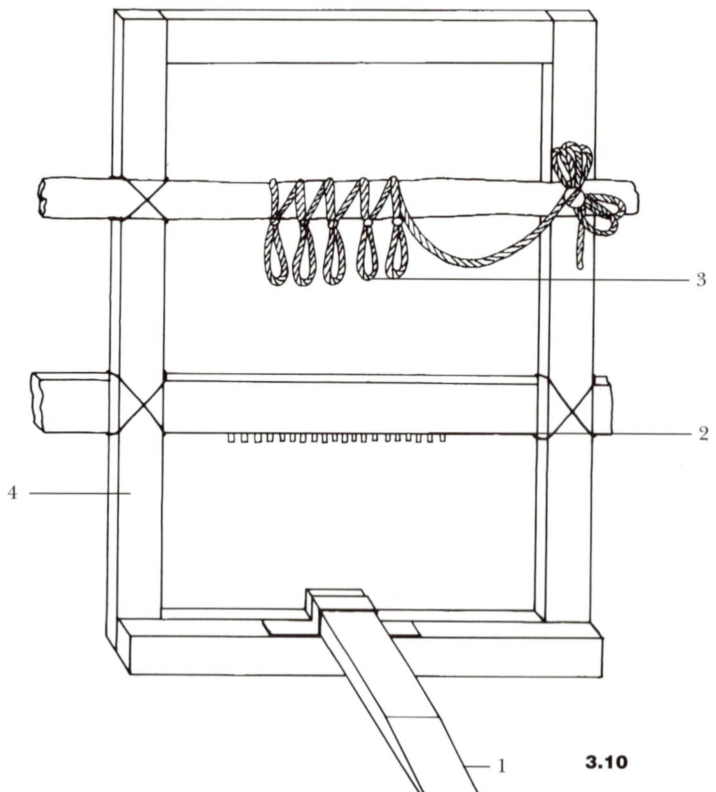

3.10

A wooden rod or wooden strip 1 in. x 1 in. in thickness and 16 in. in length, forms the top part of the frame. On this rod individual loops or half heddles are made. Each heddle is secured by a knot. These are made from one continuous length of twine. The remainder of the twine is made into a finger hank and hung on to a wooden peg positioned on the left side of the frame.

This twine is deliberately kept long to assist the weaver in making additional loops/half heddles, either when he wants to increase the width of the *nāḍā/newār* by increasing the number of ends or when he needs to replace a broken heddle.

The warp is tensioned between two poles, one at the back of the loom and one in the front. Stick shuttles and small knife-shaped wooden beaters help to complete the various parts of the loom.

Raw Material

The material used is *suttar*/cotton. Hanks of 2/10s unbleached cotton yarn used for both warp and weft is bought in Surendranagar. The weaver dyes this in a simple process, using a dye bath of hot water, common salt and direct dyes. The yarn is entered and turned around till the dye is completely absorbed. The excess dye is washed out after which the yarn is dried.

Warp Making

The colours red, yellow and green are used in larger proportions in the warp, whereas blue, black, white and pink are used in comparatively smaller quantities and can be considered to be accentuating colours.

The warping patterns for all the *nāḍās* or *newār* are similar. The width however, may differ. The selvedges are red. The inner warping pattern in the narrow tapes is done in an alternating pattern of two yellow ends and two green ends. The colours for the narrow tapes are limited to red, yellow and green while blue and white are included in the broader *nāḍās* and *newār.*

Warping Process

The warp is wound around two wooden poles fixed into the ground, the distance between the two determined by the desired length of the tapes to be woven. The warp ends vary according to the width of the tape; in present times a wider fabric used for making bags is also produced by these weavers and hence the warp ends can vary from 20 ends to 650 ends. Once the required number of warp ends is

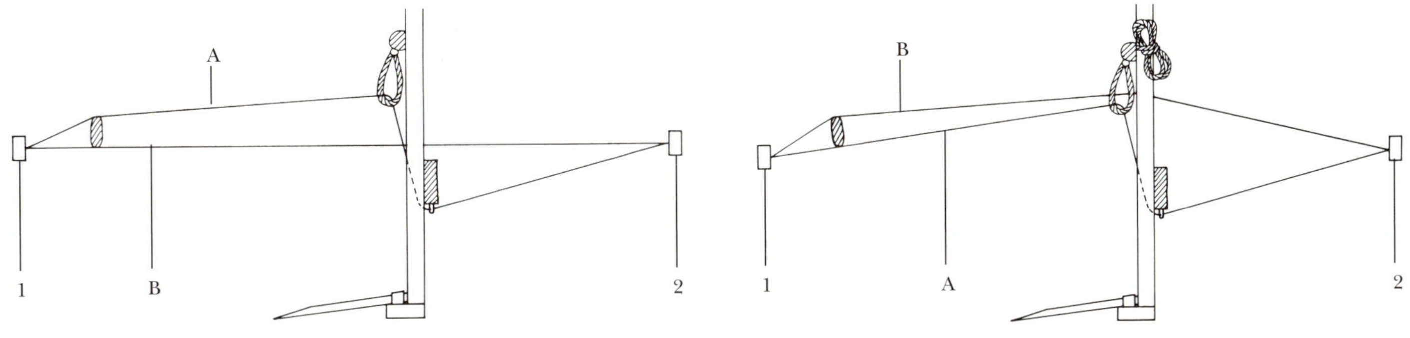

3.11 3.12

wound around the two poles, the stretched warp is cut in this position. Alternatively the ends may be held in tension by a helper. Subsequently the ends are rolled on to a sturdy wooden rod.

The drafting is done by passing one end from the back of the loom, drawing it under the wooden slab and placing it in the dent of the warp spacer. It is then taken above, and passed through the heddle and drawn forward towards the weaver. The second end passes straight from above the slab with the warp spacer, before being pulled forward near the weaver.

In this way all odd ends, passing through the heddles would create one fixed shed, whereas the even ends would pass on straight. On completion of the drafting, all the ends are held and the tension made even. These are then tied temporarily around the pole in front of the loom to the other ends at the far end of the loom. As a result of this procedure, the warp ends assume an elliptical shape, rather than being knotted or bunched up. One layer forms the drafted ends whereas the other passes from under the slab with the warp spacer.

Weaving

One shed being fixed, the first pick is passed through this shed and beaten in. The shuttles are ordinary cylindrical wooden pegs/sticks of 12 in. in length, around which the weft is wound. The weft colour matches the colours used in the selvedge (see fig. 3.11, Stage 1); the countershed is formed by pushing the lower warp ends of the first shed by hand above the warp ends in the half heddles and inserting the beater (fig. 3.12, Stage 2). The beater is turned on its edge to create a clear shed through which the stick shuttle with weft is passed and beaten in place. The weaver continues weaving in this way. A 2/10s six ply yarn is commonly used for the weft. However, for thinner *nāḍās* 3 ply yarn of 2/10s count is used. A warp-faced plain weave forms the structure of the cloth/tape.

To bring the warp forward as the weaving progresses the weaver pulls the warp towards him, adjusts the tension, and re-ties it to the unwoven warp. In the absence of a cloth beam, the woven tape is wound on to a wooden stick.

Guddar Loom: Navalgund

Located in the northern corner of Navalgund town there is a group of *guddar* weavers who migrated to Navalgund from Hebsur, a small town 15 km. away. They moved more than a hundred years ago in the hope of achieving better prospects for their woven products. They weave a type of floor covering which is warp-faced and has coloured stripes running along the length of the warp. These are woven in narrow strips 9 in. wide and 18 ft. long. Each strip is called a *hali*. Several such strips are sewn together to make a *guddar*.

The *guddar* can be considered a multiple use fabric. Apart from being used as a floor covering, a cart cover and for covering stored grains; it can be converted into double walled cube-like containers for storing ingredients used in the making of *pān*/ betel leaves (fig. 3.13). It is also used for covering of temporary structures such as *maṇḍapam* constructed for community gatherings.

Men, women and children are all involved in the process of weaving and finishing the *guddar*. Women perform this activity in their homes during their leisure time.

3.13

Stage 1: Cross section, fixed shed. (Fig. 3.11)

1. Front pole of loom
2. Back pole of loom

A. Set of warp ends

B. Set of warp ends

Stage 2: Cross section, formation of counter shed. (Fig. 3.12)

1. Front pole of loom
2. Back pole of loom

A. Set of warp ends

B. Set of warp ends

Cubical storage container made with *guddar* fabric. (Fig. 3.13)

Guddar/ kaimagga, horizontal fixed heddle ground loom used in weaving *guddar.* (Figs. 3.14a, 3.14b)

1. *Mund dandagi/* front beam
2. *Mundin guta/* front peg
3. *Yedicolu/*iron rod
4. *Anidora/* lease thread
5. *Chinicoalu/* lease stick
6. *Sutradhar/*rope which holds the heald rod
7. *Bejicolu/* heald rod
8. *Beja/*heald
9. *Hulai Hatta/*beater
10. *Ani/*shed
11. *Kodruvamani/* wooden platform
12. *Kallu/*weight
13. *Hind dandagi/* back beam
14. *Hagga/*rope
15. *Hindin guta/* back peg

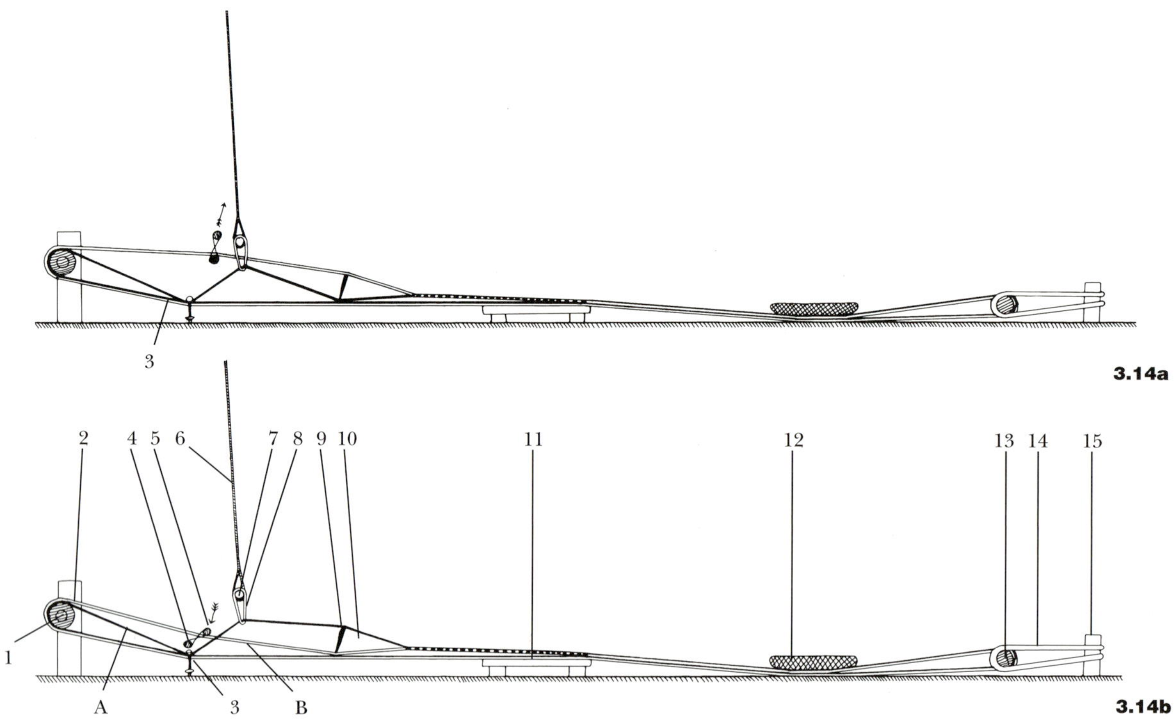

Guddar Loom

This loom is a horizontal fixed heddle ground loom locally called *kaimagga* (figs. 3.14a, 3.14b). A continuous warp is put up on this loom which is assembled each time a new strip/ *hali* has to be woven (fig. 3.15).

Four pegs are driven into the ground, which help establish the front and back of the loom. Two beams, the front beam and the back beam hold the warp under tension. The front warp beam is positioned behind the front two pegs, whereas the back warp beam is tied by rope to these pegs. This rope can be released, or tightened, facilitating the weavers in maintaining the tension of the warp.

A type of warp-faced plain weave (fig. 3.16) fabric with the weft compactly beaten in is woven on this loom. Different bright coloured stripes of varying thicknesses are spaced almost at regular intervals amidst a white ground which form the main characteristics of a *hali.*

Each *hali* has thin stripes of four dyed ends along the edges, about 1/2 in. away from the edge, forming the guideline for sewing or demarcating the seam allowance. These form part of the design. A *guddar* is made of twelve *halis* of which ten have similar patterns. The two *halis* used on either edge of this unit are called *mari hali.* These have different patterning in the rhythm of the stripes helping to contain the format of the *guddar.*

Guddar/kaimagga loom Navalgund. (Fig. 3.15)

Hali basic unit of *guddar* fabric. (Fig. 3.16)

Pre-weaving, Weaving and Finishing Processes

Cotton yarn of 3/6s count is used for the warp. Warping is done on four pegs which are fixed on to a wall. Two of these pegs are fixed at a distance of about 9 1/2 ft. apart which is equivalent to

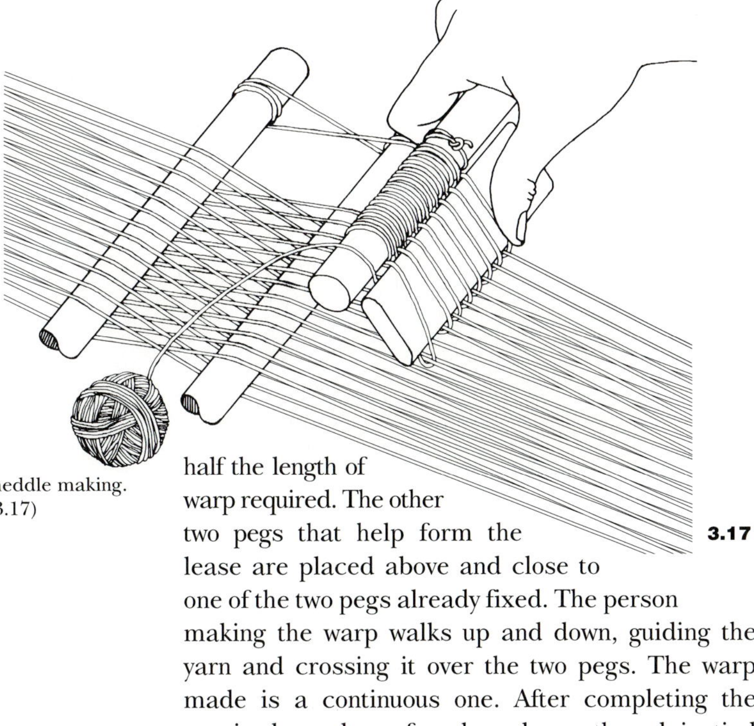

Beja, heddle making. (Fig. 3.17)

3.17

half the length of warp required. The other two pegs that help form the lease are placed above and close to one of the two pegs already fixed. The person making the warp walks up and down, guiding the yarn and crossing it over the two pegs. The warp made is a continuous one. After completing the required number of ends a lease thread is tied around the crossing warp ends.

Setting up the Loom and Weaving Process

The warp is transferred to the front warp beam and the back warp beam. The lease thread is replaced by two wooden rods/lease sticks, which are tied to each other at both ends. These are called *chinicoalu*. Half heddles/*beja* are drawn with thick twisted thread which is passed around every alternate upper layer of warp ends. A wooden heddle making guide is used to achieve equal lengths of heddles. The heddle rod helps to hold the heddles in position and the space between two heddles is maintained by carrying or wrapping the yarn around the heddle rod twice after the heddle is made (see fig. 3.17).

Once the heddles are made both ends of the heddle rod are tied overhead with a rope called a *sūtradhār*, to the ceiling of the house (figs. 3.14a, 3.14b). The loops of the heddles are held vertically and are kept taut, with the warp ends lying within the top segment of the inside of the loops.

An iron rod is passed over the warp threads (A) (see fig. 3.14a) with heddles pulled downwards and tied to the nails fixed into the ground. One of the lease rods is removed and tied over the set of warp threads (B) (see figs. 3.14a, 3.14b) which are not threaded through the heddles to the lease rod which is located under the same set of warp threads.

Whenever the *chinicoalu*, the lease rod, now lying on top of the warp, is pushed down, the set of alternate warp threads (B) go down and form a shed/*ani*. After the weft is passed through the shed, it is beaten in with an iron sword. The counter shed is formed by pulling the lease rod upwards, lifting with it the (B) set of threads crossing over the (A) set. The weft is passed through this shed and beaten in. The lease rods here are seen to be utilized in performing the three functions of:

1. Maintaining the lease.
2. Acting as pressure bar.
3. Shed rod.

The weaver sits on the warp which is supported from underneath by a wooden platform, positioning himself almost midway along the length of the warp. Behind him the warp is weighed down by a stone called a *kallu* (see figs. 3.14a, 3.14b). After weaving 6 to 7 in., the stone is lifted and the woven fabric pushed back by beating it away thereby shifting the warp forward. The weaver can thus continue to weave fabric as long as the shed can be formed and opened wide enough to allow the passage of the weft. About 7 to 8 in. of the warp remains unwoven. The heddles are unwound and the continuous woven material removed and cut to form a long strip. Several such strips are joined together by hand to form a *guddar* (fig. 3.18).

Kushti Loom

The ethnological evidence of the loom used in Parsi *kushti* weaving serves as a complement to card weaving.

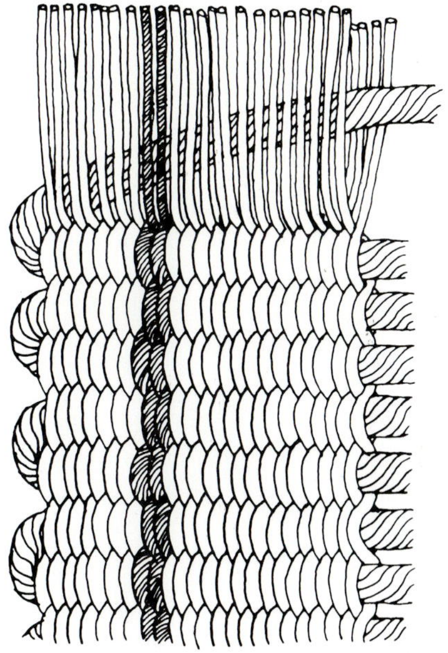

3.18

Guddar, warp faced plain weave. (Fig. 3.18)

According to Jer Kutar,[1] in earlier times only females of the priestly class, Dasturs, could weave *kushti* which was a tubular tape made of lamb's wool (fig. 3.19). Other Parsi women wove *newār, nāḍā* and *pāṭṭī*. In Punjab, each year the cotton wool of *razāī*/quilts, would be taken out, spun, and woven into these latter items, and the quilts would be restuffed. It is possible that the *kushti* loom may have been introduced by the early Zoroastrian immigrants into India.[2] Lamb's wool was spun with a spindle, called *chhatri* by the Parsis (fig. 3.20). The loom would have a continuous warp of seventy-two ends (figs. 3.21, 3.22). This would be divided into nine shafts, each shaft manipulating eight ends (figs. 3.23, 3.24).

Shed separators were made of sandalwood (*Santalum album*) or sisal (*Agave sisalana*) selected for their lightness. At the first stage, wool from two spindles would be knotted and

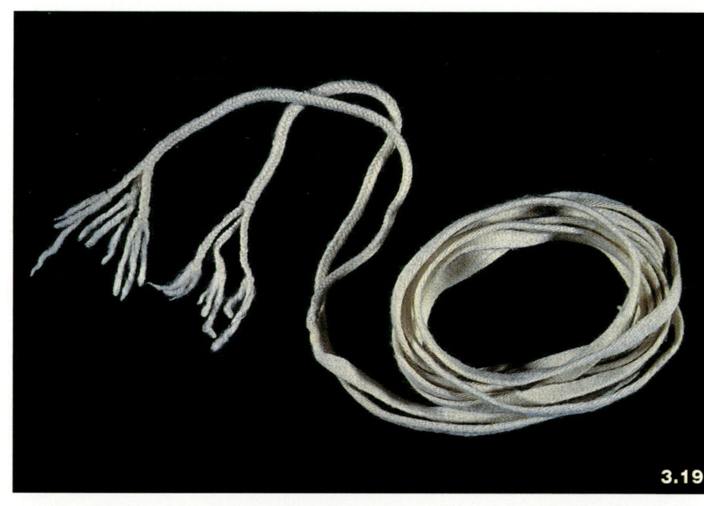

3.19

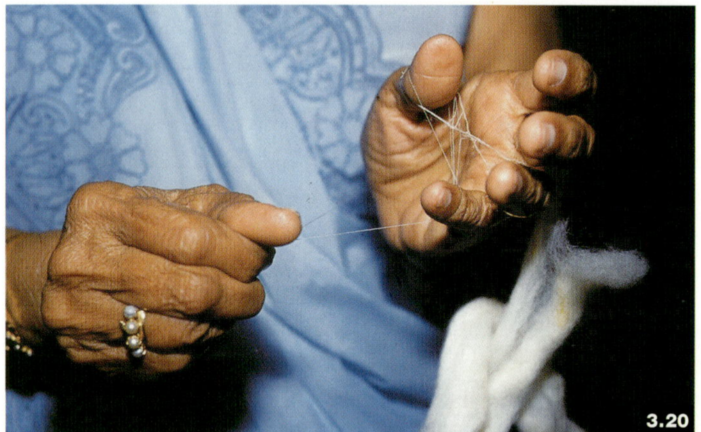

3.20

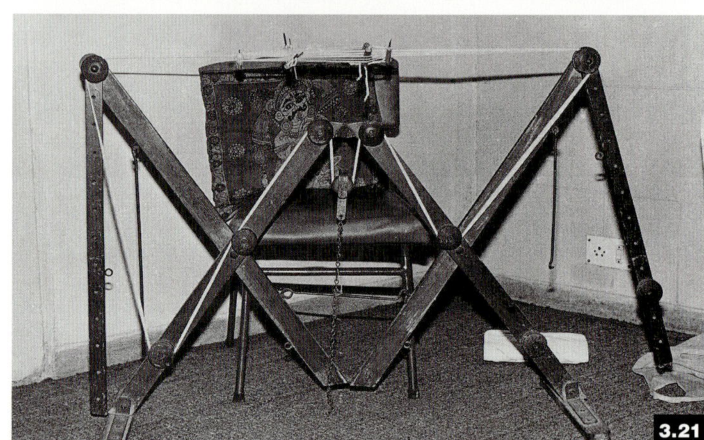

3.21

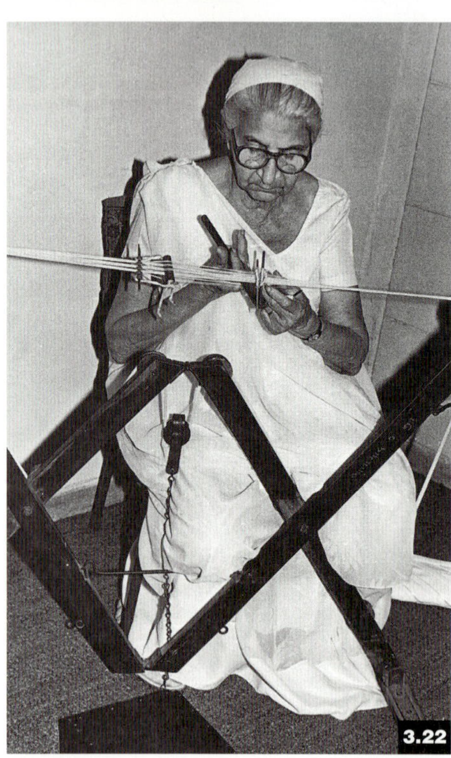

3.22

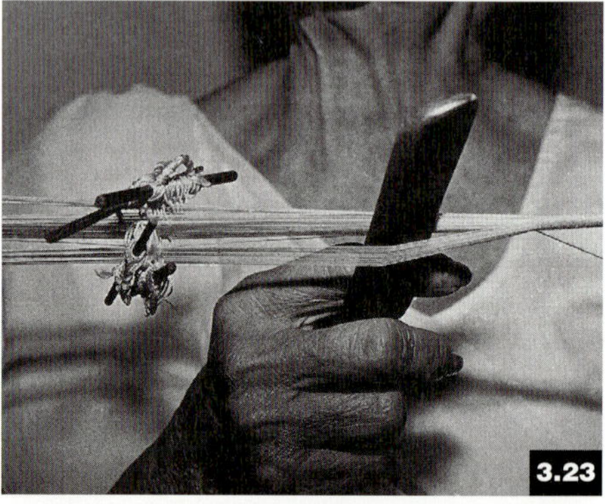

3.23

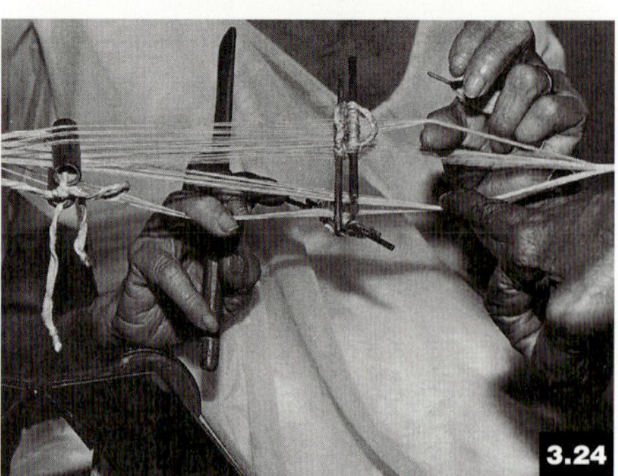

3.24

Kushti, ready for wear. Tubular tape, warp faced plain weave. Courtesy: Tanaaz Irani, Pune. (Fig. 3.19)

The wool is initially twined on the fingers before being transferred to the *chhatri*. Photographed at Exhibition, Zoroastrian Values and Traditions, India International Centre, New Delhi, 18-21 March 1989. (Fig. 3.20)

Parsi *kushti* loom. Photographed at Exhibition, Zoroastrian Values and Traditions, India International Centre, New Delhi, 18-21 March 1989. (Fig. 3.21)

Jer Kutar at loom. Note hair covering. Photographed at Exhibition, Zoroastrian Values and Traditions, India International Centre, New Delhi, 18-21 March 1989. (Fig. 3.22)

Shaft in Parsi *kushti* loom. (Fig. 3.23)

Demonstration of shed formation. Photographed at Exhibition, Zoroastrian Values and Traditions, India International Centre, New Delhi, 18-21 March 1989. (Fig. 3.24)

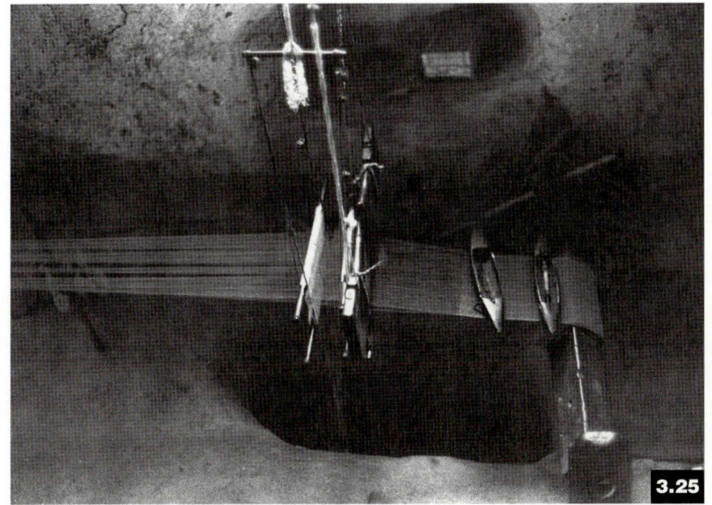

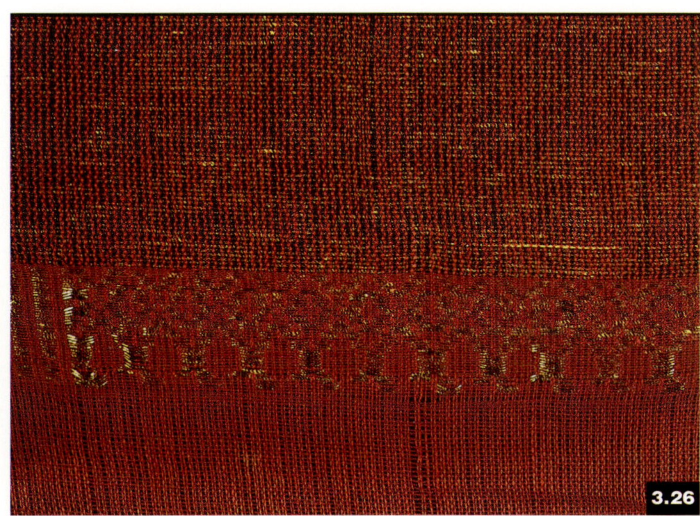

Pagaḍī loom, Yeola. (Fig. 3.25)

Pagaḍī. Detail, made to order at Yeola. (Fig. 3.26)

then formed into a ball. Twisting of this two ply yarn would then be accomplished by means of a bigger and longer spindle. On completion of weaving on the loom, a length of 2¾ in. would be left unwoven. With the help of a large sewing needle the *kushti* would be turned inside out. The free ends would be plaited for about 11½ in. after which it would be fastened together with a string. The remaining portion would then be plaited into three separate braids and left hanging. The item would then be washed. The last phase is smoking which has a bleaching effect. In a small utensil burning charcoal is placed on a bed of ash to which sulphur is added. The utensil is covered. The *kushti*, wrapped in muslin, is left on the cover for one hour, after which it is ready for use. The endproduct would be tubular in form.

Yeola *Pagaḍī* Loom

Another narrow width product of importance is the *pagaḍī* or turban. Yeola has been an important centre of turban/*pagaḍī* weaving. Momins, a caste of Muslim weavers have specialized in *pagaḍī* weaving. They are believed to have migrated to Yeola *c.* 1875 (oral communication: Abdul Gani Khalliq, Yeola).

Pagaḍīs are woven in different widths and lengths. The smallest width woven was 6 in. whereas the maximum length was equal to *sau hāth* or 100 *hāth*, *hāth* literally meaning hand. One *hāth* is about 80 cm.

Pagaḍīs are woven on a narrow width pit loom (fig. 3.25). Two shafts and two treadles are used. They are woven in plain weave with patterning in extra weft in *zarī* at their two ends. The extra weft patterning is done using the pick-up extra weft method. Thin metal rods are used to pick up the warp ends according to the required design. There may have been some patterning on the main field of the *pagaḍī* earlier, but there is no record of this. The colours most commonly used were a dull red, white, pink and golden yellow. The older men are believed to have worn red and white, while the brighter colours such as pink and yellow were worn by younger men. The names for designs were given according to the number of pick-up sticks/*tiris* used for weaving the pattern, e.g. *cheh kamḍī kā phūl* implies that six pick up sticks are used to form this design. These *pagaḍīs* are believed to have been worn by the men in Gujarat, Maharashtra, and Rajasthan, and were only made to order (fig. 3.26).

Turbans are also believed to have been woven in the weave that can be considered to be related to the one used in the making of *pītāmbaras*. One such piece is displayed in the Calico Museum of Textiles, Ahmedabad (accession number 51). This turban is woven in the double-cloth weave wherein the two layers of the fabric can be separated. There is also an interchange of the face colour with the back face colour at the two ends of the piece.

Woollen Weaves: *Dhablas* of Kutch and *Tangalia* of Chitarwada

Dhablas

Traditionally Kutch has been a centre of cotton, silk and wool weaving tradition in India. The processes involved in the making of *dhabla* have been studied and recorded from the villages of Bhujodi, Hodka and Khanderi in Kutch, Gujarat.

The womenfolk of the Rabari community in Kutch are still seen adorned in their traditional black woollen *oḍhnīs*/veils and *ghāgharās*/type of skirt while the Rabari men are dressed in their

Spinning of wool for Kutch *dhabla*. (Fig. 3.27)

traditional white costume. The *dhabla* is used by the male members of the rural population of Kutch.

A *dhabla* is a coarse woollen woven blanket known as *khato* among the Rabari community. The custom of carrying a *dhabla* by the Rabari men still prevails. The uses of the *dhablas* are various. A man may wear it like a shawl against cold or rain; sit on it while he rests on his journey; spread it on his bed as a bedspread, or cover himself while sleeping. He carries it over his shoulder wherever he goes, even in summer (*Dhablas of Kutch*, NID: 9).

The Rabaris, traditionally camel rearers, and Bharwads, herders of sheep and goats are two classes among the herdsmen and shepherds. Their main source of income is through the sale of milk, milk-products and wool.

Earlier, the woollen yarn was handspun by the people themselves, who then provided this yarn to the *vankars*/weavers to be woven into products required by them.

Oḍhnīs/Veils

The woollen *oḍhnīs* worn by the Rabari women are woven in balanced plain weave and are mainly black with maroon-coloured tie-dye patterns and extra weft ornamentation in white at the two ends of the *oḍhnī*. The extra-weft patterning is deliberately done in cotton, so that when the *oḍhnī* is dyed in red and black in dyes meant for wool, the cotton yarn does not accept the dye. It does however get slightly tinted but after a couple of washes it regains its whiteness. *Oḍhnīs* are woven in narrow widths and joined together with decorative interlacing stitches to achieve the required width.

Dhabla: Blanket with Multiple Uses

A finished *dhabla* measures between 18 in. x 90 in. to 27 in. x 108 in. These are pieces woven in narrow widths. Two identical narrow width woven pieces are joined together to achieve the total width. The main colours used in the blanket are black and white with narrower stripes of deep purple, red, magenta, orange, green and yellow. The warp stripes in the border vary in rhythm from blanket to blanket. Traditional *dhablas* have *sachchi kor* with *kangris* woven in the single interlocking technique. The blanket also has the two *chhedas*/ends and the centre or the whole blanket is colourfully patterned, by use of the extra weft technique.

Spinning

For spinning (fig. 3.27), a stone weight or a stick is commonly used instead of a spindle. The raw wool is teased by hand or combed and made into a rove. This roving is wound around the left forearm. With the right hand the spinner draws the fibres from the rove while these are being gently guided by the left hand, simultaneously providing a twist with his right hand. This end is attached to the weight. The spinner twirls the weight to spin the yarn, draws the required fibres from the left hand, raises it and twirls the weight once again to spin the yarn. The spun yarn is wound around the weight or stick and the process continues. A stick is inserted and the yarn is secured by wrapping the yarn in a figure of eight around the stick to avoid the unravelling of the spun yarn. Today, mill spun yarn is used and very rarely is somebody found engaged in the traditional method of spinning.

Dhabla Loom and its Accessories

This loom can be classified as a pit loom with throw shuttle. In present times, however, a number of these have been converted to the fly-shuttle loom. The basic structure of the loom is a large frame fixed on to the ground. On to this frame are fixed the beater and a lathe-turned wooden cylinder. The shafts are attached to the wooden cylinder with the help of two cords, one on either side, of adjustable length. The shafts may also be attached to pulleys or heddle horses (*Dhablas of Kutch*, NID: 1).

The beater consists of two strips of wood. The reed is set within the inner longitudinal grooves of this implement. The reed is made of the outermost layer of a bamboo. The number of dents formed between each bamboo split corresponds to the number of threads in the warp. Two wooden sticks of the same height as the reed are positioned between the two strips of wood to hold the reed in place. Each shaft consists of cotton heddles suspended between two wooden shaft sticks. Four shafts are used. Facing the weaver is the cloth beam on which the woven fabric is wound. The cords holding the treadles are tied to the lower shaft sticks of each shaft with the help of a long string.

Under the large frame constituting the basic structure of the loom is a pit which accommodates the treadles. The weaver sits on a narrow shallow depression in front of the loom, placing his legs within the pit. The pit is 30 in. by about 18 in. wide and 24 in. deep. The warp is rolled on to a stick and kept at the back of the loom either on the ground or hung from the ceiling (*Woven Products Kachch*, NID: 21). A fabric width of not more than 30 in. can be woven on this pit loom with throw shuttle (fig. 3.28).

3.28

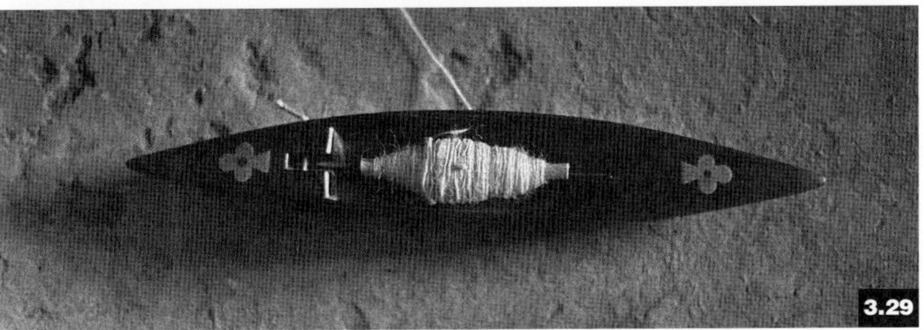

3.29

The *dhabla* loom, Bhujodi, Kutch. (Fig. 3.28)

Kathlo/ shuttle used in Kutch *dhabla*. (Fig. 3.29)

Accessories for the Loom

***Kathlo*/shuttle:** The *kathlo* is made of wood with rounded sides that taper to a point. The thread emerges from an eye on one side of it (fig. 3.29).

***Nardi*/Spool/Pern:** The *nardi* is made from the stem of a locally grown plant from which the pulp is scooped out. The thread is wound around the spool, which is placed upright in the shuttle. A piece of wire serving as an axle passes through it fitting into a hinge at one end and into a notch in the other. Above the notch is a hole which is stoppered with a small stick.

***Pankho*/ *Charkho*/Hank Winder**

***Kana*/Pick-up Stick:** The *kana* has both its ends tapering to a point. The ends are used to pick up the warp ends in extra-weft patterning.

***Kana*/Shed Sword:** The shed sword, also called *kana* is a strip of wood about 2 to 3 in. wide tapering to a point. The shed formed beneath the picked up yarns are usually held open with a shed sword turned sidewise in the warp.

***Panakh*/the Temple:** The *panakh* is employed to maintain uniformity of width in cloth in the course of weaving. It is made of two bamboo sticks with spikes at both ends. These are pushed through the selvedges of the cloth to stretch it.

***Phani*/Reed**

***Pankheriyo*/Stoppers:** These hold the upper and lower part of the sley together.

Brush

***Hantho*/Beater with the Reed**

***Kadhani*/Warping Stick**

***Phanto*/Stick to Maintain the Width of the Warp**

***Gharodi*/Pulley** (*Dhablas of Kutch*, NID: 2, 3).

Warping

Two methods of warping are used either with warping pegs or with warping frame.

Warping pegs: Wooden pegs are fixed on to the ground near the wall, from which they continue on to the wall according to the required length of the warp. These may be found either within the premises or outside the premises of the house. In the first row of these pegs there is an extra peg fixed at either the far left or the far right of the row of pegs. This peg serves the purpose of maintaining the sequence in the warping pattern and facilitates the formation of lease.

Warping frame: The warping frame has pegs fixed on to a wooden frame which is propped up against a wall while making a warp (fig. 3.30).

Warping stick/*khadani*: The warping stick or *khadani* is a stick 2 to 3 ft. long to which at one end a hollow stick is attached at an angle of 90° through which the yarn is guided while making the warp (*Dhablas of Kutch*, NID: 5).

Dyeing: Chemical dyes are bought in Bhuj from the Khatris. The dyeing process is carried out by individual weavers at their workplaces.

Sizing: Sizing is done after the process of dyeing. Starch is extracted from a locally available onion like vegetable, the *Dungra* (*Asphodelus* sp.). The bulb is simply immersed in hot water in a vessel for a few hours and then mashed. The warp is then placed in this solution. It is then stretched between two poles and additional supporting poles. These poles are fixed to the ground so that the warp dries under tension. While drying, the strands of the warp are separated and the starch is evened out with the help of a brush. At some places fingers may be used. After the warp has dried it is wound on to a beam (*Dhablas of Kutch*, NID: 17) (fig. 3.31).

Putting up the warp: The new warp ends are individually linked to the ends of the previous warp by means of twisting them together. This process, known as piecing, ensures that the original drafting order is retained. Fresh drafting is done only when a new weave or texture is required. The drafting sequence is usually that of plain weave (1, 3–2, 4) or twill (1, 2, 3, 4). Two ends are threaded per dent of the reed. The warp which was wound on the stick is unwound and removed from the cloth beam. A bamboo stick/*phanto* is fixed on the warp, which helps divide the warp into sections and also helps to maintain its width. The tension of these sections is maintained with the help of small sticks around which a thread is wound. These sticks are first tied in the centre and then to the sides. Later, the threads are brought to a point so that a triangle is formed. The apex of the triangle is tied with a long rope which goes round a pole fixed on the ground alongside the cloth beam, where it is tautly stretched and fastened to a pole next to the weaver. When this bracing rope is loosened, the warp tension is released and the woven cloth can be wound around the breast beam. Only when the cleared part between the shafts and the *phanto* gets exhausted does the weaver have to get up and clear out the other length of warp by shifting the *phanto* away from the shafts recontrolling the tension (*Dhablas of Kutch*, NID: 18).

Weaving techniques: The blanket is either plain woven or in reversible twill (*Dhablas of Kutch*, NID: 19), however, the plain weave blankets are more commonly woven than the twill weave blankets. In either case the pattern in weaving is done using the extra-weft and the interlocking techniques. These are elaborated below.

a. motifs achieved with discontinuous extra-weft patterning: In this method of discontinuous extra-weft patterning, the warp yarns are picked up with the

3.30

3.31

Making of warp on warping frame for Kutch *dhabla*. (Fig. 3.30)

Street sizing of warp in Bhujodi, Kutch for *dhabla*. (Fig. 3.31)

fingers or with slender smooth pick-up sticks in the shed formed by depressing the treadles one and three. The shed formed beneath the picked up warp ends is usually held open with a shed sword which is propped up on its edge in the warp. The patterning wefts are inserted individually and the shed changed. The pattern weft is often a loosely held weft engaging only a part of the warp, regardless of whether the extra weft patterning is done from end to end or is localized.

A ground weft in the plain weave shed is inserted to strengthen the structure of the fabric. One throw of weft in plain weave shed is made between each insertion/throw of pattern weft with throw shuttle or fly-shuttle. Alternative rows of plain weave mesh to form the ground fabric.

Motifs are woven without reference to a graph. The movement of the steps in weaving of a motif is done from memory and based on the experience the weaver has gained over a period of time.

This method of patterning provides great freedom in achieving a variety of shapes, as this in fact becomes a canvas for drawing through weaving. Nature and the living environment have ceaselessly provided the artisan with inspiration. Simplified shapes, geometric in character, of trees, animals, insects, human forms, musical instruments and so on are commonly found being woven by this method of weaving. The weavers identify these by their names.

b. Extra-weft patterning with pattern rods/*chhaḍ* (figs. 3.32, 3.33): For end-to-end extra-weft patterning of simple geometric forms such as triangles, squares, diamonds and triangular weaves, where symmetry is required, pattern rods are used and set up before the weaving is begun. The selection of warp ends according to the patterning required is done behind the shafts as part of preparing the loom for weaving.

For each step in a motif, a pattern rod is inserted. These are also inserted sequentially to allow for progressive growth in pattern while weaving. Thus, the number of pattern rods is equal to the number of steps required to form a particular motif.

After the pattern rods are arranged the weaver draws these rods one by one towards the shafts and clears the warp to form a shed. The pattern rods could be rounded sticks or flat. When flat they are propped up on their edges to form a shed. In the case of rounded sticks with a small diameter, the shed is kept open by inserting a shed sword for the passage of the extra-weft. The weaver throws in the weft pick that forms the ground weave, beats in the weft and proceeds with the second pattern rod. He continues this process till the motif is completed. On completion of weaving of each step of the motif the pattern rod is transferred to the back of the loom thereby creating possibilities for reversal in the weaving order, and the creation of different motifs. In addition, several permutations and combinations are exploited by the weaver through the desired manipulations, other than the sequential operation of the pattern rods.

3.32

3.33

Selection of warp ends prior to insertion of pattern rods in warp for Kutch *dhabla*. (Fig. 3.32)

Kutch *dhabla* loom with pattern rods at back of loom for extra weft patterning. (Fig. 3.33)

c. Extra-weft patterning with pick-up sticks/*paṭṭī*: The shed is formed by pressing pedal numbers one and three. It is in this shed that selection of ends can be done by using a pick-up stick/*paṭṭī*, 1 to 1½ in. thick to achieve extra-weft patterning.

The ground between the motifs across the width of the warp is woven in basket weave creating a speckled effect around the open areas surrounding the motifs. This method allows innumerable possibilities in the creation of a variety of forms, as the selection of the warp ends has to be done afresh each time. The desire to expedite this process and change it from a time-consuming one to a comparatively more efficient method, may have led to fresh thinking and the germination of new ideas and development of the method of the use of the pattern rods (see fig. 3.33).

d. *Kangri* achieved with single interlocking tapestry weave: Traditional *dhablas* have *sacchikor* woven in the single interlocking tapestry weave. *Sacchikor* not only carries the literal meaning of true border but it also connotes a traditional border. The borders are usually interlocked with a different coloured weft from the main field of the *dhabla*. While interlocking with the weft of the main field, the weaver staggers the meeting points of the two wefts to create the *kangri*. This is an elongated triangle made with the same weft as the border, pointing inwards from both sides (fig. 3.34). This type of interlocking where the stress points are staggered can be seen in Kanchivaram silk and cotton saris, in the *aṅgavastram* from Andhra Pradesh and a number of other traditional fabrics woven in India. It has been noted that in the traditional *Paithani* saris the borders can separate from the main field over a period of time as all the interlocking meeting points worked along one warp line, or in between two warp ends, thus creating stress along the inner edge of the border. Hence the triangle form, commonly known as temple or *kangri*, has been developed from the understanding of the structural property of interlocking.

Finishing: The finishing or making up of a blanket, such as the joining and adding of fringes is considered to be an extremely important and integral stage in completing the product.

The two identical pieces are joined together with the use of interlacing embroidery stitches. Remnants of coloured thread which are left over from the warp that cannot be woven are collected and kept carefully for this purpose. These threads or embroidery threads are then used for joining the two pieces together. Several colours may be used by the weaver or his wife while embroidering with a needle to complete the joining.

Colourful tassels are also attached to both ends of the *dhabla*. The two *chhedas*/ends are ornamented with *meri*, a type of twining, and brightly coloured *phumtas*/tassels.

One of the striking features of the *dhabla* is the colourful fringe. Two coloured weft twining is also used in the S and Z twist to lock the weft in its place. Another method of finishing the two ends which continues to be practised in Chitarwada, is use of tablet-weaving. The warp for tablet weaving is laid across the width of the ends of the blanket. Each individual warp end or a group of warp ends is treated as weft and interlaced one by one in the shed formed by each turn of the tablets. Oval tablets made of cardboard with two holes are commonly used.

Woollen *Tangalia* of Chitarwada

A special reference needs to be made to the traditional woollen products woven by the Vankar community of village Chitarwada (approximately 30 km. from Khambhat) in Gujarat, who weave the *tangalias*, *charmalias* and *lubdi/dhablas*. These are also woven in villages around the district of Surendranagar, Gujarat. All members of a weaver's family are involved in the pre-weaving and weaving processes of these traditional woollen textiles. Locally available wool is procured from the shepherds, cleaned, combed and made into roves by both the young and the older members of the family.

The *tangalia* is a traditional woollen unstitched garment worn as a wrap-around skirt secured at the waist by a string, and worn by the married women of the Bharwad, a pastoral community in Gujarat. As a part of the marriage customs of this community, the father of the bride presents her with a

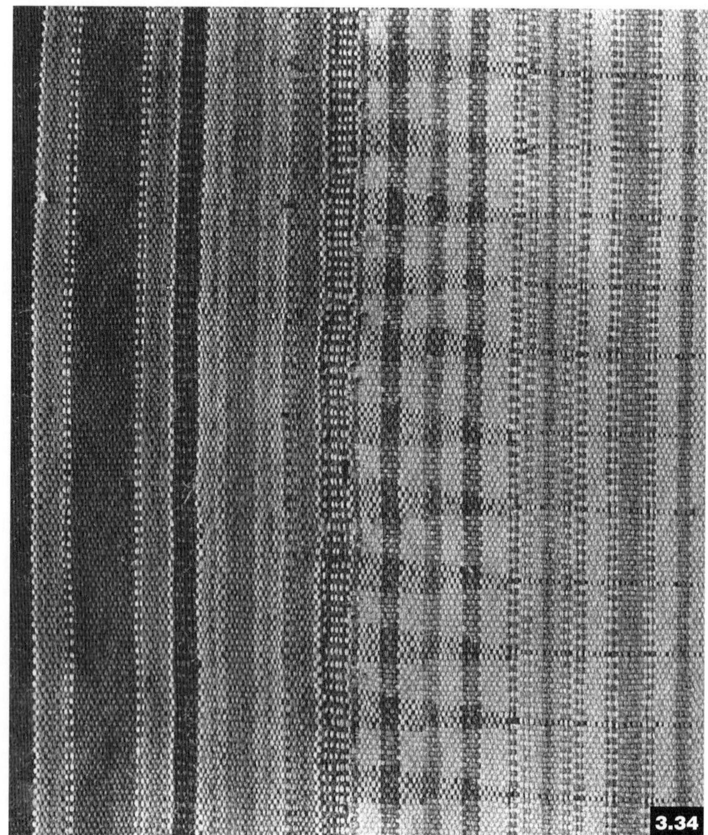

Single interlock in *sacchikor* border of Kutch *dhabla*. (Fig. 3.34)

tangalia and a *charmalia* at the time of marriage. Married women of this community use this garment for daily wear. This is worn with a backless blouse and an *oḍhnī*.

A *tangalia* measures around 9 ft. in length by 4 ft. in width. There are a number of products that can be categorized under the heading of *tangalia*.

Visual Characteristics of *Tangalia*

Tangalias are identifiable by the patterning made with dots/*dānās*. *Tangalia* are predominantly black with a patterned border along the selvedge. The two ends of the *tangalia* are elaborately patterned and are the most striking part of the garment. Narrow warp stripes in maroon, black and white run along the length of the warp forming a border. These stripes form the base and guideline for the patterning with *dānās*.

The motifs used in the *tangalias* reflect the influence of natural and manmade forms, while also echoing the religious symbolism of the weavers. The main motifs are: *jhadhva*, tree; *mor*, peacock; *sītāmaḍhī*, temple; *khajūrī*, date palm.

Various compositions are made with different combinations of trees, peacocks, and temples. A popular combination is the one with peacocks on a tree. In this type the number of peacocks can vary from one to twelve or more depending on the weaver's skill and the quality and price of the fabric. Other combinations are:

1. Trees around a temple,
2. Peacocks on trees and temple,
3. A number of peacocks grouped together.

The geometrical shapes which are used as design elements are:

1. Dots,
2. Circles with dots inside,
3. Diamonds,
4. Triangular wavy lines.

Flowers and fruits are represented by stylized circles each centred with a dot inside. Such motifs are known as buttons and are also used as decorative forms. Other contemporary forms like balloons, aeroplanes, trains, cars, elephants, tri-coloured flags and so on reflect the craftsperson's reaction to his changing environment (*Tangalias*, NID).

Rāmrāj

The *rāmrāj*, also commonly known as *tangalia* is a very decorative and colourful kind of fabric, worn only on special occasions. The motifs formed with a type of herringbone pattern are elaborately ornamented with colourful *dānās*/dots. Both the *chhedas*/ends and the body have maroon coloured stripes in the weft, and both the warp and ground weft are woven in black. *Tangalia* may also be woven with only maroon and white *dānās*.

Charmalia

The *charmalia* is identified by the maroon and black evenly spaced stripes that run along the warp in the main field. The weft used is black. These stripes are approximately 1/4 in. in width. The two *chhedas*/ends have maroon weft stripes making the ends distinguishable from the other parts of the main field. The patterning on the *charmalia* is comparatively less elaborate than that on the *tangalia*. These may also be woven with or without colourful *dānās*. A cheaper quality of this fabric is also woven with minimum patterning which appears only on the two ends of the *charmalia* (fig. 3.35).

Lubdi/Dhabla

The *lubdi/dhabla* is used only by men. Woven in wool using the balanced plain weave for the ground, the patterning incorporates a combination of warp stripes, *dānās*, extra-weft and the tablet weaving techniques.

The main ground colours are maroon, black or white with patterning done in subtle combinations such as maroon on black, highlighted in places with vivid colours and/or contrasting colours.

The patterning is done across the two *chhedas* or ends of the *dhabla* using a combination of both

3.35

Charmalia. Note red and black background. Work place of Dayabhai Karsanbhai, village Chitarwada, district Kheda. (Fig. 3.35)

Card woven panel joining two narrow width *rāmrāj* pieces. Work place of Kanubhai Savabhai Parmar, Chitarwada village, district Kheda. (Fig. 3.36).

Card weaving in *tangalia*. Demonstration of shed formation. Work place of Dayabhai Karsanbhai, Chitarwada village, district Kheda. (Fig. 3.37)

end-to-end extra-weft patterning and localized extra-weft patterning. This is done using the pattern rods and the pick up stick similar to the method followed in Kutch, whereas the patterning with *dānās* is done as in *tangalia*.

Finishing of the *Chhedas*

Two identical pieces are joined together in interlacing embroidery stitches, using brightly coloured wool. The two *chhedas* are sometimes finished with tablet-woven brightly coloured borders. These are made only on customer request as this adds to the total cost of the product (fig. 3.36).

To prevent unravelling of the weft, various methods are used. The warp ends are knotted or braided. Tassels may be added, weft twining done or tablet woven border made to meet both the practical and aesthetic purpose of making the product complete (fig. 3.37). In Surendranagar, where *dhablas* are also woven, tablet weaving is a seasonal occupation with weavers making either *dhabla-na-chheda*, or *dhabla* narrow width borders. *Paṭṭī*/narrow tapes, however, are now used only as straps in footwear.

Tablet Weaving in Surendranagar

For tablet weaving in Surendranagar rectangular tablets are made out of cardboard measuring about 3¼ in. x 1¾ in. They have two holes each measuring about 1¼ in. in diameter which are punched about ½ in. from the top and ½ in. from the bottom. A minimum of eight and a maximum of sixteen such tablets are used for making the tablet woven border. These tablets after some use become oval in shape as the corners tend to wear out with use. A two-metre long warp is made with brightly coloured wool.

An example of tablet weaving with eight tablets is given below. A special feature is that the warp fringes of the *dhabla* constitute the weft of the *chheda*. The warping is done with pairs of cards and its alignment is done through sets of two tablets. The weaver knots two separate yarns together and passes it through the top-punched hole of the two tablets and does the same for the second hole. He then turns the tablets anti-clockwise so that the yarns are sandwiched between the two tablets. He repeats this procedure with the other three pairs also thus obtaining four pairs of tablets (*Rhythm of Fingers*, NID: 5–6).

One end of the warp is tied to a bed post/a peg driven into the ground or around something that will hold the warp in place, whereas the other end is wrapped around the craftsman's big toe. By flexing his foot or leg back and forth he maintains the tension of the warp. The tablets are arranged vertically and according to the warping order. The tension is checked before commencement of weaving.

The weaver sits on the floor with his legs stretched out. The *dhabla* is spread over one of his legs. The tablet weaving apparatus comprising the tablets with warp is positioned above his legs. Moving the tablets in a clockwise direction, a shed is formed through which he passes three warp ends of the *dhabla* at a time to form one pick of the weft. He pushes the weft in place with a small stick or his fingers, changes the shed by turning the tablets once again in the clockwise direction and continues weaving. Reversal in the order of weaving to achieve a mirror image of the weaving pattern is accomplished by moving the tablets in an anti-clockwise direction. This also helps in unwinding the twist at the other end of the warp. Sometimes patterns such as diagonal lines and zigzag lines are created by the method of weft twining in between the tablet woven areas (fig. 3.38). The patterning is restricted to the width of the *dhabla*.

Once the weaver completes the weaving of the border he ties the ends near the selvedge by wrapping them with woollen yarn, finishing this with a simple knot, to prevent unravelling.

Tangalia Loom

This is a throw shuttle counter balanced pit loom very similar to the *dhabla* loom of Kutch but with two shafts and two pedals/*pavla*. The sley/*kathlo* holds the reed and is suspended from an overhead horizontal beam with the help of ropes. In front of the sley facing the weaver is the breast beam on which the woven fabric is wound with the help of a take up peg attached to this beam. This is manually operated by the weaver. The width of the loom is about 30 in. on which a fabric of maximum 25 in. width can be woven (cf. figs. 1.15, 1.16).

Pre-weaving and Weaving Processes Involved in the Making of *Tangalia/Rāmrāj*, *Charmalia* and *Lubdi*

The yarn is hand spun by the women using the *charkhā/ raintio*. Once the yarn is ready it is dyed in the required colours. Warping is done using warping pegs which are fixed into the ground forming a square. The distance between these pegs depends on the length of the warp required. The warp maker stands in the centre holding the hank-opener/*parita*, with yarn in one hand and a long stick with a hook in the other hand. The hook helps the weaver to guide the yarn around all the four pegs. While making the warp, a lease/*jogpatti*, is tied to all the four sides of the warp.

A warp of 13 metres length is made from which two identical pieces can be woven. It is sized using similar ingredients, tools, and methods as in Kutch.

When ready, this warp is pieced together end-by-end with the warp ends of the previous warp eliminating the tedious process of repeating drafting and denting. When a different set up in drafting is required or when additional shafts, new shafts, new reeds, or a change in reed are required, the drafting and denting are done afresh (*Tangalias*, NID).

Specific units of dents in the reed are identified as *visi*. As the width of the reed is constant, the number of *visi* is dependent on the total number of dents in the width of the reed, e.g. three sets of 80 dents in the total width of the reed would make the reed of three *visis* (*Tangalias*, NID).

The warp is stretched and the tension maintained as seen in the *dhabla* loom from Kutch. The *khanthalo*/shuttle is boat-shaped and made of wood. The pern/*nali* fits into the scooped out area of the shuttle. The temple/*panas* is made of two longitudinal sections of bamboo approximately ¾ in. in width which are held together by string where the two opposite ends overlap. The two opposite tips of the bamboo which are inserted into the cloth may have sharp pins attached to them or the bamboo tips may be sharp enough to be inserted into the cloth without damaging it.

3.38

Patterning in card woven ends of *charmalia*. (Fig. 3.38)

Dot wrapping on loom: Stage 1. Photographed at work place of Dayabhai Karsanbhai, Chitarwada village, district Kheda. (Fig. 3.39)

Dot wrapping on loom: Stage 2. Photographed at work place of Dayabhai Karsanbhai, Chitarwada village, district Kheda. (Fig. 3.40)

Weaving Technique, Fabric Texture and Characteristics

Woollen yarn is used both for the warp and the weft. These are woven in plain weave with the extra weft technique which is used to create patterns with the pick up stick and the pattern rod methods. This is, however, dominated by the patterning done by a technique where individual dots/*dānās* are created by tightly wrapping, almost twisting, small quantities of white, maroon and brightly coloured woollen/cotton fibres onto individual warp ends according to the requirement of the motif.

The weaver breaks about 1 to 1½ cm. of white or coloured yarn to be used for creating the *dānā*. He opens the twist in the yarn and separates the fibres using both his hands. He then creates a shed by depressing a pedal. From the raised warp ends he selects and lifts one warp end at a time with one hand. By using his thumb and forefinger to hold the fibres, he wraps these around the lifted warp end and pushes them down to the level of the woven area (figs. 3.39, 3.40). After having completed the required *dānās* across the width of the warp the ground weft is inserted, the shed changed and the pick beaten in. In the *tangalia* two picks of maroon yarn are inserted after every inch, creating stripes. As the value of the maroon colour is very close to black, these do not stand out. The stripes are also not woven from end to end but return from the periphery of the motifs created with dots.

As the yarn used for this is thicker, these are slightly raised from the ground weave thereby adding variation in the textural quality of the fabric.

The wrapped dot is used as a unit and several such units are used to create a variety of lines and motifs that are simplified forms, derived from nature and other objects seen in daily life (see fig. 3.35 showing details of *charmalia*). The extra weft technique is used to create patterns on *lubdi/dhabla*. This is done with the use of pattern rods as in Kutch. As these products are woven in narrow widths, two identical pieces are joined together using interlacing embroidery stitches so that the seam lies right at the centre of the finished piece.

Notes and References

1. *Kushti* weaving was documented at the India International Centre, New Delhi, during a Parsi festival organized at this institution on 21 March 1989. The weaver was Jer Kutar aged about 74, a resident of Jamshedpur.

2. A section of the Zoroastrian community of Persia migrated to the west coast of India following the introduction of Islam into Persia. The first stop is stated to have been Diu, after which their presence is recorded at Sanjan.The dates of these events are not definitive. The source of Parsi history in India is the *Kisse-i-Sanjan* written by Dastur Bahman Kaikobad, *c.* AD 1600. The authenticity of this work is, however, questionable. See S.H., Hodivala, 1920: 3, 37, 101; S.K. Hodivala, 1920: iii, 36, 41; Altekar, 1924–5: 25.

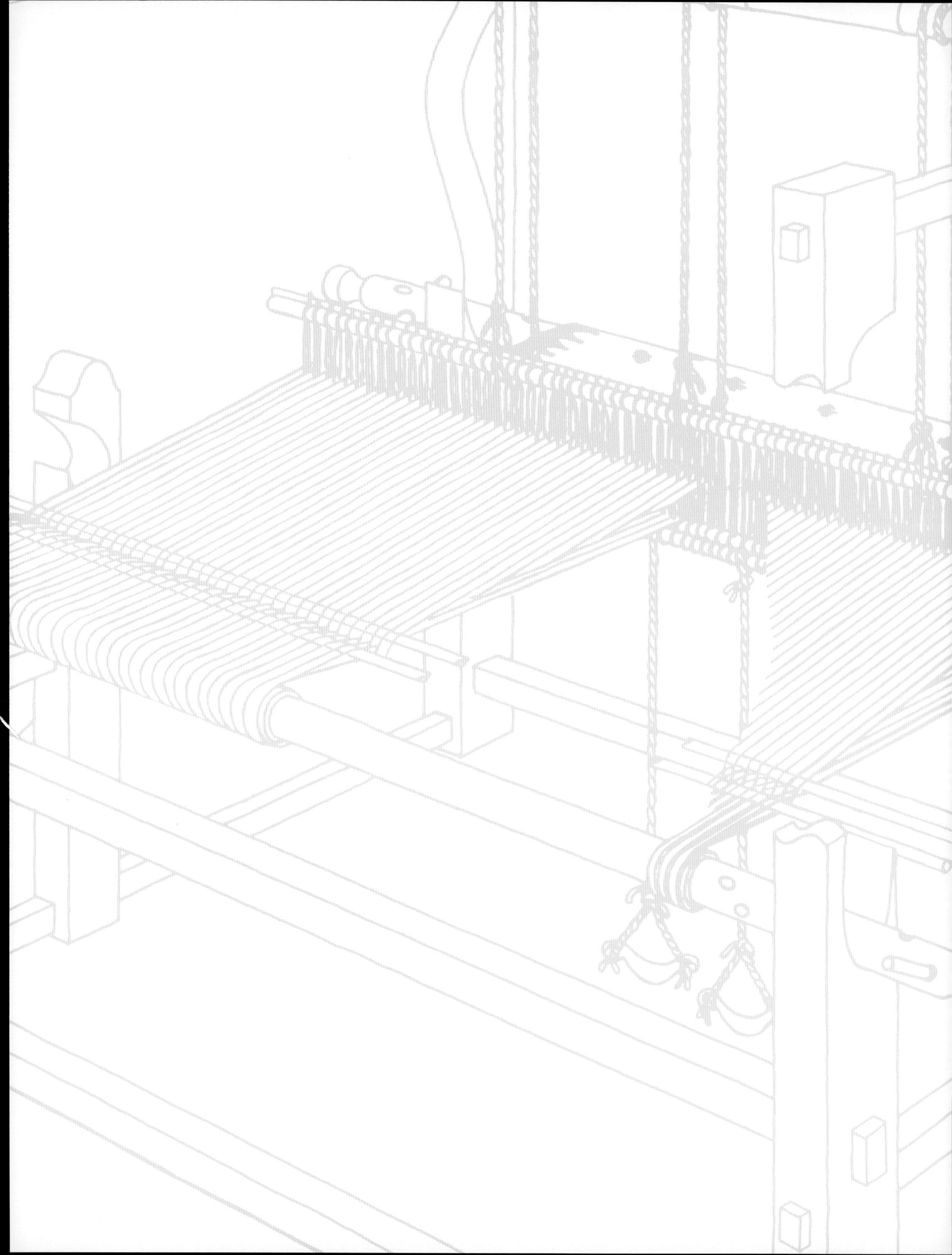

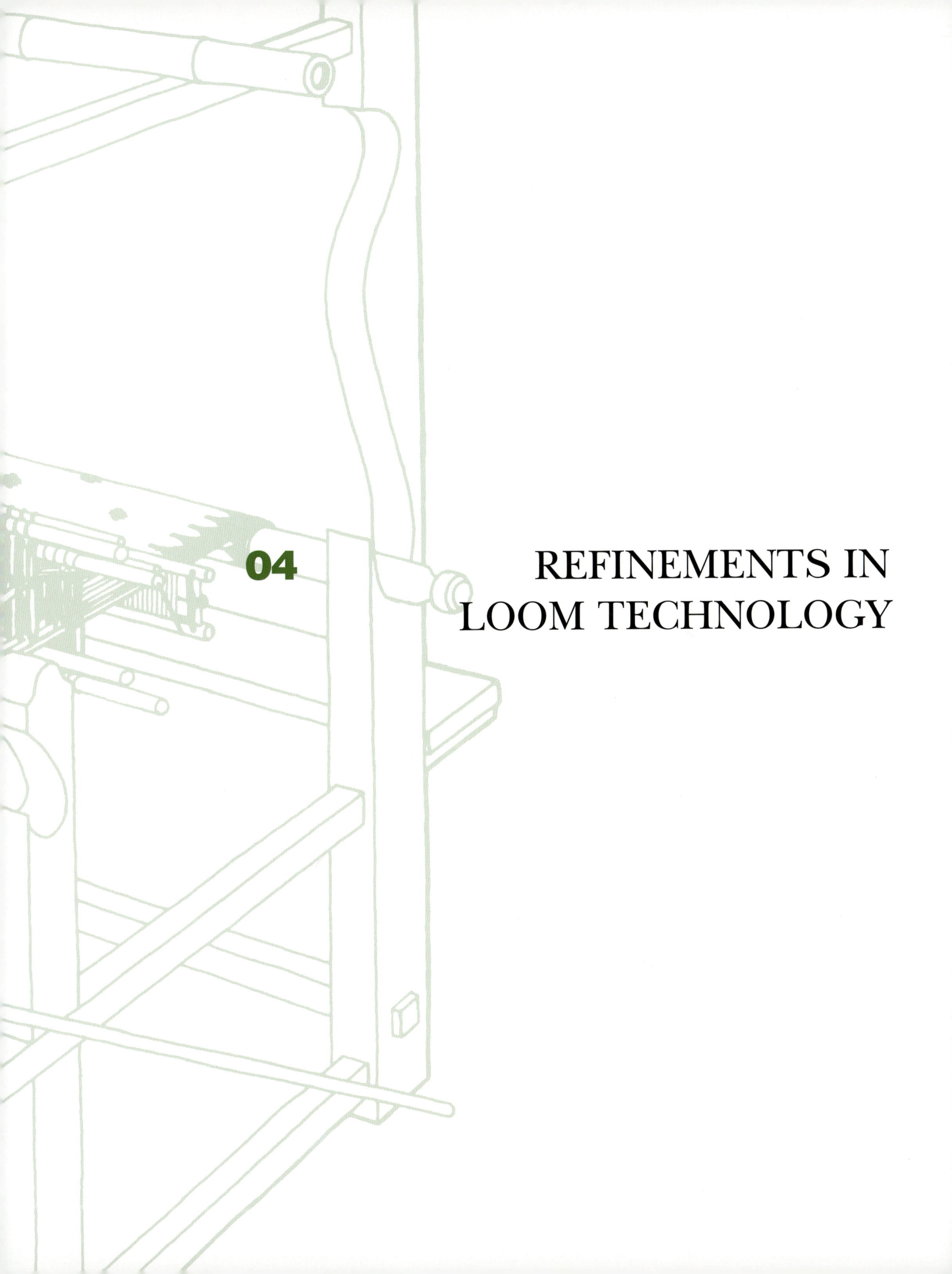

04 REFINEMENTS IN LOOM TECHNOLOGY

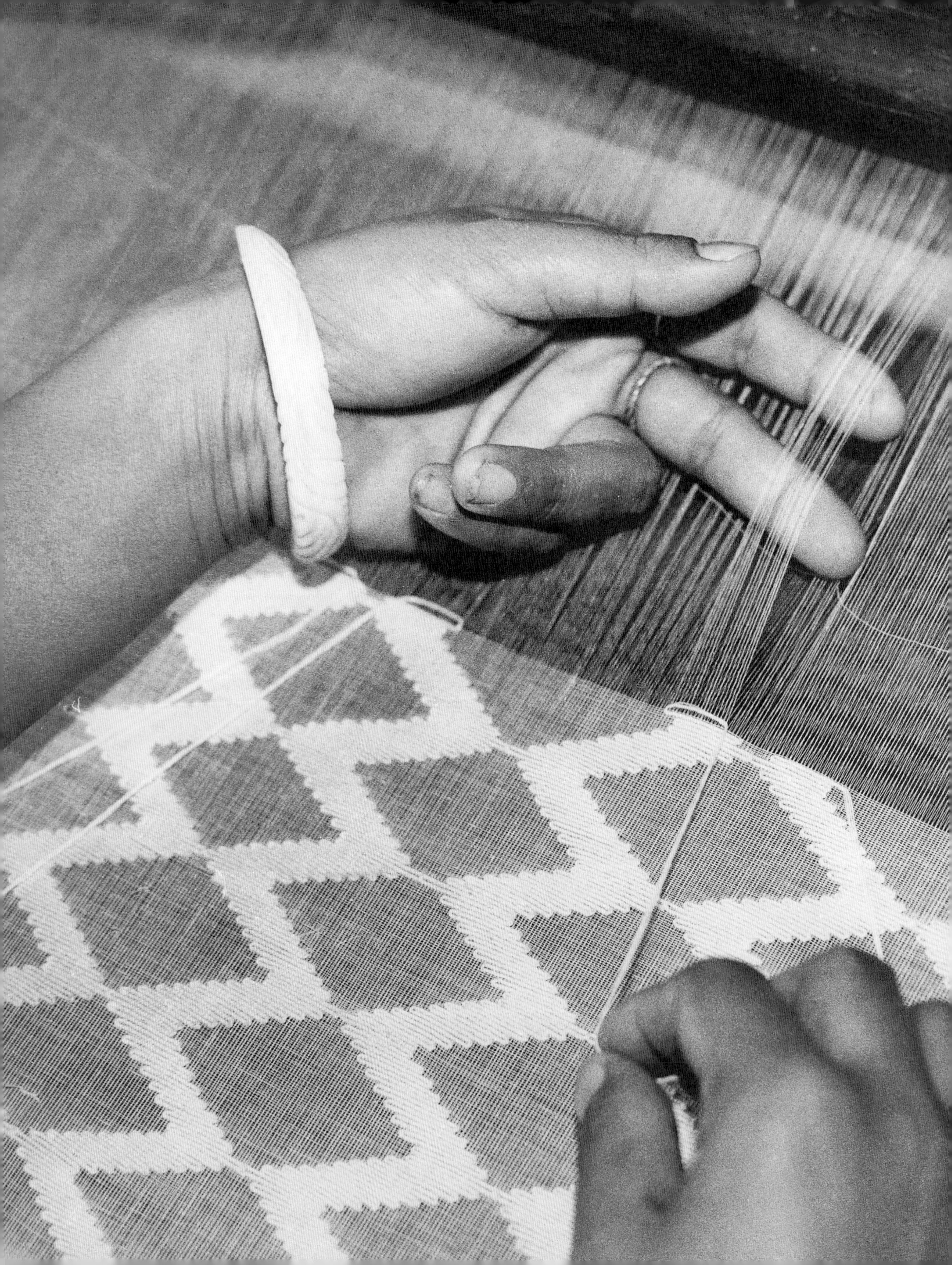

04

Refinements in Loom Technology

The existence of a large variety of loom usages and practices in India, and the absence of dated material artefacts, makes it difficult to frame a chronology for the evolution of the loom on the subcontinent. However, scholars have worked on the subject in other culture regions, and their analysis of the evolution of the loom could establish some pointers for the study of specific direction in loom development in India. The evolution of the loom is linked to the fibre being woven but this aspect will not be covered[1] owing to limitations of time and space.

If a broad synoptic view is taken, it can be stated that the earliest loom accessory was probably the shed stick. This was followed by the introduction of the heddle. The heddle is estimated to have been in use in Egypt before 2000 BC (Picard-Schmitter, 1967: 22–8). The next step would be the introduction of the shaft and treadle. As has been stated above, the treadle loom has been in existence in China since Han times. The pit treadle makes an appearance in Egypt in the sixth century AD (Endrei, 1958: 24), but the source of diffusion may well have been India (Broudy, 1979: 20–1).

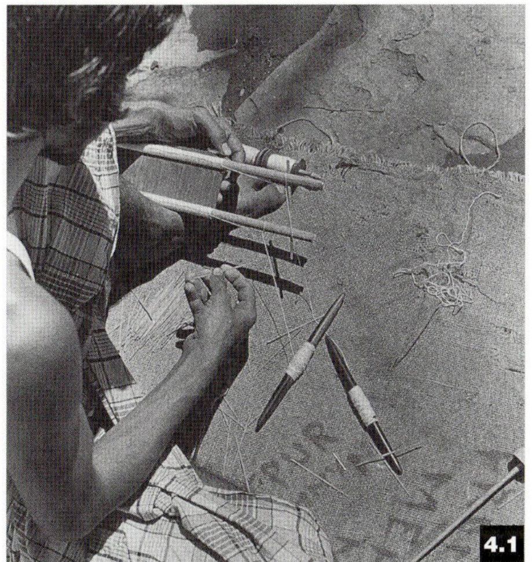

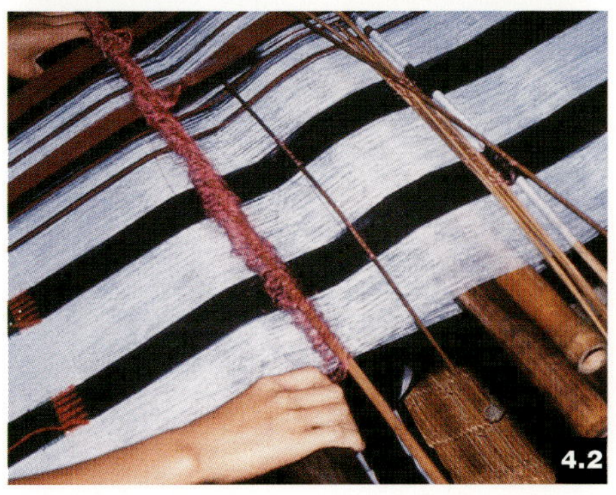

Reed making. Chanderi. (Fig. 4.1)

Loin loom. Tumnoupokpi village, Manipur. Reedless loom with half heddles and pattern rods. (Fig. 4.2)

With regard to the weft, the earliest method would have been introduction by hand. From this it would progress to being wound on a stick to form a spool leading to the next stage, when the spool would be encased within a shuttle (cf. figs. 9.27, 2.14). Then would come the batten lending itself to two purposes, that of beating in the weft and of enlarging the shed. For spacing the warp, the reed (fig. 4.1), and batten could be combined so that it functioned both as beater and warp spacer. This kind of reed beater is in evidence in many Indian looms but the reed could well have been a later entrant into the loom repertoire. Cloth and warp beams, heddle rods, shed stick; lease/cross rods are also included among parts of the loom in its earlier stages of evolution (Albers, 1965: 26–9).

Predilections for extra warp or extra weft to effect patterning have also played a role in the development of the loom. In the north-eastern loin loom tradition the predominant mode has been in favour of extra weft as has been the case in *jāmdānī*. In Gujarat and peninsular India border ornamentation has been based on extra warp but elaborate *būṭā* motifs have been organized around the play of extra weft. Stripes, as in Kanchipuram, have been based on extra warp.

Broadly speaking, on the basis of ethnological evidence to date, the evolution of the Indian loom could be postulated as follows: at the first stage the loom was bereft of treadles and heddles. The next stage was the introduction of fixed heddles, lashes, shaft and treadle. In such areas where extra warp ornamentation may have been predominant, pattern rods used in body tension looms (fig. 4.2), the Kutch *dhabla* looms and the Orissa pit loom (fig. 1.4), may have been less in evidence. Alternatively, the evolution may have occurred in the following stages:

4.3

4.4

1. Looms without heddles, reed, and treadle such as the Haryana *pañja* frame loom.
2. Fixed heddle (fig. 3.15).
3. Reedless looms with one heddle rod with half heddles and a shed rod (fig. 2.2).
4. Reedless looms with two heddle rods, with half heddles (cf. fig. 2.13).
5. Looms with shafts, with full heddle, treadles, sley/reed.
6. Looms with multiple shafts, sley and treadles.
7. Looms with two or more shafts with attachments for patterning such as the dobby, *jāla* and *aḍai* (figs. 4.71, 6.5).
8. Introduction of the improved dobby and adaptation of the jacquard mechanism to the traditional forms of harness attachments (figs. 7.44, 7.43).

The earlier repertoire of weaves may have been tabby, tapestry, extra warp, and a variation of supplementary weft laid-in or *jāmdānī* (fig. 4.3). No contrivances are required for *jāmdānī* as warp ends used in pattern areas can be manually lifted (figs. 4.4, 4.5). After stage six there may have been wider usage of metallic yarn and appearance of more convoluted designs.

Looms which may be associated with early stages of weaving technology have already been described. In the present chapter the emphasis will be on patterning with extra weft in relation to the loom. Where relevant, processes involving extra warp ornamentation will also be mentioned. Before we move to a description of the *jāmdānī* as evidenced in *moirangphee* woven on the Yongkham loom of Manipur, it would be appropriate to follow other kinds of fabric centred around the *jāmdānī* weave. These include Tanda *jāmdānī*, Dacca *jāmdānī*, the *jāmdānī* of Chanderi, of Paithan and Venkatagiri (fig. 4.6).

Wherever the term *jāmdānī* has been used in this text it must be understood as:

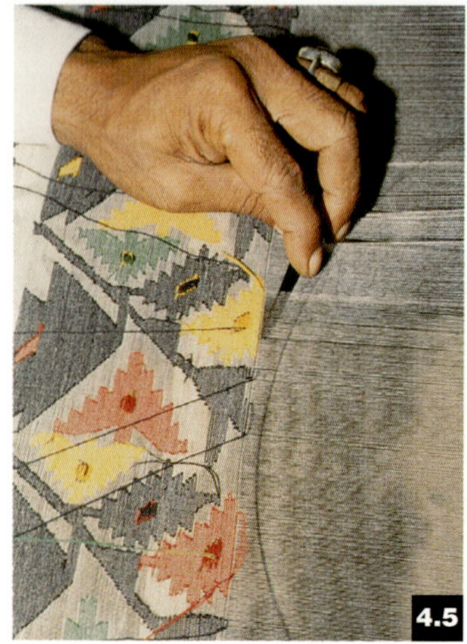

4.5

4.6

Surya. North Bengal. Ninth to tenth centuries AD. Courtesy: Museum of South Asian Art, Berlin. Note texture of upper garment and its similarity to Dacca *jāmdānī*. (Fig. 4.3)

Jāmdānī, pitloom. Naopara, Demra, district Dhaka, Bangladesh. Photographed at Craft Museum, New Delhi. (Fig. 4.4)

Discontinous supplementary weft pattern pick in colour woven within the same shed with a continuous white pick. *Jāmdānī*, pitloom. Naopara, Demra, district Dhaka, Bangladesh. Photographed at Craft Museum, New Delhi. (Fig. 4.5)

Illuminated manuscript *Aṣṭasahasṛka, Prajñaparamita, Newari samvat* 191/ AD 1071. Courtesy: Asiatic Society, Kolkata. Note lower garment. (Fig. 4.6)

> The simplest extra-weft patterning structure is one often described as 'laid-in', 'inlaid' or 'inlay'. It is produced by 'laying' supplementary wefts along with the ground wefts in the regular sheds of a ground weave. ... The ground weave is usually, though not necessarily, plain weave, but if plain, the faces of the fabric are identical except where the pattern wefts are either carried from one area to another, ... or from one shed to the next ... The thread count of the ground weave is usually more or less balanced, although the wefts may outnumber the warps. Obviously, with a warp faced ground weave, 'laid-in' patterning would not be effective. (Emery, 1980: 141)

In the places associated with *jāmdānī* weaving to which a reference has been made above, a number of woven products have been found which use the supplementary wefts along with the ground wefts in the regular sheds of a plain weave used as a ground weave. It has been noted that the thickness of the 'laid-in' supplementary wefts is almost double; also sometimes the yarn used could be thicker in count or ply than that of the yarns used in the ground weave. They may or may not be of the same colour as the groundweave fabric. In fact, the use of the supplementary 'laid-in' wefts in the same colour as the ground weave, warp and weft, has a subtle yet delicate and refined tonal quality, differentiation being achieved through the mingling of the ground weave texture with the 'laid-in' supplementary weft (or wefts) texture.

When the laid in supplementary wefts are of a different colour than the ground weave (both warp and weft) the colours of these interact with the colours of the ground weave. Depending on the outcome of the interaction between the colours and the shadows created by the yarn, the patterned areas reflect a different hue, while retaining the textural effects.

While there may be other related design aspects which will not be gone into, it is necessary to mention here the use and combination of materials such as cotton with silk, silk and cotton with gold or silver *zarī*. When these combinations are compared with other 'laid-in' supplementary weft yarns, it is found that the effect achieved is similar to shimmering reflections of light seen on rippling water.

It has also been observed that in some of the traditional woven textiles of India localized supplementary laid-in wefts have been used to weave individual shapes or motifs. These laid-in wefts are also used to create an all over surface pattern like a *jāl*/web, spiralling creepers and a variety of other forms, which are stylized versions of any form by which the craftsman is inspired.

The Banarasi *jāmdānī* saris were inspired by rich Bengali zamindars of the Raj who were landowners in Bihar, Orissa and the United Provinces. Many settled in Varanasi. Such saris were woven in Madanpura and were very much like the Daccai *jāmdānī*, which were becoming increasingly rare and expensive. Cotton is used in the weaving of both the ground and the patterned *jāmdānī*, but at Madanpura, Varanasi, *zarī jāmdānīs* have also been found.

With increased demand for *jāmdānī* saris, a faster technique for production was developed. This was the use of a throw shuttle. With this innovation cutwork fabrics made an entry into the Banarasi repertoire. The modern artist, K.G. Subramanian, earlier with the Weavers Services Centre, Varanasi, is credited with introducing this technique. A finished cutwork sari resembles a *jāmdānī* in texture. It is also woven using cotton both for the ground (plain weave) as well as for the patterning. The patterning is laid-in extra weft and the design area is woven in *dampach*.[2] However, as the ends are not manually lifted for each motif, and the design pick extends across the width from selvedge to selvedge, there are floats on the reverse side of the fabric in the non-design areas. When this fabric is taken off the loom, the unwanted floats are cut. Surface patterning in Banarasi cutwork is much denser than is the case with *jāmdānī*.

Yongkham: The Throw Shuttle Frame Loom of Manipur

Manipur with its capital at Imphal is situated in north-east India. The predominant group in this valley, the Meiteis, are members of the Kuki–Chin group. Unlike other parts of India, weaving in Manipur is done exclusively by women. There are no caste or class boundaries and the labour of weaving is held in high esteem by all sections of the community. Being part of the women's household duties, this handloom industry is monopolized by them, not from economic necessity alone but also in conformity with social custom. In fact, to know how to weave is a primary qualification for a Meitei girl. So much so that a throw shuttle loom forms part of the bridal dowry.

The Meitei community of Manipur uses a type of throw shuttle frame loom called *yongkham*, on which several different kinds of fabrics for apparel and household use are made. *Moirangphee* is one of them. *Moirangphee* is the name of a type of *chādar* which is worn as a draped unstitched upper garment. The *chādar* is draped around the shoulder with the two ends brought to the front covering the front part of the body from neck to waist. A number of variations exist in the way in which the *chādar* is draped. However, it is essentially worn to cover the torso. It is worn over a blouse and *phanek*, the lower garment.

The name of the *chādar*, *moirangphee*, is believed to have its roots in a village about 44 km. south of Imphal. According to oral tradition, a Meitei king had invaded and conquered the village of Moirang. He brought some Moirangs as captives. The Moirang king, while paying tribute to the king of Manipur for the release of his prisoners, presented some handloom cloth woven by the Moirangs. The king of Manipur was highly appreciative of their design quality and named them *moirangphee*, meaning cloth from Moirang. Today, there is no trace of weaving in Moirang (Mohanty, 1984, II: 262). The fabric, which is bordered by the characteristic temple-top shaped motifs, is called *moirangphee*.

Yongkham Loom

The *yongkham* is essentially a frame loom. The whole structure is based on the main outer frame made of sturdy wooden beams (fig. 4.7). The cloth/breast beam, *mang konaba*, and the rear beam/warp beam, *tung konaba*, of the loom are supported on the vertical poles in front and the vertical poles at the rear of the loom frame which have slots. The seat, *phisaphan*, for the weaver is attached to the front just 8 to 9 in. below the cloth beam. There is a thick bamboo rod called *nawonkok* on the top of the loom frame to which the heddle shafts are attached and balanced. This is a counterbalanced loom. The harness cords tied to the shafts are hung in balance on the bamboo and move up and down counter to each other: that is, when one shaft is pulled down by the treadle, its mate is automatically pulled up. The warp is stretched in a perfectly straight line from the warp beam to the cloth beam.

The shafts are made of three wooden sticks having thread heddles. The wooden sticks, called *nachei*, are approximately 1 in. in thickness and slightly longer than the width of the shaft. These are held parallel to each other with the help of thread heddles. The upper stick is also accompanied by a thin wooden stick called *shingmit* which is necessary to hold the heddles in position. The harness cord, called *nak-en-thouri*, ties the *shingmit* and the upper stick together. Normally each heddle shaft is supported by three harness cords. The pedals, *khunet*, are tied to the lower stick of the shaft with the help of cords. The placement of the pedals is such that they can be operated conveniently by both feet. The reed of the loom is kept hanging on the warp and does not need any other support.

The reed, *samchat*, fits into the grooves of two wooden rods that hold it in place. The dents, *samchatmaya*, are made from fine strips of bamboo. The wooden rods are tied to each other with a flexible bamboo string so that the reed is firmly held between the rods. The thickness of the rod is between 3 to 3.5 cm. and the height of the reed with the rod is approximately 11.5 cm.

To make it easy for the warp beam to be brought forward, there are other wooden poles with carved slots fixed to the loom frame at some distance from the original position of the beam. When the warp is nearing the end, the warp beam is brought forward and is kept on these poles secured into one of the slots. The tensioning and releasing of warp tension is achieved by tightening one of the two beams (as required while weaving) with the help of a bamboo rod, *hanglakchei*, which is kept inserted into one of the holes provided on the sides of the beams. After winding and tightening the warp to the required extent the bamboo rod of the cloth beam is secured to the ground, the front vertical beam or any other part of the frame. There are three to four holes provided for this on each of the beams. Nowadays, instead of the bamboo rod/*hanglakchei*, an iron rod is sometimes used. The width of the fabric is maintained by stretching both selvedges with the help of a bow-shaped temple/*shingkap*. The loom is approximately 25½ in. long, 19½ in. wide and 20½ in. high.

Tools and Accessories

Besides the shuttles/*pangan dem*, and bobbins/*langchak*, used during weaving on the throw shuttle loom, a number of other implements and tools are used. The processes of dyeing, winding, warping, denting, drafting, setting the loom for weaving and the actual weaving all require specialised sets of tools as listed:

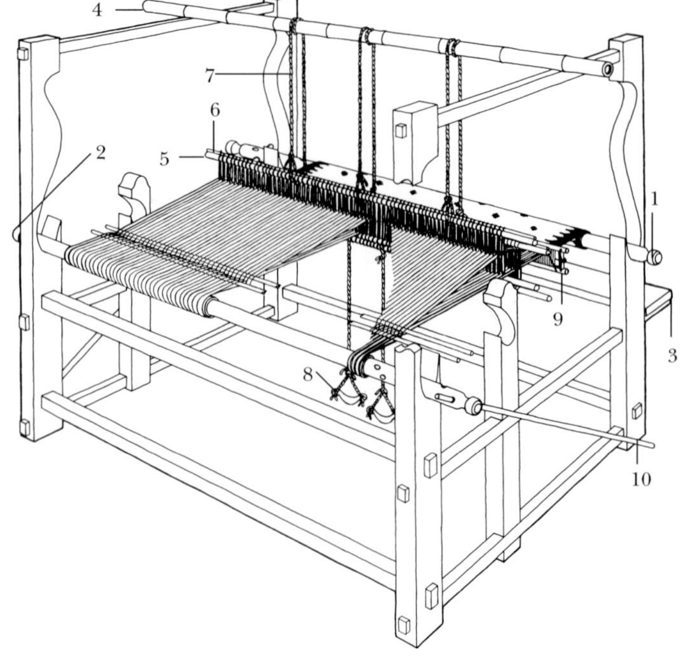

Yongkham, throw shuttle frame loom used for weaving *moirangphee*. (Fig. 4.7)

1. *Mang konaba*/ breast beam
2. *Tung konaba*/ rear beam
3. *Phisaphan*/seat
4. *Nawonkok*/ bamboo rod for heddle shafts
5. *Nachei*/shafts
6. *Shingmit*/ wooded support for the healds
7. *Nak-en-thouri*/ the harness cord
8. *Khunet*/pedals
9. *Samchat*/reed
10. *Hanglachei*/ bamboo rod used in tensioning warp

4.7

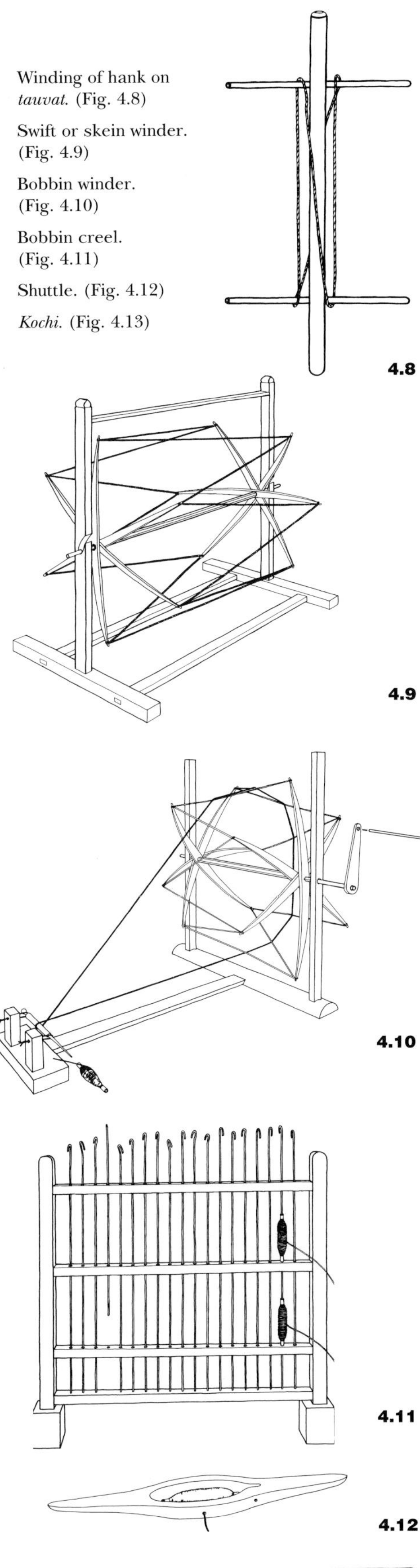

Winding of hank on *tauvat.* (Fig. 4.8)

Swift or skein winder. (Fig. 4.9)

Bobbin winder. (Fig. 4.10)

Bobbin creel. (Fig. 4.11)

Shuttle. (Fig. 4.12)

Kochi. (Fig. 4.13)

4.8

4.9

4.10

4.11

4.12

4.13

Tauvat: device for making hanks

This is an upright wooden stick which has two horizontal sticks of the same size fixed at right angles on opposite ends. This is used to make the hanks prior to dyeing the yarn. The yarn is wound on the *tauvat* in a crosswise manner (fig. 4.8).

After winding the required amount of yarn on the *tauvat*, it is taken out in the form of a hank. This hank is used by the weavers in the dyeing process if small quantities of yarn are required.

Swift

The yarn acquired from the market is in the form of hanks or skeins. These have to be transferred on to the bobbins. The hank is fixed on the swift from which the yarn is wound on to the bobbin with the help of the bobbin winder. Both the bobbin winder and the swift are made of wood and bamboo. The swift has a wooden stand on which a rotating bamboo structure is fixed. This structure is made of two sets of bamboo strips which are tied to each other with the help of strong thread. The hank is fixed on to this structure and then the yarn is transferred to the bobbin. The total height of the swift is about 25 in., the length about 22 in. and its maximum width is about 14 in. (fig. 4.9).

Bobbin winder/*tareng*

This has a similar structure to the swift and is rotated by a handle attached at one end. A string/*langhi* links the rotating structure to the part that holds the bobbin while winding yarn on it. The extended portion, called *mayot*, tapers towards the end so that the bobbin can be fixed easily on to it. The height of the *tareng* is 16 in. and it is about 30 in. in length (fig. 4.10).

Bobbin creel

The bobbin creel has a frame and a strong stand made of wood. It has several thin vertical iron rods on which the bobbins are put. This creel is utilized in traditional methods of warp making (fig. 4.11).

Wooden comb

To brush and make parallel the warp threads, a wooden comb is used while setting up the warp on the loom.

Shuttle and perns (figs. 4.12, 1.2)

Two shuttles are used while weaving on the loom. These are turned on their sides so that they move smoothly inside the shed when thrown from one side to the other. The length of the shuttle is approximately 12 in., the width 1½ to 1¾ in. and the breadth ¾ to 1 in. It has one hole at the side from where the weft thread emerges. Along the same side there is a second hole into which the peg is fitted for fixing the *mayot*/stick holding the pern. The pern used in the shuttle is made from bamboo and is hollow measuring approximately 2½ in. in length and ½ in. in diameter. These shuttles are very similar in shape to the ones used for weaving on the northern Thailand loom (fig. 1.3).

Reed hook/*Kochi*

The *kochi* (fig. 4.13), is a reed hook used for drawing threads through the reed. It is a thin iron rod having a thickness of about 3 to 4 mm. and a length of about 9½ to 10¼ in. It has a slot on one side and a hole on the other.

Shuna-ubak: Guiding plank for making heddles

This is a flat wooden strip tapering towards one end. The narrow portion serves the purpose of a handle and the wider part is used to wrap the heddle during the heddle-making process. These types of heddles are without heddle eyes and are known as *shunna akhanba*. The *shuna-ubak* is 23½ to 25½ in. in length and ¼ to ½ in. broad. The widest part is about 3 to 3.5 in. (fig. 4.14).

Jig for heddle-making

A jig is used for making heddles (fig. 4.15). This type of heddle has heddle eyes and is very rarely used. It is specially associated with *mūgā*, *tasar* and other silk yarns. These heddles are known as *shuna-akiba*. Four nails are fixed on the jig, the distance between these depending upon the length of the heddle and the distance between each knot.

Shing kap/Temple

This is a bow-shaped implement made of bamboo for keeping the woven cloth stretched to maintain its width in course of weaving (fig. 4.16).

The loom and the other accessories are made by carpenters who are experts in this field. Traditionally, bamboo was very commonly used in making the loom. In contemporary times bamboo is gradually being replaced by wood, though even now some parts of the loom continue to be made of bamboo. A few examples exist, nevertheless, of looms made entirely from bamboo. The wood used for making the loom is mainly the *Uningthoy* and *Uchan* types of Pine.

Pre-weaving and Weaving Processes

The weavers in Imphal either buy ready made dyed yarn from the market or buy undyed yarn which they have dyed by local professional dyers. The dyers are mainly located in the area near the pond called Ningthem Pukhari. The dyeing is done exclusively by men.

Almost thirty to forty years ago, natural dyes were commonly used. The dyers have practised this profession for generations and have thus gained much expertise. Both bleaching and dyeing of the yarn was carried out using natural substances. Nowadays as natural dyeing and bleaching has proved to be time-consuming, synthetic dyes are used, as they are more easily available in the market. Synthetic dyes were introduced about forty years ago. Normally, yarn is bleached before dyeing.

Preparation of Yarn: Soaking, Washing and Dyeing

The soaking process has not changed. The grey hanks (natural colour of fibre) are taken out of the tank and a single hank chain is made out of one bundle (fig. 4.17). Before dyeing or bleaching, the hanks have to be soaked in water. For soaking, a cement tank approximately 3 ft. x 5 ft. and 2 ft. in height is used. In some places, instead of cement tanks, iron drums cut in half are also used (figs. 4.18, 4.19). The drum or tank is half-filled with water and 1 kg. of soap, called *momposho*, is added to it. Soap is used to achieve uniformity in scouring. The hank chain is dipped and pressed in the prepared solution, using the feet. Then it is taken out of the solution and the hanks are separated. A bunch of three to four hanks at a time is rinsed in the solution. The hanks are squeezed out again. For squeezing, bamboo sticks are used, one of which is fixed to the two vertical pole structures while the other is manipulated by hand (fig. 4.20). They are then spread on a horizontal pole in sunlight for drying. Yarn is bleached only when light shades are desired. For dark and medium colours, the dyeing is carried out on the grey hanks and bleaching is completely avoided.

Traditional Method of Bleaching and Dyeing

The traditional method of bleaching was a long process and originally cowdung was the main natural substance used. A fine quality of cowdung was strained, stirred in water and made into a solution. The hanks were made in chains and dipped (fig. 4.17), into this solution for five to six hours. After, taking out the hanks, they were not squeezed but spread in the sunlight to dry. This process was repeated two or more times until the required whiteness was achieved. This bleaching process, called *santhitanba* is still practised but being time-consuming, professional dyers have shifted to chemical bleaching.

The cowdung used for bleaching was specially chosen. The finer quality obtained from cattle feeding on green grass was used; characterized by a greenish colour and very fine particles. This was preferred to that collected from cattle feeding on hay which was coarser and had a brownish colour.

For natural dyeing, the substances mainly used were seeds, bark, flowers, leaves and roots of trees (see fig. 4.21). Some examples which can be cited include:

1. *Urai Romba Machu*: The seeds from the tree called *urai romba* (*Coscinium fenestratum*), were crushed to a powder and filtered through a cotton cloth. The filtered powder was mixed in cold water and some common salt was added to make the colour fast. The yarn was squeezed out

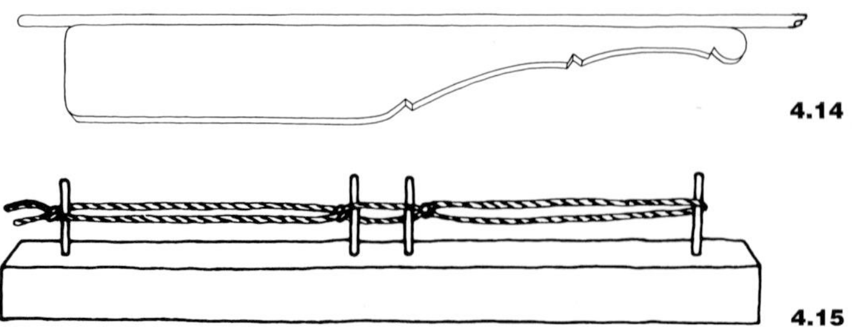
4.14

4.15

4.16

Shuna-ubak, guide for making heddles. (Fig. 4.14)

Jig for making heddles. (Fig. 4.15)

Shingkap/temple. (Fig. 4.16)

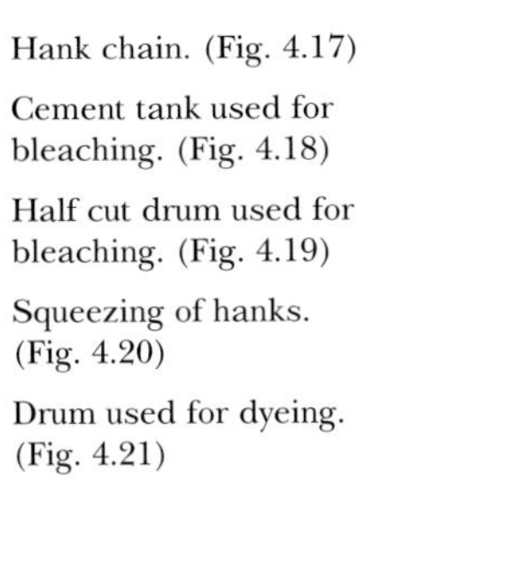
Hank chain. (Fig. 4.17)

Cement tank used for bleaching. (Fig. 4.18)

Half cut drum used for bleaching. (Fig. 4.19)

Squeezing of hanks. (Fig. 4.20)

Drum used for dyeing. (Fig. 4.21)

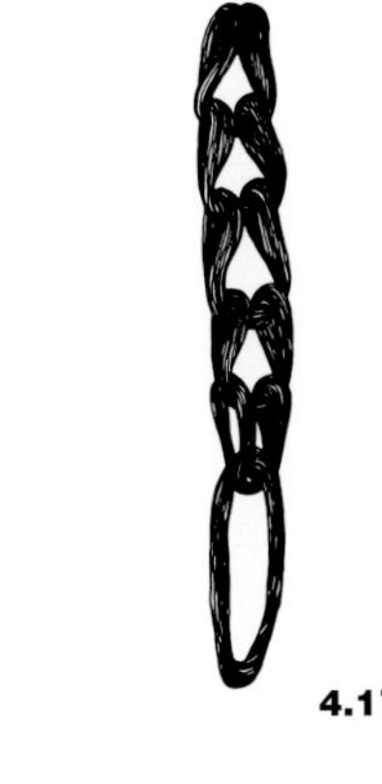
4.17

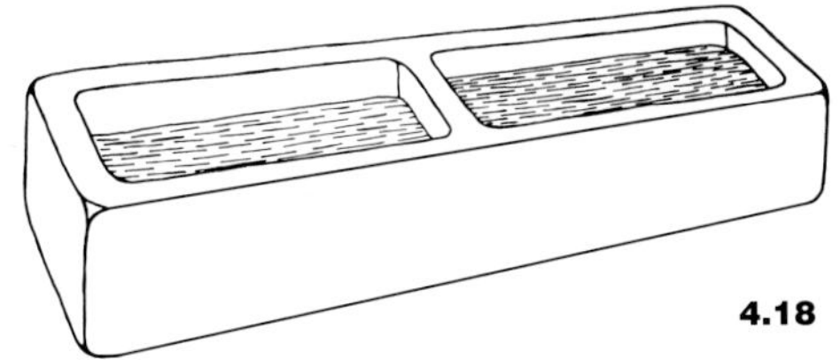
4.18

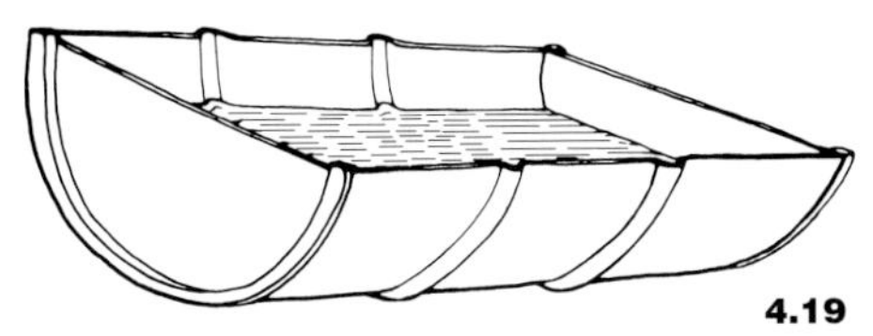
4.19

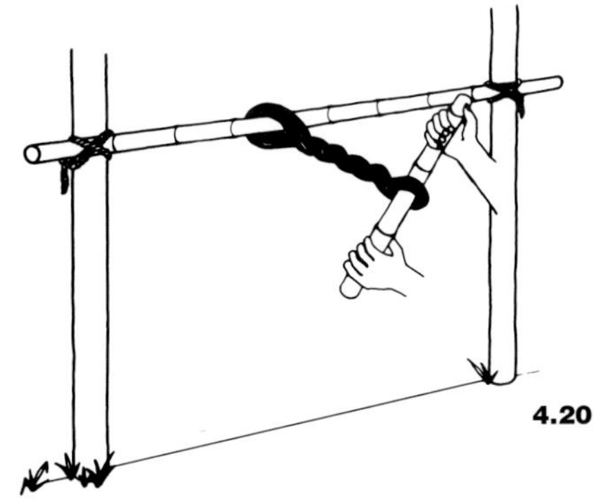
4.20

4.21

and dried in sunlight. The colour obtained from these seeds is light orange, a colour that was worn only by women during occasions pertaining to death (Mohanty et al., 1987: 120, 25).

2. *Sama Machu*: A golden yellow colour is obtained by mixing the powder from the seed of the same plant, *urai romba* with freshly ground turmeric/*Curcuma longa*, Linn; and diluted with water. After adding the salt the yarn was dipped in the dye and kept for one hour. It was then taken out and washed in lukewarm water. After appropriate squeezing it was dried in sunlight.
3. *Kuhi or Sahi Machu*: This is coffee coloured. Bark of the trunk of the *kuhi* tree, (*Lithocarpus fenestratus*), or *Pasania qachi* was stripped and soaked in water for one day after which it was taken out and the solution filtered. After adding salt to the solution, the yarn was steeped in it. The dyed yarn was then removed from the solution after a day or two, and squeezed out. Lastly, the yarn was washed and then dried in the sun. Though this process of dyeing yarn is now obsolete it is still used for dyeing fishing nets.
4. *Kamfoi Machu*: The same process as for *kuhi* was used to obtain a shade of yellow from the bark of the pomegranate plant. In Manipur, the pomegranate, *Punica granatum*, is called *kamfoi.*

Winding of the Yarn

The yarn is bought from the market in the form of hanks. This yarn is wound on to bobbins which are made of bamboo and are normally of two sizes. The bigger ones are used for the warping creel and the smaller ones are used in the shuttle. The winding of the yarn (converting of yarn from hanks to the bobbin) is carried out by using the traditional bobbin winder and the swift or skein winder. The hank is fixed on to the swift and the bobbin on the bobbin winder. An additional stick is put in the hole of the handle of the winder which is then rotated. While rotating with the right hand, the yarn is guided to the bobbin by the left hand.

Warping

In Imphal, the traditional method for making warps is still practised though drum warping is now being introduced. The weavers use a type of warping creel which is called *pheenau grak.* This comprises several parallel vertical bars. Each bar is divided into two parts each accommodating one bobbin. Hence two bobbins can be put in one vertical bar.

Generally, the preparation of the warp is carried out by people who are experts in warp making. But sometimes the weavers also make the warps themselves, especially in the case of small warps. Two vertical wooden rods are fixed in the ground in front of the bobbin stand while a third rod is fixed at the same distance from the first rod. This is fixed so that a narrow angle is formed with the first and second rods.

Each pair of threads from one vertical bar of the stand is taken around the first rod, crossing diagonally to the second rod and going around the second and the third rod. These threads cross diagonally to meet the first rod, and go around the first rod again to the second and the process is continued till the warp is made with the required number of ends. The wooden rods are taken out and the lease threads are put in the warp which is then wound on to the wooden cloth beam. This beam then continues to perform the function of the cloth beam while weaving.

As modern accessories replace old implements, the weavers of Imphal have also started using a new warping method which consists of a wooden warping drum and the warping creel stand.

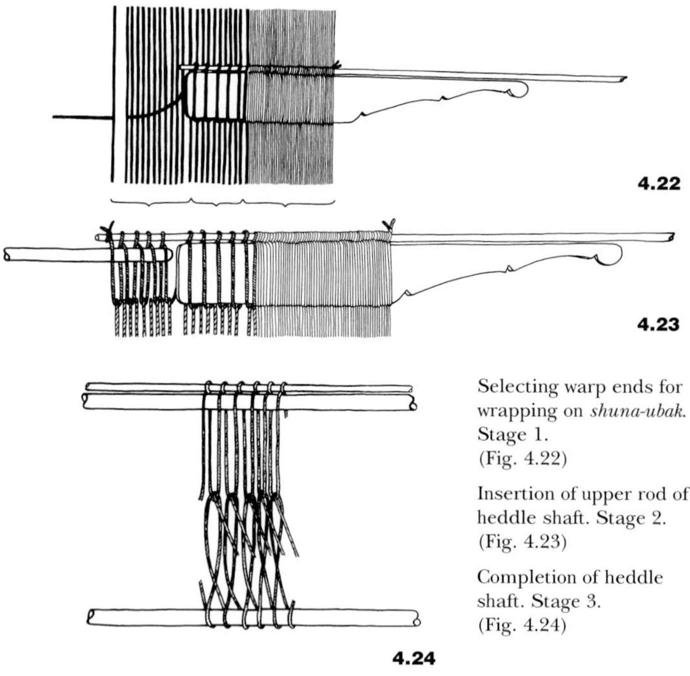

Selecting warp ends for wrapping on *shuna-ubak*. Stage 1. (Fig. 4.22)

Insertion of upper rod of heddle shaft. Stage 2. (Fig. 4.23)

Completion of heddle shaft. Stage 3. (Fig. 4.24)

Denting and Heddle Making

Before denting, and heddle making, the warp is on top of the loom frame and the ends of the warp are kept suspended from the beam. The reed is also kept suspended at a height of about a foot above the ground. Two persons sit on either side of the reed. One passes the thread, selecting it from the lease thread and another pulls it from the opposite side with the help of the reed hook/*kochi*. Normally, the process is carried out from the right to the left of the person passing the thread. In order to achieve 80 ends per inch 160s reed is used. Generally, the denting is done in either of the following ways:

1. Two ends in one dent, leaving every alternate dent empty.
2. One end in each dent across the width.

For the selvedge, the density of the ends is doubled both in denting, as well as in drafting. Once the heddles are made they last for several warps, as the weavers do not change the drafting order. Successive warps are attached by means of piecing thus obviating the procedure of redrafting.

The heddles are of two types namely *shuna-akiba* and *shuna-akhanba*, *shuna* meaning heddle. The *shuna-akiba* is a way of making heddles which have knots in the middle and at the bottom. The heddles are separated from each other, having a heddle eye in the middle through which the warp end is drafted. To make the *shuna-akiba* heddle, four nails are fixed on a wooden board (cf. fig. 4.15). The distance between the nails is determined by the length of the heddle for each *shuna-akiba*. A piece of twine is taken around the first nail and then a series of knots made, one before the second nail, a second knot after the third nail and the third knot after the fourth nail. The heddles made in this way are taken out and put on the heddle frames. The *shuna-akhanba* is a thread heddle which strings all the warp ends. It is made from two separate loops which pass through each other. The bobbin of the thread used to make the *shuna*/heddle is placed to the right of the loom. The thread is passed between the two layers of the warp separated by the lease and the end is tied to a thin wooden stick called the *shingmit*. The *shingmit* is held along the top edge with the *shuna-ubak* by the person who executes the drafting (cf. fig. 4.14). The thread is wrapped around the *shuna-ubak* along with one alternate warp end of the upper layer. The thread is taken back to the *shingmit* and wrapped around it crosswise and again taken downwards for the next heddle. The process is continued till the last warp end (fig. 4.22). While doing this, the previous heddles are shifted further along the shaft so that new heddles can be accommodated. After finishing the heddle making and drafting as required, a thick rod is inserted which serves as a rod of the heddle shaft and the *shuna-ubak* is taken out, while the *shingmit* remains as it is (fig. 4.23). This process is undertaken only for the alternate warp ends which were drafted in the heddles. Turning the beam upside down, the same process is repeated. But this time the previous half part of the heddle is also taken inside the new half part of the heddle along with the remaining alternate warp ends and the *shingmit* is not used (fig. 4.24).

After the drafting is completed, the warp ends are tied to the warp beam ensuring even tension. Finally, the whole warp is wound on the warp beam. The loom is then ready for weaving.

The Weaving Process

The *moirangphee* is either woven in full width or two half pieces. Most often they are woven in two identical half pieces in sequential order. These are then stitched together, and are referred to as centre-stitched *moirangphees*.

For weaving a full width fabric, two weavers are required. The border shuttles are inserted up to the required number of ends simultaneously in the first shed, by both the weavers, sitting on two sides of the loom. Then the third shuttle is passed through the shed from one side to the other, interlocking the border weft of the first side. At the other side, the weaver catches the shuttle before it reaches the border and interlocks it with the other border weft. This is done in the single interlocking technique, referred to as three shuttle weaving in Kanchipuram and other weaving centres in India. At the same time, the weft yarn from the other side is inserted in the same shed. The shed is changed and the body shuttle is thrown back to its previous side, through the shed where it is again interlocked with the border weft. While doing this for each successive pick, the border weft is passed one step further than the previous one. This is carried out till the triangular shape reaches its maximum required length. The border weft yarn forming the triangle is inserted to an extra length,

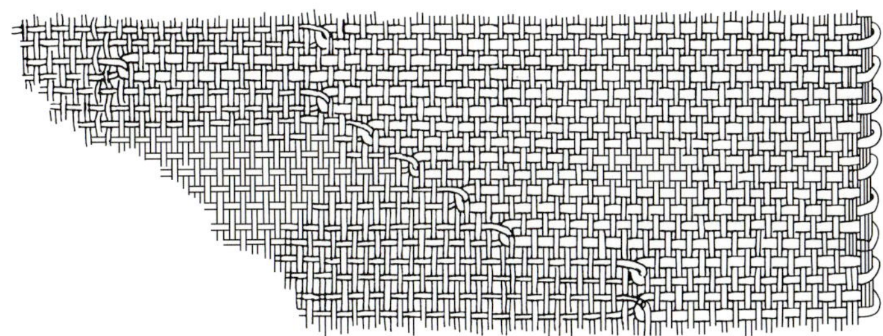

4.25

Structure of triangular shaped border motif. (Fig. 4.25)

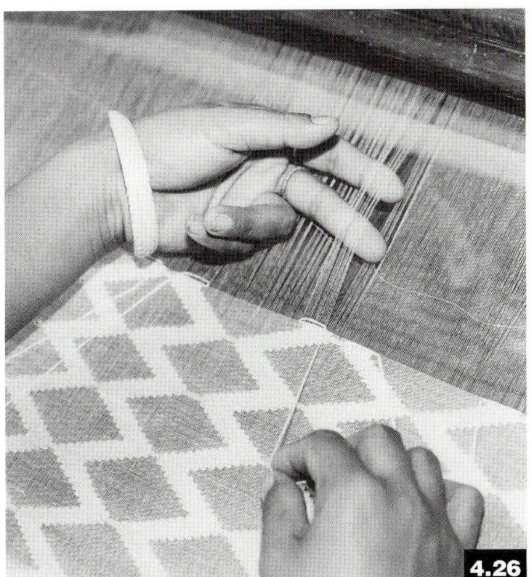

4.26

Weaving of *moirangphee*. Note manual lifting of warp for *jāmdānī* weaving. Craft Museum, New Delhi. (Fig. 4.26)

4.27a

4.27b

Structure of star motif (a). Detail (b). (Fig. 4.27a,b)

4.28

Part detail *moirangphee chādar*. (Fig. 4.28)

to give an elongated point at the peak of the triangular motif (fig. 4.25, cf. fig. 3.34). After this, again the insertion of the border weft decreases gradually till it reaches its lowest point. When this is reached, the insertion of the border weft is stopped, and for a few picks only the main body shuttle is inserted, so that a certain distance is maintained between two triangular shapes.

The fabrics which are woven in the half width are executed by one weaver. In this case, the temple-top shaped border is woven only on one side of the cloth. Two shuttles are used, one for the main field and another for the border; both are operated by the same weaver. In both methods, whether one weaver weaves or two weavers weave, a reference drawing fixed under the warp provides the guideline for the inlaying of the supplementary wefts. The count of yarn used for figuring is generally the same as for the ground warp and weft, but it is plied two or three times. Usually it is of a different colour from the ground weft. However, in certain types of *moirangphee*, supplementary weft is also inlaid with white ground weave yarn. This *jāmdānī* technique appears to be of Bengal origin in the context of Manipur (fig. 4.26). As the background is plain weave, only two shafts are used, which are operated by depressing the two pedals alternately.

Although sizing is not carried out prior to weaving, an application of starch is necessary while the weaving is in progress. About 1 to 1 in. of warp is starched before weaving.

For some of the fabrics, a comparatively thicker quality of 60s count cotton is also used. However, the use of staple yarn is gaining popularity. The use of *mūgā* silk was very common in the early days, but now as the *moirangphee* is one of the most common garments worn by Manipuri women, the use of cotton has become unavoidable. But, on rare occasions *mūgā* and silk *moirangphee* are still worn.

The fabrics, mostly *chādars*, woven on this loom are characteristically of fine quality because they are woven with the yarn of 80s to 100s count. Plain *chādars* as well as the *moirangphee* are woven on the *yongkham*. As a high degree of craftsmanship is required to weave the *moirangphee* design, only very skilled weavers make these fabrics.

The *chādars* are generally 42 to 48 ft. wide and 80 ft. long. The balanced plain weave used as ground weave forms the main fabric structure of the *moirangphee*. For weaving with supplementary wefts the yarn is supplied from small spindles to form the figures in the border as well as in the body of the fabric. These supplementary wefts are inlaid in the regular plain weave sheds that form the ground weave similar to the *jāmdānī* from Dacca. The designs are in the form of motifs or *jāl* spread all over the body or arranged at certain distances. The motifs are called *kabok* (fig. 4.27a, 4.27b). The intricate patterns developed from traditional or completely modern designs are also used in the cross border as well as along the warpwise border.

The fabric with temple top shaped design known as *moirangphee* (fig. 4.28), and also other *chādars* have one background colour on which the patterning is done with a weft of another colour. Sometimes the background is white and the designs are also woven in the same white yarn which gives a beautiful effect of two different types of translucency against light.

When *moirangphee* are worn on special occasions, appropriate colours are used. On the day of the *śrāddha* ceremony, i.e. the thirteenth day after the death of relatives or friends, the people of Manipur wear light orange coloured *moirangphee*. Lord Krishna and the *gopīs* are an inseparable part of their dance festivals when orange silk *moirangphee* are worn. No motifs are used in the background. The *chādars* with a white background are quite popular in Manipur. They have the inlaid supplementary weft patterning in other colours, and often with the same white yarn. Such fabrics have great importance on ritual occasions.

However, *moirangphee* is also a garment worn by the Manipuri women for daily use (see fig. 4.29). Originally, they were woven only in specific colours like red, green, black and blue, but nowadays an unlimited variety of colours are used.

The temple top shaped border which was used only for *chādars* is also being used in several other fabrics. The sari has recently become more popular in Imphal and the weavers have introduced this motif in saris as well as in bedsheets and mosquito nets to name a few among the items. However, many other motifs and patterns are now being used for the *chādars*, some of which are woven along with new types of all over *jāmdānī* patterning in the main field. Some variations in the temple top shaped motif are also to be seen, for instance, as in the Taj Mahal motif. The *cashi* border is a very old version of the temple top shaped motif in which the extra weft technique is used. The *cashi* border is woven inside the selvedge. Often, it is used together with the temple border.

4.29

Usage of *moirangphee* as carrying cloth. (Fig. 4.29)

Other products woven on the throw shuttle loom besides the *chādars* include:

1. *Pagung*: This is an open plain weave cloth with motifs/*kabok* in the body.
2. *Set thava*: A fancy type of *moirangphee*.
3. Curtain cloth: This is a recent development with production based in Imphal. It is used for curtains as also for mosquito nets. The fabric has a pattern of large shapes in the main field and a border across the width.

Making the Designs

To facilitate the process of weaving nowadays, the designs are either made on paper or drawn directly on the warp in fugitive colours by the male members in the family. The complicated patterns are usually drawn on paper and while weaving; this paper is fixed below the warp. Most of these are traditional. Variations in and additions to these patterns, combining the old with the new are prevalent. Nature is the main source of inspiration for new patterns, with flowers and birds as subjects. Also, the old patterns from other types of traditional fabrics are used in the fabrics woven on throw shuttle looms.

The temple top shaped border is mainly used in *chādars* with different types of motifs or patterns in the body. The motifs are distributed in the body at varying yet planned distances. The more complicated patterns and designs which are made in the cross border are called *phijin*. In the lengthwise border, parallel to the selvedge, they are called *phiban*.

Finishing

The *moirangphee* and the other fine count *chādars* are finished in a special way. After the weaving is completed, the fabrics are taken out and washed. If it is woven in two pieces, it is hand-stitched together to make one width and then washed. The stitching is called *taiba*. If the fabric is white, it is bleached before washing. The warp ends of the *chādars* are finished by applying rice starch. When the starch is still wet, a set of eight to nine ends is straightened and made parallel and then twisted by applying the starch from time to time, and later ironed carefully. These strands last well, even after several washings. The making of these strands is done either by the weaver or given to the laundryman who is an expert in this technique. The folding of fabrics is called *thuppa* (fig. 4.30) (*Moirangphee*, NID).

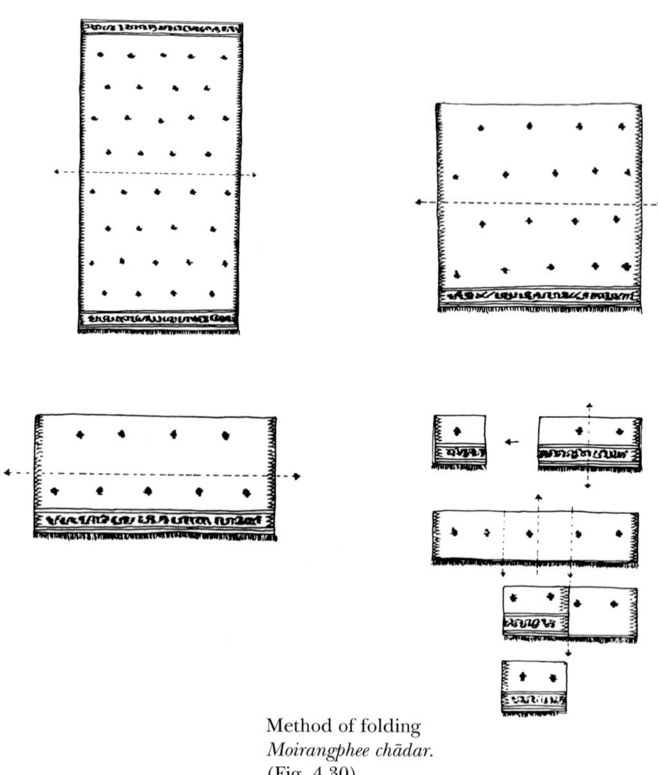

Method of folding *Moirangphee chādar*. (Fig. 4.30)

4.30

Lattice dobby. Courtesy: Panduranga Silk Handloom Society, Dharmavaram. (Fig. 4.31)

Barrel dobby used for weaving *Paithani* in Yeola. Craft Museum, New Delhi. (Fig. 4.32)

Barrel dobby, *nuapatna* loom, Nuapatna, Orissa. (Fig. 4.33)

Brocade woven on *jāla* loom, Nardipur, Gujarat. (Fig. 4.34)

The Loom: From Simple to Complex

Noticeable by its absence in relation to early Indian weaves, is satin, found in China at the turn of the Christian era, and twill, developed by Syrian weavers between the second and fourth centuries AD.[3] In Persia and China the possibilities of supplementary weft patterning in *satin* and twill were fully exploited particularly in the area of silk as this could be done without additional loom devices. *Satin* and twill may have been among the many post-Islamic interventions into the area of decorative Indian weaving. In the Han loom a combination of shaft, treadle and pattern rods comprised the components of the loom as back-up for the production of Han monochrome and polychrome textiles.

In the case of the Indian pit loom the following indices are of interest. In ethnographic textiles, mainly unstitched garment lengths made for rural markets, the width has tended to be narrow and this also extends to the sari (Chisti, Sanyal, 1989). However, in clothing items made for consumption by the higher castes, maximum widths woven on the pit loom range between 4 and 4½ ft.[4] In south India, medium to coarse cotton items ranging in count from 10 to 40 could have been woven outdoors as recorded by Europeans in the seventeenth century. However, fine weaving in the range of 60 to 100 counts would only have been accomplished indoors as any weather turbulence could cause problems. Fine count cottons required lighter beaters and thinner shuttles. In 1933, A.F. Barker (Barker, 1933: 133), recorded of wool weaving in Kashmir that the *paśmīnā* loom was carefully designed to put the least stress on warp yarn in course of weaving. The shed was designed to be shallow hence necessitating the use of a smooth and thin shuttle. Considerations were, therefore, similar in the weaving of fine cotton, silk as well as wool. A loom equipped with a reed to accommodate 100 count cotton yarn required no adjustment for transition to silk weaving. Loom ornamentation was accomplished with pick-up sticks, pattern rods (fig. 3.33), loops, multiple shafts, and, later, the lattice and barrel dobby (figs. 4.31, 4.32, 4.33).

Both in the case of the Han loom in China and the pit loom in India, the draw loom mechanism was incorporated when the limits of development within the previous system had been reached. In both cases, the draw loom, which was grafted on to existing loom structures with necessary modifications, was drawn from non-indigenous sources. The Malaysian Kota Bahru loom type appears to have occupied an intermediary position in the evolution towards the above loom. Whereas the Chinese loom seems to have been an adaptation of either the Sassanid or the Byzantine damask loom, the prototype for the Indian draw loom as manifested in the *jāla* of Varanasi was derived directly from the Persian draw loom.[5] The *aḍai* represents a separate chain of development if the Malaysian Kota Bahru loom is taken as its probable point of origin.

Brocade Weaving of Borders and *Pallus*: Gujarat

Historically, Gujarat was a great centre for the production of woven textiles including brocade, but in 1981 there were only two weavers practising the method of brocade weaving in Ahmedabad. One was specializing in weaving of borders with extra warp patterning, and the other in the weaving of *pallus*, the end pieces of the sari, with extra weft. The mechanism of the *pallu* loom in Ahmedabad is similar to that traditionally employed for the weaving of *maṣrū* in Patan, Gujarat. The items were woven on looms with multiple shafts and treadles, whereas the brocade saris woven at Ridrol, Mehsana district, Gujarat, were executed on looms with the *jāla* attachment. The *kimkhāb*, selvedge to selvedge brocade, manufactured at Nardipur (fig. 4.34, cf. fig. 7.76), Upera, Gajhariaya and Charda in the Mehsana district, Gujarat, earlier woven with the *jāla* attachment, are now woven with the jacquard mechanism.

Typology in Brocade

Fabrics in which brocaded patterns extended from selvedge to selvedge across a silken ground were called *kimkhāb*, and those in which the weft comprised picks of pure metallic wire were called *tash*.

Brocades in which the major area of the ground was of silk with certain portions worked in metallic yarn were called *bafta* or *pot thāns*, whereas silk gauze with design elements in metallic yarn was called *abrawan*. The borders and end pieces could be sewn on. These were called *abrawani*.

As was observed by George Watt at the beginning of the century, Gujarati textiles were noted for their lively colour schemes both in the field and in the floral designs, as also for their bright and pleasing tones. There was little difference between the brocades made at Ahmedabad and those at Surat. The popular colours at both centres included gold, green and dark maroon. The cross-border was embellished by a panel of flowering trees with pale green leaves; such motifs bear a striking resemblance to the enamelled motifs used in Mughal architecture. The term *Asavali* has been used in relation to this category of weaving. This kind of pattern is woven with the *jāla* attachment on the loom at centres such as Ridrol.

In earlier times brocade borders, *pallus* and borders patterned with *zarī* in extra warp were woven by weavers of the Hindu Khatri caste, whereas the *Asavali* saris were woven mainly by the Patel community. The Thakurs and Rawals were involved in the ancillary activities related to the craft. The dyeing of yarn was the task of the Muslim Khatris. All members of the family participated in the activities involved in the making of these fabrics. The primary task of weaving was carried out mainly by men whereas all the subordinate tasks related to weaving, such as bobbin winding were undertaken by women and children.

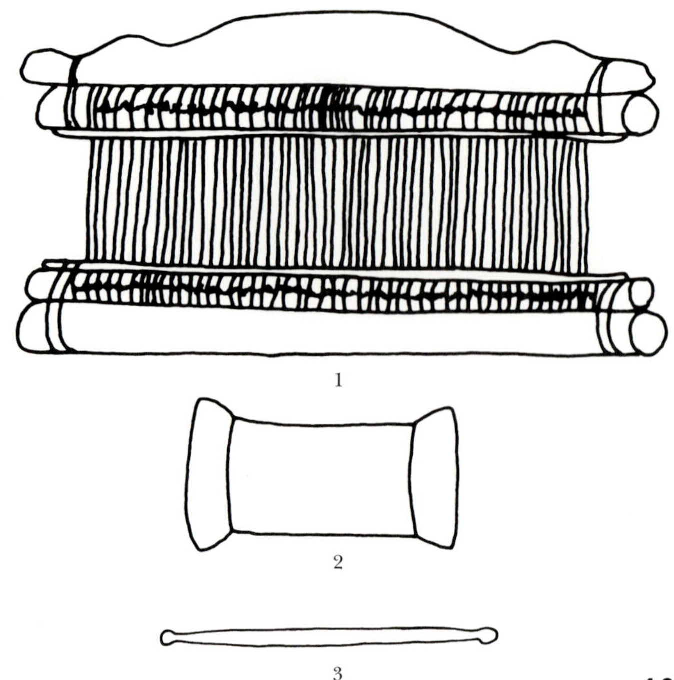

4.35

Tools and accessories. (Fig. 4.35)

1. Reed
2. Spool
3. Pern.

Ahmedabad Border Loom/*Ghoḍa Shāl*

Equipment, Tools and Accessories

The equipment, tools and accessories (fig. 4.35) include the following:

1. Cloth beam or breast beam: The woven fabric is wrapped around this beam locally known as *tor*. The end of the fabric is hemmed and a rod is inserted into the beam, which is held in place by two nails on either side of it. The whole beam is attached to the stool.
2. Warp beam: At the far end of the warp all the different coloured yarns are tied to different rods. These rods are tied to cords which pass around pulleys and have weights tied at the end.
3. Stool: Unlike the major usage in which the weaver sits on a piece of planking, in this case the weaver is provided with a stool. He sits on this stool and operates the pedals.
4. Sley/*hatho*: This holds the reed and is swung forward to beat in the weft. It is not attached to any frame, but rests on the warp. This is possible because it is small and light and about 6 in. wide.
5. Reed/*phuni*: The reed serves to keep the warp threads evenly spaced. The dividers are made from splits of the stem of the *jowār* plant. The spacing is maintained by the binding cord used while making the reed. The reed count used is 52s.
6. Spool and bobbin winder: This is used to wind yarn from the hanks into bobbins, spools for the weft and for making the warp.
7. Hank holder: This holds the hank in a stretched position so as to facilitate its unwinding and winding onto the bobbins. These are made from bamboo splits which are kept in place by interlocking strings. There are large and small hank holders for large and small hanks of *kasab* and cotton respectively. A stand/*thapu* is used to hold the hank holder upright during the unwinding of the hank.
8. Creel: This is a stand for spools which is used during the process of warp making, facilitating the smooth unwinding of yarn. It is a rod resting horizontally on two hooks in the wall. The spools are put on this rod.
9. Warping board: This is used for making the warp. It consists of two boards with nails or pegs driven in. Each board is placed at opposite ends of the wall.

In 1981 it was reported that the brocade *pallus* (fig. 4.36), and borders were used largely by the Rabari and Bharwad communities of Kutch and Saurashtra. Originally, borders were used by the more affluent Patels, Kanbis and other communities. Rabari communities of Saurashtra stitched the Ahmedabad woven *pallus* on to their maroon coloured *oḍhnīs*. The width of the *pallu* is about 40 to 42 in.

Borders Patterned with Extra Warp

Borders are woven in narrow-width patterned lengths. They were woven in imitation and tested *kasab/zarī*. The *kasab* stands out because the pattern is in extra warp. These borders were previously used as edging and trims for wall hangings and *chaklās* (square hanging), as well as for garments such as *ghāgrās/* skirts, *abhas*, a long *kurta*-like garment, and *cholīs*, a stitched upper garment worn by women. These were also used as edgings of *cholī* sleeves. The older borders were generally of a narrow width in which the motifs were clearly defined. They used to be woven in pure silk and pure *kasab*.

The *Ghoḍa Shāl*/Border Loom

The border loom is called *ghoḍa shāl* because the appellation *ghoḍā* meaning 'horse', and also 'stool' in Gujarati is based on the fact that the weaver sits on a stool above ground level.

The border loom comprises a wooden stand that houses the treadles ranging from eighteen to thirty-two in number, along with multiple shafts, each single shaft being called a *gullo*. The pedals are connected to the lower part of the shafts by cords. While the single pedal is called *pāvḍī*, the entire set of pedals including the system of tie-up below the shaft is called *jāla*. Thread heddles, called *hoyo* or *vay* are used in the shafts (figs. 4.42b).

There is a counterbalance mechanism, the *joro padhati*, which allows one shaft to go up when the other is pressed down by the treadle. Thus a positive open shed is formed. Every two shafts are attached to a heddle horse made of wooden sticks. The string, suspended from the ceiling is tied to the centre of the horse balancing the shafts. The term, *rach* is used for the entire system linking the shafts to the upper limits of the loom. The *hatho* or sley houses the reed/*phanni*, and it rests directly on the warp. The warp is held in tension, its farthest end being tied to a rope which is guided by spools or a pulley and weighed down by weights. The other end of the warp is tied to the cloth beam/breast beam (figs. 4.37, 4.38, 4.39, see figs. 5.44, 5.45, 5.46, 5.47).

Pallu woven on *Khāḍa shāl*, Ahmedabad. (Fig. 4.36)

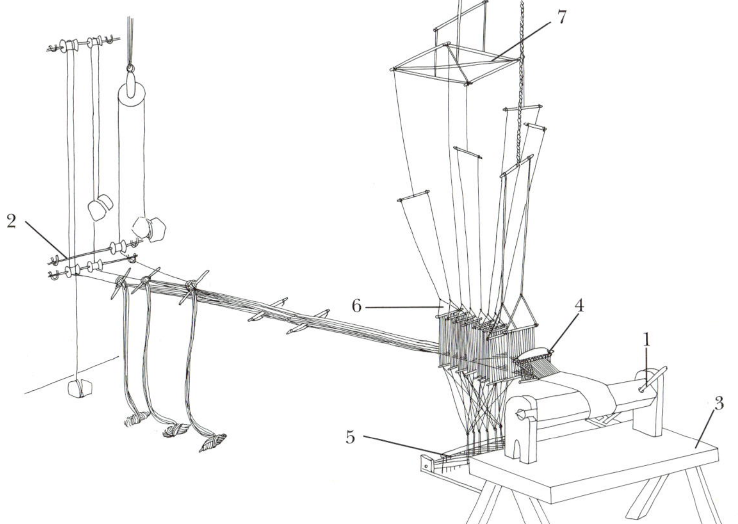

4.37

Ahmedabad *ghoḍa shāl* with multiple shafts and multiple treadles. (Fig. 4.37)

1. *Tor*/cloth beam
2. Warp beam
3. *Ghoḍa*/stool
4. *Hatho*/sley
5. Treadles
6. Shafts
7. Heddle horse

Border being woven on *ghoḍa shāl*. Photographed at work place of Manilal Somnath, Manek Chowk, Ahmedabad. (Fig. 4.38)

Rear view of pedals of *ghoḍa shāl*. Photographed at work place of Manilal Somnath, Manek Chowk, Ahmedabad. (Fig. 4.39)

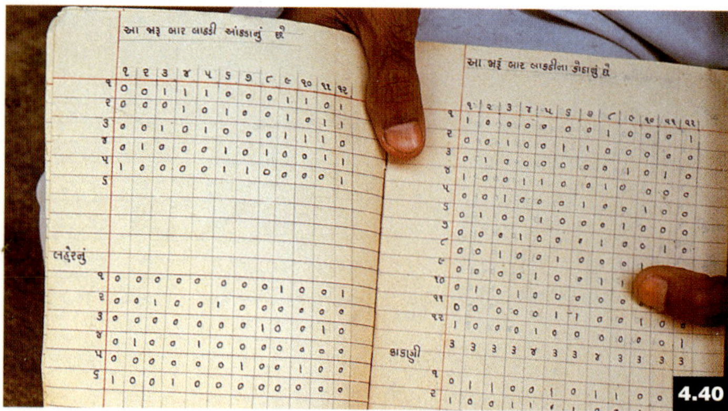

Khata/pattern book showing lifting order in Ahmedabad brocade *pallu* weaving on *khāḍa shāl.* Courtesy: Ambalal Khatri, Ahmedabad. Cf. fig 4.41. (Fig. 4.40)

Chart on paper indicating weaving order for pattern. *Gaṭṭhuā* weaving. Courtesy: Haji Muhommad Ayub, Madanpura, Varanasi. (Fig. 4.41)

Materials Used

The materials used are metallic yarn, silk and cotton. The *kasab* of count 2.60s is used for the extra warp. Previously, pure *kasab* manufactured in Surat was used but in the year 1981 pure *kasab* was found to have been replaced by imitation *kasab* known as *khoṭī kasab.* Pure *kasab* in which the gilding was done with silver or gilded silver is called *rupeṛī* (silver) or *soneṛī* (gold) *kasab. Khoṭī kasab,* made from base metals turned black with the passage of time. Tested *kasab,* silver plated copper, does not tarnish and is much cheaper than pure *kasab* (Mohanty, I, 1984: 105–15). The 13 to 15 denier filature silk used is called *mustūl.*

Suttar/cotton has replaced silk in usage for the weft. Mercerized cotton of 2/80s or 2/60s count is bought from thread shops near Dhalgarwad and Dosi-Varani Pole, in Ahmedabad. Sometimes hanks unfit for mill consumption are used, as they are cheaper.

Pre-weaving and Weaving Processes

The design is first worked out on a graph, and the system of notation is very similar to that followed for *gaṭṭhuā* (cf. figs. 4.40, 4.41). The symbol '0' means down, 'I' refers to up. The number of columns relate to the number of shafts. The weaver reads the graph in terms of the lifting order of the pedals.

Dyeing

The hanks to be dyed are immersed in a hot dye bath containing the required dyestuff with some common salt. The hanks are turned around from time to time, to allow for the even absorption of the dye. They are then taken out of the dye bath and rinsed in water to remove the excess dye content and dried.

Spool Winding

For spool winding the hanks are put on the hank holder; the spool on the spool winder which are then filled with yarn. This operation is carried out by all the female members of the family.

Winding the Warp

When winding the warp the weaver takes 8 ply yarn and knots them to the first peg on the warping board (fig. 4.42a). He winds the warp back and forth in a zigzag manner.

Drafting

The new warp ends are pieced together with the old warp ends. The warp is then brought forward through the heddles (fig. 4.42b) until the new warp is arranged and set ready for weaving. Only when a new set up in drafting pattern is required, does the weaver go through the whole process of drafting individual ends. This process is time consuming as it takes approximately three to four days to complete. Two shafts are allotted for the ground weave. The supplementary warp ends are drafted in a pointed drafting order on eight shafts, four each for the *hīrā* or the diamond motif and four for the *khajūrs,* date palm motif.

To achieve the fluidity in form demanded by the *keṛī*/mango and *popat*/parrot motif which are typical of Ahmedabad borders, individual lifting arrangements are necessary for each of those warp ends that constitute the motif. These motifs form a narrow width of the border and do not repeat across the width. An arrangement that consists of each patterning warp end passing through individual heddles that constitute a part of the motif, and need to function individually is thus made possible. These are held tautly in place as they are tied overhead by a cord and attached to the treadle below.

Weaving

These borders are woven using the supplementary warp for patterning with plain weave as the ground weave. The two treadles used in the plain weave are operated by the right foot. The treadle for the supplementary warp is operated simultaneously with the left

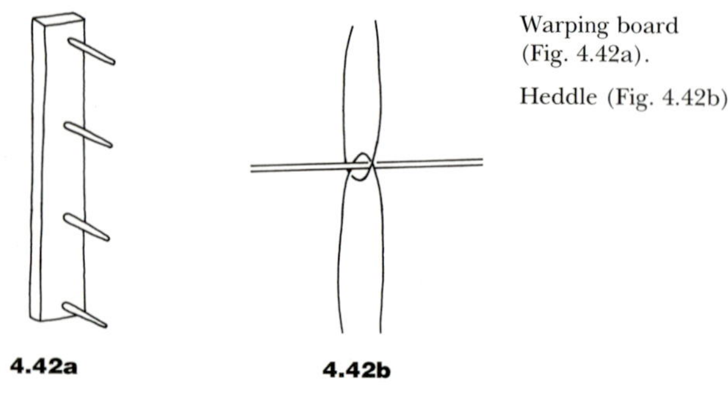

Warping board (Fig. 4.42a).

Heddle (Fig. 4.42b).

4.43

Some border designs / Different colours used (handwritten labels, Fig. 4.44)

4.44

Representative drawing of pattern in border. (Fig. 4.43)

Borders. Woven on *ghoḍa shāl*, Ahmedabad. (Fig. 4.44)

1. *Makru*
2. Warp beam
3. Shafts
4. Cloth beam
5. Beater
6. Pedal
7. *Athnoni lakḍī*

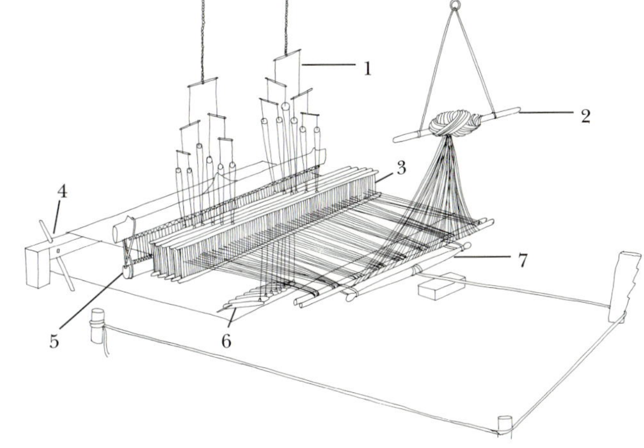

4.45

Khāḍa shāl used in weaving detachable *pallus.* (Fig. 4.45)

foot. The supplementary warp lifted remains in position for the required number of picks, and the binding takes place subsequently. On analysing a traditional border with *hīrā*, *khajūrī* and *popat* motifs, it is found that the binding points for the supplementary warp ends float over the ground weave for a minimum of six picks and a maximum of nine picks.

Patterns

The patterning on woven borders has seen little change over the past 60 to 70 years. Formerly, a silk and cotton mixture called *gul* was used along with real *kasab.* The colours usually used are *lāl*/red, *jambli*/purple, *lilo*/green, *gulābi*/pink, *kāro*/black. The red border has a pink *nakshi*/selvedge, with thin green lines running the length of the warp, acting as space dividers for the motif (figs. 4.43, 4.44).

Sizes of borders vary according to changing trends. The widths range from 2 to 4 in. The narrower borders do not have the mango motif. The border widths are identified by the number of diamond shapes in the body of the border and are available in even numbers such as:

Chār Hīrā - 4 diamonds
Chhe Hīrā - 6 diamonds
Āṭh Hīrā - 8 diamonds
Das Hīrā - 10 diamonds

Finishing

Dhauar Gundar/gum obtained from the secretion of the *dhodki* tree trunk, *Grislea tomentosa* is soaked overnight in water and strained. The border is stretched and the gum is applied evenly on its front face. Once dry, these borders are calendered, giving them a crisp feel and a glossy surface finish.

The *Khāḍa Shāl/Pallu* Loom

The loom/*shāl* is a pit/*khāḍa*, pulley, counterbalanced loom. The pit is preferred as the *pallu* is wider than the border. The warp beam is hung from the ceiling. To keep the warp spread out to the right width and to maintain its tension, bunches of warp ends are twisted around a rod called *athnoni lakḍī.* This rod is tied to a cord which is wound round two stumps at the back and is brought in front, to the right side of the weaver, where it is tied to another stump. This stump is within the weaver's reach so that he is able to loosen the warp without changing position (fig. 4.45). In the *pallu* loom the system of heddles horse and pulley is called *makru.* The basic principles on which this loom is worked are the same as those to be found in the border loom.

Unfurling of disused *gaṭṭhuā* loom. The *ankhada* tie up mechanism, may be viewed in the centre. Also in view is a heddle shaft. Courtesy: Suhail Ayub, Varanasi. (Fig. 4.46)

Unfurling of disused *gaṭṭhuā* loom. Setting up of *pāvḍīs*/pedals below the warp. Courtesy: Suhail Ayub, Varanasi. (Fig. 4.47)

Reassembling of dismantled *gaṭṭhuā* loom, Varanasi. Note similarity with Ahmedabad *khāḍa shāl*, *pallu* loom. (Fig. 4.48)

The *Gaṭṭhuā* Loom

The *gaṭṭhuā* loom marks a progression in development over the *ghoḍa* and *khāḍa shāl* looms. The word *gaṭṭhuā* is derived from *gaṭṭh* or *ganṭṭh* which means knot or tie-up. The characteristic feature of the *gaṭṭhuā* loom is the elaborate and intricate tie-up of shafts to facilitate weaving in the extra-weft technique. The *gaṭṭhuā* is a multiple-shaft and multiple-pedal loom. Although one sample of the loom has been traced, it has been dismantled and put away. The present documentation is based on a special re-assemblage which was made of this loom by Suhail Ayub, Varanasi (figs. 4.46, 4.47, 4.48).

According to the late J.N. Supakar, Varanasi, the *gaṭṭhuā* loom was prevalent before the *jāla* loom and was exclusively used by the weavers of Madanpura in Varanasi. This loom facilitates the weaving of intricate extra-weft patterns on saris and other fabrics. Small motifs are used to give textural effects which characterize fabrics woven on the *gaṭṭhuā*. These motifs may be used to create an allover pattern, to form diagonal lines or to fill areas within larger motifs. This loom operates on a complex *nachchni*/heddle horse system and has innumerable *pāvḍīs*/foot pedals which the weaver operates with his toes. The introduction of the *jāla* mechanism, followed by the jacquard, led to the decline and eventual extinction of this loom.

The differences between the *jāla* and the *gaṭṭhuā* may be listed as follows:

1. In the case of the *gaṭṭhuā* loom, the shafts that are required to be lifted simultaneously are tied together in a bunch. There are many such bunches. According to the lifting order required by the design, these bunches are lifted on to a hook while the weft is inserted. However, in the *jāla*, the repeat of the design is woven with lease threads on a miniature warp, that is, the *naksha*, placed perpendicular to the main warp. These lease threads are lifted one at a time by the draw-boy according to each lifting order.
2. The *pāvḍīs* and the *ankhada* of the *gaṭṭhuā* are replaced by the pedals and the *naksha* in the *jāla*.
3. The *nachchnis* or the heddle horses of the *gaṭṭhuā* are eliminated in the *jāla* system as the lifting order is integrated within the *jāla* attachment to the loom.

It is believed that the Banarasi weavers are descendants of weavers from Gujarat. According to oral tradition, the silk weavers from Gujarat were forced to migrate during the famine of AD 1630–2 and many settled in Varanasi.

The similarity between the border and *pallu* looms of Gujarat which operate on the multiple harness and multiple-pedal system and the *gaṭṭhuā* of Varanasi, which is based on usage of the *nachchni* and *pāvḍī*, substantiates this theory.

The flexibility of the *gaṭṭhuā* system allowed the weaver to innovate in the course of weaving and create new patterns through various permutations and combinations of lifting orders and shed-formation. In the *jāla*, the lifting order is determined according to design and no variation is possible within this order of warp lifting by the weaver in the course of weaving. The *gaṭṭhuā* system can be said to encompass the weaving order found in the *ghoḍa shāl* and the *khāḍa shāl* as discussed above. In this system each pulley, and each heddle horse is manipulated by a separate pedal. Different combinations of shaft and pedal are possible both within the lifting order of the pattern as well as by changing the sequence of the pedals. The limitation of the loom arises from the fact that the number of pedals which can be conveniently operated by the human foot is restricted as also limiting the number of shafts which can be accommodated on the loom.

The advantage of the *jāla* arises from the fact that the physical movement involved in operating the lifting order is based on skilled hand picking. Since the *jāla* is not subject to the same physical constraints as is constituted by the manipulation of the pedal in the *gaṭṭhuā* system, it offers greater freedom within the prescribed pattern.

Reassembling of *gaṭṭhuā* loom. Note *nachchni*. (Fig. 4.49)

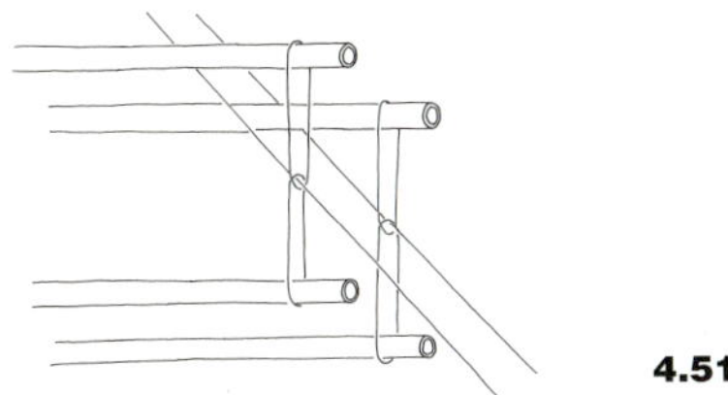

Bai/ heddles through which warp ends have been passed. (Fig. 4.51)

The reassembled *gaṭṭhuā* loom. (Fig. 4.50)

1. *Thappa batta*/ warp peg
2. *Sakada*/lease rod
3. *Kadis*/shafts
4. *Bai*/heddles
5. *Nachchni*/heddle horses
6. *Pāvḍī*/pedals
7. *Ankhada*/ design tie-up
8. *Hattha*/sley
9. *Phanni*/reed
10. *Toor*/cloth beam

4.50

This is exemplified in Gujarat where the more complex patterning of Ridrol brocade is done with the *jāla* attachment.

However, the *jāla* does not offer as wide a possibility of working in variations on the loom in the course of weaving as in the *ankhada* system. The pulley system in the Ahmedabad *khāḍa shāl* allows for the accommodation of a larger number of shafts and thus for greater variation in a possible range of weaves. The square set-up on the uppermost level of the shaft tie-ups on the Ahmedabad *ghoḍa shāl* permits four further permutations in the lifting order through manipulation by pedal of the four corners of the square tie-up. This system also ensures even shedding at the level of the shafts farthest removed from the reed.[6] It is perhaps because of the usage of the *gaṭṭhuā* that the introduction of the dobby has been obviated both in Gujarat as well as at Varanasi.

When the *jāla* was introduced to Varanasi from Gujarat, it began to supersede the *gaṭṭhuā*, as its operation was less time consuming. However, even in 1961, there were two or three *gaṭṭhuā* looms still in operation. By this time, the jacquard had come in, and the *jāla* itself had begun to give way to the jacquard. With the passage of time the *gaṭṭhuā* also disappeared.

The Loom

The *gaṭṭhuā* loom of Varanasi, is a multiple-shaft, multiple-pedal pit loom (figs. 4.49, 4.50, 4.51). The harness mechanism above the warp is called the *nachchni* while the pedals below are called *pāvḍīs*. The loom parts comprise the following:

***Thappa baṭṭā*/warp peg:** The *gaṭṭhuā* does not have a horizontal warp beam, but instead the warp is wound around a short wooden pole which, being very smooth and highly polished, averts friction and thus prevents the silk warp from breaking. This pole is fixed in a hole and is passed through a round stone slab which weighs it down, and keeps it in position (*Banaras Brocades*, NID: 5), (fig. 4.50).

***Sakada*/lease rod:** This is made of a smooth and narrow strip of bamboo or wood, and is placed behind the shafts. The lease rods separate the odd and even ends and maintain their order. A lease rod is 56 in. long. These rods are joined together with a string called a *jhutti*. The lease rod was also a tool which supported the reed in maintaining the width of the warp.

***Kādis*/shafts:** These are made of cane. The *kādi*/shaft is a long stick which is supported by another shorter, thicker stick in the centre. Usually there are 36 to 40 shafts in the *gaṭṭhuā* though the number can vary.

***Bai*/heddles:** These are made of twine or mercerized cotton. A rectangular wooden piece, half the length of a heddle, is taken and a length of twine/*jīre* is tied to one end of this piece. Then the thread to make the heddle is wrapped around this wooden piece thus forming a loop which is one half the length of the heddle. The loop is knotted on to the *jīre* so that a uniform spacing of heddles is maintained. After the required number of loops is wound, depending on the number of heddles required, the wooden piece is flipped over. The entire process is repeated in such a way that a new set of loops is formed interlocking with the previous one, thus forming the heddle (*Banaras Brocades*, NID: 7) (fig. 4.51).

***Nachchni*/heddle horses** (see fig. 4.49): This is a counter-balance mechanism which facilitates the smooth operation of the shafts. The *nachchni* system is made up of small wooden units 4 in. long x $\frac{1}{2}$ in. broad x $\frac{1}{4}$ in. thick which are strung together. Each end of the *nachchni* is tied to an alternate shaft of the loom. The system of the tie-up of the *nachchnis* is demonstrated below. The number of *nachchni* progressively decreases in the vertical plane culminating in one final *nachchna*. In the *gaṭṭhuā*, there are two such tie-ups, one at either end of the shaft.

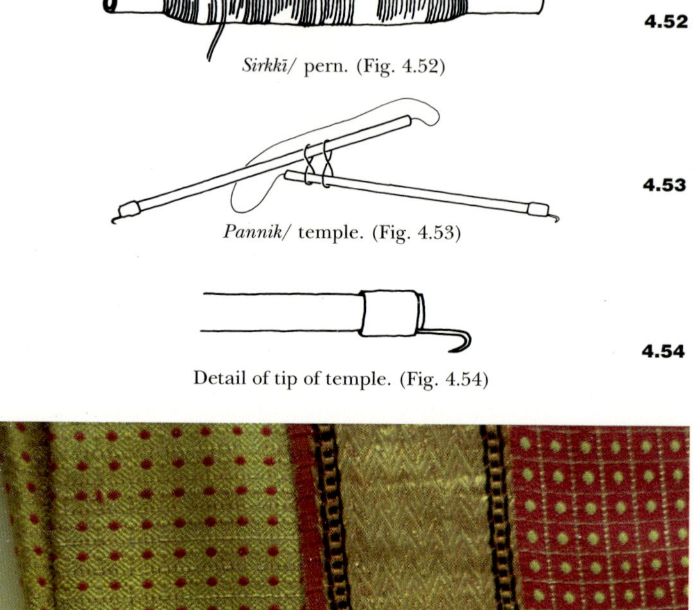

Sirkkī/ pern. (Fig. 4.52) 4.52

Pannik/ temple. (Fig. 4.53) 4.53

Detail of tip of temple. (Fig. 4.54) 4.54

4.55

Oḍhanī, detail, Banaras brocade. Woven on *gaṭṭhuā* loom. Courtesy: Bharat Kala Bhavan, Varanasi. (Fig. 4.55)

Pāvḍī/pedals: Under the loom, is a *pausar*, operated by the feet, which is made up of numerous small wooden pedals/ *pāvḍīs*. The number of *pāvḍīs* depends on the design, i.e. the number of permutations and combinations in the lifting order/ *pāvḍī kī chāl*. The number of strings joining the shafts to the *pāvḍīs* depends on the number of times the shaft has to be lifted in one repeat of the design. The tie-up of the *pāvḍīs* governs the ground weave construction and other basic weaves, used to create the patterning within the larger motifs. These weaves may be twill or variations of twill such as pointed, wavy, or diamond.

Ankhada/design tie-up: This is placed above the shafts between the two *nachchni* systems. It is a complex yet systematic tie-up of the shafts and like the *naksha* in the *jāla* system, it controls the allover patterning/motifs on the fabric.

Each lifting order of a motif is grouped together by threads to form a bunch which is tied to a wooden rod on top of the loom. There are as many bunches as there are lifting sequences in one motif/repeat.

At one side of the *ankhada*, a hook is suspended from the top. If one bunch of *ankhada* is to be lifted, it is put on to this hook. This keeps the ends in a lifted position until the weaver completes the pick.

Hattha/sley: There is a hanging sley, which is usually suspended from the ceiling. This facilitates beating-in of the weft according to the requirements of the fabric structure.

Phanni/reed: Originally, the reed was made locally in Varanasi from bamboo splits. The first step in making the reed is the removal of the outer skin of the bamboo. This is then dipped in oil after which it is boiled in water. It is now dried to become smoother and harder. After this the bamboo splits can be used to make the reed. A dent of the reed is called *gewā*. A 60s reed is used.

Toor/cloth beam: This is a wooden beam, either angular or cylindrical in shape. The cloth beam is indented into a groove along the length of the beam. It also has holes at one end through which an iron rod/ *bidhanī* is inserted to lock it into place.

Accessories and Tools

Dharkī/shuttle: This is a boat-shaped throw-shuttle made from buffalo horn/ *seengh* locally made in Varanasi. The size of the shuttle varies according to the design. The commonly used shuttle is 7 to 8 in. long, 3/4 in. wide and 1/2 in. thick. It is a type of throw-shuttle into which a pern/ *sirkkī* is fitted. This shuttle is thrown by hand through the warp shed.

Sirkkīi/pern: This is also called a *tillī* or a *kandhi*. These are small, polished bamboo sticks on which the silk and metallic wefts are wound. These are used to insert the localized extra weft (fig. 4.52).

Suā/needle: This is a wooden needle approximately 4 in. in length and is used in denting.

Pannik/temple: This is also called *maṭhī*. It is a device used to maintain the width of the weaving. It consists of two sticks of cane strung together with two loops. Each cane has a small sharp needle on one end, fixed in place with lac/ *chapada kī lāh*. This tip is 1 in. long and the cane is 34 1/2 in. long (figs. 4.53, 4.54).

Pattra/seat: This is the wooden plank on which the weaver sits.

Pigeon feather: This is used to comb the warp from below.

Pattern on paper: The predominant pattern of the fabric, which in this case is the *koniyā* or corner design, is drawn by hand on paper. This paper is pinned on to the warp from below.

Lifting order charts/weaving plan: Unlike procedures in *kani paśmīnā* this is represented by a binary code of dots and dashes, the line/dash indicating the depressing of a *pāvḍī*. The lifting orders are written in columns and are read column-wise bottom to top from right to left (figs. 4.55, 4.41).

Garment. Banaras brocade. Woven on the *gaṭṭhuā* loom. (Fig. 4.56)

Kāgaz daud bel motif. (Fig. 4.57)

Latifā motif. (Fig. 4.58)

Chānda motif. (Fig. 4.59)

Pre-weaving Processes

In Varanasi, 99 per cent of the silk used today, is obtained from Bangalore, and 1 per cent from Kashmir. Bangalore has become the main supplier only after Independence. Before 1947, the silk yarn was obtained from Malda (West Bengal), Kashmir, China and Japan.

The mulberry silk is a 2 ply twisted yarn, known locally as *katan* and is used in the warp. The Kashmir silk is mostly used in the weft.

The yarn is available in hanks of 1 kg. each. It is de-gummed with soap and water and then bleached. After bleaching, the hanks are given to the dyers. The dyeing of dark colours is more expensive than the light. Sometimes, the weaver does the dyeing himself or one of the members of his family does.

Once dyed, the hanks are taken to the bobbin-winders after which the warping is done. Generally, one warp is 27 yds. 1 ft. long, sufficient for weaving four saris. If a different coloured warp has to be attached for the same design, it is knotted to the previous warp by another person.

Weaving Process

The weaver pins the design paper underneath the warp. According to the design, the required ends are lifted by depressing the appropriate *pāvḍīs*, thus creating the *nas*/twill line. Each *pāvḍī* is held between the big toe and the second toe and is pulled downwards. Once the shed is open, the required weft is inserted in accordance with the paper pattern. The shed is then closed.

This is followed by the manipulation of the *ankhada* overhead, for further patterning. The *ankhada* consists of *guchchhās*/bunches, each bunch constituting a lifting order for one pick of the minor motifs. Subsequently, the *pāvḍīs* are worked, and a ground weave pick is inserted from one side of the warp to the other. This entire process constitutes the weaving of one complete pick, one ground pick along with extra-weft. Thus, the weaving continues.

The distribution, placement and direction of the motifs can be manipulated and altered at the will of the weaver. If the weaver so desires, only parts of the motifs can be woven, thus generating variations of the same motifs. The fabric/sari is woven front face down in the *gaṭṭhuā*.

Layout of the Sari

Designs and Motifs

The traditional motifs that were used to ornament the fabrics woven on the *gaṭṭhuā* loom may be seen on saris and other fabrics that are woven through other processes in Varanasi even today. However, weavers familiar with *gaṭṭhuā* weaving are able to differentiate between the two sets of loom usage (figs. 4.41, 4.56).

Main Field of the Sari

In order to distinguish between the larger paper designs and the smaller textural motifs that constitute the fillers, motifs can be categorized as major or minor. However it is to be noted that the illustrations of motifs which have been provided are indicative of a repertoire of patterns rather than weave structure.

Major Motifs

The major motifs are those which include the bold and large designs which are represented on the design paper. They are drawn by the master craftsmen/*kārīgars*, and weavers. Some examples of this category are:

1. *Koniā/ corner motif*: This is usually in the shape of a teardrop or an *ambi*/stylized mango motif. The central axis of the *koniā* is aligned to the diagonal of the corner of a sari.
2. *Kāgaz daud bel*: This is essentially a large floral, border motif. It is also called the *latthar bel* (fig. 4.57).
3. *Latifā*: This is said to have been named after its designer, Abdul Latif. This is a perfectly symmetrical betel-leaf or teardrop shaped motif (fig. 4.58).
4. *Chānda*: This is a circular floral motif placed at the centre of *dupaṭṭās* or on the back of *kurtās* (fig. 4.59).

5. Diagonal lines: Broad lines, 2 to 3 in. in thickness, constitute the common design of the *gaṭṭhuā* saris. Each diagonal line comprises small *būṭīs*/motifs in extra-weft, and are usually in *zarī*. The *būṭī* may be in the form of a stylized flower. Sometimes the weaver manipulates the design changing the direction of the diagonal so that a zigzag effect is achieved.
6. *Hyderabādī bel*: These are diagonal lines which run parallel to each other, interspersed with individual motifs also placed diagonally. This may be seen on borders or as an all over design on the sari (fig. 4.60).

Hyderabādī bel motif. (Fig. 4.60)

Minor Motifs

The minor motifs are the smaller motifs which are used individually, or as textural filler for the main design, or as the units which form the major motifs, as for instance, the diagonal lines. They may be woven either with the *pāvḍīs* or by manipulation of the *ankhada*.

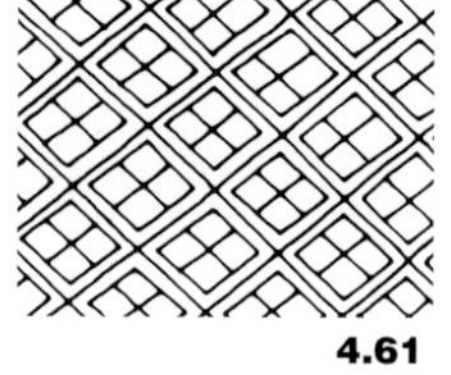

Chaudānī motif (Fig. 4.61)

Chaudānī motif, a variant. (Fig. 4.62)

Minor motifs made with the *pāvḍīs*

1. *Chaudānī*: This is one of the most popular motifs woven on the *gaṭṭhuā*. It consists of diamond-shaped motifs arranged in a half-step repeat creating a grid-like texture (figs. 4.61, 4.62).
2. *Twill lines*: Various twill lines are also woven with the help of the *pāvḍīs*. These are as follows:
   a. *Ādi Nas*: The straight twill (fig. 4.63).
   b. *Leheriyā Nas*: The zigzag or herringbone twill (fig. 4.64).
   c. *Samudrī Laher*: The wavy twill (fig. 4.65).

4.63

Ādi nas motif. (Fig. 4.63)

4.64

Leheriā nas motif. (Fig. 4.64)

Minor motifs made with the *ankhada*

1. *Chiretan/Tripankhā būṭī/3-petal flower*: This is a tiny motif with three petals and a short triangular stem (fig. 4.66).
2. *Panchpankhā būṭī*: This is similar to the *tripankhā būṭī*, and has five petals with a short stem (fig. 4.67).
3. *Jhumkā būṭī*: It is a tiny, mango-shaped motif (fig. 4.68).
4. *Basket jāl*: It is an ogival grid with *panchpankhā* or other *būṭīs*. Other small floral motifs are also made with the help of the *ankhada*. The term *jāl* refers to a web of floral forms which create an overall surface or grid (fig. 4.69).

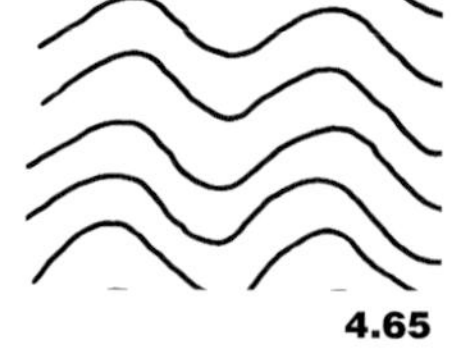

Samudrī laher motif. (Fig. 4.65)

Tripankhā motif. (Fig. 4.66)

Border

The border is weft faced in *zarī* and twill weave. It consists of two parts:
1. *Lais*—which is a solid area of twill.
2. *Mutthra*—are the characteristic silk and gold lines which outline the *lais*.

Panchpankhā motif. (Fig. 4.67)

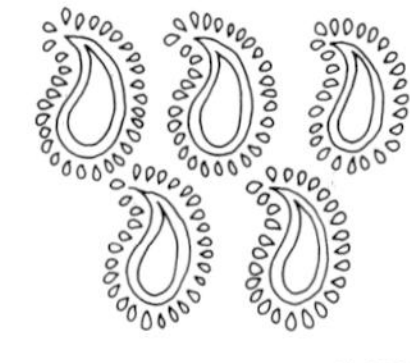

Jhumkā būṭī motif. (Fig. 4.68)

Pallu

The *pallu* is usually very elaborate and highly ornamental. It has various layouts, the most popular one consisting of bands of different designs that have been brought together. The *pallu* length measures between 2 to 2½ ft. Often, on both corners of the *pallu*, is a diagonally placed *koniā*. Sometimes, a large bel-like pattern, a stylized running creeper design is used on the *pallu*.

Colours

The span of colour used in weaving the *gaṭṭhuā* saris was extensive. Colours ranged from subtle soft ones like white, cream, straw, dull yellow, dull green and brick red

Basket *jāl* motif. (Fig. 4.69)

to intense and bright colours like magenta, bright blue, fuchsia pink/*rānī*, green, purple and bright red. Of these colours, *rānī*, cream and green were the most popular.

The Products of the *Gaṭṭhuā* Loom

1. *Sari*: The present-day dimensions of a sari are 6 yds. long and 48 inches wide. It is not known if the dimensions were radically different in earlier samples woven on the *gaṭṭhuā*.
2. *Kurtā*: These are traditional Indian stitched garments worn originally by men, and are woven according to the garment layout (fig. 4.70). Each part of the garment is woven separately on the loom itself and stitched together off the loom.
3. *Oḍhanī/dupaṭṭā*: A *dupaṭṭā* is half the length of a sari. It is draped for modesty over the head and shoulders to cover the breast; it may also be draped around the shoulder and around the head. In Maharashtra the usage differs.
4. *Sāfā/turban*: A *sāfā* is a length of fabric wound around the head.

Varanasi Throw Shuttle Pit Loom with *Jāla*[7]

In Varanasi, throw shuttle pit looms are used. As can be noted from the drawing (figs. 4.71, 4.72), the weaver's seat is placed across the pit. He sits with his feet resting on the pedals operating them as required. These looms also have overhead beams to accommodate the *jāla*, and, in recent times, the jacquard.

On the loom, the *jāla* is suspended from an overhead beam and looks like a complicated mass of threads (fig. 4.73). These threads, however, are leased. In other words, an independent thread is passed through, to keep the threads in sequence and to prevent their getting entangled and knotted. These, when operated, result in the production of silk fabrics of intricate design.

The intricacy of the designs in the textiles of Varanasi has earned considerable fame over the years. The *jāla* weaving technique has enabled the Banarasi craftsmen to produce exquisite designs because of its great capacity to manipulate different weaving orders. *Jāla* is not, and was not, the only technique, of weaving prevalent in Varanasi. The *gaṭṭhuā* preceded it, and today, the jacquard is taking over. Although, for purposes of clarity, the term *naksha* has been used for the preparatory stage of *jāla* making, the words *naksha* and *jāla* are in fact synonymous. This attachment to the loom appears to have reached Varanasi via Gujarat. The term for such a loom at Nardipur, Gujarat, is *doriyā*/thread *sharishāl*, loom (*Jāla Banaras*, NID: 58). The term *jāla* could refer to an arrangement of yarn such as that seen in a spider's web. This would draw *jāla*, as a descriptive term, close to the *doriyā shāl* of Nardipur, Gujarat. The term *naksha*, used in Varanasi, can be linked to the visual impact of the pattern as worked on the *macchan*. The fact that there is a difference in terminology

Sleeve pattern of *kurtā*. Patterns sketched on paper. Used in *gaṭṭhuā* weaving. Courtesy: Haji Muhommad Ayub, Madanpura, Varanasi (Fig. 4.70)

Banaras pit loom with *jāla* attachment (Fig. 4.71)

1. *Jāla*
   a. *Naksha*
   b. *Nathia*
   c. *Seuvarni*
2. *Paggiā*
3. Wooden bench for draw boy
4. *Akkada*
5. *Kādis*, shafts controlling ground weave
6. Reed
7. Pit of the loom
8. Wooden shaft supporting the *jāla*
9. Wooden bench for the weaver

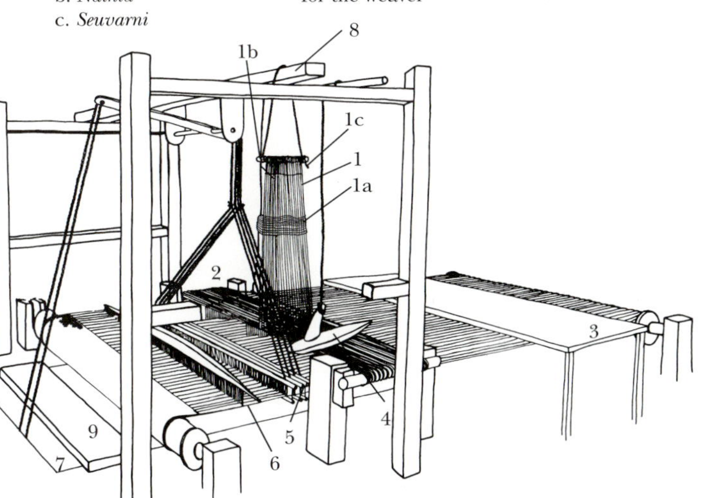

4.71

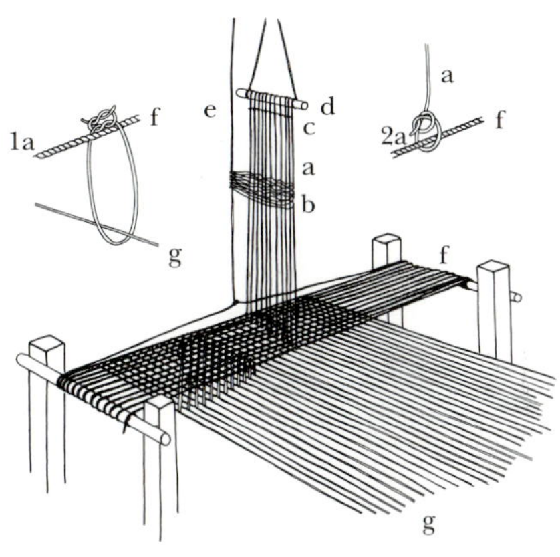

Attachment of *jāla* to *Paggiā*. (Fig. 4.72)

1. 1a. Thread connecting the *paggiā* string to the warp thread
2. 2a. *jāla* end tied to the *paggiā* string
   a. *Jāla*
   b. *Naksha*
   c. *Seurvārni*
   d. *Gulli*
   e. *Nathia*
   f. *Paggiā*
   g. Warp

4.72

Jāla. Operated by Jaffar Ali of Varanasi. Craft Museum, New Delhi. (Fig. 4.73)

Macchan (Fig. 4.74)
1. *Naksha* warp
2. *Naksha* weft/*khewā*
3. *Sāthī*
4. *Seuvārnī*
5. *Nathia*
6. *Gulli*
7. *Ḍorī*
8. *Macchan*/ Frame
9. *Chowk*/ unit of ends
10. *Jathi*

4.74

between Gujarat and Varanasi, does not necessarily point to different centres of origin. *Doriya shāl* and *jāla* can be synthesized at the visual level. If this is accepted, *naksha* can be taken as a supplementary term having its roots in an Islamicized culture.

The *Jāla*

The *jāla* is an attachment at the upper level of the loom consisting of threads which contain the complete series of lifting orders for the weaving of an intricate design. These threads are attached to their corresponding ends on the warp below. When each of the *jāla* threads is raised, its corresponding combination of warp ends also gets raised, forming a shed through which the weft is passed. The *jāla* is essentially a system of lashes and leases which enable the weaving of highly intricate patterns on fabric.

Patterning done on woven fabrics using the *jāla* technique is always in extra weft. Extra weft is a surface ornamentation where the pick floats on the fabric and is bound down to the ground cloth at intervals. In the areas where it is not required to be seen, it floats in a similar manner on the reverse side of the fabric.

The making of a *jāla* is a lengthy process involving many steps. First, a drawing is made of the required motif/design in a 1:1 scale. This is converted into a graph.

Using the graph as a guide, the *naksha* is made on the *macchan* (fig. 4.74). The *macchan* is a locally made wooden frame. *Macchans* can be of different sizes depending upon the size of the *naksha* to be made. For example, a single motif/*būṭī* in the body of a sari requires a small frame whereas a *naksha* for a *pallu* would have to be made on a larger *macchan*. Most *naksha-bandhas*/designers work on a larger *macchan*. Seated on the ground, the *macchans* are ergonomically constructed to ensure a comfortable working position (fig. 4.75).

4.75

The warp for the *naksha* is made after the *parita* (fig. 4.76), is placed on the floor. The *naksha* maker pulls out the string and winds it diagonally on two sticks driven in the ground at a distance of 40 to 42 in. The required number of ends is wound on these two sticks. The number of these ends is calculated according to the size of the design, the count of the yarn used, and the number of ends and picks per inch. As in the making of any warp, a lease thread, *ḍorī*, is inserted to prevent the ends from getting entangled.

The *naksha* is finally to be suspended from the top of a loom vertically. The *naksha* becomes a *jāla* when it has been transferred on to the loom and becomes operational. To facilitate this a bamboo rod/*gullī* is used to which all the *naksha* ends are tied. The *gullī* is connected to the central beam of the *macchan* with a plied cotton yarn/*jathi*. The other end is tied to the frame on the opposite side of the *macchan*. These ends are tied in groups at regular intervals (fig. 4.77). A second thread, known as the *seuvārni* twines around each end gripping them together and maintaining even spacing between each end (fig. 4.72).

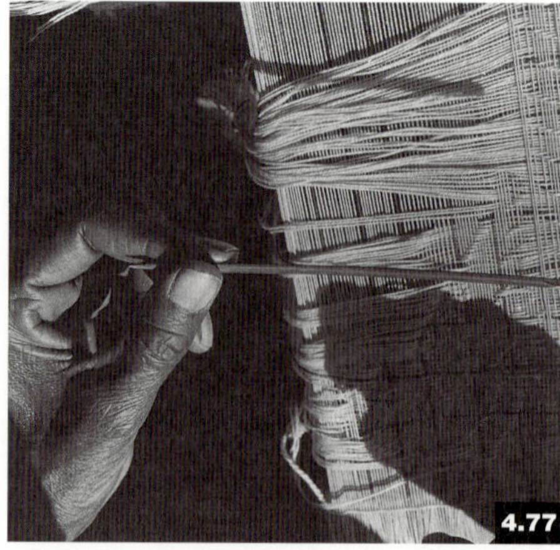

4.77

4.76

At work on a *macchan*. (Fig. 4.75)

Parita (Fig. 4.76)

Tying of *naksha* ends. (Fig. 4.77)

A third thread runs parallel to the warp and acts as a lease for the weft. This is called a *sāthī*. The *khewā*, or *naksha*/weft, threads are tied on to a fourth thread which in turn is tied along one end of the frame of the *macchan*. The *khewā* is tied in groups to enable the *naksha* maker to pick up each thread conveniently. The *sāthī* thus plays an important role in preventing the weft from getting entangled.

Each square of the graph is referred to as a *chowk*, a unit used by the *naksha-bandh* not only as a guide for the making of a *naksha* but also as a unit to calculate the price of the finished *jāla*. Each *chowk* indicates a certain number of threads on the *naksha*. The number of threads depends upon the end and pick ratio to be used. Generally, the picks are about one and a half times more than the number of ends per inch. The count of the reed used, defines the number of ends per inch.

The traced design/motif is placed in front of the *naksha-bandh* and he commences work. The motif is divided into a specific number of parts. The *naksha* threads are also divided in the same manner, and a coloured thread is used to indicate the demarcations on the *naksha* warp. For example, if the original design is divided into 20 parts, there should be 20 equal divisions. Similarly, divisions are made in the *khewā* but here coloured threads are not used in order to avoid confusion when the pattern is made. Each division on the drawing is a part of the whole motif. The *naksha-bandh* transfers each division on the drawing to its corresponding division on the *naksha*.

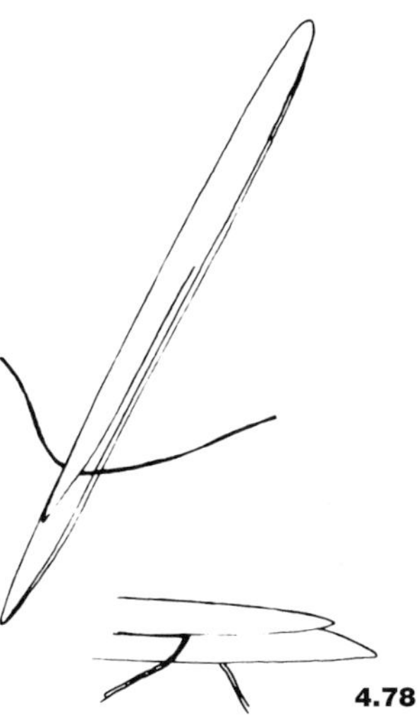

4.78

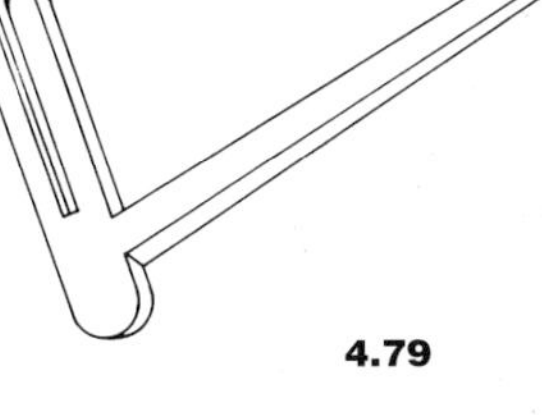

4.79

Two perspectives of *tillī*/reed needle used to make *naksha* (Fig. 4.78)

Mandha. (Fig. 4.79)

The *naksha-bandh* uses a reed needle/*tillī* to make the *naksha* (figs. 4.77, 4.78). Through experience, he understands the *tarāz*/movement of lines in the design, and can adjust the paper design as he transfers it if he detects any irregularity. Various weaves are incorporated to bring out the design on the finished fabric and also to bind it firmly with the ground weave. Therefore, the *naksha-bandh* picks the *khewā* in such a manner that different weaves will ornament different parts of the design. The outline/*hunar* is first picked out on the *naksha*. Subsequently, the colours/*mīnā* are filled in. The extra weft colours that are used to fill in areas in the motif are known as *mīnā* because of their resemblance to *mīnākārī*/enamel work. The drawing/graph indicates the number of colours that will appear on a single pick. Once all the colours that constitute a single weft is picked on the *naksha*, they are tied in a loop through which a string/*nathiā* is passed. This procedure is repeated after every set of single picks. This is done to enable the drawboy to lift each pick separately and easily.

Naksha-bandhs employ various ways of transferring their designs onto the *naksha*, each using the method most comfortable for him. Essentially, they follow one of three methods:

1. A 1:1 coloured drawing with divisions is made. Each of these divisions is then transferred on to the corresponding divisions of the *naksha*.
2. A graph is made using the grid of the drawing as a guide. The *naksha-bandh* works on his *naksha* by referring to the graph.
3. The original paper drawing is enlarged, and this enlarged drawing is placed under the *macchan*. The *naksha-bandh* picks up the design directly from this drawing. This is called the *fanni jāla*.

Once the *naksha* is made on the *macchan* it is transferred and set up on the loom. A *naksha* can be fixed on to any loom which has an overhead beam from which it can be suspended, and which has a facility for the *paggiā* attachment.

The *paggiā* is a set of horizontal threads stretched tight across the warp. It is placed behind the shafts and acts as the link between the ends of the *jāla* and the warp on the loom.

The *jāla* is suspended with the *gullī*/bamboo rod from the overhead beam. Each of the *jāla* ends is tied to a corresponding *paggiā* thread.

A heddle like attachment called a *naka* links the *paggiā* to the warp. The *paggiā* is attached to its corresponding end on the warp in a straight drafting order. This is repeated as many times as there are repeats in the design. For example, if a repeat has over 325 ends, the drafting order would be: 1, 2, 3 ... 325. The procedure would be repeated if so required.

The drawboy sits behind the loom, on a bench above the warp. He lifts one thread after the other in the *jāla*. As earlier mentioned the *sāthī*/weft lease on the *naksha*, makes the drawboy's job, a simple one. The *mandha*, is a wooden tool (fig. 4.79), used by the drawboy to keep the required threads lifted up and also to maintain a clear shed.

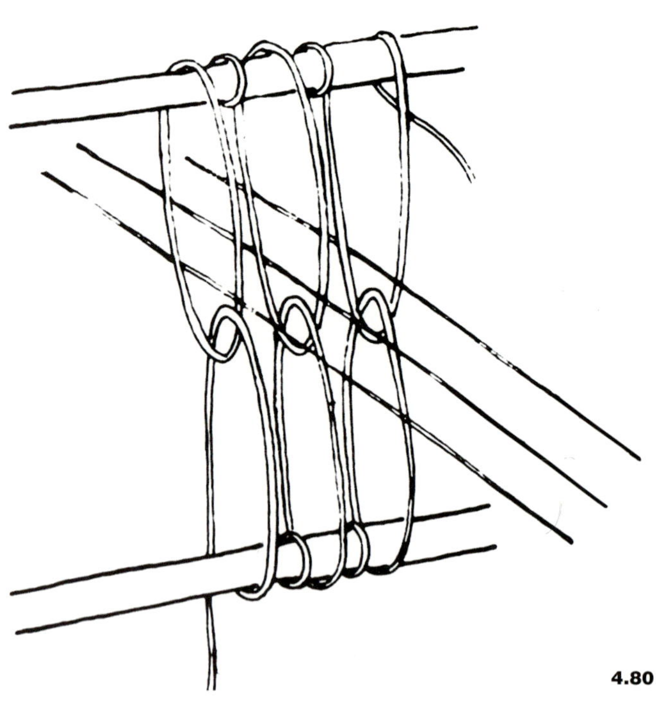

Passage of warp thread through upper section of heddles. (Fig. 4.80)

The weaver uses another wooden implement called an *akkada* (see fig. 4.71). While the ends that are lifted by the *jāla* are being woven, the *akkada* is thrust under the *paggiā*, to enable a clearer shed. When the *akkada* is placed under the *paggiā* the drawboy releases the *mandha* and starts picking out the next thread on the *jāla*. After the weaver has inserted all the required picks, he removes the *akkada* to drop the shed. This serves as a cue to the drawboy to lift the next pick on the *jāla*.

The shafts of the loom are pedal-operated. The ground weave is controlled by the lifting of the shafts and the extra weft patterning is controlled by the *jāla* or the jacquard.

From the position of the seated weaver, the warp passes through the reed, and then through the heddles. The heddles constitute two looped threads which are attached on the frame of the shafts at either end. The warp does not pass through the eye of the heddle but through the upper half of the heddle itself (fig. 4.80). This enables the end to move independently in two movements, namely, the ground cloth being woven by the lifting of the shafts and the patterning on the fabric being woven by the lifting of the *paggiā*. A ground pick is inserted, followed by the insertion of the extra-weft picks. After all the colours of one extra-weft pick have been inserted, the next ground pick is woven.

In the *jāla* weaving process, it is common to find two weavers sitting together at the loom. This facilitates and speeds up the work especially where the *jāla* picks have to be inserted. In weaving of the ground, the second weaver also helps by passing back the throw-shuttle.

Fabrics of Varanasi

Brocades

Brocades refer to those textiles where patterns are created by weaving, transfixing or inserting the pattern thread in between the warp. In plain weaving, the weft thread passes over and under the warp thread regularly. However, in brocade fabrics a combination of metallic and silk or cotton threads have to be woven. Hence, special threads are transfixed in between after skipping, resulting in 'floats' the passage of the regular weft over a certain number of warp threads. This number depends upon the pattern. The skipping resulting in floats is regularized by rearranging the heddles for each unit in patterning.

When gold and silver threads are used along with or without silk threads and are inserted either as extra warp or extra weft to create a glittering raised ornamentation, the fabric is called *zarī* brocade. Apart from this, different types of brocades from Banaras are known by different names according to the characteristics of the fabric.

Kimkhwāb: *Kimkhwāb* was very heavy because the gold and silver threads were used so densely that the ground silk was hardly visible. These are weft dominated fabrics and are very expensive because of the amount of *zarī* used in their weaving.

Khewā: *Khewā* have a silk ground with *zarī* extra weft. They are woven in the same technique as *tanchoi*; the difference being in the use of *zarī* for extra weft, while only silk is used in the *tanchoi*. *Khewā* brocades are all woven using the throw shuttle.

***Bāluchar*ī Type Saris:** The *Bāluchar*ī sari is not a traditional product of Varanasi. *Bāluchar*ī saris, a product of Bengal, were introduced into Banaras by the late Kamala Devi Chattopadhyaya, a key figure in the development of handicrafts and handloom products in the post-Independence period. Since chemical dyes were used in the *Banārasī* product, the range in colours was wider than in the Bengal prototypes. The *Bāluchar*ī sari had gone into virtual oblivion in Bengal at this time. Largely resulting from the remarkable skill of the Banarasi weaver, Kalu Hafiz, it was possible to revive the *Bāluchar*ī, unhappily not at its place of origin, but at Varanasi. The Bengal revival took place subsequent to this.

Tanchoi: *Tanchoi* was introduced to Varanasi in the early twentieth century. As pointed out by the late J. Supakar of Varanasi, the basic characteristic of this fabric is that the ground is almost totally covered with ornamentation. There is an oral tradition that the weaving of *tanchoi* was started by three brothers known as *choi*. In time, they got to be known as *trun*/three, *choi*, which led ultimately to the term *tanchoi* being used for the product.

It is clear, however, that the name *tanchoi* does not suggest a known Indian weaving tradition nor does it take its appellation from its place of origin, the two basic features of nomenclature for traditional Indian textiles. The technique is of Chinese origin. It was transplanted to form a part of the Indian repertoire at Surat. This is not surprising in view of the fact that the major portion of the textiles used by the females of the Zoroastrian/Parsi community showed a marked Chinese orientation (fig. 4.81). *Tanchoi* fabric can be recognized by a few characteristics which include:

1. Exclusive use of silk extra weft ornamentation on a silk ground of plain or *satin* weave.
2. The extra-weft ornamentation on the surface is very dense and well packed.
3. Throw-shuttle weaving is employed.

According to Haji Mohinoodin,[8] the early *tanchois* came from China and Japan. These cloths were of fine yarn woven in satin weave and had five to six colours for the extra-weft patterning. It was used in India as dress material especially in *śerwānīs* for weddings. According to the late J. Supakar the original *tanchois* had silk extra-weft ornamentation on a plain weave silk ground. The ornamentation was so dense and the floats so small and well packed, that they could not be easily seen.

Tanchoi worn by Zoroastrian women. (Fig. 4.81)

There are no traditional motifs in *tanchoi* fabrics. The only requirements from the design aspect are that the surface pattern must be dense and compact, showing as little of the ground as possible. With such a pattern, the extra-weft floats are small. In a satin weave *tanchoi*, the floats are not visible on the back face of the fabric as they are woven in with the ground fabric. These colours do not appear on the surface of the reverse side, as the ground weave is a warp satin.

The inspiration for designs of *tanchoi* fabrics is taken by the weavers from other traditional textiles of India. Traditionally, *zarī* was never used in a *tanchoi* but it is now showing its presence. Until about 1970, *tanchois* were being woven by only four or five families in Pili Kothi, a section in Varanasi, but today they are woven in all areas of the city. *Tanchois* are now woven with the jacquard and the cost increases depending upon the number of colours used in the extra-weft. If the *pallu* motif is too long to be woven feasibly by the jacquard, the *jāla* is used.

Ek-tāra Tanchoi: *Tanchoi* may also be woven with a combination of the jacquard and *paggiā* system to achieve an even greater degree of intricacy and fineness in design. This is evident in *ek-tāra tanchoi*. In the case of usual satin weave *tanchoi*, there are five ends in a dent and correspondingly five ends in one *paggiā*. However, in *ek-tāra*, while there are five ends in a dent in a *paggiā*, they are separated into two sets of two, one of three and one of two ends respectively. Therefore two picks have to be inserted consecutively as two *paggiās* have to be lifted. This allows for a greater flow in the movement of the design.

Ek-tāra tanchois are woven only on special order, as they require twice the amount of labour that goes into the making of a regular five end *satin tanchoi*. They are therefore expensive than the normal *tanchois* more because of the extra time and labour involved.

Brocades: *Khoechen* and *Gyaser*

Origin of the Terms

The presence of the Tibetan community in India has served to highlight a variation in brocade, *khoechen*, manufactured specifically for this community. *Khoechen* is the generic term for brocade in the Tibetan language (oral communication, Lobsang Dawa, Zongkar Monastery, Hunsur, Mysore). In Varanasi, the term *kochin*, obviously a corruption of *khoechen* is the trade term for the product (oral communication, Anjan Chakravarty, Varanasi). The term *ghysar/gyaser* refers to the category of *khoechen*/brocade, in which the background colour is yellow. The derivation is from

Old sample of *gyaser* from China. Courtesy: Kasim Silk Centre, Varanasi. (Fig. 4.82)

Gyaser from Russia. Courtesy: Kasim Silk Centre, Varanasi. Note Byzantine orientation in ornamentation. (Fig. 4.83)

Sri Kalachakra Mandala, Courtesy: Namgyal Monastery, Dharamsala. Motifs woven in *khoechen* and Kinnaur *dohru*. (Fig. 4.84)

Dubel, cylindrical hanging from the ceiling. Courtesy: Tibetan Monastery, Sarnath. (Fig. 4.85)

Gyaser used for draping images. Courtesy: Tibetan Monastery, Sarnath. Note usage of gold coloured drapery, associated with golden glow said to emanate from Buddha, after attainment of enlightenment. (Fig. 4.86)

ghya, famous, and *sar*, gold. Until recently the only patrons of *khoechen* had been members of the Tibetan community. According to suppliers in Pili Kothi, Varanasi, earlier the main sources appear to have been China and Russia. The material however, was used in Tibet. Although *Banārasī* also appears to have played a role in this market, the exact amount of material supplied remains unclear.

Description of *Khoechen* and *Gyaser* Fabric

The production of this fabric is the virtual monopoly of two business concerns emanating from the same family at Pili Kothi, Varanasi. According to these sources, earlier prototypes came from China (fig. 4.82), and Russia. This is not surprising as the Tibetan tradition in the pictorial arts drew inspiration from both these sources, Russia being taken as the Russia of the Orthodox Church drawing its artistic tradition from Byzantium (fig. 4.83). *Gyasers* are heavy silk brocades woven in narrow widths not exceeding 30 in. There is a profuse use of gold thread *zarī*, in these fabrics. The motifs are usually bold and are based on Buddhist themes (fig. 4.84). Each motif and colour has a specific symbolism. Hence, most designs must be reproduced exactly from samples given by the client. These samples are in extra weft brocade woven in twill and satin weaves; the motifs are always in a weft faced construction, predominantly in twill or sometimes satinette weave. The common colours used additionally in *gyasers* are red, golden-yellow, dark blue, gold, green, white and sometimes silver.

Relevance and Use

Khoechen and *gyaser* fabric are used by Tibetans primarily in their temples and monasteries in the following ways:

1. Temple hangings and temple cloth: Fabric is cut up and pieced together to make panels hung on pillars for decoration. Sometimes single piece panels are hung alongside the shrine. They are also used as covers for platforms on which sacred objects such as photographs of the Dalai Lama are kept.
2. *Dubels*: These are cylindrical hangings 6 ft. in length made of pieces of *gyaser* which are suspended from the ceiling (fig. 4.85).
3. Coverings for pillars: Items which are woven to order.
4. Drapery for temple images: As seen in Sarnath, this gold coloured fabric is used as it is taken as a reflection of the glow believed to emanate from Buddha in his state of enlightenment (fig. 4.86).
5. *Gyaser* is used for making cloth bags in which ritual objects are stored. Such objects include scarves and necklaces, symbolizing respect; bells and other musical instruments, symbolizing praise; canopies, rosary beads and the sceptre, symbolizing attendance on the Buddha; cylinders and banners, symbolizing protection and guidance. Canopies and banners could also be made of this fabric.
6. Borders for *thankas*: These are scroll portrait paintings held to be as sacred as their equivalents in wood, stone

Thangkas, framed in *khoechen* and *gyaser*. Courtesy: Tibetan Monastery, Sarnath. (Fig. 4.87)

Gyanta. Courtesy: Badruddin Kasim Silk Centre, Varanasi. (Fig. 4.88)

Dodgen, gyaser. Worn on apron by married women. (Fig. 4.89)

Double *dorje*. Courtesy: Tibetan Monastery, Sarnath. (Fig. 4.90)

and metal. Such paintings are executed on canvas or cotton cloth. On completion borders are sewn along the edges comprising strips of coloured silk or brocade (fig. 4.87). In present times the brocade used is usually *khoechen* and *gyaser*. Borders, taken from the inner to the outer edge are laid in panels of red, yellow and blue. These symbolize the separation of the sacred inner subject from the outer material world. The outer border of blue is broader than the red and yellow, having its greatest width at the base. This panel is usually divided by a vertical patch of brocade in golden yellow, orange or amber symbolizing the door to Lhakhang, the House of the Gods. However, while Chinese influences were important (fig. 4.82), there were also certain themes associated with an Indian origin.

7. As ritual garments some *gyasers* are worn during rituals or the performance of religious dances. These are called *gyanta gyasers*. *Gyaser* utilized in aprons worn on such occasions, have *yakṣa* faces and other human skull motifs on a dark background (fig. 4.88).
8. Triangular pieces of *khoechen* and *gyaser* called *dogden* are used for ornamenting the *pangden*, aprons of married women. These are aprons of daily wear and not necessarily used for special occasions (fig. 4.89).
9. As coverings for *pothī*, Tibetan manuscripts. The title of the work is written in Tibetan behind the flap.
10. Recently, *khoechen*, woven with allover patterns and self-designs are also being exported as furnishings to Russia and some West Asian countries. Such monochrome patterns are also gaining favour in the Indian market as apparel ware.

Varieties of *Khoechen* and *Gyaser*

1. *Tchingo*: *Gyaser* which fall into this category are imbued with considerable symbolic value. Those enjoying special significance are *gyaser* with motifs of *dorje*, *gyanta* and *zinko*. *Dorje* are square *gyasers* so called because of the main motif of the *vajra/dorje* on it. They are usually found in single colours with *zarī* work (fig. 4.90).

Gyanta have motifs of the head with three eyes in gold and silver thread on a black *satin* ground (fig. 4.88).

Zinko constitute *gyaser* pieces forming the top of the panels which hang in the Tibetan temples. They usually have the *yakṣa* motif, called *zeba*, symbolizing protection.

2. Apart from *gyaser* which fall into the *tchingo* category, *khoechen* and *gyaser* can have other motifs such as: *pad ma chandan, thiu gyamu, thiu sum, mor-pankhī*, dragon *gyaser*, dragon *chandan, dogden*, Russian *gyaser* and floral designs.

Pad ma is an ornamental *gyaser* with stylized lotus motifs in numerous shapes and sizes. These designs are predominantly woven in silk with minimal *zarī* (figs. 4.91, 4.92, 4.93, 4.94, 4.95).

Thiu gyamu (figs. 4.96, 4.97) are *gyasers* believed to be of Chinese origin. They are golden in appearance on account of elaborate gold *zarī* work. The motifs are flower buds. A variation of the *thiu gyamu* with all over *zarī* is of Russian origin. The gold weft forming the ground in this sample is of georgette *zarī*.

Motif, *Pad ma chandan gyaser.* (Fig. 4.91)

Motif, *Bada phool chandan.* (Fig. 4.92)

Motif, Pad ma chandan gyaser. (Fig. 4.93)

Motif, Pad ma chandan gyaser. (Fig. 4.94)

Reverse face of fig. 4.94. (Fig. 4.95)

Thiu gyamu. Gyaser of Chinese origin. Floral motif. Front face. (Fig. 4.96)

Thiu sum is a variation of the *thiu gyamu* with clusters of three buds as the motif.

Mor-pankhī (figs. 4.98, 4.99), are *khoechen* traditionally executed with a ground weft made from peacock feathers, a practice which originated in China but was later emulated in Varanasi. An attempt was made to revive a simple form of *mor-pankhī* in the exhibits woven for the Festival of India in the United Kingdom in 1987 by Kasim Silk Emporium, Varanasi. The original Chinese weave supplemented the allure of silk in brocade with the additional texture and colour provided by the judicious use of peacock feathers. The fabric was monochrome without ornamentation. The peacock feather was prepared by re-moving the fluffy part of the feather in the first stage. Individual barbs were then knotted to form a barb string. This was then woven in as extra weft in five step weft *satin* weave with silk for the ground weave; when the shed was opened, both wefts, silk and barb string were passed through. The shed was then changed and the operation repeated (figs. 4.100, 4.101, 4.102).

Thiu gyamu gyaser. Back face of fig. 4.96 (Fig. 4.97)

Mor-pankhī gyaser. Front face. NID Coll., Ahmedabad. (Fig. 4.98)

Back face of fig. 4.98. (Fig. 4.99)

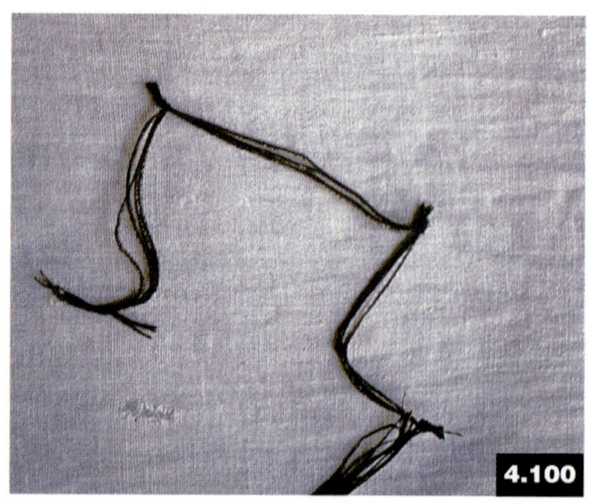
Continuous weft made by knotting individual lengths of the shaft of peacock feathers. Courtesy: Haji Kasim, Pili Kothi, Varanasi. (Fig. 4.100)

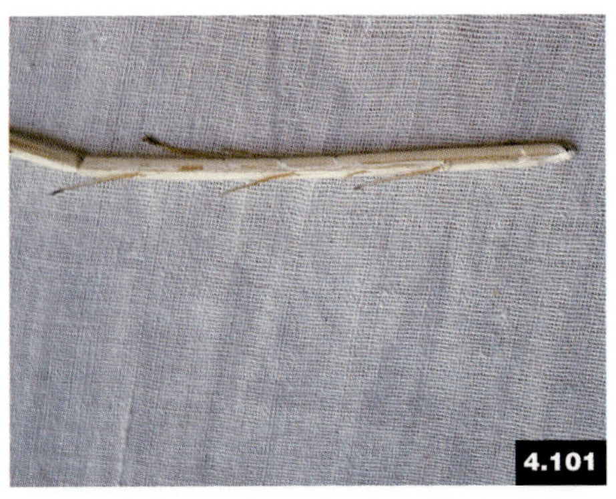

4.101

Denuded spine of peacock feather. Courtesy: Haji Kasim, Pili Kothi, Varanasi. (Fig. 4.101)

4.103

Motif in *gyaser* wheel of law/*'khor lo.* (Fig. 4.103)

4.104

Motif, conch/*Dun.* (Fig. 4.104)

4.102

Mor-pankhī. Obverse and reverse faces of fabric. Courtesy: Haji Kasim, Pili Kothi, Varanasi. (Fig. 4.102)

4.105

Motif, Parasol/*gdugs.* (Fig. 4.105)

4.106

Victory banner/*rgylmtshan* (Fig. 4.106)

Dragon gyasers are very elaborate and grand, and a single repeat can extend up to 1 yd. 6½ in.

Dragon chandan motif is often used in conjunction with other motifs like the *chandan* flower, forming an allover pattern. These are consequently called *dragon chandan.*

Dogden (fig. 4.89) are triangular pieces of *gyaser* fabric with floral motifs. They are attached to a larger rectangular fabric forming an apron at two corners. *Dogden* are traditionally worn by married women. The ground is usually georgette *zarī* with *mīnā* work, a form of ornamentation in which the extra weft figuring is executed in coloured silk yarn.

Russian gyasers as the name implies, originated in Russia. These have a marked orientation towards Byzantine tradition in patterning. There is interplay between floral motif and a geometric grid within which they are distributed (fig. 4.83).

Motifs and their Symbolism

Eight Auspicious Symbols/Tibetian: *bkra shis rtags brgyad,* *Sanskrit: aṣṭamangala*

These auspicious signs, also found in early aniconic representations of Buddha's footprints have been transferred to the area of *gyaser* ornamentation. These comprise:

1. The wheel, *'khor lo/cakra* symbolized the teachings of the Buddha. The eight spokes symbolize the eightfold path (fig. 4.103).
2. Conch with right spiraling whorls, *dung dkar/śaṅkha* (fig. 4.104). The right spiralling movement of the conch echoes the celestial motion of the sun, moon, planets and stars across the heavens. The hair whorls on the Buddha's head also spiral to the right. When such a conch is ritually blown it symbolizes, acoustically, the true proclamation of *Buddhadharma.*
3. White or yellow parasol, *gdugs/chatra* (fig. 4.105) symbolized the monastic emblem of religious sovereignty.
4. Victory banner, *rgyalmtshan/dhvaja* (fig. 4.106) symbolized the Buddha's victorious dharma radiating to the four directions and his triumph over the four *Māras.*
5. Lotus/*padma, pad ma* (fig. 4.93) symbolized purity, perfection, compassion and renunciation.

There are various kinds of lotus each enjoying a symbolic value.

a. Red or pink lotus, *padma* or *pad ma/ Nelumbium nucifera,* is the most common variety depicted.

b. Blue lotus/*Nymphaea nouchali, var. aerulea,* Sanskrit *utpala,* is held by the deites such as the Green Tara. In such cases it represents the purity and compassion of the holder. (For further details see Williams, 1932: 253–66.)

4.110

Khoechen. Svāstika motif in corner, *dorje* in the centre, eight auspicious symbols in the border. Courtesy: Tibetan Monastery, Sarnath. (Fig. 4.110)

4.111

Dragon and *pad ma* motifs. Courtesy: Badruddin Kasim Silk Centre, Varanasi. (Fig. 4.111)

4.107

Motif, treasure vase/*gter gyi bum-pa* (Fig. 4.107)

4.108

Motif, golden fish/*gser nya.* (Fig. 4.108)

4.109

Motif, endless knot/ *dpal be'u.* (Fig. 4.109)

6. Treasure vase/*gter gyi bum pa* (fig. 4.107). As the divine vase of inexhaustible treasures it possesses the quality of spontaneous manifestation and is ever full.
7. Golden fish, *gser nya/suvarnamatsya* (fig. 4.108), is a symbol of happiness, fertility and abundance.
8. Endless knot *dpal be'u/śrīvatsa* (fig. 4.109) is without beginning or end and symbolizes the endless wisdom and compassion of the Buddha.

Each individual motif within the assemblage of the eight auspicious symbols is characterized by the presence of a fillet or an intertwining ribbon. The latter represents the red pieces of cloth tied around anything which is believed to possess the efficacy of a charm. These fillets represent the rays or aura of the charm (Williams, 1932: 155).

Dorje

The *dorje* or *vajra* is one among the chief ritual objects of Tibetan Buddhism. It is also believed that Padmasambhava or Rin-po-che, the founder of Tibetan Lamaism, used his thunderbolt, *vajra/dorje*, to destroy all his supernatural enemies (fig. 4.90). At another level the *dorje* symbolizes the impenetrable, immovable, immutable, indivisible and indestructible state of enlightment (fig. 4.110).

In the double *dorje*, the southern point represents peace, the western point multiplicity, the northern point initiatory power, the eastern point fearlessness and the centre is taken as the focus of all spiritual endowments and perfections (fig. 4.90).

Dragon/ *'brug*

The dragon is not a malevolent force as in Christianity but is an emblem of strength and goodness. It represents the spirit of change and is therefore, akin to life itself (fig. 4.111). Dragons are of many kinds and are extensively used in *thankas* (Williams, 1932: 131–9; Waddel, 1979: 395, 410–14).

Svāstika/g. yung drung

This emblem is found on the stomach/chest of idols, on the eaves of houses, on fabric and many other objects (fig. 4.110). It is also supposed to be the symbol imprinted on Buddha's heart. The commonly used orthodox *svāstika* has its crampons directed towards the right. A group of heterodox Tibetans, the Bon, adopted the same symbol as *sauvāstika*, but directed its crampons to the left. The orthodox *svāstika* is the first of the sixty-five auspicious signs on the footprints of Buddha and the *sauvāstika* is the fourth.

It is also the symbol of good augury. It typifies the continuous moving or ceaseless becoming, which is called life. The *svāstika* was specially associated with the divinity of fire as representing the two crosspieces of wood, which by friction, produce fire.

Khoechen. Cloud motif. Courtesy: Badruddin Kasim Silk Centre, Varanasi. (Fig. 4.112)

Khoechen and *gyaser* at Tabo monastery, Tabo, Spiti. Circular motifs are the symbol for longevity woven at Pili Kothi, Varanasi. (Fig. 4.114)

Chatun mehto, round stylized flower. (Fig. 4.113)

Flame/Fire

Analogous to the sun which appears consumed by fire but remains whole itself, the deities consume evil with the fire of their wrath.

Clouds

These are in perpetual movement and are popular decorative elements (fig. 4.112).

Other Floral Motifs

These are purely ornamental in function and include:

1. *Thiu-gyamu*: Maroon bud (*thiu*) (fig. 4.96).
2. *Thiu-sum*: A cluster of three buds.
3. *Chatun-mehto*: Round, stylized flower (fig. 4.113).
4. *Shinglo-mehto*: Greenery (trees, creepers, leaves).

Seven Possessions of the *Chakravartin*, the Universal Monarch.

These comprise:

1. The Wheel: This is the victorious wheel of one thousand spokes, representing both temporal and spiritual sovereignty.
2. The Eight Faceted Precious Jewel: This is the wish granting gem.
3. The Precious Queen: The *chakravartin* and his spouse bring prosperity and happiness to the realm.
4. The Precious Minister: He regulates the business of the empire in the service of his lord.
5. The Precious Elephant: He is the perfect vehicle for his lord.
6. The Precious Horse: He bears the *chakravartin* with ease, often carrying the precious jewel in his saddle.
7. The Precious General: He is one who conquers all enemies and is ever righteous.

To the above seven gems, an eighth has been added in Tibetan Lamaism, that of the vase, being called in this context, *bum-pa-ter*, in which are stored the hidden riches of the three regions. These three regions, exemplified in terms of *kāma*/desire, *rūpa*/form, *arūpa*/formlessness, are the ethical transpositions in Buddhism of the three Brahmanical physical categories of *bhū*/earth, *bhūva*/heaven and *svar*/space. Taken collectively these three entities have formed the Brahmanical *Bhavanātraya* and the Buddhist *Trailokya* (Waddel, 1979: 74–84, 389–91, 397).

Triratna

The three gems—symbolizing Buddha, dharma/law and *saṅgha*/community.

The Trigram, Spinning Jewel

Trigrams are not Chinese characters but linear figures composed of three whole or divided lines. The Chinese work, *Yi-kin*, Book of Changes, was based on eight such trigrams (Waddel, 1979: 456). The unbroken line represents the male/*yang* principle, the broken ones, female/*yin* principle. Other binary poles are active–passive, heaven–earth, sun–moon, light–darkness. The trigram, Spinning Jewel, has three segments and symbolizes ceaseless change or the process of becoming. The Tibetan symbol for longevity, found in *gyaser*, may be oblong or circular. It could be either a trigram or a hexagram, but in all cases, in form, the outlines would have some relation to the Chinese character, *shou*[9] (fig. 4.114).

Coins/*Ṭaṁka*

The symbol of prosperity and popular both as amulet and ornament, the emblem is used for decorative purposes. Some coins are considered good luck charms. They are also said to ward off evil (Williams, 1932: 71–5).

Zeba

Yakṣa motif used on the *zinko* is symbolic of protection.

Colours

The five/*pañca*, primordial Buddhas are associated with five colours: white symbolized the transmutation of ignorance, blue or black that of anger, red of passion, attachment and desire, yellow of pride and green of envy or jealousy.

In Indian texts the elements also carry colour symbolism: white for water, yellow for earth, red for fire, green for ether, blue for air. The same order is relevant for Tibet but there is also a degree of flexibility. Three colours are particularly popular—red, yellow and blue. Red stands for spiritual power, yellow for sublime truth and blue for transcendental knowledge. However, yellow can be taken as an alternative to white and blue to green, but there is no substitute for red. Different combinations of colour can carry different meanings (Sinha, 1989: 17–18).

Colour can also indicate mood in images. White and yellow typify mild moods while red, blue and black are indicative of fierce ones. Generally gods/*lha*, are pictured as white, goblins/*tsan* as red and the devil/*bdud* as black. Gold is taken as an analogue for divine lustre and the light of spiritual bliss. In the double *dorje* the four cardinal points east, west, north and south can be represented through white, red, yellow and green.

Fabrication of *Khoechen*—From Yarn to Fabric

Process as practised in 1993:

1. Yarn purchased from bazaar
2. Twisted (often however, it is bought twisted)
3. Conversion of hanks to bobbins
4. Winding the warp
5. Bleaching
6. Dyeing
7. Warp stretched and threads untangled
8. Warp rolled on to the warp beam
9. Drawing of warp threads on loom
10. Weaving
11. Cutting of ready fabric from loom
12. Inspecting the fabric
13. Rolling and stacking the fabric

Raw Materials and Pre-weaving Processes

At the pre-weaving stage the following raw materials are used:

1. *Silk*: 2-ply mulberry silk called *katān* procured from Bangalore is used. Earlier silk was imported from Malda (West Bengal), China and Japan. The silk is brought according to requirements and is not stocked.

The silk is of 13/15, 14/16, 16/18, 18/20, 20/22, 26/28, denier. The most commonly used are 18/20, 20/22 and 26/28 denier in the warp.

Art silk/*kela* is also used. Rayon filaments of counts 60, 80, 120, 140 are used in the weft. The lower the count, the finer is the yarn.

2. *Metallic yarn/zarī*: finely drawn gold wire wrapped over a core of silk, now also art silk yarn. Threads used are measured in terms of the yard/*gaz* equal to 0.914 m. The various counts used are:

   1,200, 1,300 and so on up to 3,700 *gaz*
   1,500 *gaz* = 2/100s = 50 count of core yarn
   Similarly 3,700 *gaz* = 75 denier

The *zarī* is bought by the *muṭhā*, one *muṭhā* being equal to 100 to 225 gm. This *zarī* is purchased from local suppliers who obtain their supplies from Surat.

Georgette *zarī*: A highly twisted/crinkled *zarī* specific only to the *gyaser* fabric. This cannot be accommodated within the traditional categories of *kalabatoon/zarī*, and *badla* and appears to be of recent origin. The commonly used count is 32/34 denier. This needs polishing and is made only in Varanasi.

Synthetic *badla*, flat filament: This has more shine as compared to other *zarīs* and is found in both gold and silver. Silver *zarī* is called *rūpa*.

Peacock feathers/*mor pankh*: Made with four ply of peacock feather barbs. The peacock feathers are brought from Rajasthan and the weft is made at Varanasi.

Conversion of Hanks to Bobbins

The yarn is bought in hanks/*ladhī*/*karchī* available in bundles of 2 kg., each hank weighing about 20 to 25 gm. The hanks are degummed with soap and water after which the weight of the yarn is reduced to 1.5 kg. They are now ready to be converted into bobbins. Where previously *charkhās* were used to wind the bobbins, today the entire process is mechanized. The hanks are put on to an electrically operated bobbin winding machine which winds yarn on to 12 bobbins simultaneously.

Warping Process

For winding the warp the bobbins are set up on a creel/*addi* which has a capacity to hold 104 bobbins at a time. The threads are made to pass through a mechanism called *heck*, which consists of two shafts which move up and down alternately thus forming the lease/*beniya*. These threads are then wound on to a huge octagonal warping mill/*charkhā*, its circumference being 5 *gaz*/17.6 ft. The mill has a capacity to wind a 55 *gaz* length of warp.

Once wound the warp is carefully removed from the warping mill keeping the tension even by forming a chain of links to avoid entanglement.

Dyeing

Dyeing is usually done by the *koṭhīdārs*/master weavers. The dyes used today are acid dyes and are bought in the open market. Colour mixing of particular shades is done at the dye suppliers according to the specifications provided by the *koṭhīdārs*.

The wet silk hanks or silk warp are put into a hot water and stock solution. After they take on colour from the dye solution, acid is added. The materials then absorb the dye fully, after which they are put out to dry.

Stretching of the Warp

The warp is stretched out in a *galī*/lane with two people holding the ends. It is then separated out and put on a thin metal rod which is fixed into a *tur*/beam. The warp is thus stretched; each end is separated and the lease rod/*sakada* inserted to maintain the order of threads, the tension, and keep the ends separate. Once the warp is ready, it is wound manually on to the *tur* which is now ready to be taken to the loom. The average length of the warp is 26 to 27 yds. 1 ft. A longer warp requires more weaving time and the longer the weaving time, the greater the possibility of ends breaking.

The Loom/*Karghā* and the Process of Weaving

Today *gyaser* fabric is woven on the traditional *Banarasi* pit loom with a jacquard mechanism attached to it. The weaver's seat is placed across the pit at floor level. He sits with his feet resting on the pedals operating them as required. The pit dimensions are 21 in. in depth, 54 in. in length and 37 in. in width. Until quite recently these looms were operated by the *jāla* system but since the introduction of the jacquard around 1961, the use of the *jāla* weaving system has gradually declined. However, some of the features of the *jāla* have been integrated within the jacquard loom. Joseph M. Jacquard invented the loom which today carries his name. It is so constructed that each warp end can be raised or lowered at will to form intricate motifs and designs. The jacquard machine was first shown at the Paris Exposition in 1801.

When the jacquard mechanism is attached to a hand or power loom, it is called a jacquard head-motion. The motion is used in the production of intricately figured fabrics and in material of a highly compact structure with a pattern which, in a conventional harness loom, would prove extremely cumbersome. The jacquard mechanism had to be redesigned in Banaras, as it was much too heavy to fix on top of the loom of an average weaver. In 1956 the Central Government took over the Institute of Handloom Technology, Varanasi. The then Principal, J.C. Seth, requested Arjun Mistry, to design a jacquard suitable for local conditions (*Jāla Banaras*, NID: 15).

The shift from the *jāla* to the jacquard was an answer to the demand for faster and more efficient production. Also, the weaving of large repeats and intricate designs was both tedious and time-consuming on the *jāla*. Once installed, the jacquard could facilitate quicker and simpler weaving of these large repeats. More importantly, the jacquard also solved the problem of the increasingly short supply of drawboys.

The Banarasi jacquard loom is classified by a measure of *chauks*; 1 *chauk* being equal to 4 *kādis*/shafts/hooks. The shafts are divided into two sets. One set is used for the ground weave and the other for the figuring. The lifting mechanism works on a system of *pāvḍīs* and *dams*. *Pāvḍīs* are the pedals which raise the shafts on being pressed, and the *dam* is the dead weight which pulls the shaft down, once the *pāvḍīs* are released. The shed formed is a bottom-closed shed usually 1½ in. high. The jacquard mechanism is attached by means of a leather thong to a long pedal which the weaver operates by hand to set the jacquard into motion.

Once the shed is opened, the ground pick is passed. This is followed by a new shed for the extra weft, formed by the figuring jacquard.

For the weaving of the *gyaser* brocade, usually two jacquard machines are used: a larger capacity one for the figuring and a small capacity one in the front for the ground weave. In some looms there is only one jacquard mechanism for the figuring, and the ground is controlled by pedals under the loom (fig. 4.115).

The comberboard is an adaptation of the *paggiā* which was used in the *jāla* loom, though in some cases the thread *paggiā* is still used to connect the warp to the jacquard hooks. But in most cases the comberboard which consists of a series of metal rods, specifications as shown in diagram, are used to distribute the harness across the warp (fig. 4.116).

The number of shafts/*kādis* used varies according to the design. These shafts of cane/*baisal* and the heddles/*bai* are made to order in relation to the reed. Earlier, cotton thread was used but today nylon is used to extend their life. The heddles can be bought easily in the market. Earlier, the reed was made of bamboo, now metal reed is used and the common reed dentages range from 64s, 80s, 84s, to 88s. In present times, the main frame of the loom supports the sley.

Earlier a hanging sley/*hatthā* was prevalent, the sley being suspended from an overhead beam by means of a rope. The suspended half reed facilitated the manipulation of the beating. The reed was made of bamboo.

Urtu is a special tie-up mechanism used for binding specific picks as required by the design.

Gyaser weaving with combination of jacquard and *jāla* loom attachments. Note pit loom and shafts with *paggia* adaptation. Pili Kothi, Varanasi. (Fig. 4.115).

Jacquard mechanism used for weaving eight step *satin* weave, *khoechen*. Courtesy: Kasim Silk Centre, Varanasi. (Fig. 4.116)

Double warp beam in *khoechen gyaser* loom. Varanasi. (Fig. 4.117)

These looms also have a correction mechanism which enables the weaver to reverse a lifting order and 'unpick' to make necessary alterations. The chief artisan or designer creates the design and transfers it on to graph paper. This in turn, is sent to a card cutter who punches the cards according to a design. These cards are made of cardboard and are 16 in. x 2.5 in. in dimension. The number of cards per design can go up to 7,000 to 8,000 cards depending on the size of the repeat. The jacquard is, therefore, much more capital intensive, than the more versatile *jāla*.

Setting up the Warp

While setting up the warp, the warp threads of the new warp are joined to the ends of the old warp and drawn through the heddles. In the case of a new warp, the ends are pulled through the heddles by hand.

There are two sets of shafts in groups of 4, 5, 6, 7, 8 and so on in increasing order. The ends are drawn through the heddles in such a way that one set of the ends passes through the lower loops of the heddles of the first set of shafts while another series of ends is drafted through the upper half of the heddles of the second set of shafts. The drafting is straight or pointed according to the design. The pointed draft is also known as *jāldār*.

The denting specifications for an ordinary *gyaser* is 10 warp ends per dent, usually in a 88s reed. The double *tānā*/double warp is 15 warp ends per dent in an 88s reed. The picks per inch/*khewā* are 50, and the ends per inch in a 88s reed are 440.

The width of the fabric is 28 to 30 in. and the average length of the warp woven, as one length is 27 yds. 1 ft.

Analysis of the Fabric

The basic ground weave used in *khoechen* and *gyaser* fabrics is the five end warp *satin*. Occasionally, a twill weave may also be used. In special cases, as for instance in the *dogden*, the fabric is weft faced or weft dominant.

The method of patterning of the fabrics is through extra-weft. The weave used to construct the motif is usually twill, 1/2 and 1/3, or a satinette, 4 end satin. The extra weft picks float on the fabric and are bound down to the ground weave at intervals. In the areas where they are not required to be seen, they float on the back face of the fabric.

After every ground pick is woven, the extra weft picks of different colours that constitute the design are inserted according to the design specifications (peg plan) thus constituting the next weft or pick in the weaving of the fabric.

Some of the *khoechen* and *gyaser* designs require two sets of warps/*tānā* (fig. 4.117). These are fabrics with no floats or a minimal number of floats on the reverse side of the fabric. They are either:

1. completely reversible fabrics, where the ground of one surface becomes the figure of the reverse side, or
2. the weft floats are completely bound and they do not create any design or floats on the reverse side.

Finishing

Once the weaving is complete, the fabric in front of the reed is cut with a pair of scissors. While still on the cloth beam, it is inspected by the weaver. The cloth is then folded and taken to the *koṭhīdār*. He, in turn, inspects, measures and weighs the fabric and accordingly pays the weaver.

No special finishing process is required. In some cases, as in *mor-pankhī*, the floats in the back of the fabric are cut with scissors.

There is a characteristic way of rolling and storing the *gyaser* brocades. The fabric is laid out flat on a *gaddā*/ mattress on the floor. A narrow cane, the length of which is the same as the fabric width, is placed on one end and the *gyaser* is then rolled into a neat, firm cylinder.

Hierarchy of Relationships

The hierarchy of relationships involved in production and sale are:

The loom, the space and the raw materials are provided to the weaver by the *koṭhīdār*. The weaver is under his direct employ and is paid wages on completion of one warp length or piecewise. Therefore, no transactions of buying or selling are directly carried on between weaver and buyer.

The buyer places the order for *gyasers* with the *koṭhīdār*. The buyer may be an individual or may represent a temple, monastery or institution. Tibetan buyers select the fabric and place their orders at the *koṭhīdār's* establishment. Sometimes, they provide a sample, which is to be replicated. At times, orders placed by mail are accepted from trusted clients such as the Dalai Lama. In such cases, both delivery and payment are conducted through the post. For ordinary

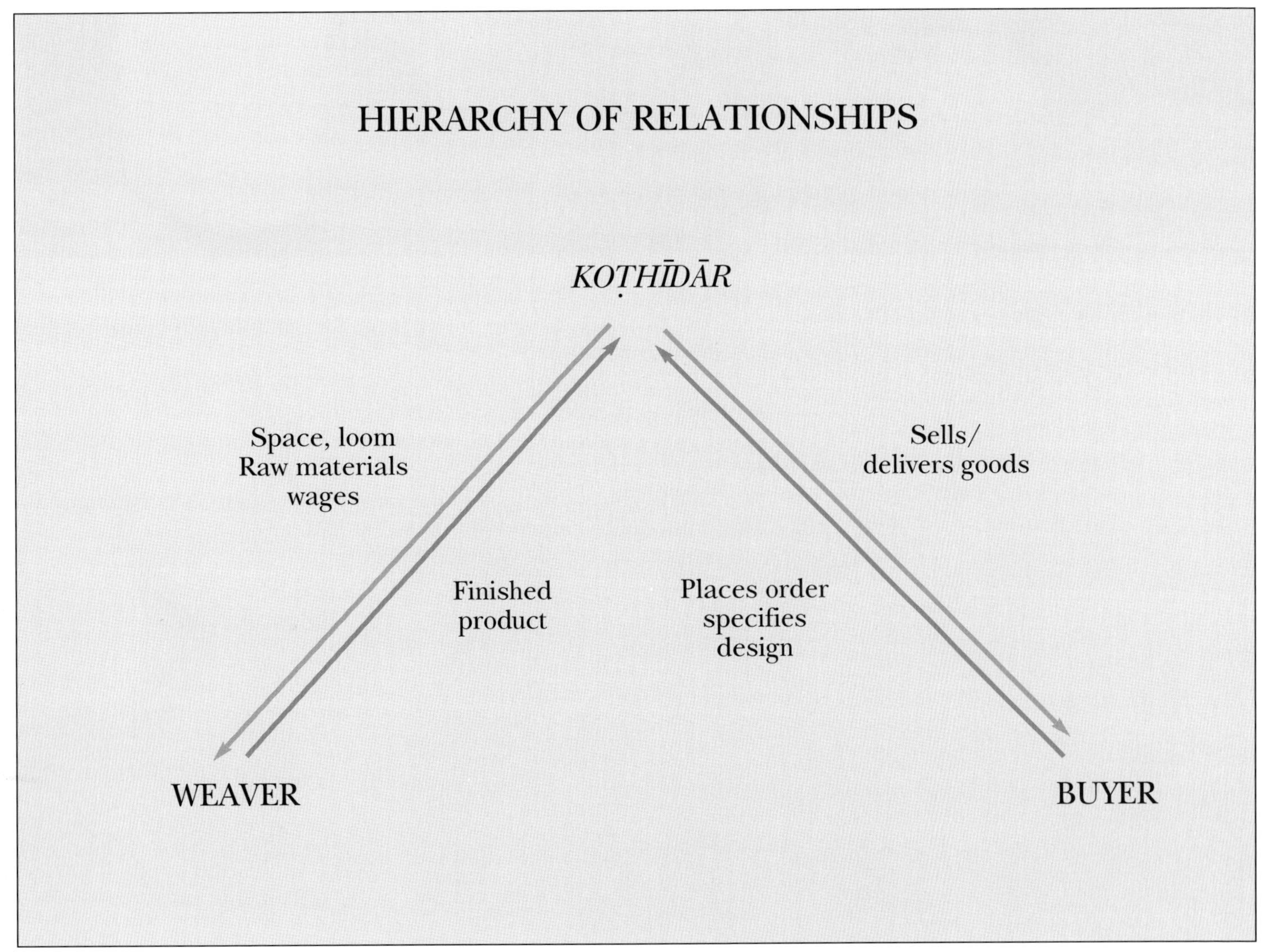

Technique of measuring *gyaser* in *kha* unit of measure. Varanasi. (Fig. 4.118)

purchases, the buyer can collect the goods immediately. Besides their establishments at Varanasi, the *koṭhīdārs* also have an outlet at Sarnath, where transactions may be conducted.

Units of Measure in Sale Transactions

Khoechen and *gyaser* are sold in *thāns* which are measured in smaller units called *mukh/kha*.

One *kha/mukh* is the square unit of the width of the fabric. It is measured by folding one corner diagonally across the fabric, to form a square (fig. 4.118).

Notes and References

1. For fibre see *Symposium, 1984*; Watson, 1977: 355–68; Emery, 1980: 1–15; for loom and fibre see Broudy, 1979: 16–22; Carroll, 1988: 15–28; Wulff, 1966: 174–5; Albers, 1965: 59–61; Endrei, 1958: 28.
2. Dam means to set, *pach* means digest. *Dampach,* thus suggests the meaning: to lay in patterning weft in the same shed as the plain weave. Treadles are used for this process in *dampach.* This sets the technique apart from *jāmdānī* where each pick of the motif is localized and manually laid in.
3. Gluck et al., 1977: 173; Wulff, 1966: 173; de Jonghe, 1985: 10. The twill weave is particularly well suited to wool. The development of this weave in relation to *kani paśmīnā* still awaits investigation.
4. *Products Tamil Nadu,* NID, *kuncham,* q.v. According to weaving expert, Shankar Kumar, Salem, in drapery material, widths extending up to 3 yds. can be woven on a pit loom manipulated with a throw shuttle. In processes documented in Salem, Tamil Nadu, the warp which is prepared at the pre-loom stage, is sized and kept in ball form.
5. Varadarajan, 1991: 208, 215, 217–18; Huzayyin, 1942: 19, n. 2.
6. The analysis of the working of the loom is based on the oral communication of R.P. Rajan, Ahmedabad.
7. *Jāla Banaras,* NID: 58.
8. Haji Mohinoodin is a master weaver whose family has been in this trade for about 125 years. He owns 65 looms of which 25 were used for *gyaser,* and the remaining 40 for *tanchoi.*
9. Waddel, 1979: 394–5; Eberhard, 1986: 298; Williams, 1932: 27.

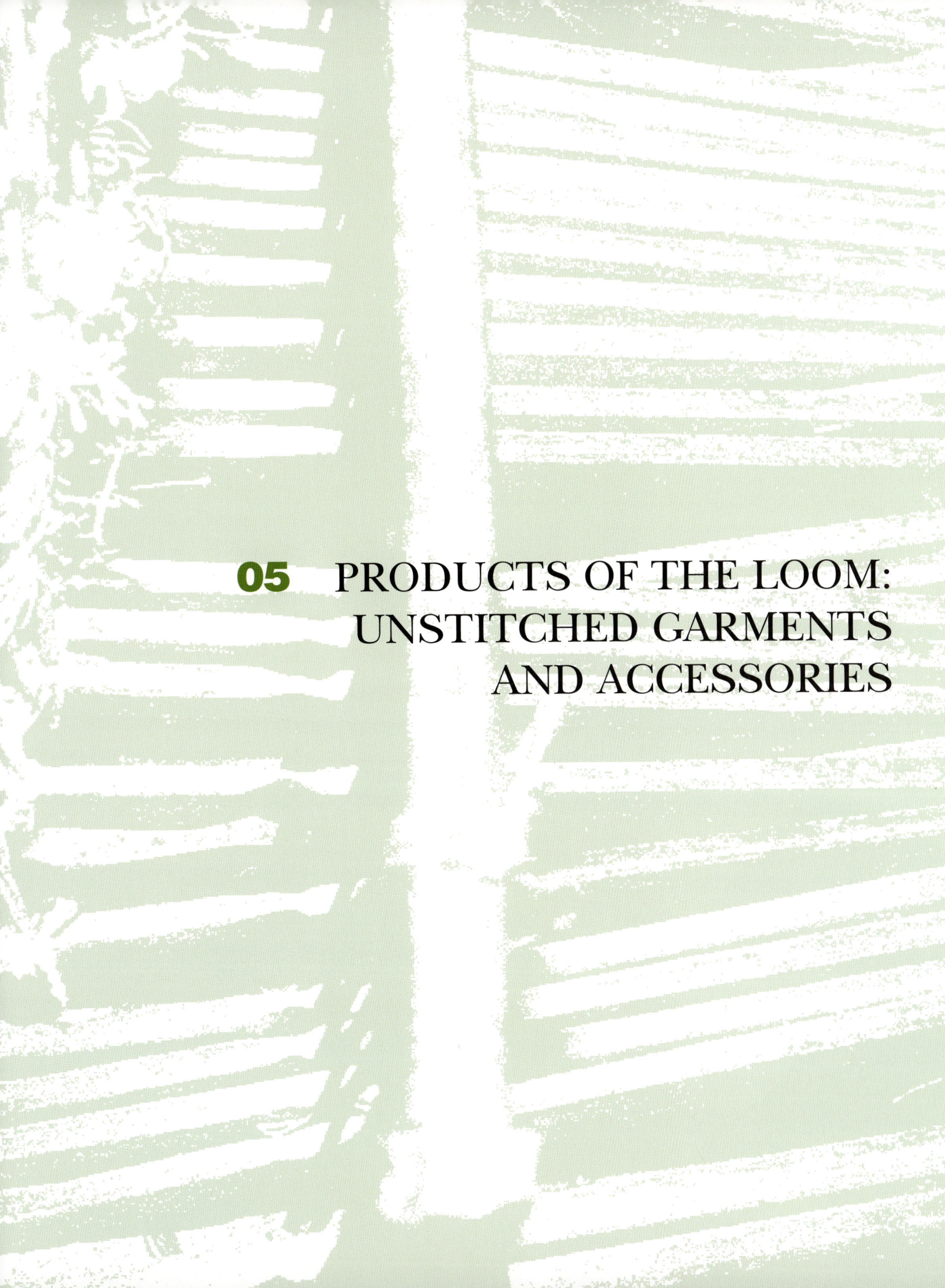

05 PRODUCTS OF THE LOOM: UNSTITCHED GARMENTS AND ACCESSORIES

05

Products of the Loom: Unstitched Garments and Accessories

Emperor Kanishka. First century AD. Courtesy: Mathura Museum, Mathura. Note stitched garment. (Fig. 5.1)

Figure of a foreigner. Second century AD, Mathura Museum, Mathura. Costumes such as this could have set trends in Indian male attire. (Fig. 5.2)

Illuminated manuscript. *Bhagwata Purāṇa*, Samvat 1705/AD 1648. Courtesy: Bhandarkar Oriental Institute, Pune. Note Rajasthani style in male costume and *pagaḍī*. (Fig. 5.3)

The diversity of detail present in Indian looms has been reflected in the range of products made on them. A very important aspect of the Indian style in clothing has been the wide prevalence, except in limited areas in the north-west (figs. 5.1, 5.2), of the unstitched garment. The *ghāgrā* comes into evidence in Rajasthan in an archaeological context *c.* AD 300 to 700. This garment, in present times, is confined to females in Rajasthan and Gujarat. The *ghāgrā* can be an unstitched garment—as stitched seams are not essential—the garment being held in place around the waist with a string (Singh, 1985: 120; see fig. 5.3). Indeed, the draw-string has played a vital role in Indian female wear. In this context, the *newār* loom, also associated with the making of *nāḍā*, drawstring tapes, cannot be ascribed to a foreign origin as has been averred by Rita Bolland (Bolland, 1968: 15). The male *luṅgī* has also lent itself to seamless wear.[1] The fabric of an unstitched garment can serve many purposes. For this reason, in contrast to Europe, there has not been a need to develop furnishing material specifically for such use. The same garment length could be put to multifarious uses by less affluent members of a community. Men could wrap the shoulder cloth around the head, spread it on the ground and sleep on it or use it as a body wrapper. The most important product of the loom has, therefore, tended to be some form of unstitched garment. The unstitched garment has readily lent itself to the loom procedure of piecing (see fig. 5.4). In this context it should be remembered that during the classical period in Greece the standard garments, the *chiton* and *himaton* (see fig. 5.5), were essentially rectangular pieces of cloth which were draped and fastened round the body (Kaplan, 1984: 27).

Piecing. Courtesy: Murali Sari Emporium. Putupaka, Andhra Pradesh. (Fig. 5.4)

Greek costume depicted on vase dated 480 BC. Courtesy: Antiqiken Museum Berlin. Note, unstitched component in costume. (Fig. 5.5)

Vedic terms relating to male and female dress varied. For male garments the terms used were *vastra*/ cloth, *vasana*/ garment, *uttarīya*/ shoulder cloth, *adhivāsa, vāsā*/ overgarment, *prāvāra*/covering, *uṣṇīṣa*/ turban and the *antarīya*/lower garment, which had to be worn in the *kachchhā* style, that is, tucked between the legs much in the form of the modern *dhotī* (see fig. 5.6). According to the *smṛtis* a garment had to be tucked in three places: near the navel, on the left side and behind at the centre back (Kane, 1941: 113–14, 672). The requirement for garment tucking for the twice born appears to be linked to aspects of chastity at times when rituals were performed. The Buddhist work, the *Milinda Pañho*, provides insight into the fastidiousness of male attire. In a description of a man enjoying the good things of life, it is stated, 'Now Nāgasena, if a layman clad in white robes, enjoying the pleasures of the senses ... making constant use of the sandalwood of Banares, of garlands and perfumes and unguents, accepting gold and silver, wearing a turban inlaid with jewels and gold. ...'[2]

The important point is that, amidst all this luxury, white garments are being worn. Generally speaking, sobriety in attire was expected of the twice born.[3] A description of the wardrobe of royalty shows greater emphasis on texture rather than colour and pattern.[4]

Garments of females varied in form, but for the elite they differed little in essence from the male garb (see fig. 5.7) as was common in most early civilizations. They revolved in the same manner around the manipulation of two strips of cloth (see fig. 5.8). For females, words used for costume included *nīvi*, undergarment, *paridhāna*, garment, and *adhivāsa* veil (Majumdar, 1934: 191). As enticement was considered integral to womanhood, and since women were excluded from the scope of Vedic learning, they enjoyed greater latitude in the range of permissible colours,[5] but the form of garment imposed in-built constraints on canons of design. The vertical fall is very important in the lower garment worn by both sexes. Since the *uttarīya* and *antarīya* had to stay in position, the edges had to be heavier than the main ground cloth. Functional considerations, wedded to the concept of symmetry resulted in the adornment of the basic fabric with two borders along the length of the warp and one or two cross borders along the horizontal ends of the garment. The ground design, if any, tended to be diminutive and repetitive.

5.6

5.8

Frieze. Birth of Buddha. Gandhara. Second to third centuries AD. Courtesy: South Asian Museum, Berlin. Note *kachchhā* in costume of maid. (Fig. 5.8)

5.7

Male figure. Sanchi. Courtesy: Archaeological Survey of India, New Delhi. (Fig. 5.6)

Female figure with turban. Sanchi. Courtesy: Archaeological Survey of India, New Delhi. (Fig. 5.7)

Terracota female figurine, excavated at Arikamedu, *c.* second century AD. Courtesy: Museum, Pondicherry. (Fig. 5.9)

Female flute player wearing *kachchhā* style silk sari. Vishwanath Temple, Khajuraho, eleventh century AD. Courtesy: Archaeological Survey of India, New Delhi. (Fig. 5.10)

Terracotta figurines found at Arikamedu between *c.* 50 BC to AD 200 (see fig. 5.9), show that in place of two separate items of varying texture serving as breast cloth and lower garment worn in *kachchhā* form (see fig. 5.10), there was also in existence a garment of uniform texture not unlike the modern sari which covered the entire body (fig. 5.11). The garment on the Arikamedu figurines bears a greater resemblance to peasant costume in Tamil Nadu and Andhra Pradesh (figs. 5.12, 5.13), rather than the costume of the more genteel folk.

There is an illustration of the lower garment worn in *kachchhā* style in A. Cunningham, *The Stūpa of Bharhut*, London, 1879, Plate XLIV, Fig. 2. Among the other sets of evidence which point to the existence of one continuous piece of cloth serving as a unisex item in costume the following may be listed:

1. A second to first century BC terracotta piece from *Pāṭaliputra*, *c.* 996 at the Patna Museum, Patna.
2. A painting dated late fifth century AD depicting a figure clothed in a garment with a swan motif at Ajantā, Cave 17, showing a palace scene with a detail delineating the subjugation of the Furious Elephant.

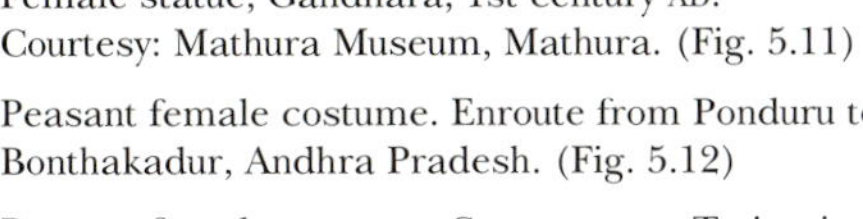

Female statue, Gandhara, 1st century AD. Courtesy: Mathura Museum, Mathura. (Fig. 5.11)

Peasant female costume. Enroute from Ponduru to Bonthakadur, Andhra Pradesh. (Fig. 5.12)

Peasant female costume. Street scene, Tuticorin. Note similarity with Arikamedu figurine. (Fig. 5.13)

3. A drawing found in J. Griffiths, *The Paintings of the Buddhist Cave Temples of Ajanta, Khandesh, India*, II, London, 1897, Plate 54, which shows the enthronement of Buddha, dated to the seventh century AD. The two male figures on the left are depicted wearing a costume of continuous cloth, draped over the left shoulder devoid of an *uttarīya*. I Tsing, seventh century AD, noted that among laymen in India, the officers and people of a richer caste had a pair of soft white cloth as their garment, while members of the poorer classes had to make do with only one piece of cloth. The former utilized wide pieces of cloth 8 ft. long. There was no girdle. Nor was the garment cut and sewn. It was simply wound around the waist and served to cover the lower part of the body (Nilkanta Sastri, 1972: 111–12).

New style of wearing *pagaḍī*, Baroda. Shahaji Chhatrapati Museum Trust, Kolhapur. Possibly woven at Yeola. (Fig. 5.14)

Pagaḍī worn by senior Maharaj, Dewas. Courtesy: Shahaji Chhatrapati Museum Trust, Kolhapur. Possibly woven at Burhanpur or Chanderi. (Fig. 5.15)

In view of the predominance of the unstitched garment as a loom product it is of interest to trace developments in this sphere. Items woven were priced not so much on the basis of the cost of labour involved as on the cost of raw material. When the European East India Companies began their trading operations in Indian textiles they discovered two terms intimately connected with this trade. The French corruptions for these two original Indian terms were *courge* and *conjon. Courge* may have been derived from Konkani *korj* or Malayalam *korja/korchchu.* While *conjon* may have been derived from Tamil *kuncham.* The Telugu equivalent was *punjam. Courge* was a mercantile term denoting sale of a given commodity in units of twenty. Unstitched garment lengths were vended by the *courge,* and Europeans had perforce to make their purchases in terms of this unit even if the individual items making up this unit varied in quality. The *conjon,* or rather *kuncham,* in present day parlance, is a section of the reed equivalent to 120 ends or 60 dents with 2 ends per dent. This unit varies in number across the width of the fabric depending on the yarn count used and the cloth count desired. The larger the number of *kuncham,* the finer the cloth. In the earlier period, at one end a red marker of cotton yarn would run through separating each *kuncham* thus making the task of counting them easier. Until recently, hand woven items had an unwoven segment between two woven areas at one end, indicating the number of warp ends used. In Gujarat the equivalent term for *kuncham* was *vissa.* One *vissa* comprised 160 ends. Gujarat weaves varied between 7 to 18 *vissas.*[6]

For the richer segments within society there would be a difference in texture between the fabrics used in turban, shoulder cloth and the lower garment. The beauty of a turban lay in the manner in which a long, narrow and fine textured piece of cloth could be transformed into a multifolded headgear (see figs. 5.14, 5.15).

It is also of some interest at this point to try and trace the origin of the word 'sari'. R.L. Turner (Turner, 1966, item 12381) provides the term, *śāṭa,* in masculine gender, as a term for a strip of cloth. In the *Mahābhārata,* the feminine form, ´ *sāṭī* is used. M. Monier-Williams (Monier-Williams, 1976 *śāṭā* q.v). on the basis of

Vāsavadattā and Cāṇakya, provides a broader range of meanings: a strip of cloth, a kind of skirt or petticoat, a particular sort of garment or gown. The strip of cloth, worn in different ways (see figs. 5.16, 5.17), possibly shrank in meaning over time until it came to refer exclusively to the female unstitched garment as known today in its many regional manifestations (see fig. 5.18).

The establishment of the Muslim élite brought about further change (see figs. 5.19, 5.20, 5.21, fig. 5.3). Marked development in the textile arts has been a feature of Islamic civilization. The efflorescence of batik in Indonesia is a case in point. In Punjab, a double cloth

5.16

Mural painting. Female costume. Veerabhadraswami Temple, Lepakshi. (Fig. 5.16)

5.17

Nine yard sari. Kannada-Telugu style. Mural painting. Veerabhadraswami Temple, Lepakshi. (Fig. 5.17)

5.18

Painting. Gallery of musicians, 1905, by Raja Ravi Varma. Courtesy: Jaganmohan Palace, Mysore. Note regional styles in female costume. (Fig. 5.18)

5.19

Mural painting. Portrait of Veerappa Temple. Veerabhadraswami, Lepakshi. Note stitched costume. (Fig. 5.19)

5.20

Mansingh of Amber, *c.* AD 1600, Mughal School. Courtesy: Bharat Kala Bhavan, Varanasi. (Fig. 5.20)

5.21

Illuminated manuscript, *Bhagwata Purāṇa. Samvat* 1705/AD 1648. Courtesy: Bhandarkar Oriental Institute, Pune. Note Rajasthani style in male costume and *pagaḍī*. (Fig. 5.21)

item called *majnū* (figs. 5.22, 5.23) probably made its appearance at this time. According to the Weavers Services Centre, Panipat, *majnū* is ascribed an origin in the Multan region. In Punjab *majnū* is differentiated from the traditional twill woven body covering called *khes*. On the other hand, in Pakistan, the product, which is called *majnū* in Punjab has been given the appellation of *khes*. According to B.M. Anand, NID, Ahmedabad, the Multan region was also the heartland of the romantic odyssey of Laila and Majnu. The fabric was named after the hero, Majnu. However, the term *khes* is a generic term for any multipurpose covering cloth. Hence *majnū* could also be subsumed within it.

Majnū, double cloth. Courtesy: Panipat Textiles and Handicrafts, Panipat. Front face. (Fig. 5.22)

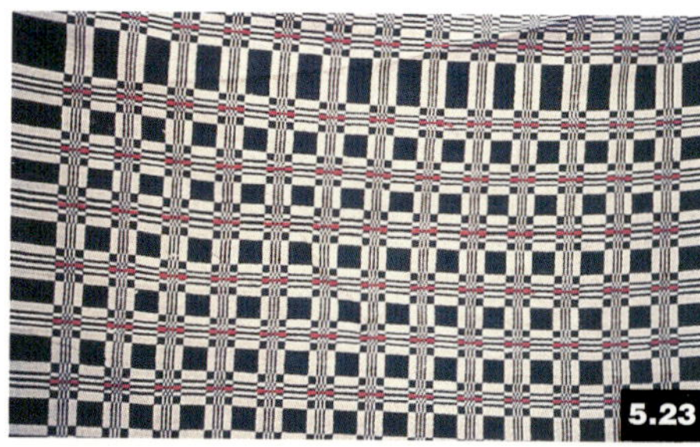

Sample of *majnū*. Courtesy: Panipat Textiles and Handicrafts, Panipat. Front face. (Fig. 5.23)

With the increasing popularity of stitched garments such as *jāmā* and *pyjāmā*,[7] the emphasis shifted from the drape and fall to the cut of the garment and the patterning in it. The play of delicate floral sprigs and emphasis on edging rather than border and *pallu* came to the fore (see fig. 5.24). The evolution of the motif known variously as *kalgā, bādām, turañj*, cone or paisley may be traced to this core.[8]

Damsel. Mughal School, AD 1625. Courtesy: Bharat Kala Bhavan, Varanasi. Note hanging in front portion of lower costume. (Fig. 5.24)

With the advent of an Islamic political and administrative ruling elite in India, courtly etiquette and dress of the new aristocracy began to be imitated and duplicated in other centres of power within the country. Rajasthan serves very well as a case in point to demonstrate this species of cultural synthesis which was taking place. Change in mode of costume was reflected in the demands being made on existing loom procedures. The curvilinear forms so dear to Islamic sensibility began to outstrip geometrical mirror image motifs in popularity even though existing weaves and loom mechanisms were better geared to the latter design repertoire. These included the extra warp weave and mechanisms such as multiple shaft and pedal and the pattern stick used on the pit loom. The addition of the Persia derived draw loom device to the existing pit loom proved an economic and labour saving solution. It also served to highlight the limitations inherent in the usage of extra warp devices from the point of view of design. It was apparent that extra weft ornamentation provided far greater liberty in securing greater complexity in form.

Another area in which Islamic impact was felt was in the increased embellishment of textiles with metallic yarn (see fig. 5.25). Although literary evidence exists from an early date pointing to the use of metallic yarn in garments, historical evidence links Sassanid Persia, AD 224 to 651, with the greatest developments in this sphere. The Chinese annals, *Liang*, mention that the Persian gold brocade sent to the Chinese Emperor Wu in the year AD 520 as something special. When the Byzantine Emperor, Heraclius, seized the Sassanian palace at Duftgird in AD 627 he found large numbers of royal silken robes woven with gold metallic yarn. The Persian method, to be followed in India, for making metallic yarn, was that of drawing and rolling precious metal into a thin wire which was then wound round a mulberry silk or linen core (Wulff, 1966: 175–7). In India, linen was substituted by cotton but mulberry silk was a far superior medium.[9] Since India was deficient both in gold and silver it is reasonable to assume that metallic yarn or *zarī* was an item developed elsewhere and absorbed by India. It would appear that both in the area of book illustration as well as in textiles the peak period for precious metals to be used for ornamentation appears to have been from the eighteenth up to about the late nineteenth centuries. Patterning by draw loom in India was a post-Islamic phenomenon. However, the new loom contrivance did not displace the earlier repertoire of weaves. Rather, it helped to increase speed in production and facilitated the usage of complex weaves.

Jehangir *c.* AD 1610, Mughal School. Courtesy: Bharat Kala Bhavan, Varanasi. Note use of metallic yarn in *pagaḍī* and *paṭkā*, also lower garment in *maśrū* showing through *kurtā*. (5.25)

5.26

Draping of *sālu* and *kachchhā* style. Stage 1. Courtesy: Bhau Daji Lad Museum, Mumbai. (Fig. 5.26)

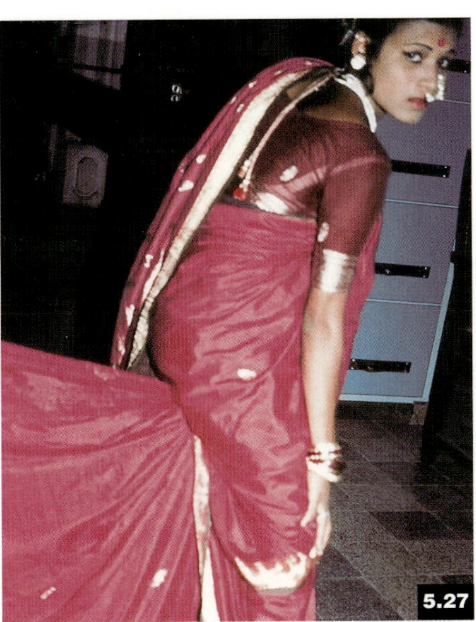
5.27

Draping of *sālu* and *kachchhā* style. Stage 2. Courtesy: Bhau Daji Lad Museum, Mumbai. (Fig. 5.27)

5.28

Style of draping sari in Bengal and Orissa. (Fig. 5.28)

Moreover, it permitted further refinement in the area of curvilinear forms.

European demand, in increasing evidence during and after the seventeenth century, was less pervasive than the Islamic one. Some existing looms were altered to facilitate the production of items with larger dimensions better suited to the needs of European drapery material. The main thrust was to be articulated in the area of cotton and within the format of dye patterned textiles. The *chintz* tradition was the most visible evidence of this impact just as the draw loom served as a constant reminder of the Islamic input. This brief survey of the Indian loom and its products cannot be rendered complete without some reference to the broad differentiation which exists between the costume range in northern and southern India. The Vindhyas can be taken as forming a notional line of division between these two regions. The north is subject to a greater degree of climatic variation and visible ethnic differentiation than has been the case in the south. Female costume has varied from the unstitched body tension loom woven hip cloths/ *sarongs* in the north-east to the Kullu and Kinnaur *dohru*/*paṭṭu*, the western Indian *ghāgrā*, the *salwār-kamīz* of Punjab as also the Kashmir *phiren*. The costume of Central Asia and Iran is in evidence in Indian sculpture from early times. The male costume depicted at Palmyra dated between second to third century AD indicates how widely prevalent the tradition of the stitched costume had become (see *Palmyra*, 1987: 62). The entry of the *phiren* and the *salwār kamīz* could be linked to Kuṣāṇa and Han times. G.S. Ghurye discusses variations in dress during the period 320 BC to AD 320 (see Ghurye, 1966: 79–106). It is possible that the Kashmir *phiren* had evolved from the *kañcuka* of an earlier period (Ghurye, 1966: 234).

The origin of the word sari which has been discussed earlier has also to take into account the term *cīra*, from Sanskrit *cīvara*, meaning cloth. *Śaṭikā* or *śāḍī* denoted a good cloth for women (Mehta, 1951: 174). Broadly speaking, the method of sari draping falls into two broad divisions: firstly by being wound round the waist, gathered in the front, with the end portion flowing over one shoulder; second, by draping between the two legs and a portion of the garment being taken from the front and tucked into the waist at the back (see figs. 5.8, 5.26, 5.27). This form is called *kachchhā*.[10] The earliest sculptural depiction of the *kachchhā* form is to be found at Bharhut dated second century BC (Dongerkery, 1909: 103). The first mode, with variations, has dominated the Indo-Gangetic delta. Gujarat, Rajasthan, Uttar Pradesh and Bihar form one unit, while Bengal and Orissa fall into another[11] (see fig. 5.28). Among the costumes of the major ethnic groups in the south, that of Kerala is quite distinct. The basic fibre is cotton and both sexes wear a tubular seamless lower garment called the *muṇḍu*. Muslims would tie the *muṇḍu* leftwards, while Hindus and Syrian Christians adopted an orientation towards the right. Among men, only royalty was entitled to wear a shoulder cloth, *veṣṭi*; other males had to remove their shoulder cloths as a sign of respect before a social superior. Hindu women were bare breasted within the confines of the home. When Nambudri and Nayar women went out to perform customary services in the homes of their *jenmies*, landlords, the *veṣṭi* was used as a breastcloth. However, like their menfolk, they had to discard the *veṣṭi* in the presence of elders and superiors (Menon, Menon, 1983: 30). All women below the Nayars in the hierarchical scale were denied the use of *veṣṭi*. Nayar women wore an unstitched cloth in the form of an undergarment beneath the *muṇḍu*. This was the *tarru* or *onnara muṇḍu*, similar to the garment worn for Kalarippayattu, the martial art of Kerala. *Muṇḍu* and *vesṭi* when worn together would have matching borders called *kara*. The *kara* could be vestigial or more prominent, the latter being a sign of status.

White was the preferred colour for Nambudris, Nayars and Syrian Christians. While the *muṇḍu* was of standard dimensions being 2.5 yds. in length and 1½ yds. in width for men, it could be worn in different ways. The *karanavan*, male elder in the matrilineal family, could fasten it under the armpits, while it was obligatory for lower caste males to double it and tuck half the width into the waist leaving the legs bare (oral communication: Professor M.G.S. Narayanan, Kottayam).

It is noteworthy that this entire range in costume emerged in a ready-to-wear condition from the loom, the preferred colour being white and the preponderant weave being tabby. Unlike Hindu women, those belonging to the Syrian Christian, Black Jew and Mappilla communities did not shun an upper covering in costume. Syrian Christian women, when appearing in public, wore a waist length jacket called *citta* or *penkuppayam*. Females among the Mappilla and Black Jew communities wore an upper garment and a veil which covered the hair but not the face (Menon, Menon, 1983: 41, 480, 530, 553).

The case of Maharashtra will be discussed later but it should be noted that regions conquered by Maratha chiefs such as Baroda, Gwalior and Indore reflected the sartorial standards set by the Peshwas at Poona. In Madhya Pradesh, the *kachchhā* form of wearing a sari is found among the agricultural Lodhi community of Chhattisgarh, the Bhils of Jhabua and the Marars of Balaghat (Chisti, Sanyal, 1989: 32–5, 42–6).

5.29

Sari *kachchhā* worn by Śaivite women, South India. (Fig. 5.29).

5.30

Standing male figure at base of pillar, Mahabalipuram, fifteenth century AD. Note presence of *nāmam* on forehead. (Fig. 5.30)

The single sari unit, varying in length from 7 to 9 yds., or 6.4 to 8.2 metres, has been worn in *kachchhā* style by Brahmin women in Karnataka, Tamil Nadu and Andhra Pradesh. The Śaivite and Vaiṣṇavite forms vary. Although the female lower garment worn in *kachchhā* form was one among other styles which have prevailed in India since antiquity, the *kachchhā* form worn by Śaivite (see fig. 5.29) and Vaiṣṇavite women in Karnataka, Tamil Nadu and Andhra Pradesh is linked with sects such as Smārtha, Śrī Vaiṣṇava and Mādhava founded respectively by Śaṅkaracārya in the ninth century, by Rāmānuja in the eleventh and twelfth century and by Madhavācārya in the twelfth/thirteenth century.

K.V. Soundara Rajan, Madras, in a personal communication pointed out that sectarian differentiation between Śaivites and Vaiṣṇavites among Tamils is to be dated to the post tenth century AD. The Alvar saint, Poigaiyar, fifth to sixth centuries AD inclined towards worship of Harihara. Even in the area of image worship, it was only towards the end of the ninth century that a degree of uniformity began to emerge. It was during the tenth and eleventh centuries that sectarian differences began to be formalized. This found reflection in the *Airāvateśvara* Temple at Darasuram and the Kampahareśvara Temple at Tribhuvanam built respectively by Rājarāja II (AD 1146–73), and Kulottuṅga III (AD 1178–1223). The division between Śaivism and Vaiṣṇavism is to be dated to the twelfth and thirteenth centuries and it was Mādhavācārya who predicated Kṛṣṇa bhakti on to the Dvaita doctrine. It was probably during this period that the female style in wearing the *kachchhā* began to differ between the Aiyars and Aiyangars, the differentiation corresponding to the *bibhūti* and *nāmam*, sectarian symbols drawn by the male on the forehead, which, by this date, had become the marks of customary identification (see also Agarwal, 1953: 75–97; Soundara Rajan, 1978: 47–8; 1981: 155, 209). A sculptural image at Mahabalipuram bearing a *nāmam* (see fig. 5.30) has been dated by K.V. Raman, Madras, to the fifteenth century. Andhra costume cannot be understood unless the geographical configuration of the present state is taken into consideration.

Andhra Pradesh can be divided into several subregions based on geographical and cultural factors. Kurnool, Anantapur, Chittoor would have affinities both with Tamil Nadu and Karnataka. The coastal areas were subdivided into the northern and southern regions. The northern unit, Srikakulam, would merge into tribal south Orissa, while Nellore in the south would find a greater affinity with Tamil Nadu. In the region comprising

Sumitra Parthasarathi demonstrating *billagoci* style of wearing sari. Courtesy: Dr V.V. Krishna Shastri, Hyderabad. (Fig. 5.31)

Sumitra Parthasarathi. Rear view. Note arrangement of border on line of divided sari. Courtesy: Dr V.V. Krishna Shastri, Hyderabad. (Fig. 5.32)

Jayalakshmi demonstrating *kasapuse* style of wearing sari. Courtesy: Dr V.V. Krishna Shastri, Hyderabad. (Fig. 5.33)

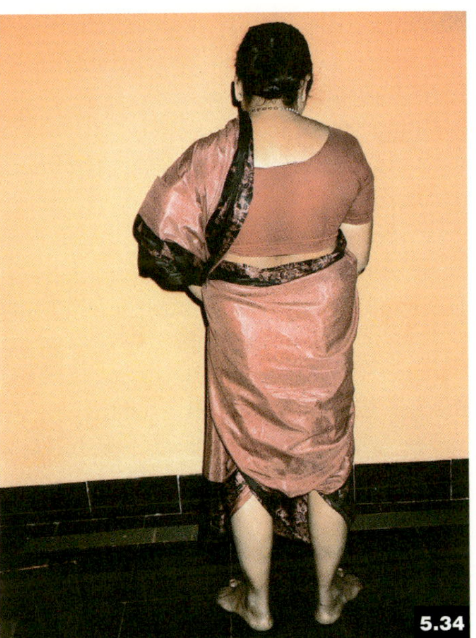

Jayalakshmi wearing sari. Rear view. Courtesy: Dr V.V. Krishna Shastri, Hyderabad. (Fig. 5.34)

Mattagoci style of wearing sari. Street scene, Hamankonda (Fig. 5.35)

Telengana and Mahbubnagar, Adilabad and Nizamabad adjoined Maharashtra while Medak existed at the crossroads between Maharashtra and Karnataka. Mahbubnagar was within easy reach of the Hubli–Dharwad design heartland.

It is, therefore, not surprising that all the sects present in these neighbouring regions should have had their counterparts in Andhra Pradesh (see Soma Reddy, 1980: 85–7, 91–4, 99, 109–11, 115). This was reflected in the field of costume. However, apart from these crosscultural influences, female costume in Andhra Pradesh had an identity of its own. There appears to have been three variations of the *kachchhā* form in Andhra Pradesh: *billagoci*, *kasakattu/kasapuse* and *mattagoci*. The term *billagoci* is derived from *billa*/tablet and *gochi*/*kachchhā* (oral communication: V.V. Krishna Shastri, Hyderabad). *Billagoci*, worn by a majority of Brahmin women, was based on a garment, about 9 yds. in length. It was essential for all auspicious community occasions (see figs. 5.31, 5.32). *Kasakattu/kasapuse* (see figs. 5.33, 5.34), was worn for all ritual activities within the home. The working costume was the *mattagoci* in which the pleats were not as neatly made as in *billagoci* (see fig. 5.35). The non-*kachchhā* form, the *godakattu*, was an innovation introduced from outside this culture zone.

Unstitched garments, particularly male wear, lent themselves to a variety of uses and, therefore, constituted the most important loom product. Mode in costume was

Gauze, raw silk, eighth century. Possibly China. Earlier taken as *byssus*. Found in Church of Notre Dame, Maastricht. Courtesy: Musee Historique de Tissu, Lyon. According to G. Vial this may have been woven on a draw loom. (Fig. 5.36)

Bāluchari. Woven at Bankura. Courtesy: late Subbo Tagore, Kolkata. (Fig. 5.37)

Bāluchari motifs in *pallu*. Courtesy: late Subbo Tagore, Kolkata. (Fig. 5.38)

Kani paśmīnā. Anavian Collection, Tehran. (Fig. 5.39)

Dhaka *jāmdānī*. *Nilambori*/Indigo midnight blue. Courtesy: late Subbo Tagore, Kolkata. (Fig. 5.40)

etched by the sensibilities of patrons and users. The greater range and variety in the costumes of the north found an echo in loom procedures. If a generalization were to be made, it could be unequivocally stated that the basic Indian fibre was cotton. There was considerable range in yarn count and the finest could be gossamer thin. India had developed *mulmul* but not gauze.[12] Gauze is linked with the silk tradition of China, its presence being attested during the Han Period 202 BC to AD 220 (Vial, 1983: 51) (fig. 5.36). Even centres like Varanasi, today renowned for its silk, had an earlier orientation towards cotton (Singh, 1985: 111). An indirect pointer to this fine tradition in cotton weaving is the *jāmdānī* tradition of Tanda.[13] It is significant that in the north, if the tribal *tasar* belt is omitted, elaborate loom ornamented silk weaving centres are also associated with the introduction of the draw loom.

Bengal was a cotton weaving centre. The silk industry was developed largely in response to European demand, and the development of the silk *Bāluchari* sari in Bengal (figs. 5.37, 5.38), is dated to the period of Murshid Quli Khan, AD 1701–26 under whom the provincial capital was shifted from Dacca to Murshidabad (personal communication: K.K. Ganguly, Calcutta). The *Bāluchari* sari was woven on a draw loom. M.C. Joshi, New Delhi, in a personal communication, has pointed out that Murshid Quli Khan had spent his youth in Persia. The *Bāluchari* draw loom may, therefore have been introduced into Bengal directly from Persia, rather than finding its way via Gujarat.

However, because of the greater ritual acceptance of silk, plain unstitched garments in silk had certainly been woven since early times as will be demonstrated later. Interlock and *jāmdānī* (see figs. 5.39, 5.40), the traditional techniques of loom ornamentation, necessitated investment in skill rather than the addition of contrivances on the loom. The unstitched garment lent itself to another labour saving work procedure on the loom. When the warp was nearing exhaustion a new warp, sometimes of a different colour, could be fashioned by twisting the new lengths to the old ends protruding beyond the reed (fig. 5.4).

Gujarat, with its West Asian linkages developed an industry in silk and *zari*. Although the satin weave, and *maśrū* (see figs. 5.41, 5.25), may have been post-Islamic developments, the term *maśrū* could perhaps be traced to *miśra*, Sanskrit for mixture. *Maśrū* has a silk warp and cotton weft. The term satin is derived from

its imputed place of origin, China. Among the nomenclatures used in medieval Europe for China was that of Zaytun and it is with this term that satin is linked. The earliest inscriptional record in which silk weaving rather than silk receives notice is the western Indian Mandasor inscription dated between AD 473 and 475 (Fleet, 1970: 80, 84–6; Fleet, n.d., 195–9). The *Laghu-Prabandha Sangraha* dated to the latter part of the thirteenth century, makes use of the term *paṭṭakul*, in the sense of fine or silken garment. The *Prabandha-cintāmaṇi*, dated AD 1305, refers to *dukūla* in the sense of fine silk (Thaker, 1970: 39, 58). In the *Māmerun*,[14] Premanand, who died before AD 1700, mentions numerous silk and brocade items such as *maśarū/ maśrū*, *char chokaḍi tāś*/cross design in metallic yarn on silk, *gajiyānī*/silk woven in *satin* weave, *asavali/cosave*, a brocade (see fig. 5.42), and *paṭolun* (fig. 5.43), double *ikat* mainly in silk, associated with the town of Patan, north Gujarat.

Due to the difference in pattern and mode, the south Indian sari developed a tradition in weaving which was quite different from that of the north. In the north, Bengal had a tradition of *jāmdānī* patterning (see fig. 5.40). Saris such as those of *Dhaniakali* were woven exclusively in tabby including the border. In Orissa extra weft patterning was introduced by means of pattern rods and shafts (see fig. 1.4). Varanasi had a tradition of *jāmdānī*. Since the tradition of brocading was introduced to Varanasi from Gujarat, it appears that the *gaṭṭhuā* loom, based on multiple shafts and pedals, also made its appearance in Varanasi at this time. The linkage between the Gujarat multiple shaft and pedal loom which has been documented in Ahmedabad (see figs. 5.44, 5.45), for *ghoḍa shāl* border loom (figs. 5.46, 5.47), and the *gaṭṭhuā* is strongly suggested through the similarity in symbols used in charts showing lifting order in course of weaving (figs. 4.40, 4.55).

Kutny, maśrū woven by members of Saurashtra community, Tamil Nadu. Book cover. Courtesy: Saraswati Mahal Library, Thanjavur. (Fig. 5.41)

Asavali sari. Gujarat. Courtesy: Bharat Kala Bhavan, Varanasi. (Fig. 5.42)

Paṭolā. Ratan chowk motif. Courtesy: Gujarat Hastha Kala, Ahmedabad. This sari is used by the Muslim Bohra community in wedding ceremonies. (Fig. 5.43)

Manilal Somnath, Hindu Khatri weaver, at loom, Ahmedabad. (Fig. 5.44)

Manipulation of pedal in *ghoḍa shāl*. Workplace of Manilal Somnath, Manek Chowk, Ahmedabad. (Fig. 5.45)

Upper portion of *ghoḍa shāl*. Workplace of Manilal Somnath, Manek Chowk, Ahmedabad. (Fig. 5.46)

Note shaft arrangement and alignment of pedals. *Ghoḍa shāl*. Workplace of Manilal Somnath, Manek Chowk, Ahmedabad. (Fig. 5.47)

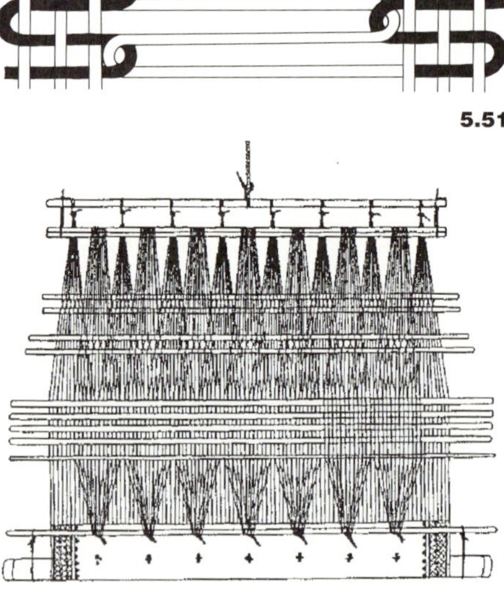

Jāla loom. (Fig. 5.48)

Sālu, woven in Varanasi. Courtesy: Rani Saheb, Udayini Raje Bhonsle, Nagpur. (Fig. 5.49)

Use of single interlock in *tope*, *Ilkal* sari. Border in extra warp *peti*/box design. Workplace of Phaniraj Konappa, Guledgud, Ilkal. (Fig. 5.50)

Single interlock technique. Weaving with three shuttles. (Fig. 5.51)

Petni, joining of the *pallu* warp to the body warp. (Fig. 5.52)

Jāla loom. Kothakota. (Fig. 5.53)

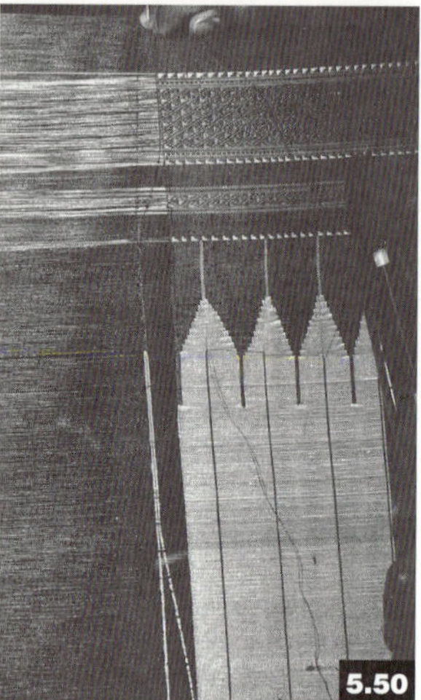

In Gujarat *maṣrū* patterning is also achieved on a multiple shaft and pedal loom similar to the Ahmedabad pallu loom (see Ahmedabad Brocades, NID: 38). A further refinement in Gujarat was the adoption of the drawloom derived from Persia. This was to be introduced later into Varanasi as the *jāla* (see fig. 5.48). It is of interest that the dobby mechanism is conspicuous by its absence in Gujarat as well as in Varanasi. In the north when the *jāla* is in evidence as a loom attachment, patterning by means of extra-weft appears to predominate (see fig. 5.49). In the south there has been a greater orientation towards extra warp ornamentation. Border variation is achieved through the three shuttle technique. In Karnataka, the *pallu* is ornamented with the *tope* motif (see fig. 5.50). The three shuttle technique (fig. 5.51), and *tope* design involved use of interlock at colour junctions. Where there was a contrast in the colour junction between the ground colour of the sari and that of the *pallu*, the joining has been achieved through the *petni* (cf. fig. 5.52). The southern traditions of ornamentation in female unstitched garments were, therefore, quite distinct from those prevailing in the north. It was for this reason that while *jāla* became increasingly popular in the north, the south preferred to refine the *aḍai* mechanism rather than adopt the Banaras type draw loom. The drawloom is in evidence in southern weaving centres such as Gadwal, Kothakota (see figs. 5.53, 5.54, 5.55), and Dharmavaram in Andhra Pradesh which have recent traditions of silk weaving. The Gadwal and Kothakota silk tradition was largely engendered by royal patronage and it was through this source that the *jāla* was consciously introduced from Varanasi. In the case of Dharmavaram, when the transition was made from cotton to silk weaving, the weaving community itself opted in favour of the *jāla*.

Jāla loom. Kothakota. (Fig. 5.54)

Two sets of six yard warps aligned along two corners of same room. Note point of overlap in loom to the right. Weaver's Complex , Kothakota. (Fig. 5.55)

Notes and References

1. The term *luṅgī* appears to be derived from Burmese *lungyi*, which, according to Conway (1992: 188), is the local term for the male loincloth and the female lower garment. Its wide dispersion on the Indian subcontinent inclusive of Punjab, and an association with Islam, raises problems of location of origin.

2. Rhys Davids, 1965: 57. The original text, no longer extant, is dated to the beginning of the Christian era. The work, as it is now known, is based on a Pali translation made in Ceylon *c.* AD 400.

3. According to the *Dharmaśāstras*, not only are white garments recommended for the householder, but there is also an injunction against the display of expensive and gaudy items (Kane, 1941: 671). The *Laghu Aśvalāya* recommends white as the colour both of the upper and lower garment worn by the male (Kane, 1953: 330). According to the *Baudhāyana Dharmasūtra* dated between 600 to 300 BC, in certain rituals the priest wore a multicoloured mantle (Bühler, 1882, 187). For continued Indian preference for white see Julien, 1847, 102.

4. Shrigondekar, 1939: 14, 89. This work was written by King Someśvara, AD 1127–38. Among the designs mentioned are those made up by lines, dots, squares and circles. These are not suggestive of a highly convoluted form of patterning.

5. Indigo blue was associated with physical attraction. While it was proscribed for Brahmin males, it was permitted to women (Vasu and Naraharayya, 1913: 430). There appears, however, to have been an ambiguity with regard to female costume. Both navel and breast had to be covered, while the lower garment had to reach down to the ankles (Majumdar, 1934: 200).

6. See Soares, 1988, *corja* q.v.; Dermigny, 1959: 439; Brennig, 1975: 246; Blancard, 1906: 232; B.N., N.A., 12087 (40); *Products Tamil Nadu*, NID, no pagination, *kuncham*, q.v.

7. The *pyjāmā* was absorbed into European apparel, the Portuguese being the first to adopt this garment. Linschoten, 1855: 200, n. 4, 206.

8. See Lawrence, 1967, 375. However, as a motif in Hindu sculpture, the *kalga* in an incipient form is in evidence in the triple crown of images such as that of Bhairava in the south-east wall of the sanctum of Babor temple, Jammu, dated to the tenth/eleventh centuries AD. See Varadarajan, 1987a: 141–9, Figs. 40–52, cf. Varadarajan, 1995: 429–30.

9. Mulberry silk is preferred to cotton and wild silk for *zarī* making. The tensile property of mulberry silk is higher than that of wild silks and cotton. It is also able to absorb temperature variations more readily during the heat generated in course of coating.

10. The *kachchhā* form for the lower garment was one among other prevailing garment styles in the north. The establishment of the Islamic ruling elite since the twelfth century reinforced West Asian trends in costume. However, among the three Hindu upper castes, women tended to wear unstitched clothing in the kitchen in the same way as men wore only the *dhotī* during meals (personal communication: M.C. Joshi, New Delhi). The acts of cooking and partaking of meals were thus imbued with ritual undertones. Such social usages may have kept alive the tradition of the sari, although in an attenuated form.

11. The earliest reference to the mode associated with Bengal and Orissa is the one made by *Rājaśekhara* dated to the tenth century AD (Ghurye, 1966: 125–7).

12. The term muslin was a generic term derived by the Europeans to cover the entire category of fine cotton weaving in India. Although the origin of this word is attributed to Mosul in Iraq, according to Marco Polo, *c.* AD 1254–1324, all cloths of gold and silk made in the territory were called *mosolin*. Dermigny, 1959: 441; Sergeant, 1972: 36. It is possible that Europeans used the term *muslin* since Indian *mulmul* may have had some similarity with the textiles of Mosul. *Mulmul*, however, is clearly of Indian origin.

13. The designs on the Dhameka *Stūpa*, Sarnath could well suggest *jāmdānī* motifs. Sukul, 1974: 43–5; Pandey, 1981: 14–17. While there is a clear reference to the presence of silk weavers in Gujarat in the Mandasor inscription there is no equivalent data for Varanasi. The many references to fine weaving could relate to fine cotton weaving. For Banarasi weaving see Krishna and Krishna, 1966: 11–17; Sukul, 1974: 43–5; Pandey, 1981: 14–17.

14. At the time of the first pregnancy, gifts of costumes and textiles had to be made by the father of the bride to the family of the groom. Premanand describes this ceremony in his *Mamerun*. The *Varṇaka-Samuccaya* provides further details. *Paṭṭakula*, woven in Patan, may have been a synonym for silken *dukūla*. *Patolun*, in the sense of the double *ikat* in silk manufactured at Patan, the *paṭolā*, appears to have entered Sanskrit in the twelfth century AD. *Atalasa*, silk in satin weave, was woven at Vidyapur, referred to as Vijapur, in Mehsana district. *Maśrū* was woven in Broach, modern Bharuch (Sandesara, Mehta, 1959: 29–30, 46–7, 56).

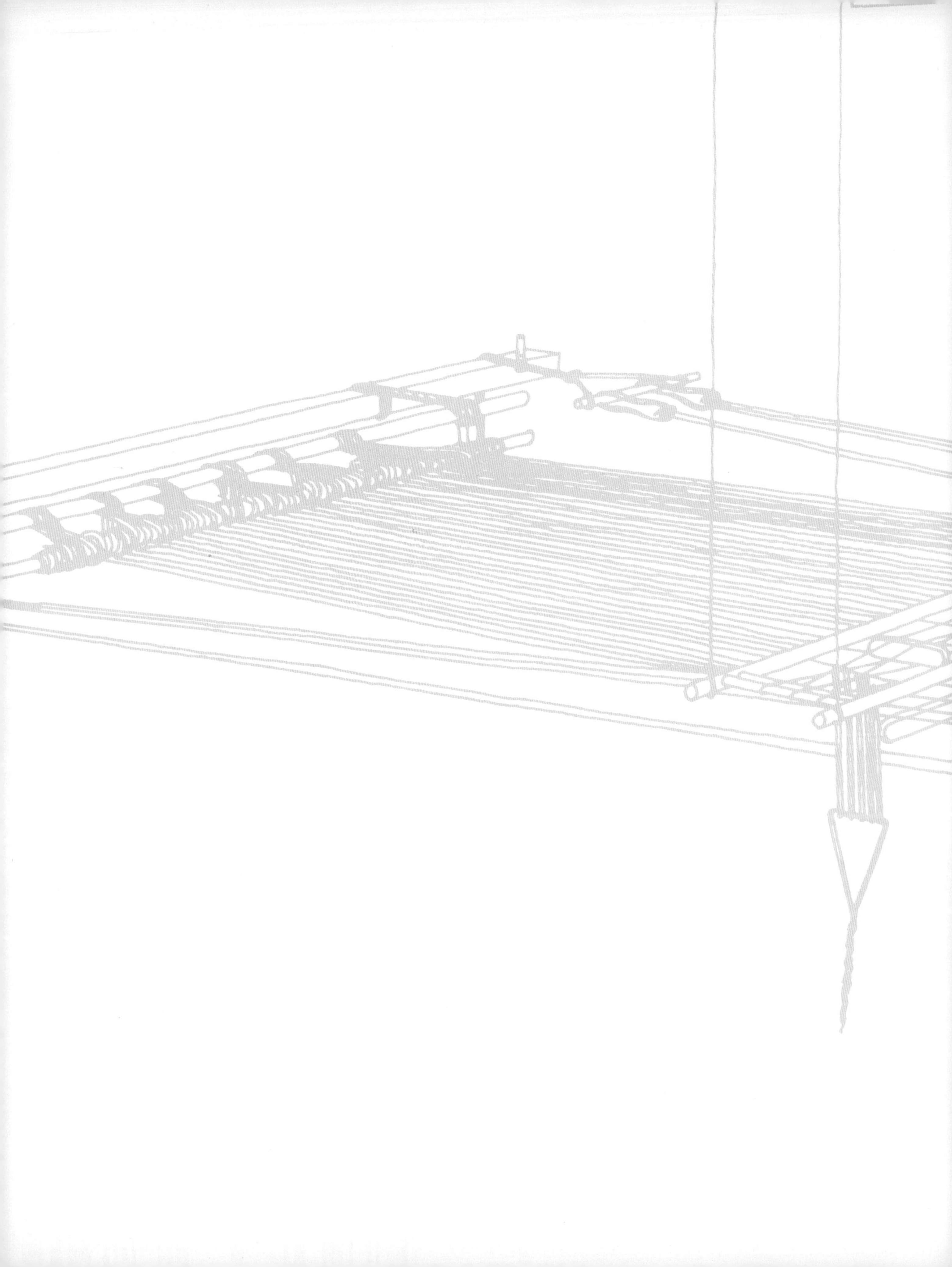

06 DECCANI HORIZONS: KANCHIPURAM FABRIC AND THE *PAITHANI* RANGE

06 Deccani Horizons: Kanchipuram Fabric and the *Paithani* Range

While any attempt to trace the ethno-historical evolution of the loom in relation to fibres used, demands an all India coverage, south India has played a significant role in terms of regional homogenization of ethnicity. In terms of loom-cum-product differentiation Paithan and Kanchipuram have played an important role. In both nuclear areas, qualities inherent to that queen of fibres, silk, have been exploited to the full. In this context a curious feature is to be noted: that although silk of *Bombyx mori* was produced in north-east India, and a more limited quantity in Kashmir, the weaving of silk in the Indo-Gangetic valley was engendered and moulded more by Islamic demand and taste rather than by the pre-existing cultural base.

Several explanations are possible for references to figured silks in early texts as delineated by V.S. Agrawala and Moti Chandra.[1] First, the items could have been imported. Second, the patterning could have been achieved by means of the *ikat* technique with the fibre being either imported *B. mori* or a very fine quality light toned *tasar*. The following quotation in the *Mānasollāsa*, 'where threads are tied, it should be well coloured', can be taken as a mid-twelfth century reference to the practice of *ikat* (Shrigondekar, 1939: 89). Third, patterning could be achieved through the *jāmdānī* technique (fig. 6.1). Finally, it should be emphasized that it was the diaphanous quality of silk which appealed to Indian taste rather than multi-layered, textured patterning. The latter category has been present but scarcely in a dominant mode. This is clear from a careful reading of early descriptions of textiles which may be found in the works of V.S. Agrawala and Moti Chandra.

Peninsular India comprising modern Maharashtra, Karnataka,[2] Andhra Pradesh, Tamil Nadu and Kerala, while sharing the cotton orientation in textiles prevalent in the north, appears to have developed an independent, possibly Gujarat derived, silk tradition, without, however, having any *B. mori* silk producing centre within its own geographical domain. Coastal peninsular India has participated in maritime trade networks since antiquity and the procurement of Chinese silk would have posed no problem.[3] Access to imported Chinese silk yarn combined with the special silk dyeing properties of the waters of the Palar River appear to have drawn considerable numbers of silk weavers to Kanchipuram by the year AD 985 as is demonstrated by the Madras Museum Copper Plates of Uttama Cola, *c.* AD 970–85. The four quarters of the town of Kanchipuram were referred to as Kambulanpadi, Atimanappadi, Kanjagapaddi and Erruvalichcheri. These were mainly occupied by weavers comprising two sections of *paṭṭaśālins*, silk weavers. They were patronized by the king and appointed to manage the temple, a mark of the high esteem they enjoyed despite their status as members of the weaving community (Krishna Sastri, 1920: 267–8).

The diffusion of silk weaving in Tamil Nadu also appears to be connected with the southward migration of a group of silk weavers originally hailing from Gujarat called Saurashtras in south India. A reference has already been made to the emergence of Gujarat as a centre of *B. mori* silk weaving.

Jāmdānī patterning on Venkatagiri sari. Workshop of Vellur Raman Aiyer, Venkatagiri (Fig. 6.1)

According to the evidence of *Periplus of the Erythraean Sea*, dated *c.* first century AD, Roman demand for Chinese silk was met by exports channelled through Broach and other ports on the west coast such as Kalyan and Sopara. In AD 437–8, a group of silk weavers who had left their native place, Lāṭa,[4] consecrated a temple to *Sūrya*, the Sun god, at Daśapura or Mandasor during the reign of King Kumāragupta (AD 415–55). A.L. Basham links the departure of the silk weavers to the shrinking Roman demand for silk, since the ineffective Roman Emperor, Valentinian III, was unable to roll back the Vandal and Hun attacks on the Roman Empire.[5] The late Professor R.N. Mehta (personal communication: Ahmedabad), did not accept this view and attributed the migration to unstable conditions generated consequent to the disintegration of *kṣatrapa* dominion in this region.

The Sun Temple appears to have fallen into a state of decay and was repaired by the same guild of silk weavers in AD 473–4. The inscription, composed by Vatsabhaṭṭi at this time, suggests that the demand for silk was still inadequate, as several silk weavers had turned to other professions (Basham, 1983: 97, 104). The next episode is the migration of silk weavers southwards following the Ghaznavid sack of Somanātha in AD 1024. This group, the Saurashtras, had an association with urban centres in Gujarat such as Gohilvāḍa, Aṇhilvāḍa Patan, Bhuj, Rajkot and Wadhwan. Their initial focus for settlement was Devagiri founded by Bhillama V in AD 1188 and made the capital of the Yādava kingdom by Jaitugi I, AD 1193–1210 (Verma, 1970: 64–5). The attacks on Devagiri waged by the Khiljis in AD 1294 and 1307 coincided with the decline of Yādava power. In AD 1318 the kingdom was annexed to the Delhi Sultanate by Qutubu'd Din Mubarak Shah.

In assessing the movement of Saurashtras what has to be kept in mind while interpreting their oral tradition is the aspiration of the community in the south to be accorded recognition as Brahmins. It is perhaps in order to reinforce this claim that oral testimony links migratory movements with Muslim attacks. The late Dr H.C. Bhayani, Ahmedabad, in a personal communication, has categorically stated that on linguistic grounds he would support both a Brahmin origin for the community and an eleventh century date for the movement outwards from Gujarat. Whatever the reason, the Saurashtras left Devagiri in the fourteenth century. They moved on to the Vijayanagar domain where they adopted their own calendar based on the Saurashtra Vijaya *Saṃvat* beginning with AD 1346. According to verses recited during the engagement ceremony of Saurashtras, it was related that from Vijayanagar the community moved towards Rameshwaram stopping at Thanjavur, Madurai and Paramakkudi on the way.[6] The two most important silk weaving units in Tamil Nadu are centred around Kanchipuram, Chingleput district, and Kumbakonam, Thanjavur district. Salem has tended to specialize in men's undyed wear, silk *vesṭi* and *aṅgavastram*. Dyed silk sari weaving was introduced into the urban complex only in the late 1960s. Saurashtras, also called *Paṭṭunulkaran* in Tamil, meaning silk thread workers, are settled in sizeable numbers at all three centres. Despite the association with Gujarat, there is no overt association with *ikat* dyeing at these places. It is only in Madurai that Saurashtras have practised *bāndhanī* on cotton, the saris thus ornamented being called *sungaḍi*.

Tamil Nadu: The Kanchipuram Tradition[7]

Among the varieties of fabrics made in Tamil Nadu, the following may be cited:

Chettinad saris

These are silk saris, the present trade name being 'puja saris'. These were earlier usually of 30 to 40 in. width so that anklets worn by women could be seen. Colours included madder, golden yellow, deep green and black. Narrow borders, with an intermixture of ground colour are achieved by single shuttle, while broader borders in contrasting colour are woven in the three shuttle technique. Today these saris are made at Karaikudi. A variation in pattern is achieved by tie dyeing of the warp so that the centre of each square in the checked pattern has a small dash.

Madurai saris

These saris, called *sungaḍi*, to which a reference has already been made, can be simple plain weave woven invariably in cotton by Saurashtras. There are also the Kodambakkam striped saris with cotton weft, art silk warp, introduced *c.* 1960. These have borders with *rudrākṣa* design.

The Paramakkudi saris

These are no longer in production and point to the *ikat* tradition of the Saurashtras in Tamil Nadu. These were three shuttle saris with weft *ikat* repeating the border colour.

Around 1910, Tribhuvanam had developed a tissue half sari or full *janglā* sari, obviously an *aḍai* version of the Banarasi *janglā* pattern.[8] In Varanasi the *janglā* pattern was achieved through the *jāla* loom device. The heaviest silk saris are those associated with Kanchipuram manufacture, while the items of Kumbakonam and Thanjavur rank next in order of preference. Silk weaving was introduced at Arni *c.* 1930 to offer a cheaper substitute of the Kanchipuram variety. Arni appears to have been a cotton weaving centre prior to this date.

Traditionally *charkhā* reeled silk, silk reeled from hand operated spinning wheel, of 18 to 28 deniers was used for both warp and weft in the Kanchipuram product (cf. figs. 6.2, 6.3). The warp is made of three or four filaments twisted together with about 24 to 30 twists per inch. The weft, *sapuri*, has 8 to 12 twists per inch, and, having fewer twists, it is more lustrous and, therefore, tends to dominate over the warp. Variations in S and Z twist have also been utilized as an element in surface design on a monochrome ground. The ends per inch in the warp usually contain 38 *punjams* spread over 48 in. This works out to 96 to 100 ends per inch. The number of picks per inch is about 70 to 80. *Zarī* obtained from Surat, has about 2,200 to 2,400 metres per *laḍī* weighing 28.35 gm. Saris[9] are made in three lengths: 6 yds., worn by the increasingly fashion-conscious married and unmarried women; 8½ yds. worn by married women with traditional tastes in Andhra Pradesh and Karnataka. Married women in Tamil Nadu with traditional tastes wear a sari length of 9 yds. 1 ft. The width varies between 44 in. and about 48 in.

Designs are derived from natural and animal forms and motifs in temple architecture. At the same time these are combined with *kollam*, auspicious designs traditionally drawn in rice paste and vermilion at domestic portals each morning (see Archana, no date). Motifs in the ground of the saris have tended to be small, repetitive and more geometrical in form.

The significant feature of this style is that the borders and *pallus* are in different colours, both in warp and weft, to those of the body of the sari. They are called *korvai* or three shuttle saris (fig. 6.4). The term *korvai* is derived from *korkarrathu*, to thread through. The analogy is that of threading flowers together. *Korvai* is thus among the rare names based on the technique used. Among these, the sari called *Kankan Rudrākṣaṃ* has a special role. The body is black, occasionally embellished with red *mallimoggu* in extra weft. The border, in three shuttle, can be mustard yellow, red or green. This sari is to be worn during the *kankan* ritual performed either in the fifth or seventh month of the first pregnancy. The ceremony can also be called *Valaikappu* or bangle. If it is performed in the fifth month, seven glass bangles are worn on the right arm and eight on the left. During the seventh month, nine are worn on the right and eleven on the left.

The three shuttle technique (fig. 5.51) solves the problem of weaving at colour junctions along the weft. The *petni* method (fig. 5.52) was devised as a solution to ensure maintenance of continuity along colour junctions in the warp. In Kanchipuram, for about an inch along the warp ends where the warp colour is changed, both sets of warp ends are woven simultaneously. The body warp is then cut as a result of which the raw ends sticking out at each end of the *petni* reflect the body and *pallu* colour respectively. In Kumbakonam the warp of the *pallu* portion is tie dyed. To remove any disjunction in outline when the *pallu* portion is reached in weaving, the irregularity in outline is straightened by warp ends being looped and tightened through bamboo sticks until a straight line is achieved. After weaving for one inch the bamboo sticks are removed and the loops cut. The Kumbakonam process called *porai*, results in a lighter join than is the case with the Kanchipuram *petni* (see *Products of Tamil Nadu*, NID; Mohanty, 1984: 156–60). Since the Saurashtras, settled in significant numbers at Kumbakonam, brought with them the Gujarat

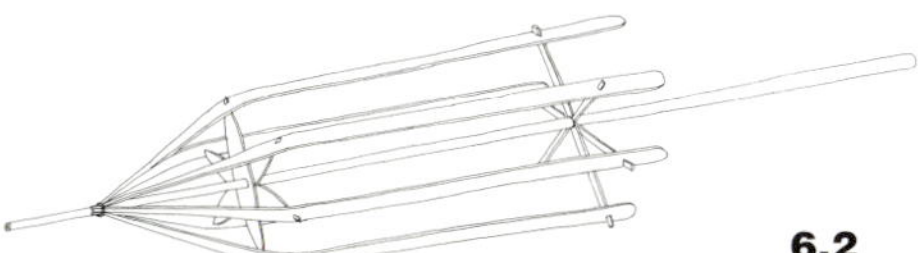

The swift/*parivattam* (Fig. 6.2)

Transferring the yarn from *parivattam* to pern. (Fig. 6.3)

Temple used in weaving *Kanchipuram* type sari on *aḍai* loom. Surajkund Crafts Mela, Gurgaon. Note three shuttle border and *petni*. (Fig. 6.4)

tradition of tie dye, it is possible that the *porai* method was an innovation introduced by them while the *petni* may represent an earlier usage.

Terms in Patterning

Korvai: borders in three shuttle technique.

Kal Erangadadu: sari with main field in checks, borders in weft stripes which run through from selvedge to selvedge.

Oosi vanam/**sparkler:** running through like a fine needle; fine closely woven longitudinal stripes in variegated colour.

Athi vazhaippu/**like lines on a plantain bud:** thick longitudinal stripes in two colours.

Attu mozhi/**eyeball of the goat:** checked effect in two or more colours.

Puppli or Sabhalankaram/**two colour checks:** 1 to 2 in. in thickness. In the centre of each square a small motif may be introduced either in coloured silk or *zarī*.

Pulainkottai/**seed of tamarind:** checks of two colours 1/4 to 1/2 in. in size.

Kasakasa/Khas Khas/**poppy seed:** tiny checks in two colours.

Muthu Kattam/**pearl checks:** discontinuous checks in silk or *zarī* woven in extra warp and weft.

Veldarī/**wavy:** stripes in twill weave with a dot in each curve running horizontally in extra warp at intervals of about 1 in. These stripes may be in *zarī* or in one or two colours. *Vel* also means creeper. The wavy stripe in twill weave could also represent a simplified form of creeper.

Thuthirippu/**flower of radish,** ***Raphanus sativus*:** dedicated to the god, *Gaṇeśa*, woven in pairs in contrasting colour or in stripes.

Mallimoggu/**jasmine:** bud design in extra weft.

Pettu/**border patterns:** two stripes define the width of the border within the *kal erangadadu.* The first *pettu* stripe is woven half an inch away from the selvedge. Motifs comprise:

Rudrākṣaṃ/**seed of** ***Elaeocarpus*** **sp.**

Samandi/**marigold:** flower dedicated to Śiva as Naṭārāja.

Jadinagam/**thin jasmine buds:** crown of serpents of Śiva.

Mīnākṣī/**fish:** shaped eye of the goddess Mīnākṣī at Madurai.

Madum/**lamp:** niche in triangular shape.

Śakuntalā: creeper with a four petal flower.

Vasisht/**circular shape:** with *thuthirippu* and wavy lines on both sides.

Pallu ornamentation was designed to balance the border pattern. Popular motifs included Gajalakṣmī, royal processions, *haṃsa* and *yali,* mythical swan and composite lion, horses, peacock, deer and parrot. There could also be *zarī* or coloured stripes.[10]

Around 1940 Kanchipuram silk saris began to find an increasing market in the north, a much prized item being the product called Temple Sari. In fact, no item with such a name exists in the southern repertoire. However, in the 1940s and 1950s temple authorities in the south began to auction old saris including those gifted by *devadāsīs,* temple dancing girls. These items were usually richly ornamented with *zarī.* Several among these were *munre bajam,* three portions, a style popular in the First World War period. The width of the sari was divided into three equal sections; the two borders comprising two-thirds of the sari width. It was perhaps this form of ornamentation which gave rise to the term, Temple Sari.

In weaving the Kanchipuram style sari the extra warp mechanism is called *jungu,* the extra-weft *aḍai.* If the weaver himself manipulates the extra weft ornamental device, the latter is called *būṭā jungu.* If there is an edge to edge design, a board is used to keep the lifted ends in position. A helper is now required, and the mechanism is called *aḍai. Aḍai* means, 'keep to one side'. The helper is called *sida pillai* in Telugu and *nesava pillai* in Tamil. The weaver concerned makes both the *aḍai* and the *jungu* devices. *Jungu* is a loom part while *aḍai,* denoting 'keep on side', is a verb (personal communication: B.K. Ravi Chandran, Saurashtra weaver, Thanjavur).

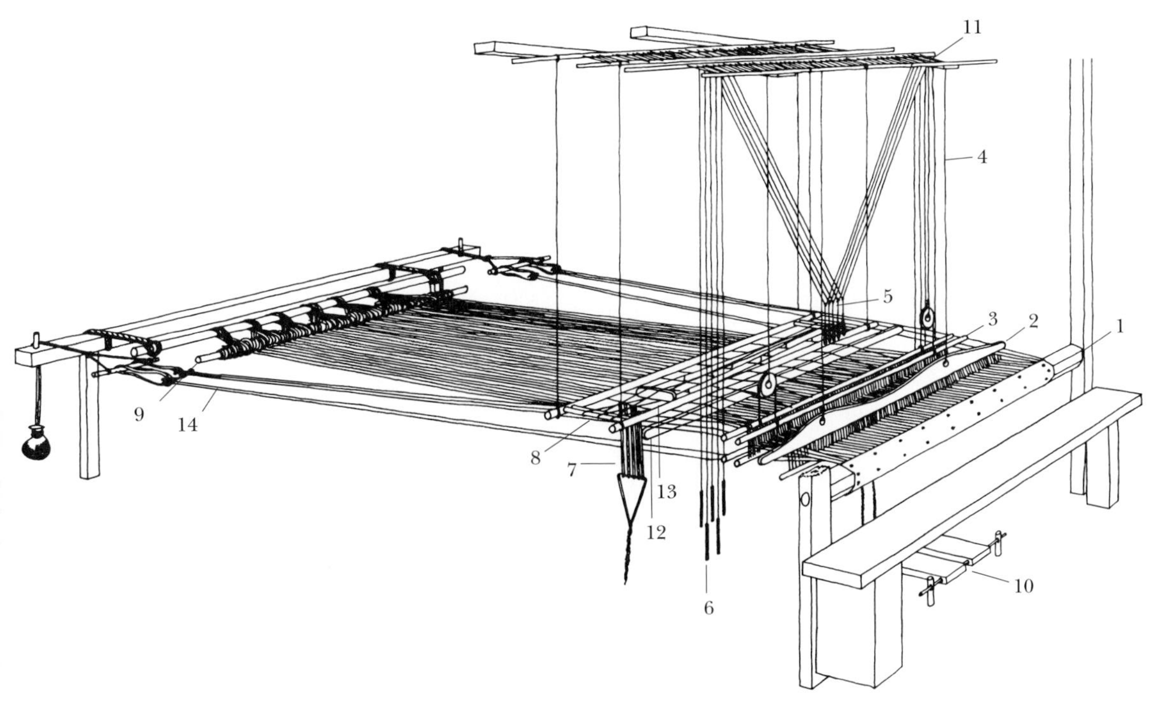

6.5

Kanchipuram loom with *aḍai* for crossborder and extra warp. (Fig. 6.5)

1. Cloth beam
2. Sley
3. Shafts
4. Harness, extra warp ends
5. *Jungu*
6. Lingoes
7. *Kuchu*
8. Cross threads for cross border *aḍai*
9. Warp beam
10. Treadles
11. Reed functioning as comberboard
12. Shedding plank
13. Small shedding planks/*Rakka palaka*
14. *Zarī* ends/extra warp patterning

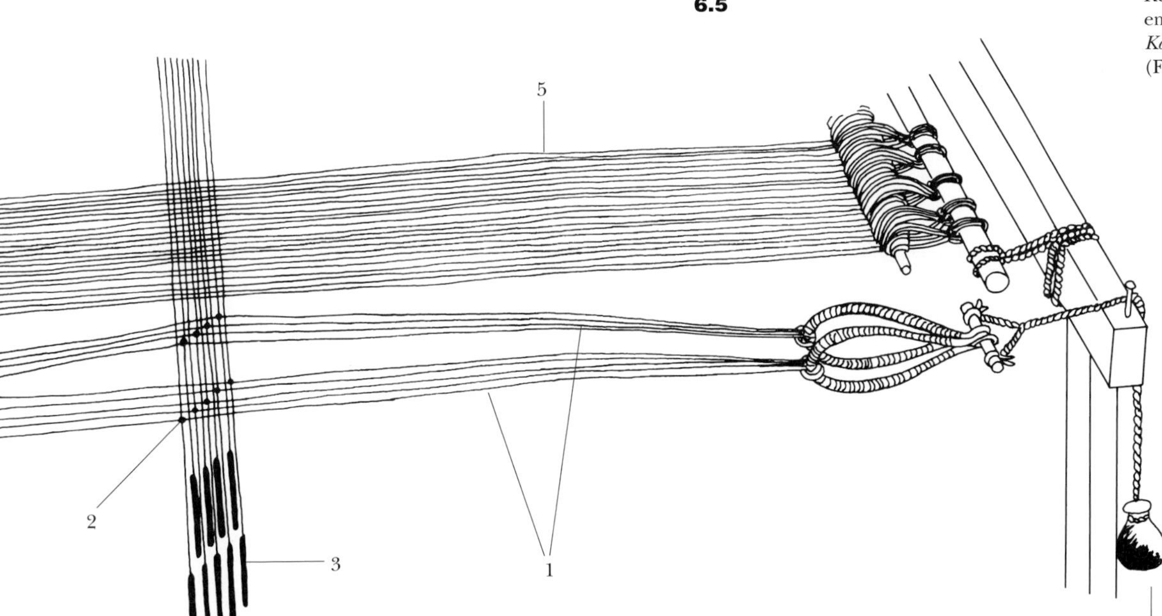

6.6

Regular silk warp and *zarī* ends. Detail. Back of *Kanchipuram aḍai* loom. (Fig. 6.6)

1. *Zarī* ends (for extra warp patterning)
2. Glass bead
3. Lingo
4. Weight
5. Body warp of silk yarn

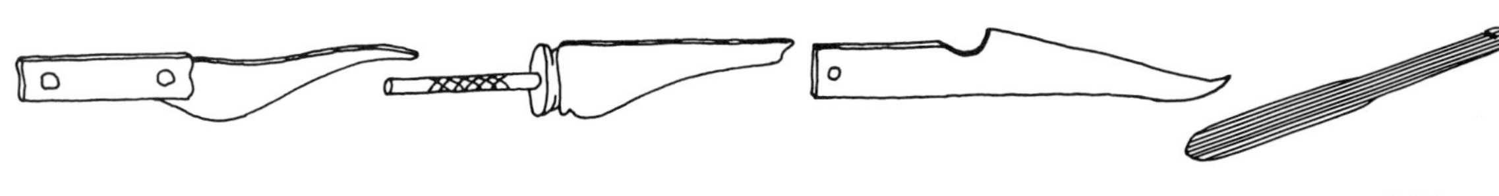

6.7a 6.7b 6.7c 6.7d

Knives used for cutting threads (a, b, c). Reed hook (d). (Fig. 6.7)

Aḍai Device

Like the *jāla,* the *aḍai* could be considered to be a system pre-planned on the loom for manipulating the required warp ends to create the shed for patterning. The device to which the word *aḍai* has been given refers to a system of lashes and leases set up on the loom for manipulation of warp ends. This is for achieving extra warp patterning, traditionally used for borders, as also for the extra weft/cross/border patterning used for the end piece, *pallu* of the sari, and the extra weft patterning of ground motifs (figs. 6.5, 6.6, 6.7).

This method works on the principle of the stationary centre closed shed with a rising and falling top. Separate *aḍai* have to be prepared for extra warp border designs, crossborder designs and for body *būṭās* (Ramaswamy, no date: 2).

Preparation of the *Pettu Aḍai* for Extra Warp Border Patterning

For the preparation of the *pettu aḍai* for extra warp border patterning, the number of warp threads in a repeat of a design, which interlace with the weft threads in the complete repeat of the design, is ascertained from graph paper (Mohanty, 1984: 1, 139). With the help of the graph the number of individual ends to be lifted for every pick is decided according to the marking on the graph. The blank squares in the prepared graph indicate the lifting order of the ends (see fig. 6.8). In the first stage, coloured twine, in this case, yellow twine is wound around two pegs fixed on a wooden plank or ground. The length of the twine varies from 3 ft. 3 in. to 6 ft. 6 in. and is dependent on the number of ends required per design (fig. 6.9). The length is increased from 3 ft. 3 in. to 6 ft. 6 in. according to the pattern requirements, and is so adjusted as to avoid overcrowding. The number of times the yellow twine is wound is directly related to the number of ends in a design. These are removed from only one of the pegs, one layer after the other, tying a knot to each one to form individual loops (fig. 6.10). A rod is inserted through these loops and tied to one side of the rectangular frame. Alternatively, instead of detachable rods which can be lifted off the frame and attached to the loom, the frame itself can be directly placed on the loom. The knotted ends are bunched together and tied to its opposite side (fig. 6.11). To determine the sequence a lease stick is inserted separating the odd and the even yellow twine (fig. 6.12).

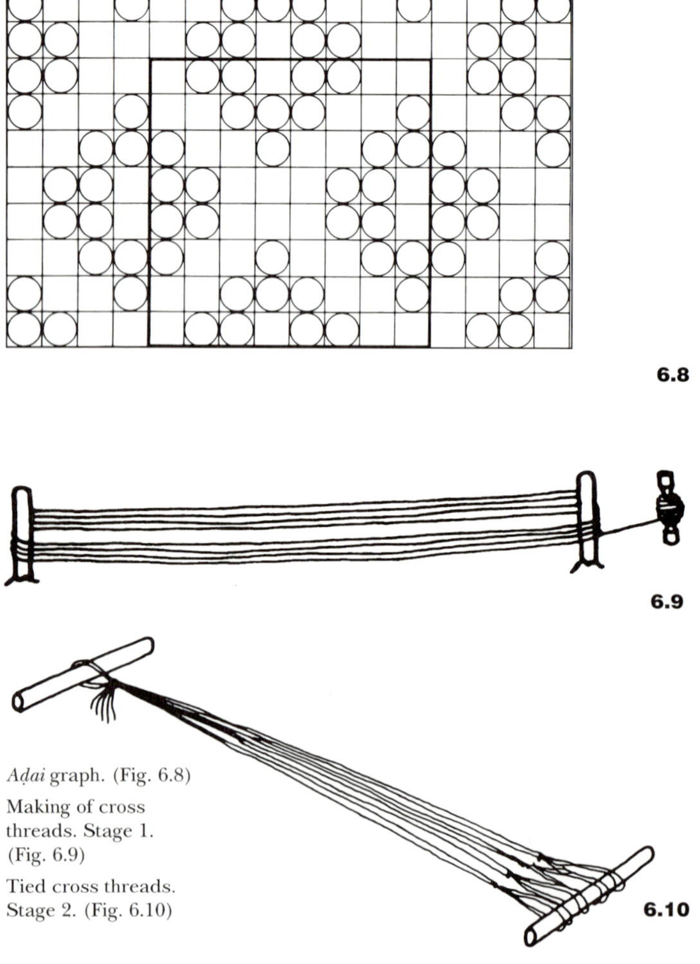

Aḍai graph. (Fig. 6.8)

Making of cross threads. Stage 1. (Fig. 6.9)

Tied cross threads. Stage 2. (Fig. 6.10)

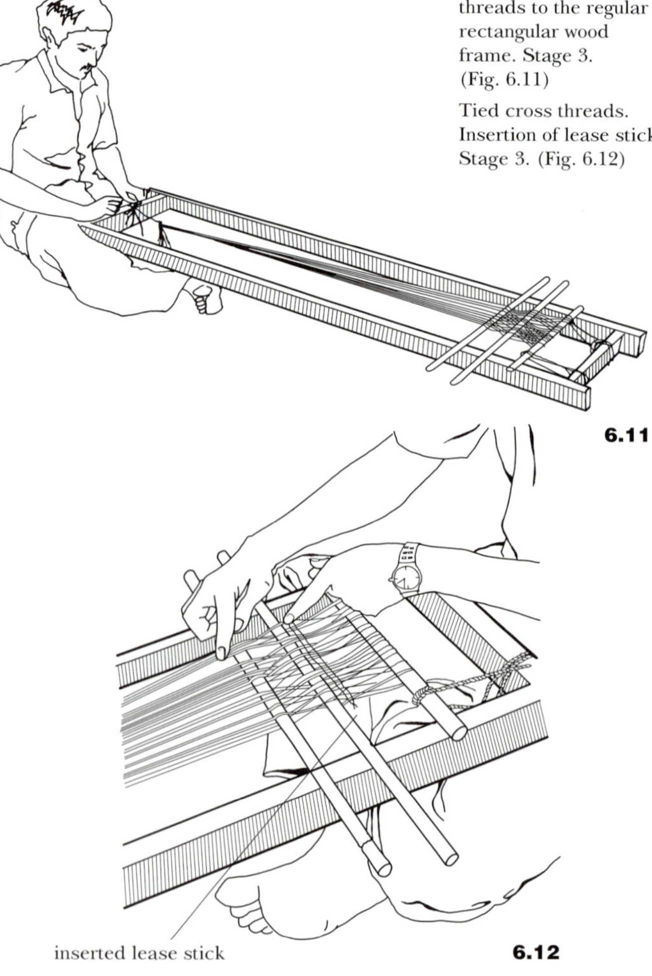

Transfer of tied cross threads to the regular rectangular wood frame. Stage 3. (Fig. 6.11)

Tied cross threads. Insertion of lease stick. Stage 3. (Fig. 6.12)

Arrangement for Lifting/Weaving Order

The number of the lengths of yellow twine on the rectangular frame are equal to the number of ends in a design. To plan the lifting/weaving order, the *pettu aḍai* maker reads each row of picks one by one in sequential order. He selects the blank squares on the graph and translates these by lifting the yellow twine from left to right on his finger (fig. 6.13). The 3rd, 4th, 5th, 8th, 11th, 12th, 13th and 16th lengths of yellow twine are picked up for the first pick. For the second pick the 2nd, 3rd, 5th, 6th, 10th, 11th, 13th, and 14th lengths of yellow twine are picked up, and this process continues for each pick till the end of the design repeat for the lifting order is completed. A rod is inserted into this selection as he withdraws his fingers.

Around each of these lengths of lifted yellow twine, individual loops with 10s twine the twine being continuous, are formed with the help of a rectangular plank locally called the *assa-palaka*, approximately 9 in. to 12 in. x 6 in. in size (fig. 6.14). This plank acts as a guide to equalize the size of each loop. Having completed tying of loops on each length of yellow twine for one pick, these are slipped off the guiding plank (fig. 6.15) and the outer ends are tied to a thick cord using the reef knot (fig. 6.16).

These bunches are called *jungus.* The next stage determines the harnessing for each extra warp/ornamenting end. If each length of yellow twine is lifted sequentially, one finds that it corresponds to the graph wherein the repeat is of eight ends and eight picks.

When the first yellow twine, which corresponds to the end number one is lifted, one finds that there are loops on pick numbers 1, 2, 7, and 8. These are made into a group and attached by a 10s twine to form a loop at the opposite end to the *jungu.* Hence the loops on yellow twine number 1 are coupled with a small loop each, and finally each loop is individually connected with a pair of harness twines. This procedure is followed for all the twines on the frame (figs. 6.8, 6.17, 6.18). The yellow twine is now removed completely.

6.13

Planning of lifting order from the graph paper. (Fig. 6.13)

Making of individual loops with aid of *assa-palaka.* (Fig. 6.14)

Representational drawing. Tying of loops. (Fig. 6.15)

Wooden board and frame used in preparatory stage of *aḍai.* (Fig. 6.16)

Representational drawing of tying of loops from outlined areas of graph. (Fig. 6.17)

Representational drawing of coupling of loops with a small loop and connection with harness twines. (Fig. 6.18)

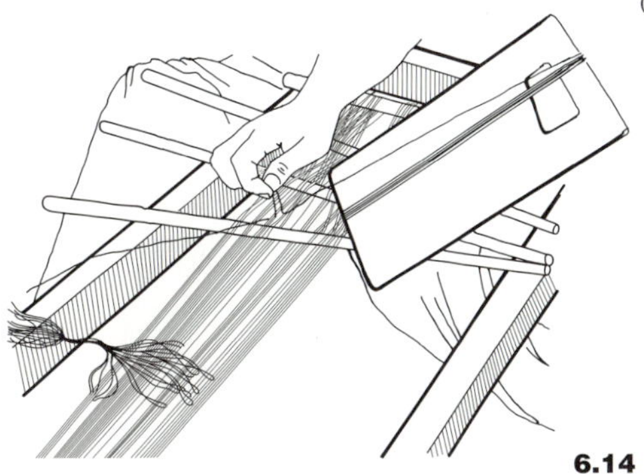

6.14

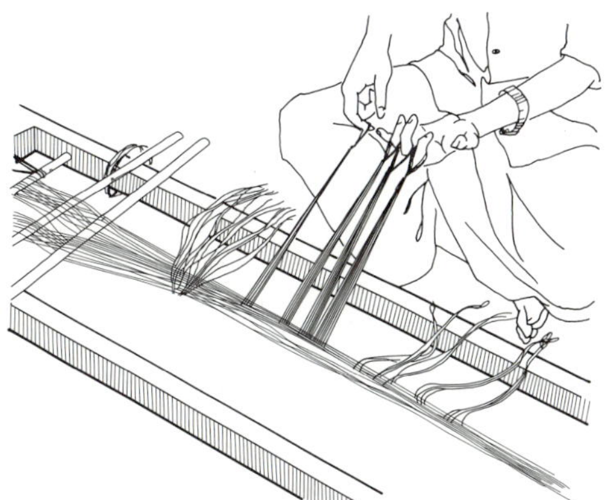

6.15

6.16

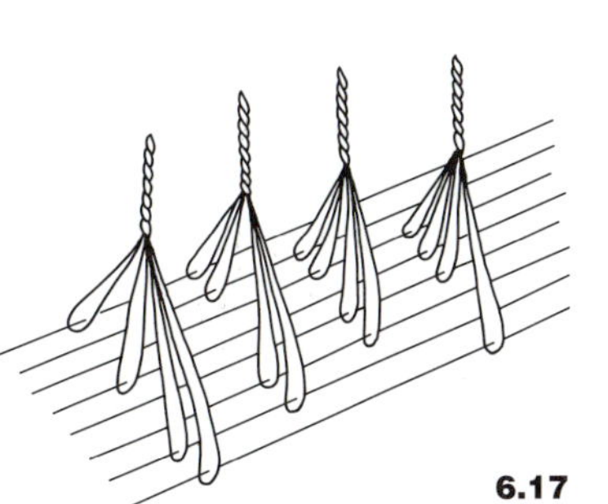

6.17

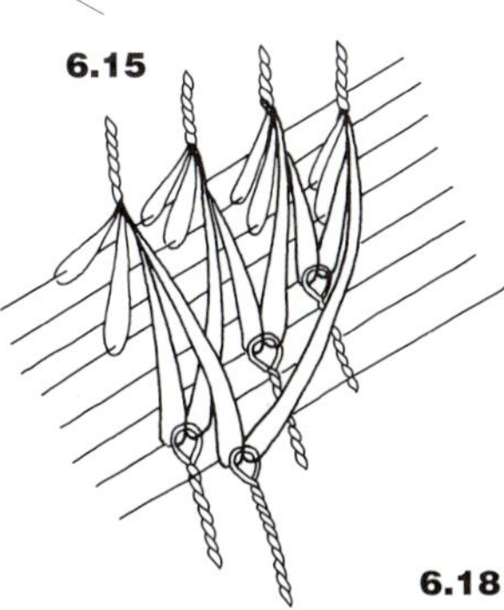

6.18

Harness mounting.
(Fig. 6.19)

1. Rod
2. Reed
3. Stand
4. Harness
5. Varnish healds
6. Mail eyes
7. *Jungus*

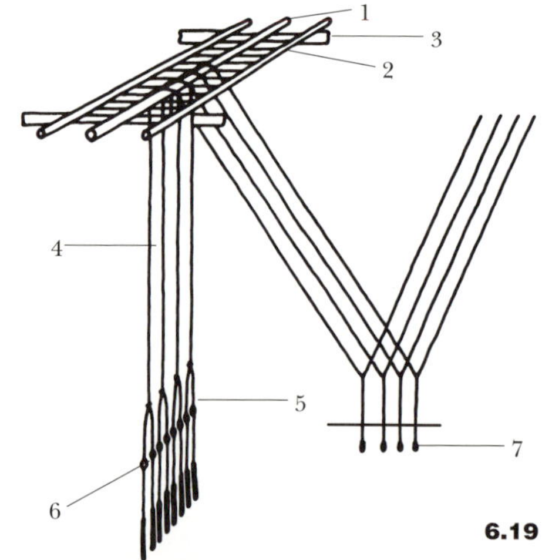

Aḍai loom. Supplementary warp harness passing through reed to maintain order in sequence. Courtesy: Kalakshetra Craft Foundation, Madras. (Fig. 6.20)

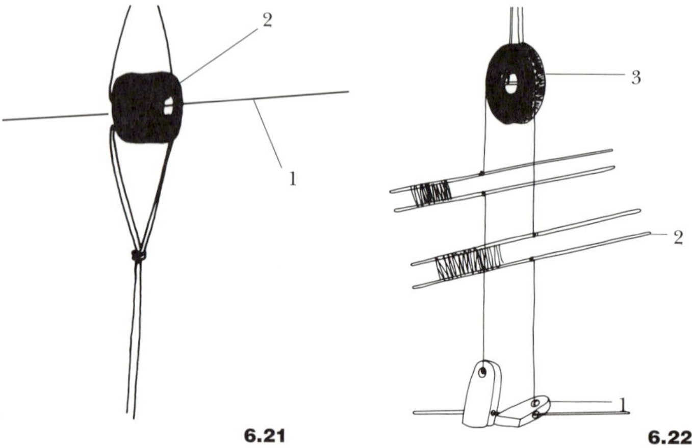

Eye of glass bead through which *zarī* extra warp ends are drafted. (Fig. 6.21)

1. *Zarī*/silk extra warp patterning end
2. Glass bead used as heald

Illustration to demonstrate the tie up of shafts for the ground weave. (Fig. 6.22)

1. Treadle
2. Headle shaft
3. Pulley

Mounting of Harness on the Loom

The harness is next mounted on the loom. Each pair of harness twine lengths is now divided into two, and each individual harness is passed through the dent of a reed placed on either side/overhead above the loom. The reed helps maintain the order, avoids crowding and provides space between each harness for efficient operation (figs. 6.5, 6.19). This is sometimes further stabilized by drawing these harness twines through a comberboard placed on either side of the loom and almost at the level of the *jungu* behind the loom (fig. 6.20). Each harness twine is attached to two varnish heddles with weights and glass beads, the holes of the beads (traditional) being used as mail eyes. It is through these eyes that the extra warp *zarī* and silk ends are drawn (figs. 6.21, 1.12).

The ends for the ground plain weave are drafted (fig. 6.22) through four shafts, with cotton heddles operated by treadles, whereas the *jungu* now placed in front of the weaver is operated sequentially to lift the extra warp ends. A leasing cord helps maintain the sequence of the *jungus*.

Aḍai for Crossborder

The term 'crossborder'/*mundi* is used to refer to the extra weft ornamentation done across the width of the sari to mark the end of its length. In Kanchipuram, for the end piece of the silk sari which is essentially and traditionally the most striking and elaborately ornamented part of this unstitched garment, the system of *aḍai* for crossborder weaving is traditionally employed.

This system is the same in principle as the one devised for the extra warp patterning. After the completion of loop making for one horizontal line on the graph paper, all these loops are brought together and a knot is tied from the top in such a way that there is a loop after the knot. In the weavers terminology, the device of loops which assists the weaver or the person who operates it for maintaining the lifting order of warp ends for extra weft patterning is also called *kuchu.* For the purpose of this document, the term *kuchu* has been retained in order to make a clear distinction between the loops that are used for the patterning with extra warp and extra weft. To maintain the order of each *kuchu* a strong thread or cord is tied to one end of the frame, functioning as a lease. The *aḍai* maker passes the free end of this thread, threading it in each loop of the *kuchu*, one by one in the order of succession. The second end of the thread is then tied on to the frame. If this is not done, the *kuchus* can be bunched together on the thread and secured. This order facilitates the work of the person who will operate these from the side of the loom, behind the shafts. This device is lifted and fixed behind the shafts and the *kuchu* bunches are moved to one end of the frame.

For the warp ends to be lifted specifically for extra weft patterning they need to be coupled with the crossborder twines running across the warp individually or in groups. For crossborder design, the entire body warp across the width has to be grouped for drawing through the cotton heddles locally known as *vadi*. This is done in accordance with the number of picks accommodated in the extra warp design portion (Ramaswamy, no date, 5). The number of warp ends that pass through each loop is related to the ratio between the number of ends and picks that was planned for the design. In order to maintain the proportions in the design, it is essential to work this out at the planning stage itself.

Once this is planned then, for example, if the design demands that three warp ends need to work as one unit, one lash/loop should be formed around every three warp ends and linked with the twine above the warp ends. Each twine on the *aḍai* frame can be said to work as a shaft. Hence a loop is formed around the first three warp ends and tied to the first twine, the four, five, and six warp ends tied to the second twine and this process is continued until the repeat in the design is completed. It is continued in the case of a normal straight drafting order across the warp, leaving only the borders (figs. 6.23, 6.24).

The *aḍai* for the *būṭās*/motifs is made in the same way as the crossborder *aḍai*. The difference occurs only in the placement of the loops. As the motifs are scattered over the main field of the sari, the loops need to be placed only in those particular areas. In the case of a design requiring crossborder *aḍai* and the *aḍai* for *būṭā* these are planned and made side by side on one frame and mounted on to the loom in the same way as for the crossborder. In Sri Kalahasti, Andhra Pradesh, the *aḍai* for *būṭā* patterns is placed behind the sley.

In Venkatagiri, Andhra Pradesh, the *aḍai* for the extra warp patterning is not attached through the reed overhead and down near the weaver as in Kanchipuram. The loops and *jungus* for extra warp are tied directly to the threads placed across the width of the loom. The *jungus* are held in place by a wooden rod (fig. 6.25).

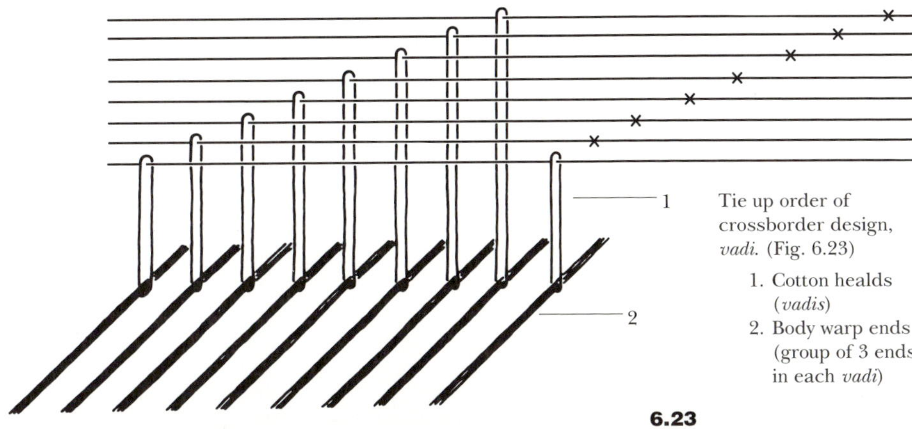

Tie up order of crossborder design, *vadi*. (Fig. 6.23)

1. Cotton healds (*vadis*)
2. Body warp ends (group of 3 ends in each *vadi*)

6.23

6.24

Working of *kuchu*. Workshop of Perneti Venkatesvarlu, Devanga, and Kalahasti. (Fig. 6.24)

6.25

Placement of *aḍai* at Venkatagiri. Fly shuttle used. Workshop of Vellur Raman Iyer, Venkatagiri. Note Influence of *Maheshwari* sari. (Fig. 6.25)

Aḍai loom. Darasuram. (Fig. 6.26)

6.27

Perns. (Fig. 6.27)

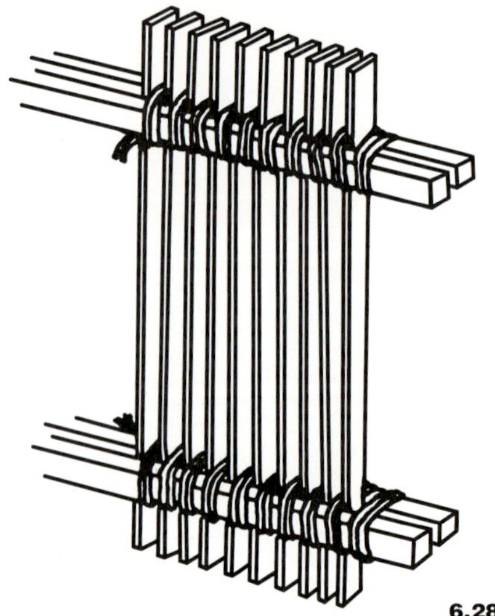

6.28

Reed. (Fig. 6.28)

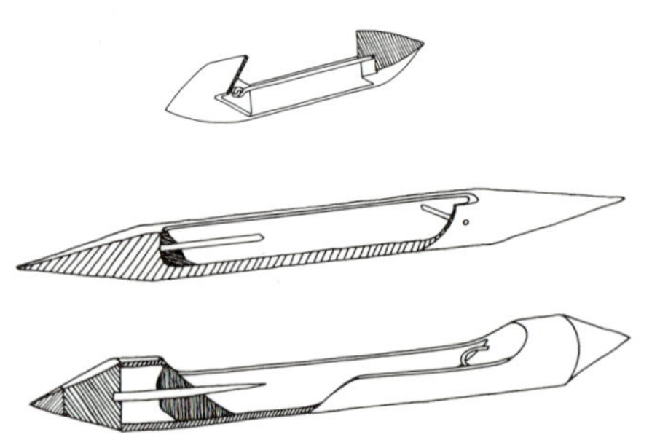

6.29

Wooden shuttle. (Fig. 6.29)

Weaving of a Sari with *Pettu Aḍai*

Solid coloured borders with extra warp patterning and solid coloured *pallu* with extra weft pattern form the distinctive visual feature of the saris from Kanchipuram. The meaning given to the term solid is that both warp and weft are of the same colour. These Kanchipuram items are extremely popular due to their quality of material, patterning and weaving. The texture of the fabric, compared to that used in other centres where silk saris are woven, is dense and thick. The warp is made of 2 ply 20–2 denier filature silk, weft of 2 ply, twisted 20–2 denier *charkhā* silk, while 3 ply pure gold/metallic yarn/*zarī*, is used for the extra warp patterning.

The weaving is begun only after confirming that all warp ends, the systems for manipulating them, the loom, loom accessories, weft yarn and all that is required for weaving are in order and in perfect working condition (figs. 6.26, 6.27, 6.28, 6.29).

The weaver first operates the treadle on his right to form a shed constituted by the lifting of the shafts for weaving of the plain weave. He simultaneously pulls the first *jungu* and hooks it on to the groove created by the strip provided on the top part of the sley (fig. 6.30). As the borders are required to be in a solid colour, two shuttles with the border weft colours are required. One shuttle with the weft in the required colour for the body is also needed. These are operated individually. When starting, the two border wefts are made to move in the same direction, whereas the body weft moves in the opposite direction and faces one of the border wefts (fig. 5.51).

One border weft interlocks with the body weft and meets at the other end with the other border for one pick. For the second pick, the two wefts facing each other interlock, the border weft moves out whereas the body weft moves in and meets with the border weft which has also moved in. This weaving with three shuttles is called *korvai* as noted earlier (see fig. 5.51). The *jungu* remains raised for two picks of ground weave. The second shed for the second pick is raised by pressing the treadle on the left. Beating in of the weft is done after every pick.

Aḍai. Note how *jungu* will be pulled. Workshop of K.J. Kamson, a Saurashtra weaver, Darasuram. (Fig. 6.30)

Insertion of the three *rakka palakas* between the selected twines. (Fig. 6.31)

Aḍai loom. Lifting of cross border *aḍai* to form a shed for extra weft. Courtesy: Kalakshetra Craft Education Centre, Madras. (Fig. 6.32)

Aḍai loom. Courtesy: Weavers Services Centre, Bangalore. (Fig. 6.33)

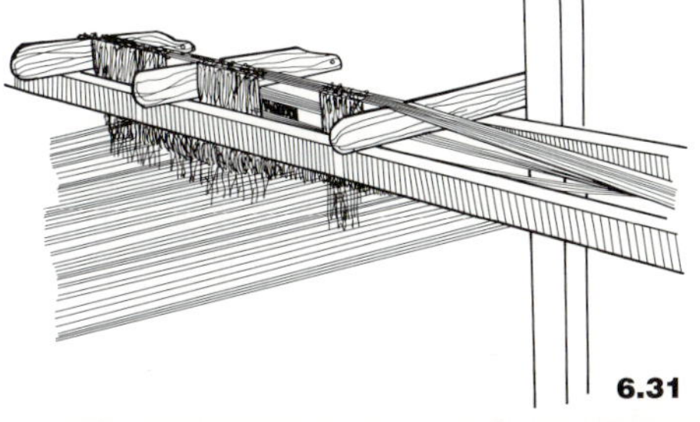
6.31

Weaving of Extra Weft by Operating the *Kuchu* for the Patterning to be done Across the *Pallu* and Motifs

To achieve this weaving of extra weft patterning across the *pallu* and motifs, another person who sits at the side of the loom and near the *kuchus* operates these loops, while the weaver continues to operate the *jungu* and inserts the weft. As the *kuchus* have been tied to maintain the order, the first *kuchu* is pulled by hand, separating the twines on the frame and bringing down the twines that are looped to the *kuchu*.

The weaver inserts three planks/ *rakka palakas* which are propped up on their sides between the space created by pulling the *kuchu*. These three *rakka palakas* are inserted at about three equal spaces across the width of the warp and in between the selected twines (fig. 6.31) thereby lifting only the selected warp ends (figs. 6.24, 6.32). Another plank called *palaka* about 4½ in. broad, 55 in. long and approximately ¾ in. thick is inserted into this shed. This is followed by the turning of all the three *rakka palakas* on to their flat sides. The *palaka* or shedding plank is now propped up on its edge and the extra weft *zarī* or silk thread is inserted and beaten in. The shedding plank now lies flat in the same shed. The next step in the operation of the *jungu* is to lift the extra warp ends and depress the treadle for forming the ground weave shed. The ground weft is inserted from right to left. At this stage, border colour and *pallu* colour are the same across the width of the warp, and all the ends are put back in their original position and the weft beaten in. At this point, the crossborder/*pallu* and the border are the same colour. Different colours in a solid plane in the crossborder are achieved through a process of interlacing a separate warp with the already existing ground warp. This is locally called *petni*.

6.32

The shedding plank, still in the same shed as for the first extra weft pick, is again turned on its edge and an extra weft thread (*zarī* or silk), is inserted from left to right. The shedding plank is removed and the extra weft pick beaten in. Each *jungu* as we know was earlier allowed to operate for two ground weave picks. The same *jungu* as operated is now pulled and the left treadle operated to make the second shed for the ground weave. With the shed formed the weft for the ground is inserted and beaten in. This completes the first step in weaving of the total design using the *kuchu* for extra weft patterning, the *jungu* for extra warp patterning and the shafts for achieving the ground weave to lock both the extra warp and the extra weft.

This process is repeated with the second *kuchu* and *jungu* and the ground weave shafts till the design repeat is completed. The sequence for weaving of the motifs/*būṭās*/*būṭīs* in extra weft is similar to that of crossborder weaving. However, the weaving of motifs and the crossborder cannot be done simultaneously as only the *kuchu* from one system can operate at a time for the extra weft patterning. Although theoretically it appears feasible to interchange units of the crossborder sequence with those of the ground motifs, in practice this is difficult due to the confusion that would be created in the simultaneous manipulation and lifting of warp ends.

6.33

The weaving of the motifs is done using several individual discontinuous wefts resulting in localized extra weft patterning. Small spools made of sticks wound with the patterning weft are used for weaving the motifs (fig. 6.33). The number of spools is dependent on the colour changes and the number of motifs.

Petni

The predominant usage is that the *pallu* colour is different and in contrast to the body warp. The border warp and *pallu* warp may be the same colour. To attain a different colour in the *pallu* warp, the weaver attaches another warp according to the size and colour required by drawing the *pallu* warp ends through the same heddles holding the body ends. He then wraps them individually around each body warp end and pulls the body warp ends towards him near the woven fabric (see fig. 5.52). This releases the tension of the warp enabling the weaver to draw the *pallu* warp ends along with the body warp ends. The *pallu* warp ends are tied to a rod. Now we have double the number of warp ends both in the heddle and the reed. The weaver continues to weave the fabric for approximately an inch, thereby firmly interlacing the *pallu* ends and integrating them with the fabric. The body warp ends are cut-off and the *pallu* weaving is continued. This process is called *petni* (fig. 6.4). A similar process is used in the weaving of the *pallu* in the *Paithani śālu* of Paithan.

Comparison between *Banārasī Jāla* and Crossborder *Aḍai*

In principle both these systems are based on the preplanned tying of lashes and leases according to the required design off the loom, as a part of the pre-weaving process. These are then transferred to the loom to facilitate the manipulation of warp threads to achieve the extra weft patterning while weaving the fabrics.

In Varanasi, fabrics are patterned by using the method of selvedge to selvedge extra-weft patterning with or without discontinuous supplementary wefts. Discontinuous supplementary wefts may also be used to weave individual units of the pattern area, in which several such units are woven to form the overall patterning on the fabric. Compound weaves are also used for the *kimkhāb* variety of fabrics as is evident in the *khoechen* category. Banaras fabrics are known for their intricate and elaborate patterning using mainly gold, silver *zarī* and coloured silk. These fabrics are well known as brocades of Varanasi. In these brocades, the patterning wefts on many occasions are woven in different weave structures while combining with the ground weave structure of the fabric. From a graph of the design, the *naksha bandh* translates each detail according to the design directly in the *naksha* threads, as a weft would be inserted (Perheentupa, NID, 1982: 1).

Plain weave is predominantly used in Kanchipuram fabrics as ground weave. The *pallu* of the sari is patterned with the use of extra weft with plain weave as the ground weave. The process of making the *aḍai* is much simpler as there are not as many complex and compound weave structures used in Kanchipuram, as is the case in Banaras brocades. While the weaving is done with one patterning weft that binds with the ground weave, in Varanasi, the usage of many picks to constitute one weft makes the process of *naksha*-making more laborious and time-consuming.

In making the *aḍai*, the *pettu aḍai* maker reads each row of picks on the graphs in sequential order. He selects the blank squares on the graph and ties individual loops around each cross thread corresponding to the blank square on the graph. As per the design, all the loops that represent one pick are bunched together and tied to a *kuchu*. For every successive pick, this process is carried out. A lease thread is passed through the *kuchus* to maintain their order. These cross threads with loops are mounted on the loom and the cross threads linked with the warp ends.

The *paggiā* threads—horizontal threads placed across the *jāla* loom—serve the same function as the *aḍai* cross threads placed across the warp on the Kanchipuram loom. Both these are connected to the warp by loops. The number of cross threads (one *aḍai* frame placed on the Kanchipuram loom) and the *paggiā* threads (on the Banaras loom with *jāla*) are linked with the warp ends with loops to correspond with the lifting order of the warp ends for each pick in a design repeat. In both cases, the cross threads and the *paggiā* can be considered to be operating as additional shafts specifically used for the patterning.

The *paggiā* threads are linked to the *jāla* which is placed behind the shafts and is hung vertically. In case the design is not complex the weaver can operate the *jāla* and weave the fabric on his own. However, in case of the design requiring more complicated lifting orders or if more than one *jāla* is required, a drawboy is essential for operating the draw system. In the same way the weaver may operate the *aḍai* mechanism for the extra warp border and simple *būṭī* ground design, but for a complex crossborder design the *kuchus* are operated sequentially by a helper standing at the side of the loom.

Usage of the *jungu* and *aḍai* has obviated the introduction of the barrel dobby. It is obvious that, with regard to design, there was a lesser infusion of Islamic elements than was the case in the repertoire of the *Banārasī jāla*. Patterning was developed for the adornment of unstitched garments with an orientation towards check patterning in tabby weave and use of extra warp for border and extra weft for creation of texture and ornamentation in the body and *pallu* of the sari. The traditional range of motifs in Kanchipuram silk saris bear a remarkable similarity to those outlined in the *Mānasollāsa* which are based on white and multicoloured straight lines and *chakra rekhā*/circular motif similar perhaps to the *rudrākṣa*. These lines could be fine or the width of two fingers, and, 'lines and dots may be densely or sparsely distributed' (Shrigondekar, 1939: 89). Patterned cottons, the *kalamkari*, for which there was considerable external demand, could have better served the needs of stitched garments than such silken ware. However, no ornamented cotton could compete with the opulent lure of the latter in a courtly setting.

The Paithan Range

The silk tradition of Paithan followed a different course from that of Tamil Nadu and appears to have been less open to Saurashtra permeation. The loom products known as *Paithani* were closely linked with the mores and values of the Maratha court brought into being by Shivaji, and carefully fostered by the Peshwas. Shivaji's dates are from AD 1627 to 1686. The relevant dates for the Peshwas are AD 1713, when Balaji Vishwanath was appointed Peshwa, while the point of termination, for the purpose of this subject, was AD 1802, when Peshwa Baji Rao II signed the Treaty of Bassein with the English. However, even though the Maratha power went into decline, such had been the nurturing of *Paithani* by the Peshwas that the tradition endured. Ultimately the death-knell was to be rung, not by a political authority but by the revolution in fibre technique after the 1930s. This was the era of velvets, lace, chiffons and georgettes which even the belles moulded by the Brahmo Samaj movement could not resist. There was no place for the *Paithani* repertoire in this new orientation in taste.

The group of weavers who seem to have played a role of some importance in the earlier period were the Salvis, credited with the introduction of the *ikat* tradition into Gujarat. Since the Salvis of Gujarat were Jains[11] (fig. 6.34), whereas Saurashtras claim Brahmin origin, it would be difficult to establish any connection between the two groups. If the link between the Salvis of Patan and the Paithan culture area were accepted it would mean that Salvis were not only weavers but also silk dyers who specialized in *ikat* as well as plain dyeing of silk. They were not associated with the practice of *bāndhanī*. This would explain the presence of the *ikat* technique in the *Paithani* repertoire as shown in a very fine example available in the collection of the Prince of Wales Museum, Bombay (figs. 6.35, 6.36).

Patron deity of Salvi Community, Nemeshwar Temple, Patan. Image is dated to the Kumarpala period, AD 1143–72. (Fig. 6.34)

Indigo black cotton, *ikat*, in stripes with *zarī* extra weft in *pallu*. Woven at Yeola or Paithan. Courtesy: Trustees, Prince of Wales Museum, Mumbai. (Fig. 6.35)

Paithani. Yeola. Cotton *zarī jāmdānī* weave in body. Double interlock weave in *pallu*. Courtesy: Trustees, Prince of Wales Museum, Mumbai. (Fig. 6.36)

Early History of the Paithan Bhakti Movement and the Moulding of Maratha Ethos

It would be difficult to dissociate the development of the *Paithani* repertoire in textiles from the broader forces which were shaping Maharashtrian society. The importance of Pratiṣṭhāna, later known as Paithan, goes back to the Sātavāhana period, *c.* second century BC to third century AD (Deglurkar, 1971: 15). Pratiṣṭhāna was a stage in an important trade route leading from the lower Krishna valley to Bharuch on the west coast. Somewhere along this route a bifurcation occurred with one branch leading eastwards to Pāṭaliputra, modern Patna, while the second proceeded northwards via Mathurā to Taxilā, Gandhāra, Balkh and onwards to the Central Asian routes. It was also linked with important settlements such as Tagara (Ter), Junnār, Karahkaṭa (Karad, district Satara), Nasik, Govardhana, Vaijayantī (Banavasi, North Kanara), and Brahmapur (Kolhapur).[12] This accessibility to urban centres and trade routes must have provided considerable incentive to existing crafts in Paithan.

After the downfall of the Sātavāhanas, Paithan lost its political importance. It re-emerged as a religious centre in the medieval period. The decline of the Sātavāhanas coincided with recognizable religious and social transformations in the period *c.* fourth century AD to the year 1000. Important ruling dynasties during this period included the Vākāṭakas, early fourth to fifth centuries AD, and the Rāṣṭrakūtas, AD 753–973. This is the time when the older Purāṇas were revised. Vedic rituals involving animal sacrifice were replaced by the *Pañcamahāyajñas* advocated by the *Smṛtis.* The latter comprised *brahmayajña, devayajña, pitṛyajña, manuṣyayajña, bhūtayajña.* These developments were pan-Indian but cultural manifestations differed in diverse parts of the subcontinent.

Maharashtra played a special role. It represented the southernmost limit of languages based on the Devanāgarī script but at the same time it evolved its own particular identity. On the one hand there was communal participation in *yātrās* or religious pilgrimage to *tīrthas,* sacred places. There was also the domestic ritual revolving around performance of *vratas,* the acquiring of desirable objectives through performance of austerities. Sectarian beliefs, religious cults and Purāṇic ritual all played a role in processes of socialization. The Viṭṭhala Temple at Pandharpur, which had become an important place of pilgrimage by AD 1250, narrowed down differences between Śaivites and Vaiṣṇavites bringing both within the confines of a common cult. This is a feature very special to Maharashtra and the enduring popularity of pilgrimages made to this temple prevented the outbreak of any schismatic rift between Śaivites and Vaiṣṇavites.

By the tenth century Pratiṣṭhāna was referred to as Paithan in Apabhraṃśa and it is from this term that the modern name Paithan is derived. Terms such as Mung Pratiṣṭhāna and Brahmpurī Pratiṣṭhāna were also used with reference to Paithan.[13] The *Brahma, Liṅga, Kūrma* and *Padma Purāṇas* referred to Pratiṣṭhāna as a significant place of pilgrimage, and Paithan assumed the position of Dakṣiṇa Kāśī, the southern Varanasi. The Purāṇic rites and rituals thus engendered led to a whole cycle of cultural activities centred around temples. Among the items generated by these activities were those of textiles.

A large number of Yādava records refer to donations made to temples providing for *aṅgabhoga,* bodily enjoyment, and *raṅgabhoga,* theatrical diversions, for the deities. Bodily enjoyment appeared to be connected with offers of food, oil, cloth, flowers, sandal and betelnuts (Morwanchikar, 1978: 112). Since there were as many as one hundred temples at Paithan under the Yādavas, there was considerable demand for cloth used in ritual and by the mid-twelfth century there was a specific association between Paithan and silk weaving. Brahmins, moreover, recognized the suitability of *Paithani* fabrics as items in *dāna* (Morwanchikar, 1978: 122, 273; Gopal, 1964–5: 102). This may be contrasted to Bengal usage where *garad* was acceptable for ritual purposes but the same did not hold for *Bāluchari* silk.[14]

Religious activity in Maharashtra was strengthened by the literary climate, and the development of Marathi as a script spilled over into the Konkan and Kanara regions. The earliest Marathi inscription, dated AD 1012, is found at the Akṣī Mahālakṣmī Temple on the Konkan coastal strip, while the AD 1117 inscription at Śravaṇa Beḷagoḷa marks its intrusion into a Kannada-speaking area.

The increasing emphasis on ritual as demonstrated in the *Caturvargacintāmaṇī* of Hemādri, written some time between AD 1260 and 1309, caused a reaction which found expression in the Bhakti movement. Hemādri, much venerated for his learning, listed two thousand rituals which had to be performed annually. The Bhakti movement was itself influenced by another sect exemplified in the Mahānubhāva cult. This cult had been founded by Cakradhara. In about AD 1278, Govinda Prabhu, who had been initiated into the movement by Cakradhara *c.* AD 1185, began to play an important role in the shaping of the Bhakti movement in Maharashtra. The Bhakti movement, apart from its religious and social role, was important for the impetus provided for the development of Marathi into a literary language (Tulpule, 1960: 173–6; 1979, 311–17, 325, 329, 338; Verma, 1970: 265–8).

Associated with the Bhakti movement was the popularization of its ideals with the help of the performing arts leading to the *kīrtana sampradāya* (Verma, 1970: 268–9). V.K. Rajwade, a historian who is not overly sympathetic to the Bhakti movement, nevertheless gives full credit to the Bhakti poets for preservation of the Marathi language. Eknath provides many references in his poetry to actors, performances and the relationship between the actor and his role. This would suggest that apart from *kīrtana sampradāya,* the

Bhakti movement contributed to all forms of the performing arts of the time (oral communication: G.P. Deshpande, New Delhi). The fact that there was a close relationship between the performing arts and the textile crafts needs no emphasis. While the Islamic intrusion led to the extinction of the Yādava Kingdom with the introduction of Persian as the language of administration, the vocabulary of Sufism could be well understood by those stirred by the Bhakti movement.[15] A new element had thus been absorbed into the cultural vocabulary.

It is against this background that the court culture cultivated by Shivaji is to be viewed (Kulkarni, 1969: 168–85). Unlike the Rajasthani courts which had adopted features of the Mughal sartorial mode, the Maratha elite patronized existing local cultural traits, and in the area of textiles, Shivaji is stated to have possessed a large number of expensive fabrics of Mungipatan or Paithan.[16]

Māṇikyasundarasuri *c.* AD 1422: provides a graphic description of Paithan, referred to as Pahīthāṇapur Paṭana:

> Among those countries, Mahārāṭhadeśa is praiseworthy. In that country Pahīthāṇapur *Paṭana* is situated. There is a high temple of Sarvajñadeva (a Jain *tīrthaṅkara*), which can compete with Kailāsa mountain in height. Many wonders exist in this town. Eighty-four shopping centres are throbbing with sound and bustle. Among these are listed cotton cloth merchants, *paḍa*, silk cloth merchants artisans knowing the art of weaving *paṭolan* one who dyes in indigo persons removing seeds from cotton ... narrow strip weavers possibly *pagaḍī* weavers, one who twists three threads into one *tragaḍīya*, experts in dyeing cloth or articles in deep red colour, *majiṭhā mañjiṣṭhā/madder, vankara* weavers there were a number of *Sadāvrata* institutions free meals were provided by donors who had taken vows where a number of people took their meals; there was a public water reservoir where all people drank water, there were step wells where to bathe and enjoy were natural; there was a big lake full of lotuses; there were gardens where cool breezes blew; such was the world renowned Pahīthāṇapur, where injustice had been stamped down to the lowest recesses. (Dalal, 1924: 94, 95)

The importance of Paithan as a centre of textile production is obvious but a significant omission is that relating to the existence of any manufacture or sale of *zarī* unless the *tragaḍīya*, twisters of three threads into one, can be taken in the sense of makers of *zarī* (see Nissim, 1909: 6, 7). The mention of *zarī* arises because if an assumption is to be made regarding the weaving of *Paithani* it is essential to find some ancillary reference to *zarī* in the above description of Paithan. Although *zarī* may be better adapted for ornamentation in silk, in India it has been extensively used in loom-derived patterning in cotton.

The repertoire of *Paithani* weaving, indicative of an early tradition in loom ornamentation, is related to the kinds of costume worn in Maharashtra. A notable feature which moulded the development of an array in unstitched costume was the belief that silk had to be worn when performing functions and duties such as worship of household gods, partaking of food, celebration of marriage and undertaking rituals such as *upanayana* (cf. fig. 6.37). For the poor, a cotton *cholī* with silk border or a sari or *dhotī* with silk border was acceptable (Edwardes, 1976: 57). In the case of Maharashtra, the development of the *khan* and *karvati kathi* traditions may be viewed against this background of the acceptance of a partial presence of silk in costume during performance of ritual activities.

Women of all castes wore the sari in the *kachchhā* form. Brahmin women showed one border, *paṭṭā*, at the back. Chandraseniya Kayastha Prabhus, and Pathara Prabhus showed two *paṭṭā* at the rear. However, Maratha women could exercise the option of draping the same length of the garment either in *kachchhā* form (figs. 5.26, 5.27, 6.38), or by having additional *nirya*, pleats, in the front. Worn in the latter manner it was called *gol nesana* (fig. 6.39). This form resembles the modern way of wearing the sari. The existing form of draping a 6-yds. sari came into increasing evidence in the nineteenth century and is believed to have been popularized by members of the Tagore family in Bengal (cf. fig. 6.40).

6.37

6.38 6.39

6.40

Upanayana ceremony of Raja Jayachamaraja. Painted in 1980 from a photograph dated 1934. Courtesy: Palace Museum, Mysore. Note costumes. (Fig. 6.37)

Kumud Damle, Bombay, wearing a 9-yard sari in *kachchhā* style. All Hindu communities in Maharashtra can wear their sari in this style. Courtesy: Dr Bhau Daji Lad Museum, Mumbai. (Fig. 6.38)

Nine-yard sari worn in non-*kachchhā* style by women of the Maratha community in Maharashtra. Courtesy: Dr Bhau Daji Lad Museum, Mumbai. Note similarity with contemporary style of draping sari. (Fig. 6.39)

Frieze. Note drape in sari. Terracotta temple, nineteenth century. Maukhira, Birbhum-Burdwan border, West Bengal. Courtesy: Satyasri Ukil, Delhi. (Fig. 6.40)

According to D.G. Godse, Bombay, Maharashtrian costume of an earlier period is best expressed in the garments delineated in the Ajantā mural paintings. These did not reach right down to the feet. Godse associated the wearing of the 9-yds. sari in the *kachchhā* form reaching down to the feet with the Brahmin group or communities with pretensions to Brahmanical status. M.S. Mate associates this form of costume with women of high lineage of the Yādava period (oral communication: M.S. Mate, Deccan College, Poona). Peasants, to this day, wear the *kachchhā* form with the garment being draped well above the calf.

Among the reasons advanced for the wide prevalence of the *kachchhā* form in Maharashtra, one suggestion is that the hilly terrain, combined with the local availability of horses provided the possibility for women to ride on horseback. To do so, draping the sari in *kachchhā* form was essential.[17] This does not seem reasonable, as it appears highly unlikely that all females were able to ride on horseback. It is possible that among a variety of costumes, as depicted in the Ajantā paintings, for some reason, the *kachchhā* became the most widespread in Maharashtra. It is possible that the women of high lineage of the Yādava period, to whom Mate makes a reference, had set the pace. This tendency of the elite to establish norms is very clearly demonstrated during the Peshwa period. Moreover, the dissolution of the Bahmani Sultanate based at Gulbarga in the first half of the sixteenth century created conditions of widespread instability with several villages becoming deserted. The migratory movements which now ensued over uneven terrain may also have encouraged the wearing of the *kachchhā* form (oral communication: G.P. Deshpande, New Delhi).

Under the Peshwas, while a great deal of patronage was extended to *Paithani* ware, there was also demand for running lengths of material which could be made into stitched garments befitting sartorial standards set by the Islamicized elite (fig. 6.41). Even in the case of women, Chapekar tells us that some wore *chūṅdārs* and *aṅgarakhā* (Chapekar, 1937: 199, 201). Further, *c.* 1813, it would appear that a certain category of dancing girls, certainly not those connected with temples, wore satin trousers and an upper garment in muslin the full skirt of which was called *peswāz*. The ensemble was completed by the addition of a *selā*, so much a part of the *Paithani* repertoire, but in this case probably of a diaphanous variety (Broughton, 1813: 55).

Painting of Shivaji by Khatolkar. Aundh Palace, private collection. Courtesy: Raje Sahib Rajwada, Aundh. Note the *selā* around waist of Shivaji and the Holkar (Indore) and the Scindia (Gwalior) style of wearing turbans. (Fig. 6.41)

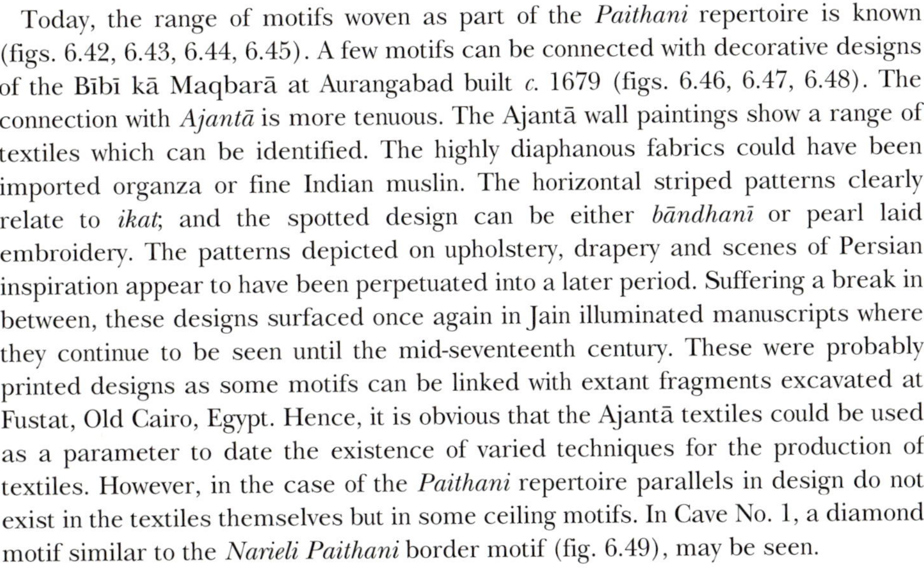

Today, the range of motifs woven as part of the *Paithani* repertoire is known (figs. 6.42, 6.43, 6.44, 6.45). A few motifs can be connected with decorative designs of the Bībī kā Maqbarā at Aurangabad built *c.* 1679 (figs. 6.46, 6.47, 6.48). The connection with *Ajantā* is more tenuous. The Ajantā wall paintings show a range of textiles which can be identified. The highly diaphanous fabrics could have been imported organza or fine Indian muslin. The horizontal striped patterns clearly relate to *ikat*; and the spotted design can be either *bāndhanī* or pearl laid embroidery. The patterns depicted on upholstery, drapery and scenes of Persian inspiration appear to have been perpetuated into a later period. Suffering a break in between, these designs surfaced once again in Jain illuminated manuscripts where they continue to be seen until the mid-seventeenth century. These were probably printed designs as some motifs can be linked with extant fragments excavated at Fustat, Old Cairo, Egypt. Hence, it is obvious that the Ajantā textiles could be used as a parameter to date the existence of varied techniques for the production of textiles. However, in the case of the *Paithani* repertoire parallels in design do not exist in the textiles themselves but in some ceiling motifs. In Cave No. 1, a diamond motif similar to the *Narieli Paithani* border motif (fig. 6.49), may be seen.

Paithani sari, Paithan. Detail. Section of main field of body and *pallu*. Courtesy: Calico Museum of Textiles, Ahmedabad. Art work: Rajendra S. Patel, NID. (Fig. 6.42)

Double interlock, possibly *Chanderī*. *Pallu*. Detail. Courtesy: Calico Museum of Textiles, Ahmedabad. Art work: Rajendra S. Patel, NID. (Fig. 6.43)

Border. *Dupatta*. Reverse face. Courtesy: National Museum, New Delhi. (Fig. 6.44)

Paithani. Mid-nineteenth century. Courtesy: National Museum, New Delhi. (Fig. 6.45)

Motif, Bībī kā Maqbarā, *c.* AD 1660, Aurangabad. This motif bears some similarity to the *Paithani* repertoire. (Fig. 6.46)

Inlay work. Bībī kā Maqbarā, *c.* AD 1660. Note similarity with *Paithani* repertoire of motifs used in border. (Fig. 6.47)

Inlay work. Bībī kā Maqbarā, *c.* AD 1660, Aurangabad. This has a similarity with *Paithani* repertoire of border motifs. (Fig. 6.48)

Paithani sari. *Narieli* border. Courtesy: Museum, Marathwada University, Aurangabad. (Fig. 6.49)

Different dates have been advanced for the Ajantā paintings and the issue is still under discussion. Professor T.V. Pathy, Marathwada University, Aurangabad, has advocated the seventh century as a terminal date. Hieun Tsang, *c.* AD 642, made a reference to Ajantā and it is known that Buddhism was practised as a living faith in Maharashtra until the eleventh century. By the thirteenth century it had disappeared. The Ajantā caves appear to have been lost to view until 1819. Between 1849 and 1855 the British East India Company commissioned Major R. Gill to make copies of the wall paintings (*Khandesh*, 1880: 480, 481). This is also the period of high patronage of *Paithani* ware. It is possible that motifs drawn from the Ajantā ceilings were integrated into the *Paithani* repertoire of design after the rediscovery of the Ajantā cave paintings.

Maharashtra Costume

The *cholī* or blouse worn by women in Maharashtra has traditionally been made of a material called *khan*. It is interesting that it was only in the category of *khan* that a specific weaving is to be identified for usage as a female blouse/*cholī*. This demonstrates another basic aspect of understanding of the aesthetics of clothing. In the manufacture of *khan* we see an articulation of the perceived need to balance the main garment, the sari, with its coordinate, the *cholī* in female wear. *Khan* was woven at Thana, Maharashtra. Ilkal and Guledgudda in Karnataka formed pivots of a thriving *khan* producing centre. Sari weaving was introduced here only in *c.* 1910 (fig. 6.50). Thana had a tradition of weaving in pure silk and J. Nissim notes in 1909 that *khan*, a silk cloth, was sold in lengths of 1¼ cubits. Approximately 22½ in. was used for making female bodices. This material was woven in lengths of 10 cubits, called *khanala*, from which eight *khans* could be cut (Nissim, 1909: 12). In Ilkal, the warp is silk, the weft being midnight blue indigo dyed cotton. Extra warp design is worked by means of a barrel dobby[18] (fig. 6.50). In Ilkal, three shuttle weave is restricted to the *tope* in the *pallu* in the case of sari weaving.

The male dress comprised a *dhotar*, lower garment, and an *uparna*, shoulder cloth (fig. 6.51). The *dhotar* was usually sold in units of two, the *dhotarjode*. The *dhotar* would be made of cotton with a cotton or silk border. It was usually 50½ in. in width and 5 yds. in length. The word *saura* was used to refer to a *dhotar* woven in mulberry or *tasar* silk with a silken border. The same male garment woven wholly in silk but having a *zarī* border was called *pītāmbara*. Originally yellow in colour (fig. 6.52), by the early twentieth century a range of other colours such as white, green, red or purple began to be used. In Maharashtra women also wore *pītāmbara*. Male *pītāmbara*, called *mardānī pītāmbara*, was 10 cubits or about 5½ yds. (figs. 6.37, 6.53, 6.54), while female *pītāmbara*, *zenānī pītāmbara*, was 15 to 16 cubits or 7½ to 8 yds. in length (Nissim, 1909, 12, fig. 6.55). *Pītāmbaras* for men have narrow borders. Those for women are wider (Watt, 1903: 310).

6.50

6.51

6.52

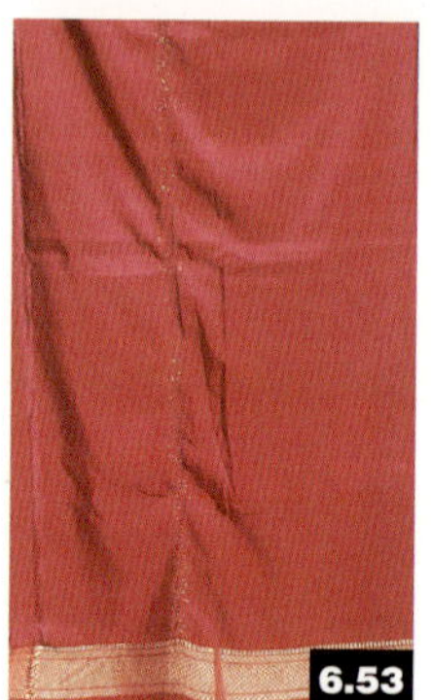

6.53

6.54

6.55

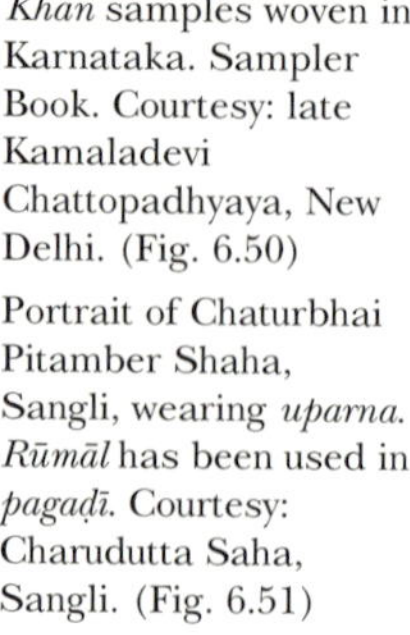

Khan samples woven in Karnataka. Sampler Book. Courtesy: late Kamaladevi Chattopadhyaya, New Delhi. (Fig. 6.50)

Portrait of Chaturbhai Pitamber Shaha, Sangli, wearing *uparna*. *Rūmāl* has been used in *pagaḍī*. Courtesy: Charudutta Saha, Sangli. (Fig. 6.51)

Illuminated manuscript, *Bhāgvata Purāṇa*, Samvat 1705/ AD 1648. Courtesy: Bhandarkar Oriental Institute, Pune, Note yellow *pītāmbara* worn by Kṛṣṇa. (Fig. 6.52)

Pītāmbara. Woven at Yeola. Courtesy: S Chitnavis, Nagpur. (Fig. 6.53)

Mardānī pītāmbara. Courtesy: Lalita Gouri Pant Pratinidhi, Aundh. Note how zig-zag motif in border is transmuted into *Narieli*. (Fig. 6.54)

Zenānī pītāmbara. Courtesy: Bharat Kala Bhavan, Varanasi. (Fig. 6.55)

6.56

Dupaṭṭā. Courtesy: Deepak Shah, Pune. Note *akroti* border, *jāmdānī* woven *būṭīs* in ground. (Fig. 6.56)

6.57

Painting. Tanjore, eighteenth century. Courtesy: Palace Museum, Laxmi Vilas Palace, Vadodara. Note *Paithani paṭkā*. Maratha style costume. (Fig. 6.57)

Black *pheṭa*, *Paithani*. Courtesy: Mrinalini Puar, Laxmi Vilas Palace, Vadodara. (Fig. 6.58)

Possibly used for turban. *Paithani akroti* end piece. Courtesy: Deepak Shah, Pune. (Fig. 6.59)

The shoulder cloth belonged to the *uparna* or *aṅgavastra* category. The texture of the *dupaṭṭā* (fig. 6.44) could extend to form a lighter version of the *uparna* (fig. 6.51) which could be draped around the arm by the nobility. It could have borders and two crossborders (fig. 6.56). *Selās* like *dupaṭṭās* were expensive shoulder cloths. There were three types: *nijatana* or *khesa selā* which functioned as a shoulder cloth; *ubekarita eravi*, which served the dual purpose of a shoulder cloth as well as a *chādar* or sheet, and the *sameta selā*, which, placed on the shoulder when going to court, was a sign of status. However, richly elaborate *selās* could only be worn by royalty. Whereas *selās* worn at the time of Shivaji could scarcely be considered luxury items, they became progressively more so under the Peshwas (Bhave, 1976: 263, 264). A *selā* could also be worn around the waist so that a sword could be tucked into it (fig. 6.41). The *paṭkā* or waist cloth, being worn over a stitched male garment, was an item which, when first introduced, must have been considered somewhat exotic in the *Paithani* repertoire although it resembled the *selā* worn around the waist (fig. 6.57). It needs to be stressed, however, that all these terms are used in a generic sense. They referred only to a particular garment by name and such garments could be manufactured in a number of ways over a large geographical area. Only a detailed description of texture, technique and materials used could help to define its specific centre of production.

Since etiquette demanded that all Hindu males should cover their heads when outside the home, a range of head coverings had been created. Terms for headdresses included *mandīla*, *rūmāl*, *pheṭa* (fig. 6.58), *tivaṭa*, *pāgoten*, *pagaḍī* (fig. 6.59), and *topī* (Chapekar, 1937: 58). *Mandīla* and *rūmāl* can both be understood as gold-edged handkerchiefs. According to Deshpande, the *mandīla* was, apparently, worn by Marathas while Brahmins, *sonārs*/goldsmiths, and *shimpis*/tailors and cloth merchants, preferred the *rūmāla*. When tied on the head it would have two projecting points (fig. 6.51), one in front and the other at the back. The *feṭa* or *pheṭa* is worn a little tilted on the head, one side being higher than the other which covers the

6.58

6.59

ear (Biswas, 1985: 62). *Tivaṭa, pāgoten, pagaḍī* all referred to the turban and could range in length from 22½ to 30 yds. (fig. 6.60). Brahmins wore the *pagaḍī* while non-Brahmins were marked by the *shinde shāhī pagaḍī*. The Marathi term *tivata* was progressively replaced by the Persian, *pāgoten* (Chapekar, 1937: 58; Desai, 1980: 156, 159). Popular colours were copper red and purple and Paithan was one of the centres of production during the period of the Peshwas. Later, Yeola also became an important centre (figs. 5.15, 6.61). In present times, Muslim weavers, Momins, appear to have been linked with the weaving of *pagaḍī* cloth, while Hindu weavers have undertaken the production of female wear. However, if a silk *pagaḍī* with *zarī* were required the sari weavers would prepare it.

Within this range of wear a special category, opulent in appeal and generally associated with Brahmins, also came to be identified as *Paithani*. This category appears to have become a symbol of status even among the local Islamic elite. Muslim brides were expected to include a *Paithani oḍhnī* among their bridal attire. The term referred to an unstitched garment in which there was a heavy gold border running along each selvedge, with equally lavish crossborders. The design could be achieved by varying the structure of the portions woven in metallic yarn or by introducing coloured silk yarn in double interlock (figs. 6.62, 6.63). The ground material could be cotton or silk. It is very difficult to date the introduction of this category of *Paithani*. It probably evolved over a period of time. Loom requisites are very simple and perhaps its development matured dramatically during the period between that of Shivaji and the first Peshwas. *Paithani* saris, unlike *śālus* were originally always woven in lengths of nine yards.

According to V.B. Kolte, Nagpur, there is no reference to *Paithani* or *śālu* in *Mahānubhāva* literature dated between thirteenth to fifteenth centuries AD. There is, however, a reference to *Kṣīrodaka pītāmbara malganthi*, double colour *pītāmbara* with heavy gold border, in the *dhavala* or marriage songs composed by the poetess of the *Mahānubhāva* cult, Mahadamba, *c.* AD 1285 (figs. 6.101, 6.102, 6.103).

Portrait of Srinivas Rao, Maharaja Raja Rishi. Courtesy: Raje Sahib, Rajawada, Aundh. (Fig. 6.60)

New style, Vadodara *pagaḍī*, possibly woven at Yeola/Chanderi. Worn by Manishah or Siledar. Courtesy: Shahaji Chhatrapati Museum Trust, Kolhapur. (Fig. 6.61)

Paithani sari. Double interlock in *pallu*, and, partially in the body. Courtesy: Calico Museum of Textiles, Ahmedabad. Art work: Rajendra S. Patel, NID. (Fig. 6.62)

Border of *Paithani pallu*. Courtesy: National Museum, New Delhi. (Fig. 6.63)

Eknath, the sixteenth century poet saint of Paithan makes an oblique reference to the excellence of weaving prevalent in his time in the stanza, 'just as when a cloth, kept folded inside a coconut, owing to the very fine spun yarn, assumes great length when unfolded, so also my exposition of the original stanza has assumed such lengthy proportions' (Soman, 1973: Chap. XIX, 317). However, this appears to relate more to a tradition of cotton weaving rather than to that of silk. The earliest pictorial representation of the *Paithani* design repertoire is in the *paṭkās* depicted in an equestrian portrait of Shahu and in another painting entitled 'Shahu with his Prime Minister'. These were painted during the reign of Shahu, AD 1708–49, and are to be found in the Marathwada University Museum, Aurangabad (Mate, 1982: 139).

By the mid-eighteenth century the repertoire had developed greatly. In correspondence between the Peshwa and Nana Fadnis dated 1765, 1766 and 1768, there are references to *dhotarjoḍe* of Paithan, *Paithani dhotarjoḍe* with *veli* design, and *Paithani dupaṭṭās* with *asavalli* and *narieli* designs (figs. 6.58, 6.49).[19] Under Shivaji, court life betrayed little ostentation. With the expansion of the Maratha empire under Peshwa Baji Rao (AD 1720–40) and Balaji Baji Rao (1740–51), there was far greater access to the luxuries of the north and south. The demand for textiles by the Maratha *sardārs* was not only reflected in their purchases of fine textiles from other parts of the subcontinent but also in the patronage they provided for the development of local textiles including those of Paithan. A small indication of the growing ostentation is that whereas under Shivaji marriages were simple and lasted for about 2 days, under the Peshwas they could extend up to 16 days.

Paithani male wear comprised *selās*, *dupaṭṭās* and *pītāmbara*. The specificity of the term *Paithani* lay in the existence of heavy *zarī* borders and crossborders with motifs done in the double interlock technique (figs. 6.62, 6.63, 6.45). Certain *dhotīs* such as the *karvati kathi* of Umred, Vidarbha, were included in this category as the three shuttle technique of the borders involved practice of the interlock. For females it included the 9-yds. *lūgaḍa*, *dupaṭṭās* and *selās*, the lines for specificity being the same as those pertaining to male wear. With regard to the *śālu* it should be remembered that it could be procured from a number of centres such as Varanasi and Kanchipuram and did not necessarily have to be 9-yds. in length (fig. 6.64). However, if it had a *Paithani* border and *pallu* and was 9-yds. in length it could qualify to be called *Paithani*.

It is therefore clear that a *Paithani* could be either a fine cotton or a silk unstitched garment. What gave it specificity was the character of the border and *pallu*. Because its development is linked with the ascendancy of the Peshwas, it began to assume a Brahmin connotation but it was certainly also patronized by the Maratha *sardārs*,[20] as well as by the Muslim elite. *Pītāmbara* would be worn by males after the bathing, for performance of *pūjā* and at meals. When officiating at public religious ceremonies, rulers would also wear *pītāmbara*. During the wedding ceremony it was customary for the bride's father to offer a *pītāmbara* to the groom, while the groom's family would offer a *lūgaḍā* and *śālu* to the bride. *Pītāmbara* would be worn by men and *Paithani* by their womenfolk on occasions such as marriages, *upanayana*/thread ceremony, Dussehra, Rām Navamī, Lakṣmī Pūjā, Diwalī and days of particular *kula dharma*, feast days of the family deity. Women would wear *Paithani* for the first celebration of Satya Nārāyaṇa Pūjā after marriage. In the month of April/Caitra during the Caitra Gaurī festival, the image of Gaurī would be placed on a brass swing behind which a decorative *śālu* would be hung. At court *darbārs*, while the *aṅgarakhā* and *chūṛīdār* could be of brocade, *Paithani selās* and *dupaṭṭās* would be worn. When gifted by the king, army officers could drape the *selā* around the waist.[21] *Selās* formed part of the costume of court officials like the *bhāldār* and *chopdār* (fig. 6.65), who announced the names of dignitaries participating in palace gatherings. To the discerning patron

Painting, *Nala-Damayanti* scene, by Raja Ravi Varma. Courtesy: Sri Jayachamendra Museum and Art Gallery, Jagmohan Palace, Mysore. Note texture of sari. (Fig. 6.64)

Portrait of H.H. Maharaja Khanderao Gaekwad (AD 1856–70). Detail of court official. Courtesy: Palace Museum, Laxmi Vilas Palace, Vadodara. Note usage of *paithani*. (Fig. 6.65)

the appeal of the shoulder cloth lay in its diaphanous character. While the fall of the lower garment was important it had, at the same time, to be made of a heavier material. Stitched garments also existed. These were worn by warriors and members of the service communities.

Among women, widows were forbidden to wear *paithani*. At the time of marriage women of Maratha ruling families would cover the face with the *dupaṭṭā*. On ceremonial occasions including processions, women would wear the *selā* to cover the exposed portion between the *choli* and the *lūgaḍa* at the waist. Usually the portion of the *selā*, which would cover the right arm, was not as heavily ornamented as the other end which could also be draped over the shoulder[22] (figs. 6.66, 6.67, 6.68). The preferred colours in *lūgaḍa* and *śālu* were red, pink and violet.

The importance of the textile industry in Paithan is reflected in some of the place names which exist even today within the town. These include Salivada, weavers' quarters, Taragalli, where wire drawing in the preparation of metallic yarn was done, Rangarahati, the place where dyers work, Pavategalli, inhabited by members of the Sonār caste who prepared metallic yarn and Mominvada where Momins, a Muslim sect, wove narrow turban lengths (Marawade, 1979: 18). Items made for Muslim patrons included *dupaṭṭās*, 5 yds. in length, 48 in. in width, used for ceremonial occasions, and *oḍhnīs* for everyday use, were 2 yds. in length and 30 in. in width. Since *Paithani* motifs tended to be more geometrical and floral rather than animal, catering to an Islamic clientele would pose no problem. Another item in which Momin weavers specialized was the range of detachable *pallus* which were then sewn on as desired (fig. 6.69).

Following a period of diffusion of *paithani* production at manifold centres,[23] *c.* 1884, there are indications of decline. Stage costume reflected the changing tastes of the time. The birth of Marathi theatre is associated with the staging of the mythological drama, *Sita Swayamvara*, by Vishnudas Bhave at Sangli. The popularity of mythological drama was reiterated in the performance of *Shakuntal* in 1880 by Annasaheb Kirloskar. K.P. Khadilkar showed how historical themes could be imbued with mythological appeal in plays such as *Sawai Madhavravacha Mrityu* (1906), *Manapamana* (1911) and *Svayamvara* (1916). This was also the period of the flowering of Bal Gandharva on the stage (*Marathi Theatre*, 1961: 2, 11, 24, 34, 37; Nadkarni, 1988: 43, 69). From 1898 to the 1930s *paithani* 9-yds. saris were worn in scenes drawn from Indian mythology. *Paithani selās* were used in productions of *Sanshayakalol*, first produced by Deval *c.* 1910, and continued in the repertory up to 1932.

Portrait. Mother of Saiyaji Rao III. Painted by Hansaji Raghunath, AD 1878. Courtesy: Palace Museum, Laxmi Vilas Palace, Vadodara. Note high *choli*. *Paithani* sari. *Choli* cloth possibly from Yeola or Thana. (Fig. 6.66)

Selā. Courtesy: Kusum Karmakar, Sangli. Front view. Two ends of *selā* have been gathered in folds as in sari. The more ornate end is draped over the right arm or over the left shoulder. (Fig. 6.67)

Selā. Courtesy: Kusum Karmakar, Sangli. Side view (Fig. 6.68)

Ground, *Banārasī* brocade, *paithani* border and *pallu* attached to ground. Courtesy: National Museum, New Delhi. (Fig. 6.69)

The *kachchhā* form is also much in evidence in the paintings by Raja Ravi Varma, 1848–1906. Although hailing from Kerala, Ravi Varma spent a major portion of his time at Malavali, a distance of 56 km. from Poona. Further, calendar art printed in Maharashtra between the 1920s to the 1940s, depicted virtually all female deities in *śālus* and *Paithanis* draped in *kachchhā* style. After 1933 costumes changed along with the thematic content of plays portrayed on stage. It was at about the same time that imported georgettes and chiffons became increasingly popular. In the *vastra haraṇa* scene of Draupadi, Bal Gandharva wore imported material equivalent to eight *Paithani* sari lengths.[24]

In 1950 because of famine conditions weavers migrated from Paithan to Surat, Yeola and other centres of silk weaving. Between 1950 and 1963, there was no weaving in Paithan. However, from 1963 to 1966 there was a small production unit run by the Panchayat Samiti (elected body at village level) under the control of the Zilla Parishad (district administrative unit). Between 1966 and 1973 the programme functioned under the Directorate of Industries, Government of Maharashtra. After 1973 a decision was taken to revive production in a programme placed under MSSIDC (Maharashtra Small Scale Industries Development Corporation), and this scheme was continued (information: Gulam Muhammad Gram, Design-cum-Demonstration Centre, MSSIDC Ltd., Paithan).

As mentioned above, the *Paithani* repertoire is in evidence in other centres in Maharashtra and in territorial units governed by Maratha *sardārs*, members of the leading feudatory families. These include Yeola, Poona, Burhanpur, Nagpur and Thanjavur. *Chanderī* (fig. 6.43), represents a parallel tradition while *Paithani* weavers introduced the repertoire to Hyderabad. *Paṭkās* woven in techniques associated with *Paithani* are to be found in Ahmednagar and Ahmedabad. Maheshwar represents yet another ancillary tradition. Traditions in Dharwad–Hubli, Narayanpet, Gadwal, Kothakota, Armoor, Amarchinta and Dharmavaram are also of interest in terms of variations in style, technology involved and clientele to which these centres have catered.

Yeola

On account of the intermittent state of warfare between the Peshwa and the Nizam of Hyderabad, boundaries were constantly shifting and Paithan remained a possession of the Asaf Jahi dynasty of Hyderabad for considerable stretches of time (*Aurangabad*: 1034). The development of Yeola as a centre for the production of *Paithanis* was an outcome of these circumstances. However, Yeola was a more catholic centre than Paithan. The ornamentation done in double interlock weave may not have been as well executed as at Paithan (fig. 6.70) but there was greater range in items woven (figs. 6.71, 6.72, 6.73). This included the weaving of *Himroo shāwls*, a twill weave in which the ground weave warp and weft are cotton, and extra weft

Paithani type. Woven at Yeola. (Fig. 6.70)

Cholī fabric. Woven at Yeola. (Fig. 6.71)

Cholī fabric. Woven at Yeola. (Fig. 6.72)

Cholī fabric. Woven at Yeola. (Fig. 6.73)

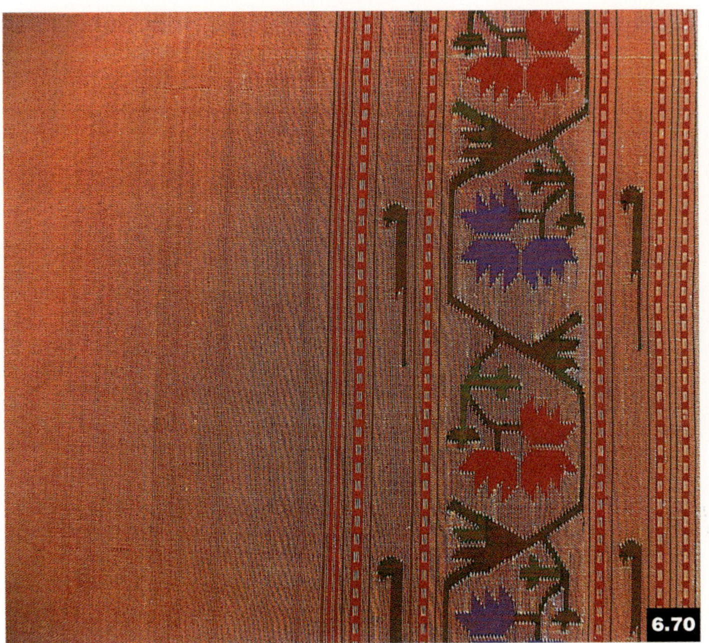
6.70

6.71

6.72

6.73

selvedge to selvedge patterning in silk and *zarī* are executed on the *jāla* loom. The drafting is such that there are no loose floats on the reverse side (figs. 6.74, 6.75). Material for *cholīs* was also produced. This could be broadly accommodated within the *khan* repertoire (figs. 6.71, 6.72, 6.73, cf. fig. 6.66).

The industry in Yeola dates back to the beginning of the eighteenth century. It is interesting that weavers were brought from Gujarat rather than Paithan to start their weaving activity at Yeola. Raghoji Naik, a forefather of the presentday Patil, who was officiating *c.* 1909, induced a Gujarati baniā called Shamdas Valji, to bring his own group of silk weavers and start production at Yeola. The incentive was that he would be given a monopoly in the weaving of all items of this category. Newcomers wishing to enter into the profession of silk weaving had to pay a fee of Rs. 350 to Shamdas Valji. This monopoly continued until 24 June 1864 when it was struck down by an order of the High Court (Mahajan, 1980: 29; Watt, 1903: 310). In the meantime, because of some dispute with Asaf Jahi officials, a group of *zarī* workers left Paithan and settled down at Shevgaon, district Ahmednagar. Leaving this group, Ramchandra Sabharam Bhaswande moved to Yeola bringing some artisans with him and on 6 January 1837, he opened a *zarī*-making unit at Yeola. *Pītāmbaras* of Nasik and Yeola enjoyed wide popularity. The most popular colours were red, purple and yellow white, green, blue and white *pītāmbaras* were also occasionally made (figs. 6.53, 6.54). The price for these ranged between Rs. 15 and 40 in 1906. *Paithani* saris, less frequently woven, 8 to 9-yds. in length and 4 yds. in width, could be sold at between Rs. 30 to 70 (figs. 6.64, 6.76). Border designs for *Paithani* saris comprised stripes, called *gomī* (fig. 6.77), *kaṭārī* from *kaṭār*, dagger, and *muthede* (Edwardes, 1976: 34).

6.75

6.76

6.74

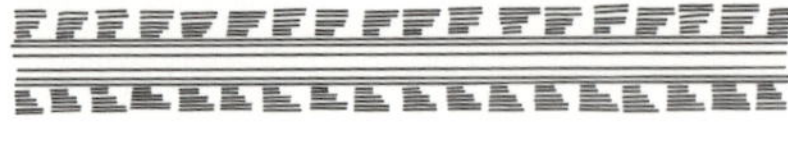

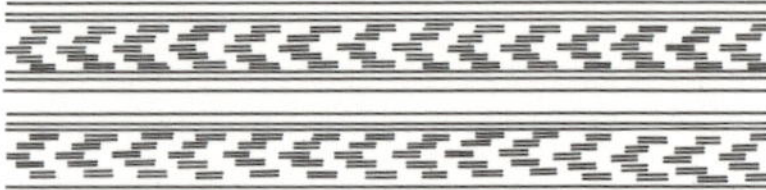

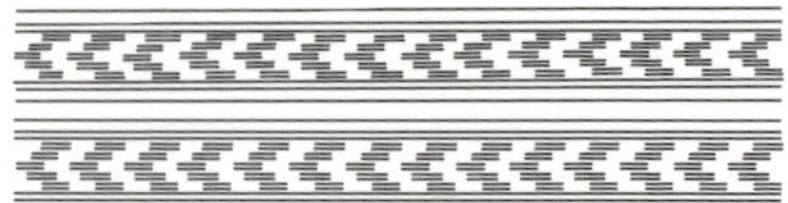

6.77

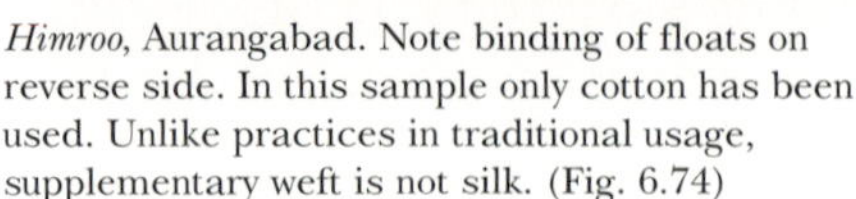

Himroo, Aurangabad. Note binding of floats on reverse side. In this sample only cotton has been used. Unlike practices in traditional usage, supplementary weft is not silk. (Fig. 6.74)

Jāla loom used in *himroo* weaving. Aurangabad. In present times the jacquard has virtually replaced the *jāla*. (Fig. 6.75)

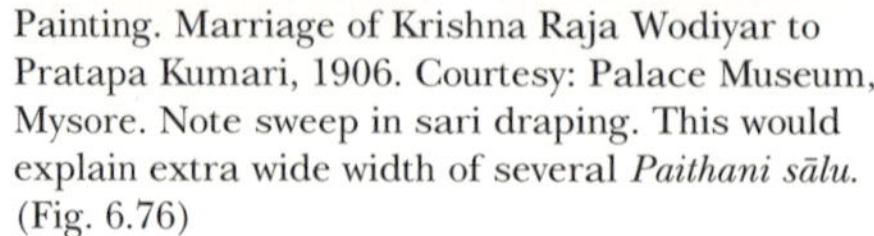

Painting. Marriage of Krishna Raja Wodiyar to Pratapa Kumari, 1906. Courtesy: Palace Museum, Mysore. Note sweep in sari draping. This would explain extra wide width of several *Paithani sālu*. (Fig. 6.76)

Gomi, border motif. (Fig. 6.77)

The Loom: Paithan and Yeola

The loom traditionally used for the weaving of *Paithani* saris in Paithan (figs. 6.78, 6.79, 6.80) and Yeola was a counterbalanced pit loom with four shafts for ground weave. A number of small shafts on either side of the loom placed behind the ground weave shafts are used for the patterning of the borders in extra warp. The heddles in these small shafts are weighed down by lingoes or weights (fig. 6.81). The number of small shafts is dependent on the designs. These small shafts may be attached to treadles for operation or may be raised and lowered manually. The shafts are hung from heddle horses or pulleys (figs. 6.82, 6.83, 6.84). The warp is held in tension by one end being attached to the warp beam and the other end to the cloth beam. The warp beam is held in place as a rope links it to the wooden pole at the back of the loom (fig. 6.80). One end of the rope is tied to the centre of the warp beam and passed around the groove on the wooden pole, after which the rope is tied to the loom. This rope, when let out releases the warp when the warp has to be taken forward for weaving (fig. 6.79). The take-up peg on the cloth beam assists the weaver to wind or release the fabric and to fine tune the required tension of the warp, for weaving (fig. 6.79). The sley houses the reed/*phani* (fig. 6.85), which is hung overhead (fig. 6.80).[25]

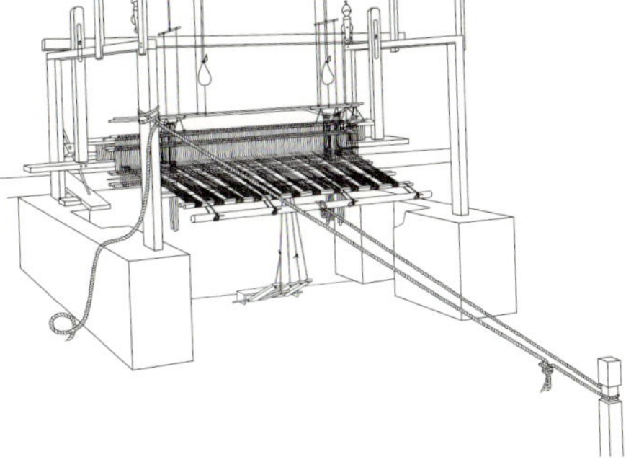

6.79

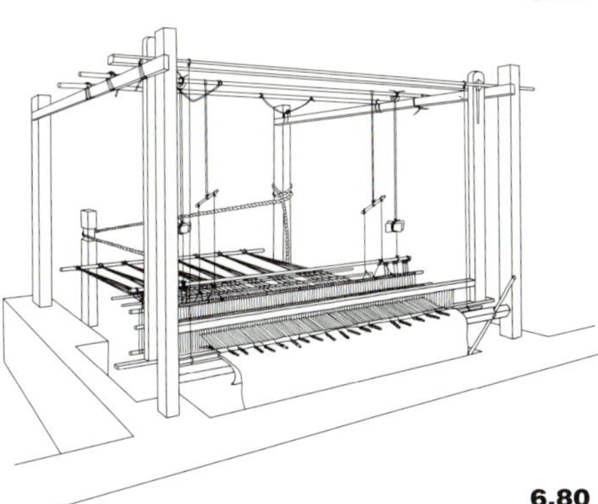

6.80

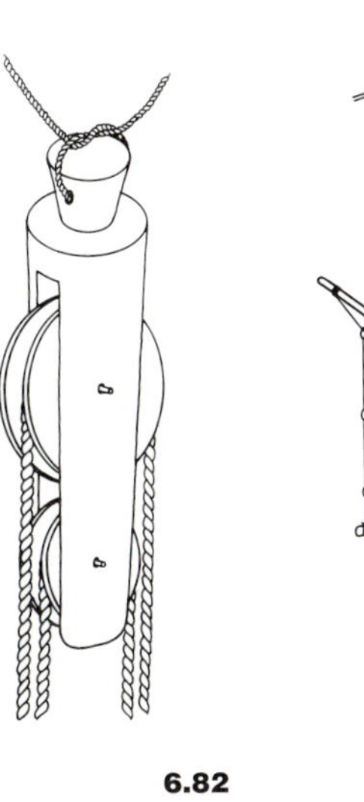

6.82

6.83

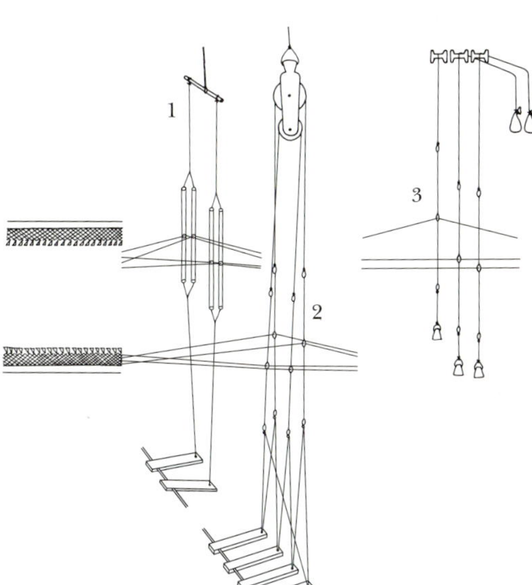

6.81

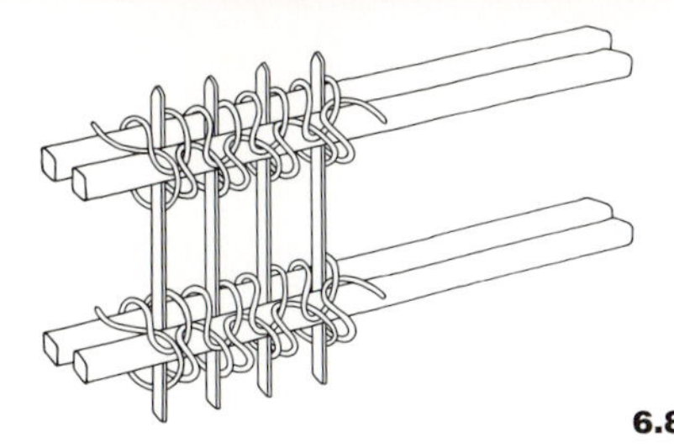

6.85

Paithani loom. Paithan. (Fig. 6.78)

Loom for weaving *Paithani* fabrics. Rear view, Paithan. (Fig. 6.79)

Loom for weaving *Paithani* fabrics. Front view, Paithan. (Fig. 6.80)

Cross section of shedding (Fig. 6.81):

1. Tie up for lifting shafts for ground weave
2. Extra warp
3. Heddles for extra warp weighed down.

Pulley. (Fig. 6.82)

Extra warp shafts. Detail. (Fig. 6.83)

Orissa *ikat* loom. Craft Museum, New Delhi. Note shafts for supplementary warp border pattern and tie-dyed warp for *pallu*. (Fig. 6.84)

Phani construction. Drawing of how reed is made. (Fig. 6.85)

Paithani sālu. Woven at Paithan. *Akroti* motif in *pallu*. Courtesy: Deepak Shah, Pune. (Fig. 6.85a)

Paithani type. Possibly woven at Chanderi. Courtesy: Mrilani Puar, Laxmi Vilas Palace, Vadodara. (Fig. 6.86)

6.85a

6.86

Raw Materials

As has been mentioned, Paithan was a well-established centre for weaving a variety of products. The double interlocking tapestry weave was an important constituent in the weaves used. From the wide spectrum of products seen in museums and private collections, it is clearly evident that cotton was the predominant raw material for the ground field of the fabrics. The use of silk for the main field of the fabric may have become more pronounced at a later date. *Zarī* has always been extensively used in the patterning, while silk has been more selectively utilized for the same purpose.

Working with the existing weaving techniques and maintaining the traditional character of the products over a period of time resulted in a sustained qualitative development of the Paithan repertoire. The craftsman made use of more elaborate and highly refined stylized forms inspired from nature as well as geometrical motifs drawn from his living environment. As has been stated, *Paithani* became popular in other Maratha dominated weaving centres such as Yeola, Burhanpur, Ahmednagar, Poona and Chanderi. It is recorded also that the *Paithani* types of saris were woven in Ahmedabad. Each centre developed particular characteristics but these are not uniformly present in all products to serve as indicators of provenance. In addition, since Maratha *sardārs* were the major patrons of the ware, the question of the transplantation of weavers from Paithan to other centres of production has to be taken into account.

The two main traditional designs from Paithan are the *Asavali* and *Akroti*. While both these were inspired by flowers, when set in the format of a *bel*/creeper in a border, the *Asavali* pattern has an overall curvilinear character, while *Akroti* (fig. 6.85a) is distinctly geometrical in nature. These designs were repeated as a continuous border along the length of the fabric or formed a framework around the *pallu*/end pieces of the textile, within which large motifs of flowering trees similarly stylized were woven. The inside of the frame could also be left plain yet woven with *zarī*. The patterning was done in silk, gold and silver *zarī* whereas the body of the textile was of cotton. These designs were also woven to demarcate the *pallu* with the *narieli* border (fig. 6.49), woven in extra warp that continued along the two edges as borders. The main field of the fabric could be plain in monochrome colour or be striped. It could also be embellished with the *mohar*/circular motifs, so called after the coin of this name. These were either woven, spread evenly all over the main field, or combined with circular motifs densely spaced over the areas close to the *pallu* (cf. figs. 6.35, 6.85a).

In course of time the *Paithani* sari, though costly, became a popular item in the marriage ceremonies of the aristocracy and laymen. Other regional centres also began weaving *Paithani*. Yeola was one such centre which later became known for its *kayari*/mango motif *pallu* in *Paithani* (Morwanchikar, 1978: 162; fig. 6.86).

Processes Involved in the Making of *Paithani* Textiles

The processes involved in the making of *Paithani* textiles are *zarī* making, dyeing, bobbin winding and warping, weaving and finishing.

Zarī

For making *zarī* a bar of silver was covered with a thin sheet of pure gold. This bar was sufficiently heated and then beaten till the metals were mixed well together. This compound was heated again and drawn through holes of different sizes in a steel plate. Generally, these were drawn to the fineness of human hair. However, for attaining even finer results the process of beating and drawing had to be repeated as many as forty times. The resulting wire was then beaten flat with a specially polished hammer and wound on silk and then polished. The trade secret apparently lay in the composition of the metal and the hammer finish. It is believed that the hammer was so well polished that its metal did not mix with the gold and silver in beating. Hence, the *zarī* composition remained pure and even with use it never became dull.[26]

Dyeing

To ensure evenness in dyeing, the yarn was bleached and freed of all impurities. It is known that traditionally natural dyes were used. The most popular colours were red, yellow, black, blue, violet, green, orange and pink (fig. 6.87).

Ethnological field derived information indicates that red was obtained from cochineal, locally known as *kirmingi*, believed to have been imported from France, and deep maroon from the fruit of a cactus locally called *sabar che bonda* that grew abundantly around Paithan. Red and yellow were obtained from *kardi* flowers (*Carthamus tinctorius*) available locally. Apart from the dyeing properties of the *kardi* flowers, it was also believed that the substance made the fabric moth-resistant and fragrant. Black was obtained from a combination of *hardo* (*Terminalia chebula*) myrobalan and either *hirakashi*/ferrous sulphate or iron filings, iron acetate and blue from *nīla*/indigo. Green was obtained by dipping a yellow dyed fabric into the indigo vat.

Bobbin Winding and Warping

The ordinary *charkhā* is used for bobbin winding and warping. The yarn is supplied from a bamboo swift, *phirkī* (fig. 6.88), which holds the hank. A wooden frame with pegs is used for making the warp (fig. 6.89). The length and the width of the warp are determined according to the endproduct required. The *pallu* warp is made separately.

Weaving

The following weaves and methods of fabric construction were found to have been used:

1. Double interlocking tapestry weave

6.87

Possibly woven in Pune. This shade of pink is a very popular colour in Maharashtra. Courtesy: Deepak Shah, Pune. (Fig. 6.87)

Phirkī/bamboo swift. (Fig. 6.88)

Warping frame. (Fig. 6.89)

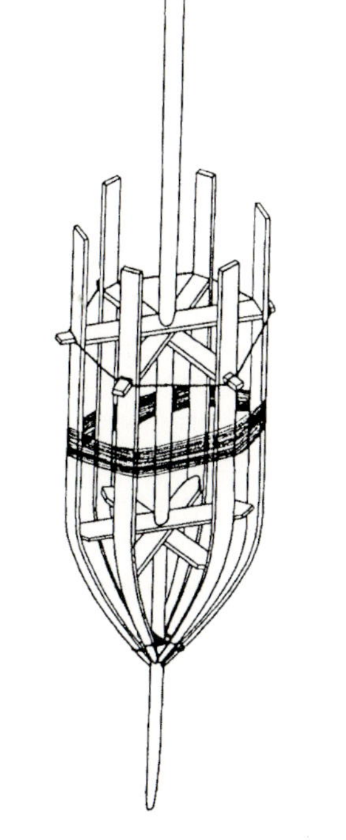
6.88

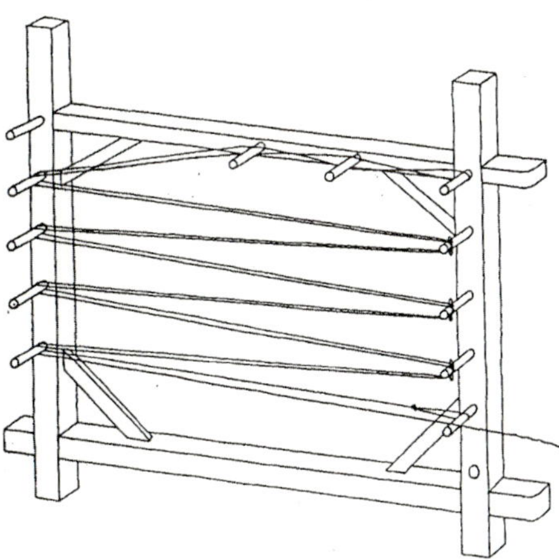
6.89

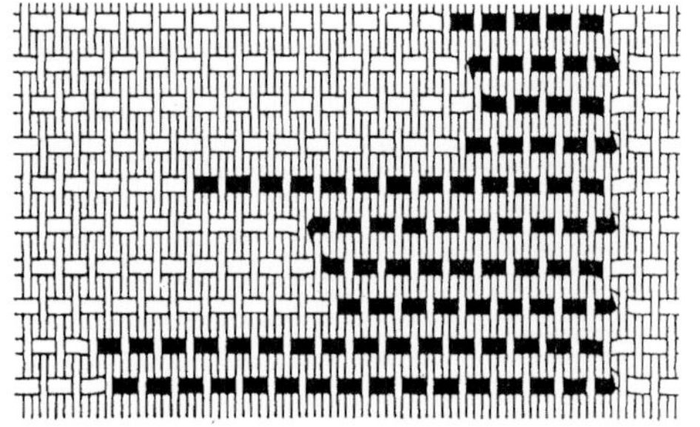

6.90

Front face. Double interlocking of wefts in weft face plain weave to achieve colour patterning in *Paithani* sari. (Fig. 6.90)

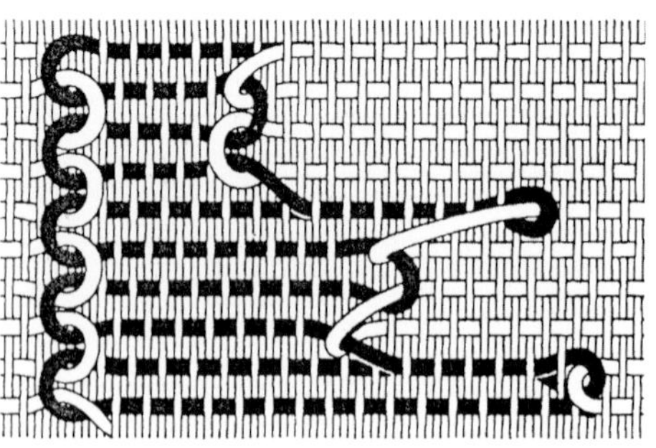

6.91

Opposite face of fig. 6.90. (Fig. 6.91)

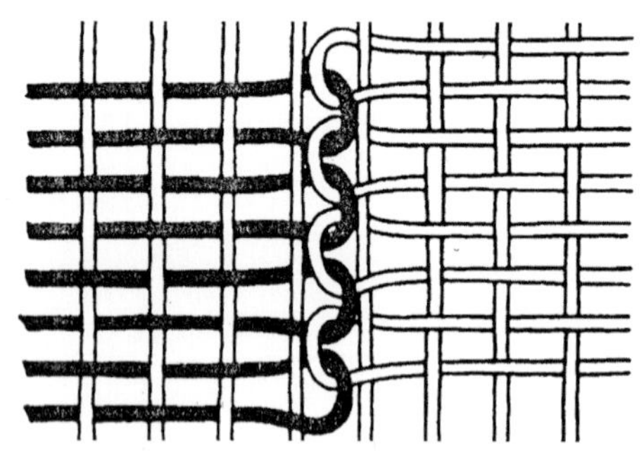

6.92

Double interlocking tapestry weave. (Fig. 6.92)

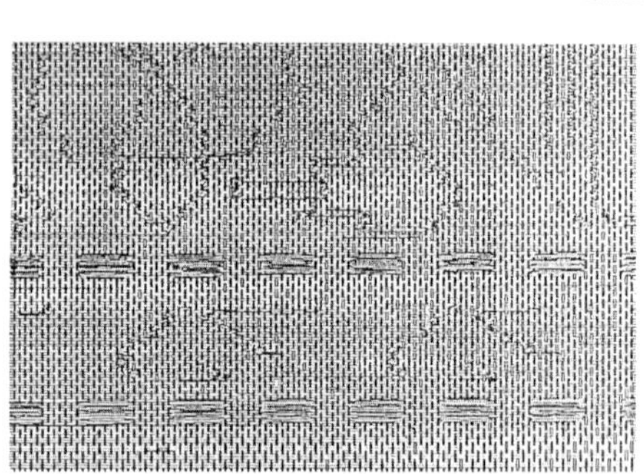

6.93

Paithani border, front face, woven with double interlocking wefts in weft face plain weave. (Fig. 6.93)

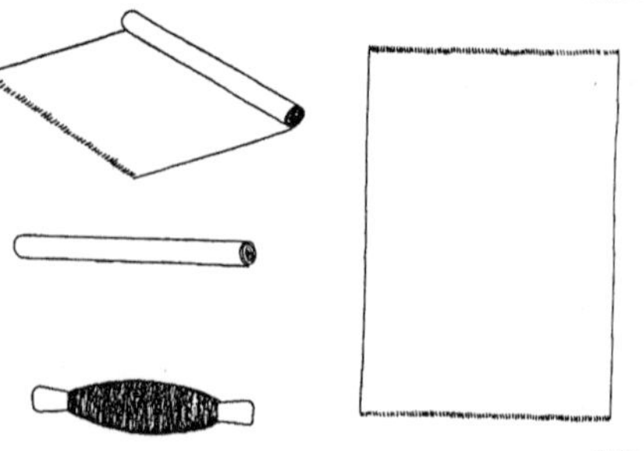

6.94

Kaḍka/quill spools made from cloth. (Fig. 6.94)

2. Single interlocking tapestry weave
3. Extra warp
4. Extra weft
5. Supplementary wefts inlaid along with the ground wefts in the regular sheds of a plain weave ground (*jāmdānī*)
6. Plain/tabby weave

The weaving and finishing processes as recorded during the year, 1980 are described here (figs. 6.90, 6.91, 6.92, 6.93):

The double interlocking tapestry weave: This weave was used for the patterning on the *pallus* to form a frame. Such patterning was also done along the borders of the sari, for *dupaṭṭās* and other items. Separate borders with traditional designs using this technique were woven also as separate *pallus*. A number of quill-like spools made from cloth, locally known as *kaḍka*, on which the silk yarn and *zarī* are wound by hand (fig. 6.94), are passed through the shed of the plain weave ground according to the paper design placed underneath the warp. Common spools may be used, which could be repeated within a restricted work area. The body ground weave weft interlocks with that of the border weft in single interlock along one warp end or in between two warp ends. A shuttle, *dhoṭā*, made of buffalo horn is used for the passage of the weft for the body of the fabric (fig. 6.95). The interlocking is extremely skilfully executed. The long floats on the reverse face of the fabric caused by the use of common colour spools (in a closer area of work) while weaving are trimmed with a small tweezer-like tool called the *chimṭā* as a part of the finishing process to complete the product (fig. 6.96). A stone, *mohrā*, found on the banks of the Godavari river, is rubbed on the *zarī* borders as a part of the finishing process to bring out the lustrous quality in the gold and silver.

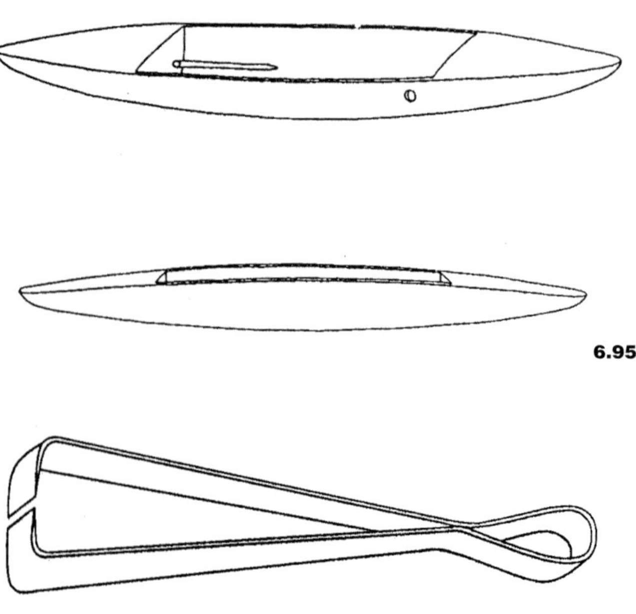

6.95

Shuttle/*dhoṭā*. (Fig. 6.95)

6.96

Chimṭā, a tweezer like tool. (Fig. 6.96)

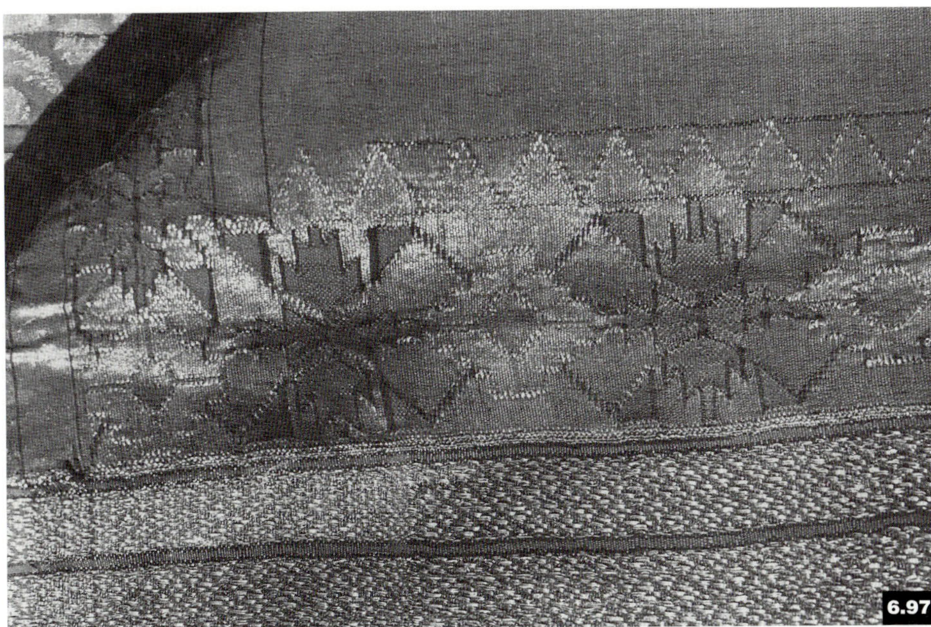

Single interlocking tapestry weave: While the borders are being woven in the double-interlocking tapestry weave, these are linked with the body fabric at each pick and between two warp ends running along the length of the fabric with the single interlocking technique. Single interlocking tapestry weave is found to be sometimes used in the weaving of the *pallu* where the triangular shapes are formed by staggering the interlocking points. This is done by using two-ply *zarī* wefts which are worked in the main ground of the *pallu* which are, in turn, linked with the single ply *zarī* weft working inwards from the border in one pick. The meeting points of the two wefts at each pick are staggered in the process of weaving, creating the triangles (fig. 6.97, cf. fig. 3.34).

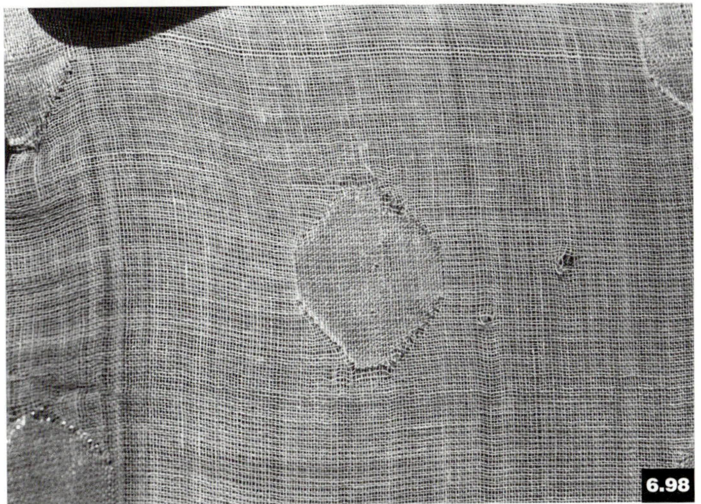

Extra warp: Patterning with extra-warp is used only for the borders. The warp threads for the extra warp patterning are drafted through the eight small shafts on either sides of the warp. These are placed behind the four shafts that are drafted and operated to weave the ground plain weave. The heddles on these small shafts are weighed down with lingoes.

These shafts are sometimes lifted or rather pushed up by hand. As these have weights on them they stay in this position till they are pushed back. These may also be linked to the pedals. The number of small shafts for the extra warp is dependent on the number of steps in a repeat of the design. A throw shuttle is used from end to end when the textiles have borders which are woven in extra-warp (figs. 6.79, 6.81, 6.78).

Extra weft: A patterned stripe in extra weft using *zarī* thread sometimes marks the beginning or end of a *pallu*. Pattern rods are placed at the back of the shafts, as is done in the weaving of woollen *dhablas* in Kutch, in accordance with the pattern required to be woven (fig. 3.32).

Selā border. Obverse. Courtesy: Sashikala Pandit, Kolhapur. (Fig. 6.97)

Jāmdānī woven gold *būṭī* on off white *Paithani* sari. Courtesy: C. Chitnavis, Nagpur. (Fig. 6.98)

Ground *haṃsa* motif woven in *jāmdānī*. Courtesy: C. Chitnavis, Nagpur. Possibly woven at Yeola. (Fig. 6.99)

Supplementary wefts: Laid-in along with the ground wefts in the regular sheds of a plain weave ground (*jāmdānī*). The *Asavali* design for example has the *mohar* motifs woven in the main field of the fabric, as done in *jāmdānī* in the *Moirangphee* fabrics of Manipur (cf. figs. 4.27a, 4.27b, 6.98).

A number of design variations were woven using this technique. These included geometrical grids and stylized birds, used singly or placed within an individual unit of the grid. The use of *zarī* is prominent in this technique and has been usually applied for the main field of the fabrics. This type of patterning was done on cotton and silk against a plain ground weave (fig. 6.99) to balance the double interlock elsewhere.

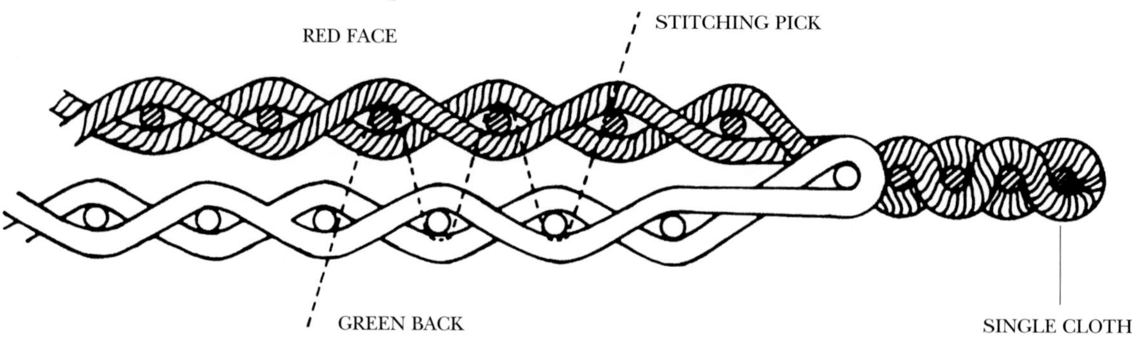

6.100

Pītāmbara in stitched double cloth technique. (Fig. 6.100)

Sometimes wedge-like inserts are woven in to create a part of the pattern both while using laid-in supplementary wefts and also while patterning with the double interlocking tapestry weave.

Plain/tabby weave: The dominant weave is the plain/tabby weave in these fabrics.

Joining the *Pallu* Warp to the Body Warp

The *pallu* is woven in a different colour warp from that of the body warp, and a separate warp is made for the *pallu*. The warp threads of the *pallu* are drafted and dented along with the body warp which would make two ends in a heddle and four ends in a dent. The *pallu* warp ends are either tied on to a stick near the woven fabric or pinned to the woven fabric. The tension is corrected and both warps are woven together so that the *pallu* warp and the body warp are interlaced with the weft. These are woven together for about 2 in. after which the body warp is cutoff and the weaving of the *pallu* continued (fig. 5.52).

Fabric Characteristics and Structure of *Pītāmbara*

While the Paithan type of *Paithani* was also being woven at Yeola, it had a definite tradition of its own. The most typical product of Yeola is the double-faced *pītāmbara*, each face of the fabric having a different colour. The main visual characteristic of the *pītambara* is that while the borders are patterned with the use of extra-warp with plain weave as ground weave, the patterning is done with gold *zarī* on a red ground. In one sample the main field of the body on one side was in a solid green while the other face of the fabric was in a solid red (the same colour as the borders). One end of the fabric had a crossborder woven as stripes using gold *zarī*. This fabric was compactly woven. *Pītāmbaras* were also woven with borders on either side of the fabric, running along its length, patterned with the use of double interlacing tapestry weave. The design repertoire used in these was essentially the traditional designs from Paithan.

Stitched double cloth forms the main weave structure used for the main field of the doublefaced *pītāmbara*. Analysis done of one of the *pītāmbaras* in the Reserve Collection of the Calico Museum of Textiles, Ahmedabad, India, accession number 212, shows that mulberry silk yarn is used for both the warp and the weft. The warp yarn is twisted, 2 ply of this yarn of approximately between 2/80s to 2/100s equivalent to cotton count is used. There are two series of warp threads. One series of red warp threads forms one face of the fabric. The border warps are also red (but these are woven in plain weave to form only one cloth on both faces of the fabric). The second series of green warp threads forms the other face of the fabric. These warp ends are placed between the two inner edges of the borders to form the ground. Three series of weft threads are used to form one repeat of the weave: red weft for the red face of the fabric and the borders; green weft for the green face of the fabric. This weft does not weave with the border warp. It is worked only in the area within which the green warp threads are placed; and a single ply red weft is used as a stitching weft.

The approximate count of the yarn used for the wefts is nearly the same as the warp yarn. However, the weft yarn has no twist. The red face (including borders) and the green face are woven with 2 ply weft yarn and the red stitching weft yarn is single ply.

After every two picks woven for one face of the fabric, the stitching pick is inserted. This pick stitches every alternate end of the face warp and the back face warp. This makes an extremely firm joining, to make both the layers one. Three shuttles holding the three different weft yarns are probably used for the weaving of this fabric. This is followed by the insertion of the two picks for the other face of the fabric and the process is repeated (fig. 6.100). As this stitching weft is much finer than the other wefts, it gets covered by them. This stitching weft does not show up on either of the faces of the fabric, making it distinctly green on one side and red on the other in the areas where the stitched double cloth is woven. Four shafts were used to weave this stitched double cloth (figs. 6.101, 6.102, 6.103).

The major item of production at Yeola consisted of the weaving of 9-yds. saris, *pītāmbaras,* fabrics for blouses, borders, *devavastras*/temple cloths and *pagaḍīs.* A pit loom with dobby attachment for the extra-warp patterning was in use till the year 1950 after which it was replaced by the Banaras type jacquard.

The castes of Hindu weavers weaving all the above products except the *pagaḍīs* include the Khatri, Sali, and Koshti. *Pītāmbara* is also known as *mukte* or *sauala* as it is used at the time of some ceremonial occasions or ritual, especially those celebrated by the Brahmins (fig. 6.53). These fabrics were used as unstitched garments such as men's *dhotīs* and women's saris.

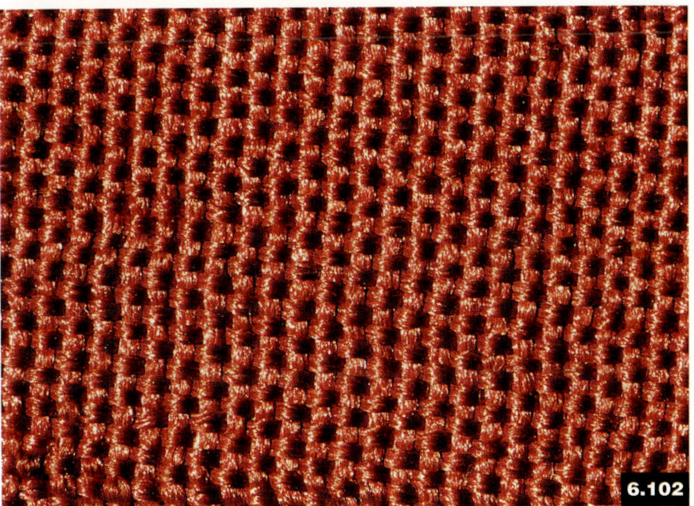

Stitched *Pītāmbara* woven at Yeola. Courtesy: Calico Museum of Textiles, Ahmedabad. (Fig. 6.101)

Stitched *Pītāmbara.* Detail of red face. (Fig. 6.102)

Stitched *Pītāmbara.* Detail. Green face. Courtesy: Calico Museum of Textiles, Ahmedabad. (Fig. 6.103)

Pune type *Paithani*. Note *nariel kalaśa* motif and the stitches running down the central section of the *pallu*. The stitching was possibly introduced to facilitate quick assessment of the gold content in the *zarī* used. Courtesy: Srimati Kusum Karmakar, Sangli. (Fig. 6.104)

Up to 1896 Yeola dealers sold most of their products at the annual fair held at Maheji, taluka Pachora, East Khandesh, on the banks of the Girnar River. This fair lasted for four months during the cold season. Merchants from Khandesh and Berar would buy Yeola Paithani along with other commodities. After consecutive bad seasons the Maheji Fair lost its importance and the Yeola merchants began to sell at Yeola itself. The *zarī* of Yeola was of very fine quality and there was brisk demand especially during the wedding season for Yeola *Paithani* ware and turbans in several Deccan towns (figs. 6.104, 3.26). The more important distributing centres included Ahmednagar, Sangamner, Nasik, Lasalgaon and Malegaon in Nasik district; Dhulia, Jalgaon, Panola, Chopda in Khandesh, Amravati, Akola, Khamgaum in Berar; and Jalna, Nander and Ingoli in Hyderabad state. Cotton turbans (silk being only woven to order), varied between 20 and 50 yds. in length. Here also the weavers were mainly Salis, Khatris and Koshtis (Watt, 1903: 310; Nissim, 1909: 7–12; figs. 6.61, 3.25). Fine cotton items were also made at Nasik (Birdwood, 1880, 225). In 1904 Chinese silk was supplied to Yeola and Poona from Bombay, while Persian silk was channelled through the same source to Poona and Nasik. Poona and Ahmednagar also procured Bengal *charkhā* silk.[27]

In the 1980s, while the majority of the *sāhūkārs*, middlemen, were Brahmins at Paithan, at Yeola they were Marwari Jains from Patan or merchants of Gujarati origin. The only items commissioned by the *sāhūkārs* were those in the Paithani range. Weavers making other ware undertook the marketing operations themselves. This included Banarasi brocade done on the *jāla* and introduced *c.* 1955 (oral communication: R.S. Morwanchikar, Aurangabad; S.M. Patel, Messrs. Gangaram Chhabildas, Yeola). When the decision was taken in 1973 to revive *Paithani* weaving at Paithan, many weavers were recruited from Yeola on the basis of newspaper advertisements (oral communication: N.T. Nagpure, Yeola). Despite the fillip that has been provided for the revival of *Paithani* in recent times, an unfortunate fall-out of the introduction of the Gandhi cap during the National Movement has been the gradual phasing out of *pagaḍī* weaving.

In the early twentieth century, a 9-yds. *Paithani* sari would have about 22 *tolās* of silver (251 gm.) and 6, 8, 12, 16 *māsās* of gold being the equivalent of 5.7, 7.6, 11.4, 15.2 gm. respectively. Saris with 8, 12, or 16 *māsās* of gold were respectively called *Aṭhmāsī, Bārāmāsī* and *Solāmāsī*. In special cases the gold could go up to 18 or 21 *māsās*, i.e. 17.1 or 20 gm. *Paithanis* used by royal families could contain as much as 36 *māsās* of gold, i.e. 34.2 gm. (Kokad). A 6-yds. *Paithani* required about 600 gm. of silk and 150 gm. of *zarī*.

In 1940 a 9-yds. *Paithani* cost between Rs. 80 and 150. In 1963 the price for a 6-yds. sari was Rs. 200 going up to Rs. 1,400 in 1975. When working the richly decorated *pallu* portion, a weaver would weave between 3/4 to 1 in. in twelve to fourteen hours. A *Paithani* sari could involve a work period of as much as two to three months. Obviously revival would entail some degree of substitution in raw material and product diversification (Kokad).

Notes and References

1. Agrawala, himself, makes a distinction between *netra* and *piṅga*, on the one hand, and *stavaraka* on the other. *Netra* and *piṅga* could indicate monochrome and polychrome *jāmdānī*. *Stavaraka* was Persian brocade. Agrawala, 1969: 103, 105. See also Moti Chandra, 1973: 114, 128.

2. Hyder Ali, AD 1761–82, was the first person to attempt the introduction of *B. mori* silk in Karnataka. *Karnataka*, 1968: 799–800.

3. See Gupta, 1991: XII, XIV. By the close of the seventh century the use of the direct sea route between China and south India had diminished the earlier role Sri Lanka had played as an entrepôt. Nilkanta Sastri, 1955: 603–4. The Pallavas had encouraged cultural contact with South-East Asia whereas the Cholas laid greater emphasis on trade avenues. Hall, 1980: 168.

4. The term, *Apāranta*, had earlier covered the whole of the western seaboard of India. By the fifth century AD, the term *Lāṭa*, as a territorial unit, began to supplant *Aparānta* despite the fact that it embraced a much smaller geographical entity. Between the fifth to the fourteenth century epigraphical records show *Lāṭa* as occupying the territory between the river Mahi in the north and river Damana in the south including the modern district of Kheda, Bharuch, Surat and parts of Thana. Sankalia, 1960: 325–9; Apte, 1959: Appendix C.

5. Basham, 1983: 95, 103. Silk weaving was practised in Rome and there is an extant sample of Roman woven silk dated by H. Granger-Taylor to *c.* mid-second century AD. Variation in colour patterning, usually projected in the form of bands, was achieved through the tapestry technique. Men tended to wear white (Granger-Taylor, 1987: Fig. 2, 16–21, 23). The description of silk woven by the guild of silk weavers mentioned in the Mandasor inscription, dated AD 473–4 has been rendered as 'variegated with the arrangement of different colours' (Fleet, 1970: 85), and 'pleasant to the touch, lovely to the eye, with varied stripes of different colours' (Basham, 1983: 99). Since Gujarat does not have a major tradition in tapestry weave but does have one in double *ikat*, it is quite possible that, apart from silk yarn and silk cloth from China, locally fabricated *paṭolā* may also have been exported to Rome through the port of Broach, or Barygaza as the Romans called it.

6. Venkatarama Sastriar, 1971: 14; I.R. Dave, 1976: 18, 26, 28; oral communication: T.L. Narasimhan, Ayyampet. The language of the Saurashtras bears out the migratory route outlined in the oral tradition of the community. In the 1970s Saurashtras could be found in all the major silk weaving centres in Tamil Nadu, clustered around the three centres of Madurai, Salem and Kumbakonam.

7. This section is largely based on a note commissioned by the National Institute of Design, Ahmedabad. Entitled, *Short Note on Silk Handloom Weaving in Tamil Nadu*, the author is T.S. Natesan, No. 15, Hanumantha Rayan Koil Street, Park Town, Madras 600003. Attention is also drawn to *Products of Tamil Nadu*, NID.

8. For Banarasi *janglā*, irregular wild flower pattern, see Yusuf Ali, 1974: 88 n. 467.

9. Terms which have been used in regional languages to denote the sari include Kanara, *shiali*; Marathi, *lūgaḍa*; Tamil, *pudawai*; Telugu, *chira*. Balfour, 1976: 536.

10. For patterning as described, see Mohanty, 1984: I, 157–9; II, 367–74.

11. After examining the *kula devīs* and *devatās*, household/clan goddesses and gods of the Salvis of Patan, K.B. Dave concludes that this group must have had its origins outside Gujarat in the Deccan and Rajasthan. The Salvis of Patan manufacture *paṭolā*, a double *ikat*. *Paṭolā* is included in the *Mānasollāsa* among the fabrics worn by royalty. It was also much prized in Indonesia where it was imbued with a magical connotation. According to K.B. Dave, when *Jayasiṁha Siddharāja*, AD 1094–1143, conquered Devagiri, modern Daulatabad, he is said to have brought back Salvi weavers from *Pratiṣṭhāna*. These Salvis comprised 84 families and married into a group of *Brahmakṣatriyas* called *vīrajos*. B.J. Sandesara asserts that it was the minister of *Kumārapāla* (AD 1143–74), Cāhaḍa, who brought Salvi weavers from Baniberapur in Mārwār to Paṭan. The Salvis were absorbed into the ranks of *vaṇikas* in the caste hierarchy and entered into marriage relations with *Leua Paṭidārs*. In discussing the origin of the *Bhulia ikat* weavers of Sambhalpur, Orissa, R.N. Mehta infers that their original home was Patan, Gujarat. In this connection, whereas K.B. Dave attributes both the Deccan as well as Rajasthan, at different points in time with the origins of the Salvis of Patan, B.J. Sandesara is of the view that the point of diffusion was to be found in Marwar, Rajasthan. See Dave, 1964a; Dave, 1964 b; Sandesara, 1953: 54–6; Mehta, 1961: 64–6, Shringondekar, 1939: 88, *slokas* 17, 20; Moti Chandra, 1973: 123; Gulati, 1951: 8, 10. However, according to the family of Chhotulal Salvi, Patan, 700 Jain Salvi families were brought to Patan from Jalna. They were *paṭolā* weavers. They claim to have brought the image of Nemināth, now installed in Paṭan, with them at this time. At first sight, the close association with mulberry silk weaving by a group of weavers of Jain affiliation appears surprising. H.C. Bhayani, Ahmedabad, has pointed out that by the twelfth and thirteenth centuries the court life of Rajput ruling princes was the model followed for depicting the image and lifestyles of the deities among Jains. Earlier wool was favoured on ritual occasions and the principle of *ahiṃsā* demanded the exclusion of vegetal adornments in ritual. With the acceptance of a courtly model for the gods, silk began to be worn and flowers entered into ritual practices. It is against this backdrop that the silk weaving of the Jain Salvi community at Paṭan is to be viewed. With regard to the origin of the Salvis of Gujarat, after sifting all available material A. Bühler and E. Fischer (Bühler, Fischer, 1979: 1, 250, 254, 326, 332), prefer to leave the subject open.

12. Morwanchikar, 1978: 182–9, 211; Fleet, 1901: 537–9. There is mention of silk in the *Gāthā Saptaśatī* as rendered by Basak (Basak, 1971: 16). However, until the date of this text is firmly established it cannot be used as evidence to attest the antiquity of silk usage in Maharashtra.

13. *Aurangabad 1977*: 1033; oral communication: R.C. Dhere, Department of History, University of Poona.

14. Bengali women usually performed their domestic ritual wearing a *garad* or undyed *maṭkā* silk sari with a red border and were cremated in the same garment (oral communication: Ruby Pal Chowdhury, Calcutta). Although the users of *Bāluchari*, always woven in sari form, were Hindu women, its technique was of Islamic origin. *Jāla* was introduced into Bengal by the Nawabs of Murshidabad and remained somewhat of an exotic intervention. *Bāluchari*, made through usage of the *jāla*, was accepted as a part of festive or ceremonial wear but not in ritual. This may be compared with the *Banārasī* sari, intimately associated with marriage but not with any other domestic ritual.

15. Tulpule, 1979: 344–6, 353, 377 n. 365; Sherwani, Joshi, 1974: 194–9.

16. Kulkarni, 1969: 170. Mungi is located in Shergaon taluk, Ahmednagar district, on the banks of the river Godavari. *Varṇakramī*, 189 and map 8/7, 107.

17. Padhye, 1933–4: 397–9.

18. Gupte, 1886, 34; workshop of Rathi and Kanade Co., Guledgudd, Karnataka.

19. Sardesai, 1933: 52–3. The letters written by the Peshwa to Nana Fadnis are dated 1.4.1765, letter no. 94, 4.12.1766, no. 95; 7.12.1768; no. 97. The term *khīrodaka* in relation to *pītāmbara malganthi* which is mentioned in the same text can be taken either in the sense of shot colour, or in the sense of a double-faced *pītāmbara* presenting one colour on one surface and a second on the other.

20. Among the ruling families those of Sangli, Miraj, Ichalkalkaranji, Jamkhandi and Aundh were Brahmin. Members of lower castes aspiring to Brahminical status would also hope to own and display of *Paithanis.*

21. There was a belief that the thickness of the *selā* could even provide protection against sword thrusts. For *kula dharma* see Thoothi, 1935: 55–6.

22. The *selā*, as an indication of respectability, has been linked by G.K. Bhat (Bhandarkar Institute, Pune), with the usage of *avaguṇṭhana*, of an earlier time. From the period of Kalidāsa, dated by this author to between *c.* first century BC, to the period of Harsha, seventh century AD, women of lineage wore a diaphanous veil extending from the head to the shoulder called the *avaguṇṭhana.* The encyclopaedic definition of *Paithani, śālu* and *lūgaḍa* are as follows: *Paithani*, silken sari with broad *pallu* with designs made at Paithan. *Śālu*, a very expensive *lūgaḍa* in silk and *zarī*. *Lūgaḍa*, lady's garment, in colour with a border. The word *lūgaḍa* was derived from Sanskrit, *dukūlam*. *Dukūlam* was transferred to *Prakrit* as *dugūlam* and *lūguḍam*. It was from this Prakrit term that the Marathi term *lūgaḍa* has been derived. M.S.K., 5.

23. *Aurangabad*, 713. Even in 1867 J. Forbes Watson had remarked that the product of Paithan was preferred to that of Varanasi in Central India and the Deccan, because the former was more durable and had better properties of washing and cleaning. Forbes Watson, 1867, 140.

24. Oral communication: Kamalabai Gokhale, Poona. Georgettes and chiffons were initially popularized by the Parsis. Unlike *Paithanis* which could be washed easily, the new fabrics required dry cleaning. By 1898 three dry cleaning concerns were operational in Bombay: Jahangir M. Desai at Dhobi Talao, Diamond Jubilee Washing Company at Thakurdwar and Leach and Weborny at Kala Ghoḍa. After 1940, branches were opened within the city and after 1951 they began to be set up outside Bombay. Prior to these dates washing would be sent from all over India by registered post to Bombay. Oral communication: Deboo, Leach and Weborny, Bombay.

25. The illustration of the *Paithan* and Yeola type of loom was based on the looms used at the Maharashtra Small Scale Industries Centre at Paithan in 1980. Black and white photographs were taken at the centre. The looms were modified by being raised above the ground. All other aspects of the pit loom on which the Paithan fabrics had been originally woven were retained. The treadles for the pit loom were originally housed in the pit below.

26. Information obtained from Bani Mia, *zarī* maker, Zargally, Paithan, aged 101 years, 1980.

27. Edwardes, 1904: 1–2. For sorting, twisting and reeling see Edwardes, 1904: 3; Gupte, 1886: 35; for bleaching, dyeing and sizing of silk see Edwardes, 1904: 3–4; for the warping process see Edwardes, 1904: 4–5; for joining and fitting into heddle and reed, and tying on beam, Edwardes 1904: 5.

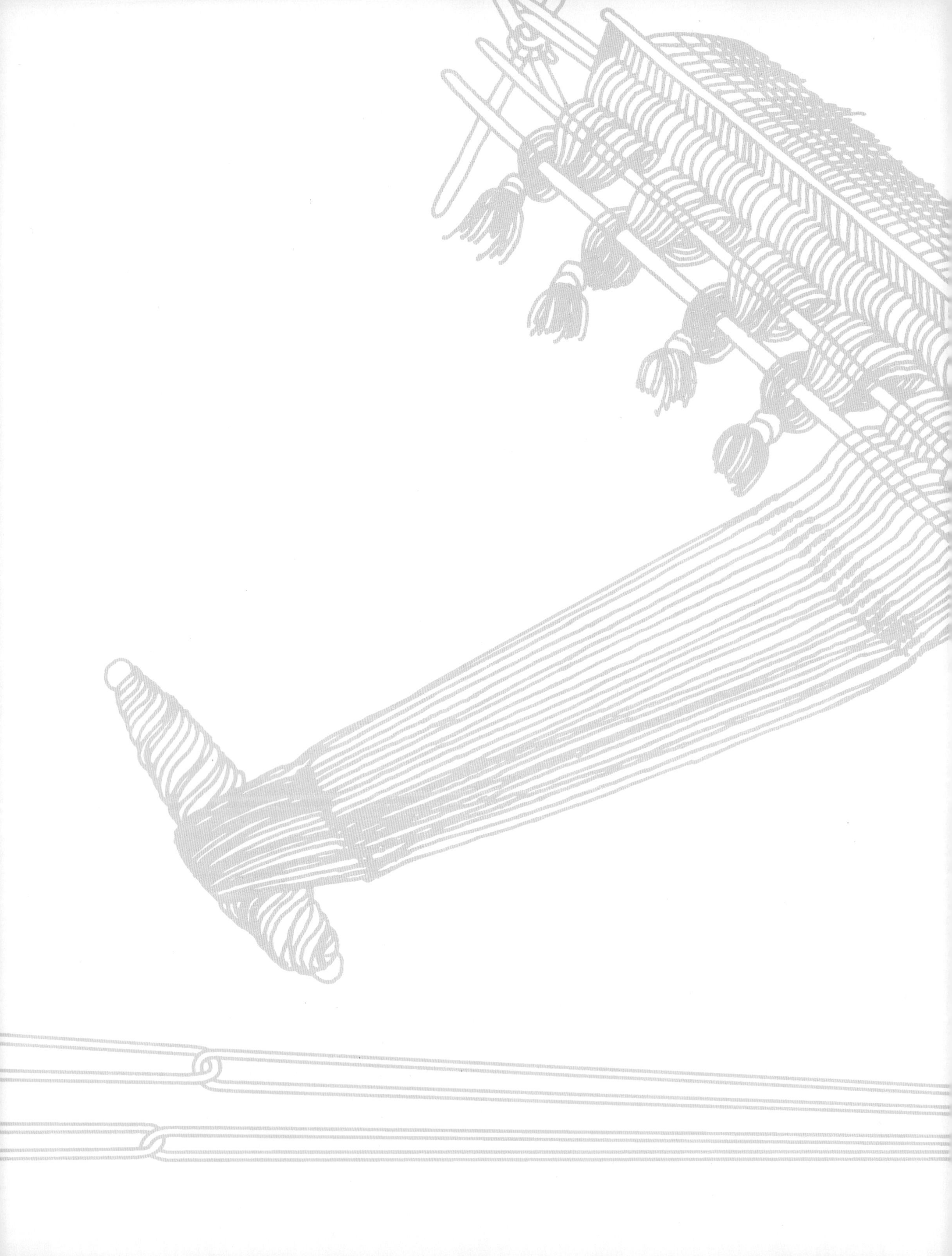

07 MARATHA FEUDATORIES AND THE SPREAD OF THE *PAITHANI* REPERTOIRE

07 Maratha Feudatories and the Spread of the *Paithani* Repertoire

Paṭkā. Double interlock. Ground cotton. Attributed to *Chanderī*. Courtesy: Sawai Man Singh II Museum, Jaipur. (Fig. 7.1)

The town of Chanderi was probably founded *c.* tenth to eleventh centuries AD. From the point of view of textile production, the important influences would have stemmed from the court of Malwa in the fifteenth century and the Marathas in the opening decades of the nineteenth century.[1] Chanderi, although also a centre of silk weaving, was particularly known for cotton weaving. In the seventeenth century, Sironj, 77 km. from Chanderi was celebrated for its fine cottons. It is possible that when trade routes were disrupted leading to the decline of Sironj, a portion of the weaving community migrated to Chanderi. A specific characteristic of Chanderi is that women undertake skilled weaving operations and the price of the fine cotton yarn used was stated, in 1905, to be worth three times its weight in silver.[2]

Chanderī. Double interlock. Cotton ground, Seventeenth century AD. Detail of fig. 7.3. Courtesy: National Museum, New Delhi. (Fig. 7.2)

Chanderī is associated with three traditions of weaving: double interlock (see fig. 7.1), *jāmdānī* and supplementary weft. The earliest example of double interlock in India is the hanging with medallion motifs presently in the collection of the National Museum, New Delhi. H. Goetz has dated it to the late seventeenth century. Referring to this artefact classified by him as a shawl, Goetz states, 'The general type is that of *Chanderī* work' (double interlock). He defines stylistic influences present as being an amalgam of Indo-Muslim and Rajput (Goetz, 1, 1955: 22–7; see figs. 7.2, 7.3). Several double interlock *paṭkās* with cotton ground in the collection of the Maharaja Sawai Man Singh II Museum, Jaipur can also be attributed to *Chanderī* on stylistic grounds (see figs. 7.4, 7.5). In the area of double interlock in silk, there is a reference to such an item in G. Watt (Watt 1903: 288, 289, 308). The green silk *rūmāl* with a delicately woven medallion in double interlock in the centre at the Calico Museum of Textiles, Ahmedabad, accession number 161, may also be attributed to *Chanderī*. Since double interlock is included within the weave repertoire of pre-Islamic India, it occasions no surprise to find its practice in such a developed form at Chanderi where it appears to have evolved independently of Paithan (figs. 7.6, 6.63).

Hanging with animal motifs. Double interlock. Cotton ground. *Chanderī*. Seventeenth century AD. Courtesy: National Museum, New Delhi. (Fig. 7.3)

Paṭkā. Cotton ground. Attributed to *Chanderī*. Courtesy: Maharaja Sawai Man Singh II Museum, Jaipur. (Fig. 7.4)

Cotton *dupaṭṭā*. Border, reverse face. Note ridged surface of double interlock weave. Courtesy: Sawai Man Singh II Museum, Jaipur. (Fig. 7.5)

Chanderī cotton *dupaṭṭā*. Double interlock. *c*. AD. 1780. Courtesy: Sawai Man Singh II Museum, Jaipur. (Fig. 7.6)

7.6

Detail of *Chanderī* sari woven on throw shuttle loom. Cotton and *zarī*, Krishna Patel Coll., Ahmedabad. (Fig. 7.7)

Pagaḍī, cotton. Chanderi. *Chheda*/end piece in supplementary weft *zarī*, plain weave, Krishna Patel Coll., Ahmedabad. (Fig. 7.8)

Jāla woven. *Chanderī* sari. Detail. Cotton and *zarī*. Note *pallu* ornamentation. Krishna Patel Coll., Ahmedabad. (Fig. 7.9)

Traditional Fabrics from Chanderi

According to oral tradition it appears that the art of dyeing was not practised by the people of Chanderi and, therefore, only natural off white and white cotton fabrics were produced earlier. There were plain borders, and plain gold borders (see fig. 7.7).

The weavers previously wove only cotton muslins of very fine quality reputed to be as fine as Dacca muslins. Cotton was locally grown and yarn up to 200s and 400s count is said to have been spun. The women used their nails to comb the cotton prior to spinning the yarn. The cotton was slightly brownish and of an average 2½ in. staple length (Mohanty, 1984: 36). Later the cotton for spinning the fine yarn was brought from Malwa and the spinning was done in Chanderi by a community of spinners called Katere, who spun yarn as fine as 200s to 260s count. This yarn was used for the warp as well as for the weft. To this day the Katere community of spinners exist but they earn their living by selling firewood.

The development of saris using *zarī* can be attributed to the late Maharaja Madhavrao Scindia, who, in the year 1908, established an organization-cum-training centre at Chanderi to develop the activity of weaving in a scientific manner to ensure production of quality products. However, this industry faced many problems as handspun yarn was replaced by mill yarn at the initiative of the colonial administration: According to oral tradition, the use of silk yarn in the warp is said to have occurred when the import of very fine count cotton was banned (Mohanty, 1984: 36).

The traditional products woven, other than saris, are *dhotīs, dupaṭṭās, oḍhnīs, pagaḍīs, sāfās, rūmāls* and yardage fabric for apparel (fig. 7.8). Today, fabrics with silk yarn of 13/15 denier in the warp and cotton weft of 140s to 200s count yarn, patterned in silk, silver and gold *zarī* are woven.

According to Ram Narain Mishra the 80 year old master weaver and merchant, there was earlier no weaving in the brocading technique. The fabrics were woven plain with or without simple borders, and *pallus* were devoid of *būṭīs* or other woven motifs. According to him, a certain Madan Baksh of Varanasi came to Chanderi and introduced *jāla* and *būṭī* weaving in the early part of the twentieth century (Mohanty, 1984: 36; see fig. 7.9).

These statements are useful for tracing the evolution of the present product range. However, there also appears to have been a more exclusive area of production, possibly engendered through courtly patronage, of superlative items executed in double interlock. A sample seen by Krishna Amin-Patel at Chanderi during a field trip in 1982 was woven in the double interlocking tapestry weave. It bore a close resemblance to the *Asavali* pattern in the *Paithani* saris. The art of weaving with the double interlocking tapestry weave to create elaborately patterned products using silk, *zarī* and cotton yarn is believed to have been taken to its peak of refinement by the patronage of the rulers of Chanderi. There is no record as to when the use of the double interlocking tapestry weave had been initiated. However, the weavers of Chanderi clearly mentioned that, by the year 1930, there was a decline in its popularity mainly because of the high level of skill, time and labour required, which in turn were reflected in the cost of the product.

The Loom

The loom on which the fine fabrics are made is a simple counterbalanced pit loom to which the Banaras type *jāla* is attached (see figs. 7.10, 7.11). Thus the lifting of warp ends to achieve patterning with both the extra warp and the extra weft with the plain weave as ground weave is planned and accomplished. The throw of weft may be achieved by the throw shuttle or the fly shuttle. Originally, a throw shuttle counterbalanced pit loom was used. The main difference between the Banaras *jāla* and the one used in Chanderi is in its placement. It is placed in front of the shafts because the warp thread is drawn through the closed heddle eye and not through the open heddle loop. After this the warp is attached to the *jāla*. It is not possible to adequately lift the threads by the *jāla* for the shuttle to pass through if the *jāla* is placed to the rear, as the closed heddle eye would prevent the thread from being lifted adequately. If, however, the *jāla* is placed in front of the heddles this difficulty is overcome (Mohanty, 1984: 37). In present times, apart from *jāla*, the dobby and the jacquard attachments are also in evidence.

Pre-weaving Process: as Recorded in 1982

Warp-making and Sizing

These activities are carried out in the open. Two sticks, about 3 to 4 ft. in height, are crossed and fixed in the ground. Two similar sticks are fixed opposite these sticks at the required distance of the width of the warp. This procedure is repeated at regular intervals to cover the length of the warp. Normally a warp for six to twelve saris is put up at any single time.

The person who does the warping walks up and down the length of the warp with a *pherna*, a warping stick or a warp-end guiding stick used to guide the warp threads while making the warp. At the same time he crosses the bobbins containing the warp yarn which he holds in each hand around the sticks to form leases. After the warping is completed, lease sticks are put in to maintain the order, width, and tension of the warp (see fig. 7.12).

Sizing follows the process of warp making. The bulbous root of *koli kanda*, *Commelina nudiflora*, is boiled and mashed to form a solution. This solution is then applied with the help of a brush to the warp kept stretched under tension. While the warp is drying, the threads of the warp are separated and the starch evenly spread with a special brush (fig. 3.31). After the warp is dry, it is wound on to a stick or rolled on to the warp beam.

Chanderī jāla loom. Weaver Abdul Gani. Craft Museum, New Delhi. (Fig. 7.10)

Detail of *jāla* attachment to *Chanderī* loom. *Jāla* made by weaver Abdul Gani. Craft Museum, New Delhi. (Fig. 7.11)

Warp making process. Chanderi. (Fig. 7.12)

Heddle making. Chanderi. (Fig. 7.13)

Chanderi *jāla* loom. Operaton of pedal by weaver Abdul Gani. Craft Museum, New Delhi. (Fig. 7.15)

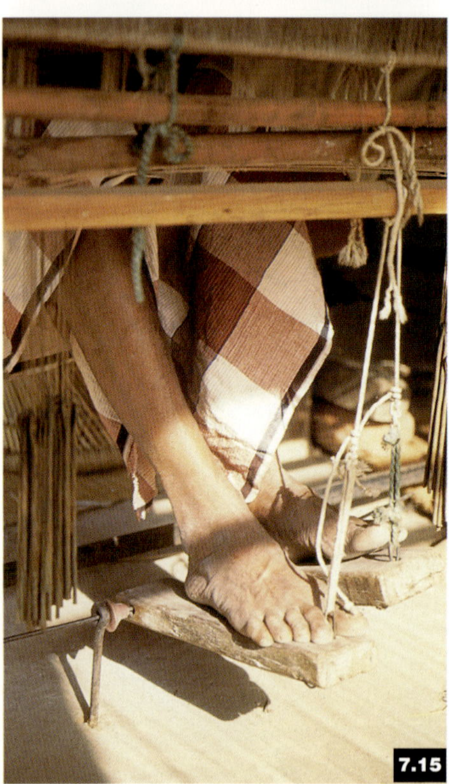

Drafting

Drafting is done only when new heddles (see fig. 7.13), and shafts are put up. Instead of repeating the process of drafting for each warp, the piecing procedure is followed while gum and ash are applied to the fingers to help in the process. Such procedures are common to almost all the traditional weaving centres in India.

Denting

Denting is done with the help of a locally made reed hook called *tarbharni.* The reed used for the border is much more dense than the reed used for the body fabric. The reeds are made of bamboo (see fig. 7.14).

The count of a reed is measured in the number of *pañjās* in one *girah* (2 in.). For example, a reed of 36 *pañjās* means 36 x 5 or 180 dents in a *girah.* The reeds used are usually of 36 to 40 *pañjās* (Mohanty, 1984: 37).

The two communities among the weavers at Chanderi are the Momin and the Sali who are again divided into three categories according to their degree of skill and experience. These comprise:

1. Master weavers who are specialized and are engaged in the production of superior quality goods. They have their own looms and employ weavers.
2. Skilled weavers who take up work where fine weaving is required. For example, weaving of a motif, with pick up sticks, and the weaving of the *pallu* is done by them.
3. Those generally employed for plain weaving.

Normally, a single weaver works on the loom (figs. 7.10, 7.11, 7.15). However, when more intricate work is required such as weaving of motifs for the main field of the fabric or in the *pallu* with the pick up method or *jāla* and the use of localized extra weft patterning, two weavers are engaged for weaving such fabrics on one loom. Two weavers are also required for weaving on wider throw shuttle looms and when three shuttles are used for weaving. Of these, two are used, one for each of the borders and one for the body, as in the three shuttle weaving done in Kanchipuram.

Denting. Chanderi. (Fig. 7.14)

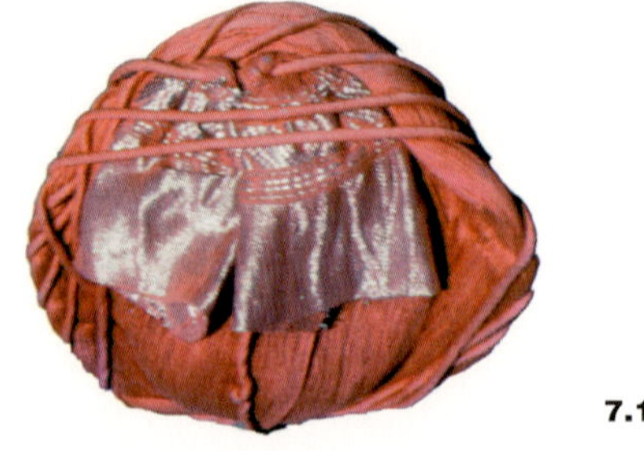

Pagaḍī worn by junior Raja, Dewas. Courtesy: Shahaji Chhatrapati Museum Trust, Kolhapur. Possibly woven at Chanderi or Burhanpur. (Fig. 7.16)

Women are employed for executing the skilled and intricate work required in the weaving of patterned areas with the discontinuous laid-in supplementary wefts which are woven in the same shed and along with the ground weave weft. This has been noted in *jāmdānī* or in the weaving of fabrics requiring patterned areas to be picked up by hand with pick up sticks for the discontinuous extra weft patterning.

Apart from the extremely fine muslin fabrics woven in cotton, diaphanous silk fabrics were also woven. The fine cotton muslin and diaphanous silk *pagaḍīs* and *sāfās* were woven in plain weave with their two ends patterned with *zarī* used in plain weave or in combination with continuous extra weft *zarī* to create the localized extra weft patterning (fig. 7.16). This was possibly done with the use of pick up sticks.

It is interesting to note that one of the ends of a *pagaḍī* is made up of two separate layers. The main fabric of the *pagaḍī* divides to form two separate fabrics. On analysis,[3] it has been found that the density of the warp and weft is consistent throughout these two layers and the main fabric, suggesting that an additional warp is put up for weaving the two separate fabrics which are linked to the main fabric in one horizontal line across its width. The analysis demonstrates that one end of the main fabric warp has loops. These loops are linked with loops of the warp of the two separate fabrics. These have been clearly woven separately after their linking with the main warp as each has its own selvedge (see fig. 7.17). The weavers thus had the knowledge and skill to use compound weaves including that for double cloth. The double cloth was not woven in tube form but in two separated layers. How this *pagaḍī* was worn is not clear. However, one possibility is that while one end was tucked in at the initial stage, this end with two *chhedas* was left free so that the decorative part of the *pagaḍī* could be seen.

Poona

Poona had no tradition of silk weaving and was a centre of coarse cotton cloth production. However, at the end of the eighteenth century there is some evidence of silk weaving. It received a fillip when the Peshwa induced some weaver families from Paithan and Yeola to migrate to Poona at the beginning of the nineteenth century. Makers of metallic thread migrated from Aurangabad, Paithan and Karanji from the territories of the Nizām. During the years 1883–4 there were as many as 1,100 to 1,300 looms in Poona of which only about 400 to 500 were devoted to cotton production. Except for two families which had come from Madras, the remaining weavers were from Sholapur, Indapur, Narayanpet, Paithan and Yeola. Muslim weavers came from Malegaon in district Nasik. As in Paithan, Hindus produced *Paithani* saris while Muslims concentrated on the production of turbans.[4] Some turbans were adorned with heavy metallic yarn tassles called *jhirmili* (Gadgil, 1945: 103–4; Nissim, 1909: 10, 11). Poona production served the needs of Satara, Sholapur, Pandharpur and other trade centres (figs. 7.18, 7.19). However, the China silk used at Poona was of cheaper quality than that used at Yeola (Edwardes, 1976: 34).

Stylistically, also, the later Poona products betray a hybrid style (fig. 7.20). The cone motif was linked more to the form of *kalaśa*/pot, *nariel*/coconut (figs. 6.104, 7.18, 7.21) and extra-weft began to supplant the double interlock weave.

By the end of the nineteenth century a decline had set in at Poona, and this centre began to be associated with the production of cheaper and lighter silks as compared to Yeola. Poona now began to produce larger numbers of cotton goods with silk

7.17

Pagaḍī, Chanderi. Two layered at one end. Krishna Patel Coll., Ahmedabad. (Fig. 7.17)

7.18

Pune type *Paithani*. Supplementary weft. Note *kalaśa*/pot *nariel*/coconut motif. Courtesy: Vatsalabai Potnis, Kolhapur. (Fig. 7.18)

7.19

Pune type *Paithani*. Courtesy: Kusum Karmakar, Sangli. (Fig. 7.19)

7.20

Pune type *Paithani*. Extra weft. Courtesy: Vatsalabai Potnis, Kolhapur. (Fig. 7.20)

7.21

Nariel/*kalaśa* motif painted on a hand cart. Patdi village, district Surendranagar. (Fig. 7.21)

borders. This was the style associated with Dharwad cottons (see fig. 7.22). The firm, Gagalbhai, Hathibhai of Poona, sold its last *Paithanis* in the 1930s (figs. 7.23, 7.24). The comparative prices at the time between *Paithani* and 8-yds. *Banārasī* was Rs. 125 for the former and between Rs. 80 and 90 for the latter.

Burhanpur

Burhanpur was founded in AD 1400 by Nasir Khan of the Faruqi dynasty of Khandesh. In subsequent years sovereignty shifted to dynasties of Mughal, Maratha and Asaf Jahi appellation resting finally with the British after 1860–1 (Grant, 1870: 125–8). In the seventeenth century Burhanpur was not only an important hub in internal trade routes but was also a centre for the production of fine cottons and *zarī*. It was at the initiative of the Emperor Akbar, *c.* 1600, that the *zarī* industry was started at Burhanpur by a group of people of Pathan origin. Silk weaving was initiated in the eighteenth century. The *Paithani* double interlock formed part of the Burhanpur repertoire (figs. 7.25, 7.27). A very fine interlock Burhanpur silk sari may be found in the collection of the Central Museum, Nagpur. The distinctive feature is that the inner end bears the typical striped pattern associated with textiles of Maharashtra (see figs. 7.26, 7.27, 7.28).[5] After products of the British mills began to carve a niche in the Indian market, Burhanpur continued to prosper as it catered to the sector consuming unstitched garments to which the mills of Manchester could cater to an extremely limited degree.

7.22

7.23

7.24

7.25

Street scene, old Hubli, Karnataka. Note *Dharwad* sari and *khan.* (Fig. 7.22)

Pune type. *Paithani. Ganga Jamna pallu.* Courtesy: Deepak Shah, Pune. This style was duplicated in Kolkata and Raichur. (Fig. 7.23)

Sālu. Possibly woven in Pune. Courtesy: Deepak Shah, Pune. Note influence of Banaras brocade. (Fig. 7.24)

Paithani sari. Possibly Burhanpur. Calico Museum of Textiles, Ahmedabad. Art work: Rajendra S. Patel, NID. (Fig. 7.25)

7.26

Typical striped end as in Pune and Nagpur cotton saris. Detail of Fig 7.27. Courtesy: Central Museum, Nagpur. (Fig. 7.26)

7.27

Double interlock, *Paithani,* type of sari woven at Burhanpur. Borders seperately attached. Courtesy: Central Museum, Nagpur. (Fig. 7.27)

7.28

Dharwad type. Courtesy: Deepak Shah, Pune. This typology was duplicated in Pune and Nagpur. (Fig. 7.28)

Nagpur

Nagpur is located in an area where a hybrid cotton, *Gossypiuim herbaceum* was grown. It also had access to the *tasar* grown by Adivasi tribals in the Chanda district. Cotton weaving of a fairly coarse nature was practised here (Blennerhassett, 1916: 42). Rajapur Barsi or Nagpur was under the Gonds when the invasion during the reign of Shah Jahan took place in 1637. The Gond ruler of Nagpur accepted Islam. Raghoji Bhonsla was active in eastern India during the reign of Peshwa Baji Rao I (AD 1720–40). After the death of the last Gond king, Chand Sultan, the succession was disputed. One of the claimants, Ratan Kunwar, obtained Raghoji Bhonsla's help offering him one-third of the territory as the price for his support. This cession was made in 1735. As relations between Raghoji Bhonsla and the Peshwa were strained, when the former wished to encourage weaving, instead of getting weavers from central Maharashtra, he brought *tasar* weavers, the Halwas, from Bhagalpur. The latter were then encouraged to settle in Nagpur. This marked the beginnings of the Nagpur textile industry under the aegis of the Bhonslas.

Front face. *Umred* cotton *uparna* with *karvati kathi* three shuttle border. Courtesy: Govind Rao Arvikar, Nagpur. (Fig. 7.29)

Umred uparna. Ruiphul border. Courtesy: Govind Rao Arvikar, Nagpur. (Fig. 7.30)

Karvati kathi loom with *aḍai* attachment. Nagpur. Suraj Kund Mela, Gurgaon. (Fig. 7.31)

In *tasar* weaving the warp was five ply and the weft three ply. In Bhandara, when the borders were in contrast colour, there was recourse to three-shuttle weaving. If sufficient *tasar* silk was not available, the body was woven in cotton, *tasar* being reserved for the border (Dewar, 1904: 11, 12). At a later stage, the textile industry was developed to meet the needs of the Maratha court. Weavers at Umred perfected the weaving of the three-shuttle *karvati kathi dhotī* and *uparnas* for which there was wide demand in regions dominated by Maratha culture (figs. 7.29, 7.30). The *aḍai* mechanism was used for ornamentation (figs. 7.31, 7.32). The weaving of these *dhotīs* was started *c.* 1775 when Mudhoji Bhonsla built a large fort. The borders were woven in cochineal dyed silk and the waters of the local well were considered very favourable for fixing this colour. Brahmin widows wore saris of red, saffron or pink cotton with a silk *karvati kathi* border. Raja Raghoji II, 1775–1816, settled some of the best weavers of famine-struck Paithan at Nagpur in 1803 and also encouraged the migration of weavers of the same calibre from Jainabad and Burhanpur. The *Paithani* repertoire may have been introduced at this time (see fig. 7.33). The local Koshti weavers now learnt the art of weaving *zarī* borders on unstitched garments (Harnetty, 1991: 456–7).

Pulling of *jungu*. Courtesy: Grama Swarajya Sangam, village Bonthalakadur, district Srikakulam. The *aḍai* mechanism on this loom is very similar to that utilized for *karvati kathi* in Vidarbha. (Fig. 7.32)

There was a difference between weavers who concentrated on *dhotīs* and *uparnas* and those who made 9 yard saris. The *dhotīs* and *uparnas*, made at Nagpur, were prized for their fineness. The yarn for these was spun by Dhers, the weavers being Koshtis. In 1827, the weavers who were originally from Jalna and Paithan, made lighter versions of *Paithani pītāmbaras* for both men and women. Silk *dupaṭṭās* with *zarī* borders were made for Hindu women of rank. These were woven both by Hindu Koshtis and Muslim Momins. This represents a departure from practices followed in Paithan and Yeola. *Selās* and turbans were also woven. Other items included *rūmāls* and *khan*.

Paithani type sari. Possibly woven in Nagpur. Courtesy: Sawai Man Singh II Museum, Jaipur. (Fig. 7.33)

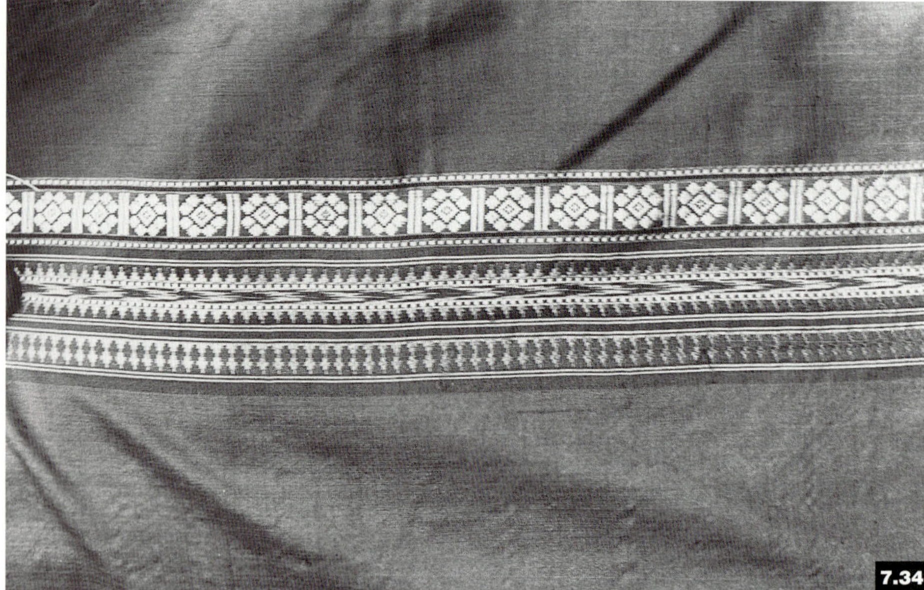

Jyot sari, Nagpur. *Ikat* motif on silk border. Ground cotton. Courtesy: Weavers Services Centre, Bombay. (Fig. 7.34)

Black *Maheshwari* sari. Courtesy: Vatsalabhai Potnis, Kolhapur. (Fig. 7.35)

The range manufactured for adults was also made in smaller sizes for children (Jenkins, 1923: 44–8). Of the saris woven at Nagpur those which resemble the Karnataka repertoire of stripes and checks are called *salai* and *choukaḍa*. *Salai* had small stripes while the checks in *choukaḍa* were 1 in. x 1 in. Borders, as in the Karnataka tradition, were in silk and the cotton used ranged from 80 to 100 counts.

The opening of the Empress Mill in Nagpur in 1867 caused a setback in the demand for handlooms. Already at low ebb in 1920, the handloom industry in Nagpur was virtually wiped out after 1930. At the height of production either a barrel dobby or the *aḍai* would be used as a loom adjunct in weaving where brocading was required. A special category comprised the *jyot* saris woven in Nagpur which included a simple *ikat* procedure in the border (see fig. 7.34). This *ikat* usage is also in evidence in some samples ascribed to Yeola.

Maheshwar

Maheshwar was located within the possessions of the Maratha state ruled by the Holkars. It was Ahilyabai Holkar, 1767–95, who started the weaving industry of Maheshwar. In 1948 the possessions of the Holkars were merged with the Indian Union. In the area of weaving, the revival of the *Maheshwari* tradition in post-independent India is closely linked with the efforts of the erstwhile Holkar ruling family.

Ahilyabai is said to have invited weavers from Gujarat to start the industry, which, from its initiation, appears to have been geared to meet the needs of royalty. The speciality of these saris was the border pattern. With the death of Ahilyabai, the moving force was lost and although the demand continued there was little innovation and by 1920 the industry had declined sharply.

The specific characteristic of *Maheshwari* saris was the checked pattern and stripes achieved through variation in intensities of colour or through varying yarn densities and skipping dents in denting. The check pattern would draw this item closer to the Karnataka repertoire (see fig. 7.35). The earlier tradition was that of pure cotton. The *pallu* portion is patterned by horizontal bands in colour along the width. This is a part of the Poona–Nagpur repertoire (Kokad; Dubey, Jain, 1965: 4, 7, 36–40).

Other Centres of Silk Weaving

Karnataka

E. Balfour noted (*c.* 1884) that in the western, central and southern Indian belts, saris presented a patterning in myriad forms of stripes and checks (Balfour, 1976: 831). According to K.V. Soundara Rajan, Madras, the checked tradition is to be connected with the *Vīra Śaiva* cult. Whereas the three shuttle technique could probably be associated with the Tamil Nadu repertoire, the layout in indigo shot colours, checks, stripes and *tope pallus* could perhaps be associated with Karnataka. With regard to loom attachments, work practices particularly in relation to the weaving of *khans*, demonstrated why the dobby mechanism grafted so well on to the existing looms. Moreover, since this part of north Karnataka merges into Maharashtra the transition from the extra warp shafts found in the *Paithani* loom to the barrel dobby found in Karnataka would have been a natural progression in terms of technological refinement.

Ilkal

Many factors must have influenced the development of the handloom industry in Ilkal, such as its topography which is considered to be conducive to weaving. Ilkal is said to have received its name from the Kannada word *lakallu* meaning slope. It is situated in a valley with hills all around and has a level of humidity which facilitates weaving. It is a predominantly cotton growing area, and homespun cotton is readily available. In addition, the mineral content in the stream which flows near Ilkal helped to fix the colours of the natural dyes used in the pre-synthetic dye era. The area was prone to vagaries of nature like floods and drought. This made dependence on agriculture precarious, leading people to take up weaving to supplement their earnings.

In the late nineteenth century and in the beginning of the twentieth century, 20s count cotton yarn was used for the manufacture of coarse *gachi*/striped border saris, and the various types of coarse cloth which were woven in and around Ilkal. *Ilkal* silk saris also gained prominence during this period. Later, Rajappa Yenni and Shankrappa Sakri took the initiative to introduce *Ilkal* saris to other markets as, for instance, in Maharashtra and the northern parts of Karnataka state. Rajappa Yenni, who had married into a family settled in Poona, used to carry a few saris for sale on his visits to his inlaws. He found that *Ilkal* saris were well received. He then introduced the *gomi* border to make them more attractive.

During First World War, there was a sudden spurt in the demand for handloom products, since mill fabrics from Great Britain were not available. Apart from this the spinning mills in India, which had been supplying yarn to Great Britain, sold their yarn to the handloom industry, reviving the Indian handloom industry. *Ilkal* weaving was also the beneficiary of this boom. Around 1930, fly shuttles replaced throw shuttles. This increased the production capacity.

Unlike the first World War, the second World War had disastrous effects on the handloom industry in India in general, and Ilkal in particular. The import of silk from China was stopped. Cotton yarn became scarce as the textile mills were consuming all the yarn produced indigenously. As a result, the handloom industry was forced to manufacture coarse cotton cloth using only waste cotton. However, the 1950s heralded a new era in *Ilkal* weaving. The Raibagis, one of the leading yarn dealers of Ilkal, introduced art silk yarn, which gradually replaced pure silk yarn, the price of which had been steadily increasing. In 1968, the Khatwas, another leading business family in this line, introduced synthetic *zarī* which considerably reduced the cost of production.

Dharwad

Dharwad is an industrially under-developed area. Within existing units of production, textiles constitute the largest group. Since there has been a degree of mechanization in the area, this may have created the climate for notable refinements in loom mechanisms such as the lattice dobby. Among the small-scale industries not dependent on power supply, the handloom weaving of cotton and woollen goods is the most important. The products are mainly *khans* saris and woollen *kamblis*.

Handloom Weaving

A census of handloom carried out by the provincial textile controller in 1946, revealed that the district had 12,965 handlooms and ranked as the third biggest centre in the Bombay Presidency. The Hubli taluka took the lead. Dharwad followed along with Rannebenur and Ron. However, the situation has changed now. In these weaving establishments, as in most family concerns, all the members of the family join in the work. When the family cannot cope with the volume of work, outside help is employed on a piece-rate basis. Earlier, throw shuttle looms were used, whereas now these have been converted to fly shuttle looms. These are manufactured locally along with accessories like heddles and reeds. Mill spun yarn is used today for weaving of *khans* and *Ilkal* saris, and these are two important products. Hubli and Gadag are the main centres for saris which are woven of different materials such as fine, medium and coarse cotton (fig. 7.22), silk and art silk.

Gajendragadh and Shirol are well known for their *khans* which also use the same raw materials. Earlier indigo and other natural dyes were used but due to the non-availability of raw materials and the high cost of additional labour, only chemical dyes are now used. The coarse saris are sold locally, mainly to farmers. *Khans* are sold both locally and outside Dharwad district. Weaving communities comprise Muslims as well as Hindus. Among the latter, Devangas, Lingayats, and Kurvinashettis form the major groups. The weavers may be broadly classified into those who own looms, those who rent looms and the wage workers.

The pre-weaving processes are generally carried out by women and children. On an average, a weaver's family has eight to ten members. There are usually three looms in a house. A weaver weaves four to five saris per week. Compared to weavers' houses in other parts of India, theirs are not small, but they have inadequate ventilation and lighting. The looms are kept in the largest space available where the weaving operations are carried out. For small jobs like pern winding and hank opening, the cooking area and the verandah are often used.

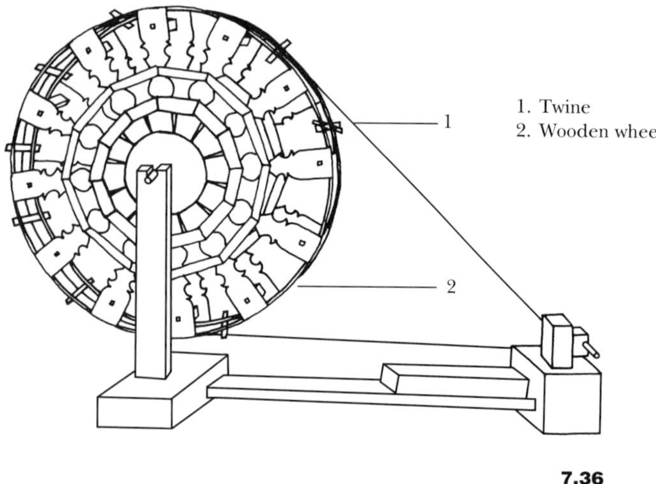

Charkhā/bobbin/pern winder. (Fig. 7.36)

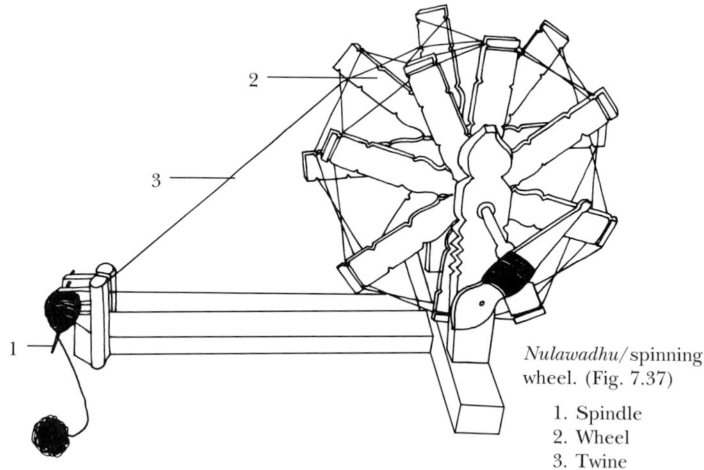

Nulawadhu/spinning wheel. (Fig. 7.37)

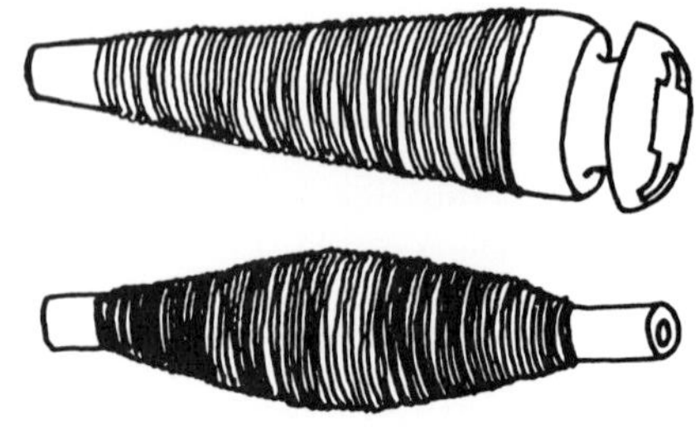

Perns. (Fig. 7.38)

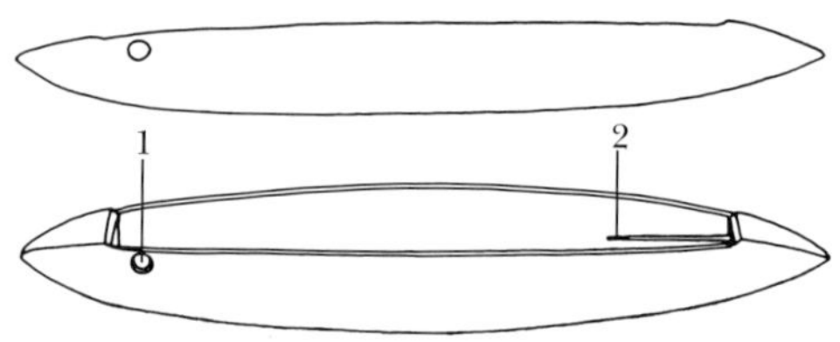

Shuttle made of buffalo horn. (Fig. 7.39)

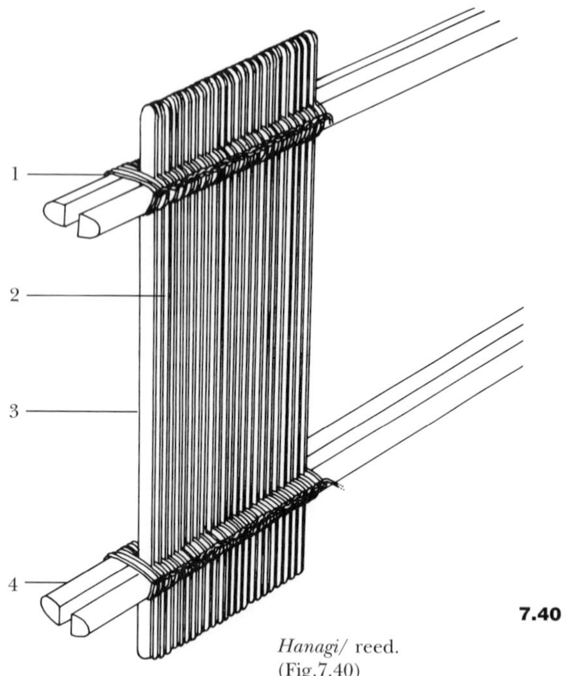

Hanagi/ reed. (Fig.7.40)

Tools Required for Pre-weaving and Weaving Process

The tools required by the pit loom weaver for the pre-weaving and the weaving process are:

1. The *charkhā* used as yarn winder (fig. 7.36) has a wooden wheel at one end and a peg at the other. The peg is connected to the wheel by a continuous loop of strong yarn. When the wheel revolves, the peg on which the bobbin or pern is mounted also revolves, thus enabling winding of the yarn.
2. The spinning wheel/*nulawadhu* (fig. 7.37) also has a wooden wheel at one end and a peg at the other. A continuous loop of strong thread connects the wheel to the peg. The spinning wheel is used to spin woollen fibres into yarn but in the present context, it is used by the weavers of cotton saris to convert yarn from hank into bobbin.
3. Perns (fig. 7.38) are made from the stems of plants or wood on which the weft yarn is wound, and are approximately 2 to $2^{1}/_{2}$ in. in length. These wound perns are then soaked in water. The wood swells up when soaked, preventing the yarn from slipping off.
4. Shuttles/*khandki* (fig. 7.39) are made of wood or buffalo horn the latter being preferred for three shuttle weaving. The shuttle is about 7 to 9 in. in length. It is handy and smooth for intricate weaving and is especially useful for weaving *pallus*.
5. The reed/*hanagi* (fig. 7.40) consists of thin splits of bamboo which are sandwiched between two long bamboo strips at the top and the bottom. These are tied firmly with twine to hold

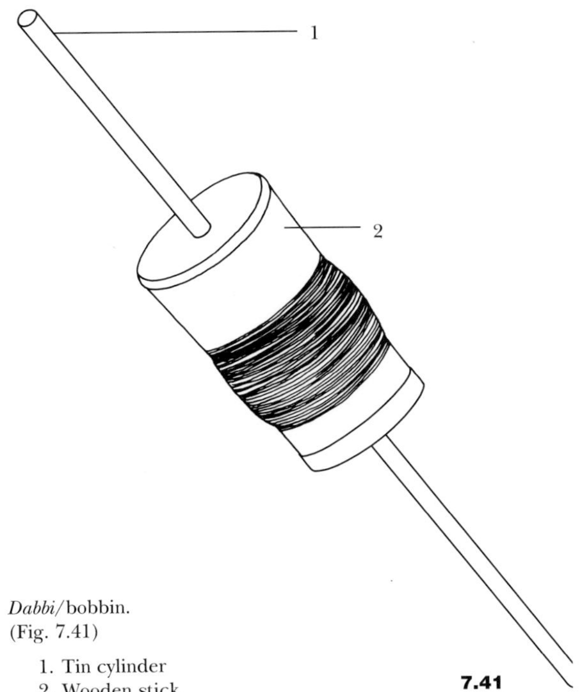

Dabbi/bobbin.
(Fig. 7.41)

1. Tin cylinder
2. Wooden stick

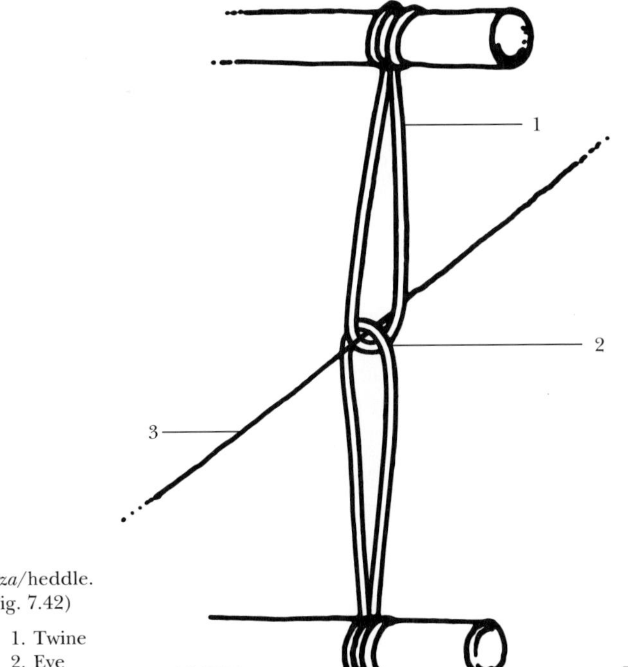

Beza/heddle.
(Fig. 7.42)

1. Twine
2. Eye
3. Warp end

the thin splits in place and the space required between each one is the width of the twine. Nowadays bamboo strips are being replaced by thin metal splits.

6. Bobbins/*dabbi* (fig. 7.41) are made of cylindrical tins. Both sides of the cylinders are closed leaving a hole on either side. A stick is inserted right through the holes. The bobbin winder holds the stick upright and rotates it manually to wind the yarn from a hank before making the warp.
7. Hollow wooden pipe/*hoonerse* is a device used for warping. The warping thread is passed through the pipe to guide it while warping.
8. The heddles/*beza* (fig. 7.42) are made of twine. The twine is first crossed between the fingers, forming a figure of eight for the required number of ends. Three bamboo sticks are passed through this bunch of threads, two in one loop and one in the other and then separated over the width required. The warp ends are then passed through the reed. After the reed is fixed, it is mounted on to a wooden frame and is ready for heddle making. Women generally carry out this task. The twine is passed through all the ends on the top layer and wound over a wooden block. After completing the entire length, the same process is repeated on the reverse side, interlacing with earlier loops.
9. Two kinds of dobbies are used, the barrel dobby and the lattice dobby.
   a. The barrel dobby (see fig. 7.43) is a circular structure with indentations on it. On one side of the dobby there are small rods which denote each pick. A lever gets attached to these rods at each pick to prevent the dobby from going

Barrel dobby, locally made. (Fig. 7.43)

1. Indentation in cylinder
2. Jack lever
3. Paul lever
4. Toothed area (ratchet)

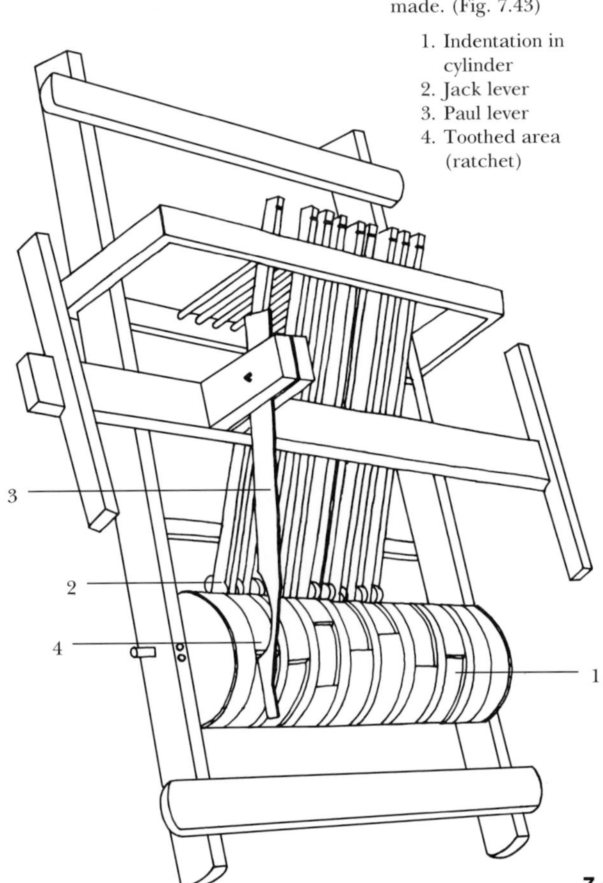

backwards. The extra warp ends are tied to jack levers/ *nakshiko* which do not lift when they come in contact with the depressions. The others which are in contact with the raised smooth surface are lifted to form a shed. This type of dobby can be used only for a particular type of design. Different designs with smaller repeats each have a separate barrel dobby.

b. The lattice dobby (see fig. 7.44) works on the same principle as the barrel dobby but is used for weaving of *khans* and *paraspet* borders. As the lifting order is more elaborate, a barrel dobby is not practical. Hence a pegged chain or lattice is mounted on the barrel dobby or the cylinder.

10. The beater/ *ṭhokrā* is a wooden frame to which the reed is attached.
11. The sizing brush/*kunchagi* (fig. 7.45) is a rectangular structure made of roots of shrubs and bound by coir. It is used for sizing the warp.
12. The warping frame (fig. 7.46) is rectangular and made of wood. On the horizontal sides three pegs are fixed, at uneven distances as shown in the illustration. The pegs on the vertical sides are fixed at equal distances from each other. The yarn is wound around the pegs according to the length required. The yarn is crossed for forming the lease on the pegs fixed on the horizontal side of the frame.
13. Weights (fig. 7.47) are used to keep the extra warp or supplementary warp in tension. These are either bags containing sand or grain, or chunks of metal with a hook at one

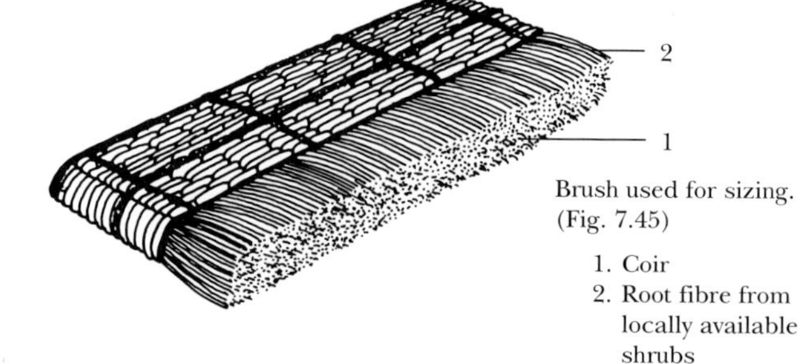

Brush used for sizing. (Fig. 7.45)

1. Coir
2. Root fibre from locally available shrubs

7.45

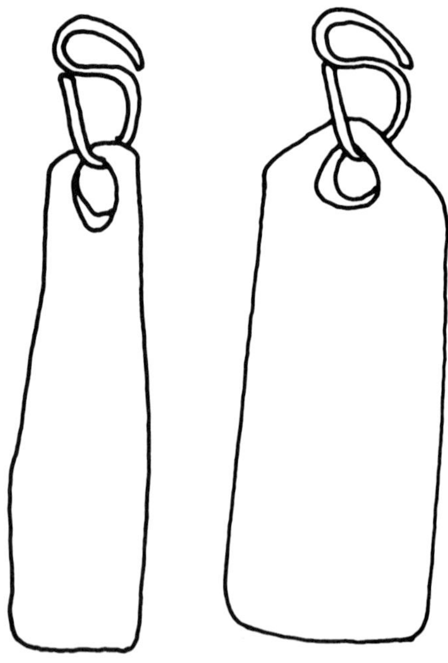

Warping frame used in pre-weaving process. (Fig. 7.46)

1. Pegs
2. Wooden frame
3. Warp
4. Lease

7.46

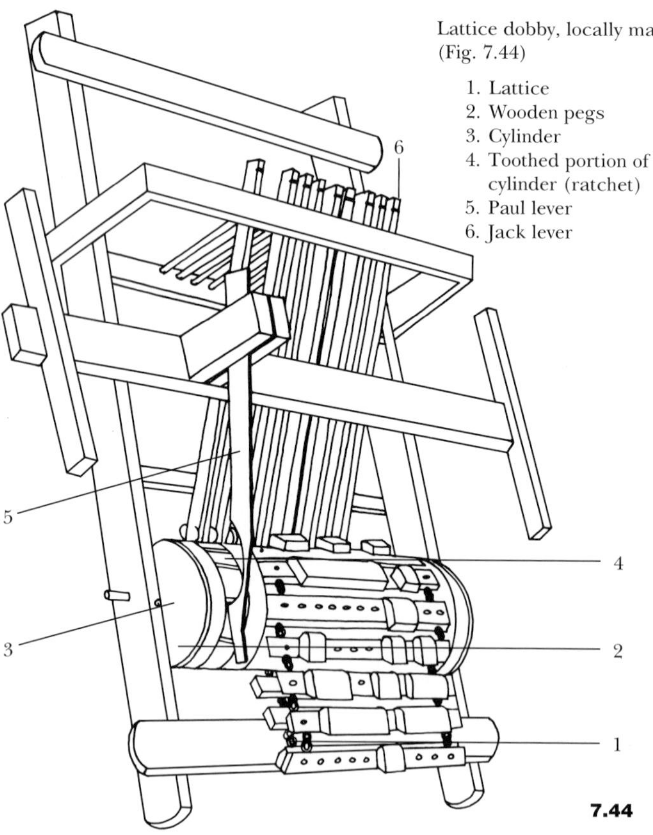

Lattice dobby, locally made. (Fig. 7.44)

1. Lattice
2. Wooden pegs
3. Cylinder
4. Toothed portion of cylinder (ratchet)
5. Paul lever
6. Jack lever

7.44

Weights used to keep supplementary warp in tension. (Fig. 7.47)

7.47

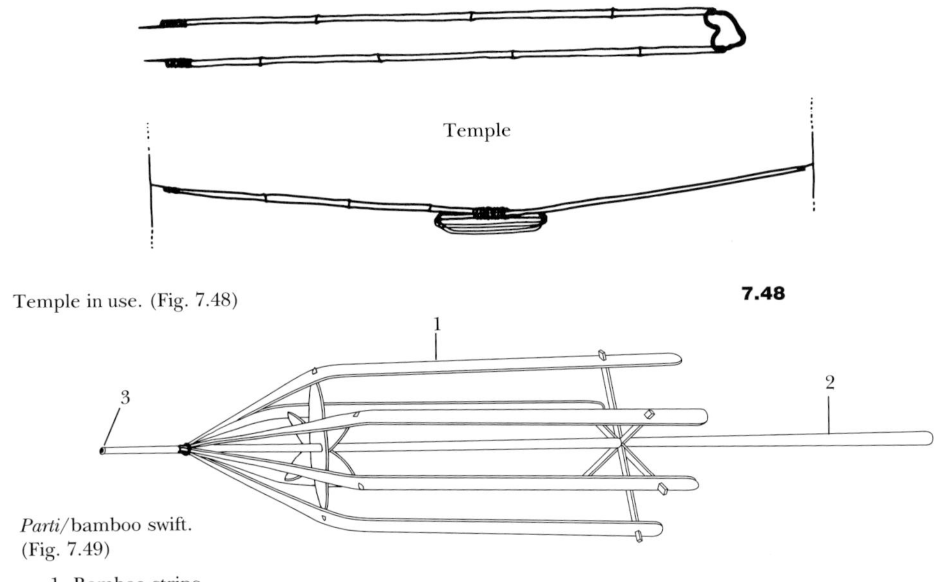

Temple in use. (Fig. 7.48)

Parti/bamboo swift. (Fig. 7.49)

1. Bamboo strips
2. Bamboo stick
3. Pivot point

Making of *sogadu* solution. Stage 1. Photographed at work place of Hanumantha Appa Reddi, Old Hubli. (Fig. 7.50)

side, which can be attached to anything. As an extra warp requires a separate beam, which is not in use here, these weights help to keep the tension even.

14. The temple (fig. 7.48) is made of two thin bamboo sticks which are tied together with twine. A sharp pointed metal is attached on one side each. This is then opened out and adjusted to the width of the fabric. The pointed sides are pierced into either side of the fabric to maintain its even width while weaving.
15. Bamboo swift/*parti* (fig. 7.49) is a device used for winding the yarn from the hanks before transferring it to bobbins. It is made up of bamboo splits supported by spokes from the inner side. The central bamboo bar forms an axis. It is locally called *parti.*

Pre-weaving Processes

These processes have been documented at the workplace of Laxman at Ilkal. The yarn has to undergo many preparatory processes before it can be woven into fabric. These used to be done manually, but today machinery has taken over, speeding up and facilitating the entire process. The master weaver who supplies the yarn, buys ready made dyed yarn from the market or obtains unbleached yarn and has it dyed by the local dyer.

Bleaching

This is done only when a pure white is desired or light pale colours are to be dyed.

Preparing the Dyes

Silk and cotton yarns in hank form are dyed in synthetic dyes in present times. A 70 gm. bundle contains 17 hanks. Forty bundles are dyed in a day. When this craft was documented in the year 1985 natural indigo dyeing was still being practised in Hubli, Laxmeshwar and Shiror. Very few weavers now use the expensive indigo dyed yarn due to the long and tedious dyeing procedure. Indigo also is not easily available. It has to be bought in the form of cakes from Kadappa in Andhra Pradesh. Prior to the actual procedure of vat dyeing it is necessary to describe the kind of vat used and two ancillary processes, that of making of *sogadu*, and the boiling of *Cassiatora* seeds.

Vatting Area

The vatting area is an enclosed space set apart in the work place where terracotta pots are arranged in parallel vertical alignments, 6 ft. in length, each row being separated from the next by a distance of 1 to 2 ft. Dried goat dung is piled up between the vats and then compressed down to a height of 1 to 2 ft. below the rim of the vessels. A mud and cowdung mixture is then applied ensuring a smooth surface. This plastering is redone every four years. When the dyer is working in the area for dyeing, the foot has to be placed in the space between two pots. This mud and cowdung mixture maintains evenness in temperature particularly during the winter months.

Preparation of *Sogadu*

The next stage is the preparation of *sogadu.* An earthen pot of the appropriate size is turned upside down and a section is removed from the lower portion. A perforated metal disc is placed at the mouth of the inverted pot. The disc is lined with some gunny sacking and on this a layer of charcoal chunks, approximately 1 cu. in. in size, is spread as a filtration agent. Fullers earth and lime/*chūnam* mix is then laid on this. The inverted pot is now placed above another vessel and the water is poured into the pot. The water which is filtered in this way is called *kara mira*/alkaline water (see fig. 7.50). The process of percolation lasts from four to six hours. When the colour of the solution being filtered changes from yellow (see fig. 7.51) to white, the process is discontinued and the remaining water is discarded. This solution is called *sogadu.*

Preparation of *Cassiatora* Seed Solution

The preparation of *Cassiatora* seed solution is done in a copper vessel. Three parts of *Cassiatora* seeds, called *nīlī bīja*, are cooked with one part jaggery. The mixture is stirred with a stick. Since jaggery requires a higher temperature for dissolving than the seed this admixture ensures a smooth paste. The process takes about six hours and the paste has to be used immediately. Straining is not required. Because of the admixture of jaggery, the solution can be used both for silk, an animal fibre, as well as for cotton (see fig. 7.52).

Preparation of Indigo Vat

For the preparation of indigo, the proportion of ingredients used are: 1½ kg. of indigo, 1½ kg. of *Cassiatora* paste, ½ kg. of lime and 2 : 1 of the *sogadu* solution. The last ingredient is *kali,* the residue from a live indigo vat. The amount used is left to the judgement of the dyer.

The required amount of indigo is reduced to a dry powder form with a stone pestle. This is mixed with a little water and left overnight for softening. It is then reduced to a paste after

7.51

Completion process of making *sogadu* solution. Photographed at work place of Hanumantha Appa Reddi, Old Hubli. (Fig. 7.51)

mixing it with the cooked *Cassiatora* seed. *Kali* is put into an empty vat along with the indigo paste and water and is mixed well with a stick. Lime is next dissolved and finally the *sogadu* solution is added. The following day a little more lime is added and the solution is stirred two to three times during the day. This process is continued for three to four days when the addition of lime is discontinued. The stirring goes on for another three to four days. During the dyeing process the starchy character of *Cassiatora* provides a temporary connection between the indigo dye and the fibre. The lime acts as a fixing agent (Holder, 1896: 40). The vat is usually ready in about a fortnight. Since the whole process is uncontrolled, the experienced dyer knows what corrective measures need to be taken from time to time. The dyer tests the vat by dipping his hand into the solution and assessing the colour on oxidation.

Preparation of Yarn for Dyeing

The yarn is taken to a plugged cistern, obviously a modern innovation. Sufficient water is added, as also *sogadu* solution in quantity equal to 4 per cent of the weight of the yarn. The hank is first pounded with the feet, then taken out and beaten on a stone. It is then squeezed out by hand in the cistern. This washing cleans and softens the yarn. However, to prevent weakening of the yarn through oxidization of the alkaline solution, and also to avoid patchiness in colour, the yarn is taken wet for dyeing (see figs. 7.53, 7.54, 7.55).

7.52 7.53

Boiling of *Cassiatora* seeds with jaggery. This paste is added to the indigo dye vat. Courtesy: Paniraj Konappa, Guledgud. (Fig. 7.52)

Hank washing. Pouring water into cistern and mixing with *sogadu* solution. Photographed at work place of Hanumantha Appa Reddi, Old Hubli. (Fig. 7.53)

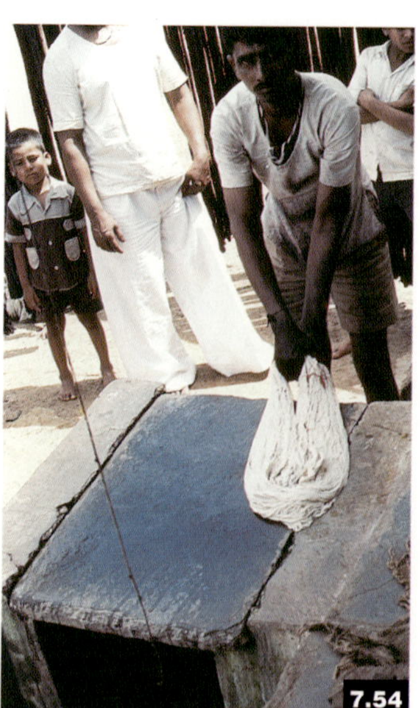

7.54

Beating of hank. Photographed at work place of Hanumantha Appa Reddi, Old Hubli. (Fig. 7.54)

7.55

Note rotary motion of hands in handling yarn. Photographed at work place of Hanumantha Appa Reddi, Old Hubli. (Fig. 7.55)

Dyeing

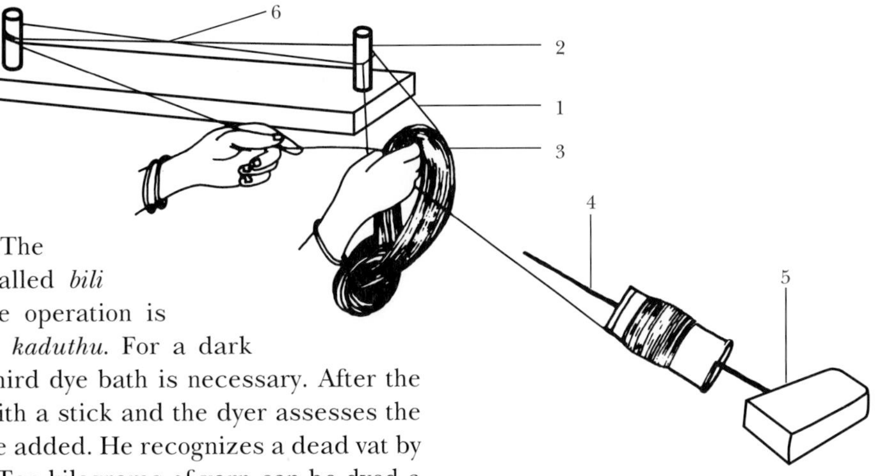

The froth on the surface of the vat is pushed to one side. The yarn is submerged in the vat by a rotary action of the hands, the hank being quickly circulated five times. It is squeezed dry in the vat and the same action is then repeated in a second vat. The hank is then sun-dried. The resulting colour is a light shade of blue called *bili kanth*. If a medium shade is required, the operation is repeated, the resulting colour being called *kaduthu*. For a dark blue-black shade, *karuthu nīlī* or *kanth*, a third dye bath is necessary. After the dyeing operation is over, the vat is stirred with a stick and the dyer assesses the proportion of lime and *sogadu* which are to be added. He recognizes a dead vat by its smell. At this point a new vat is started. Ten kilograms of yarn can be dyed a deep blue from 1 kg. of indigo.

The dyed hanks go through the following processes before weaving:

Pre-weaving process: looping of *pallu* warp with body warp. (Fig. 7.56) **7.56**

1. Body warp
2. Wooden peg
3. Body warp
4. Bobbin
5. Weight
6. *Pallu* warp

Winding: The hank is fixed on to a stand which has three sticks fixed on to a triangular base. The yarn is passed through a hook above the bamboo winder and wound on it.

Weft winding: Once on the bamboo winder, yarn for the weft is wound on perns with the help of a *charkhā*.

Bobbins: Used for warp winding.

Warping: The horizontal warping drum is used in Bijapur. This process simply involves combining yarn from several bobbins to form a warp of the required length and width.

The maximum periphery of the warping drum can be about 30 yds. and the minimum about 6 yds. But the majority are in circumference on which 16 yds. for three sari lengths are wound. The total number of ends is calculated and the width of the warp is made accordingly. The lease is made at the starting point by the insertion of a thread. The warp is half the length of the sari but with double the number of ends. Later, after the weaving of the *pallu*, the upper layer of the ends is cut and pulled behind to obtain the total length of the sari.

Sizing: The sizing is usually done by the hank-sizing method. This is done before sunrise to ensure that ends do not stick to each other through quick drying.

For silk yarn sizing, *saras*, a type of natural gum, is used. This is soaked in water overnight. In the morning the yarn is moistened with this solution and spread out for drying.

For cotton, flour of *jowār/ sorghum vulgare, katthā/ acacia catechu* and groundnut oil are used. *Jowār* flour is made into a smooth paste with the help of water. *Katthā* and groundnut oil are added, as they are believed to give strength and shine to the yarn. This solution is boiled for ten to fifteen minutes before being applied on the yarn.

The warp is stretched on to two sticks resting on a stand. A sizing brush is used to separate the ends stuck together. Entangled ends are separated with the help of a stick.

Looping: Of *pallu* warp with body warp (see fig. 7.56). This method of attaching a separate warp for the *pallu* has many advantages.

A different coloured yarn can be used for the *pallu* without dyeing by a tie-dye method and if the body is of cotton, silk yarn can be used for the *pallu*. The warp for the body once dried is arranged without extra labour in the form of a figure of eight. The lease is on either side of the loops of the figure of eight.

For winding the *pallu* warp a plank with two wooden pegs fixed on it is used. The distance between the pegs is planned according to the size of the *pallu*.

The yarn which is used for the *pallu* is taken from a bobbin or a bamboo swift. The end from this bobbin is passed through one section of the wooden pegs. The warp for the *pallu* is wound around the pegs in the form of a figure of eight.

The winder takes one round of the *pallu* warp over the pegs and leaves one loop (formed due to the warp being in the form of a figure of eight), of the body warp loose.

Piecing/*kichu* (see figs. 7.57–8, 5.4): The term refers to the attachment of the new warp ends to the corresponding ends of the exhausted warp, and is normally done off the loom by the female members of the weavers' family. Portions of the old unwoven ends of the sari warp are left on the loom after the sari has been cut out.

Each end of the new warp is attached to the corresponding end of the old warp by twisting them together. Because of the fineness of the yarn and the firm twist, the yarns do not separate. However, sometimes ash is taken in between the thumb and the first finger to get a firmer grip on the yarn. The shedding rods assist in separating the individual ends of the old warp.

The new warp threads once joined to the old warp threads are stretched along their length for even tensioning.

Variety of Products

The main products are saris, and *khans*. *Dhotīs* are also woven in some areas.

Saris

There are two main types of borders produced in the saris, *gadidadi* and *paraspet*. The saris are woven in plain weave with extra warp borders. Occasionally, stripes, *paṭṭa* and checks/*tirki* are also used. The borders and *pallu* are always in red/*kempa* in contrast to the body colours such as green and indigo blue. Red, different in shade from the border and *pallu* can also be used. Now, all colours are used because chemical dyes are available.

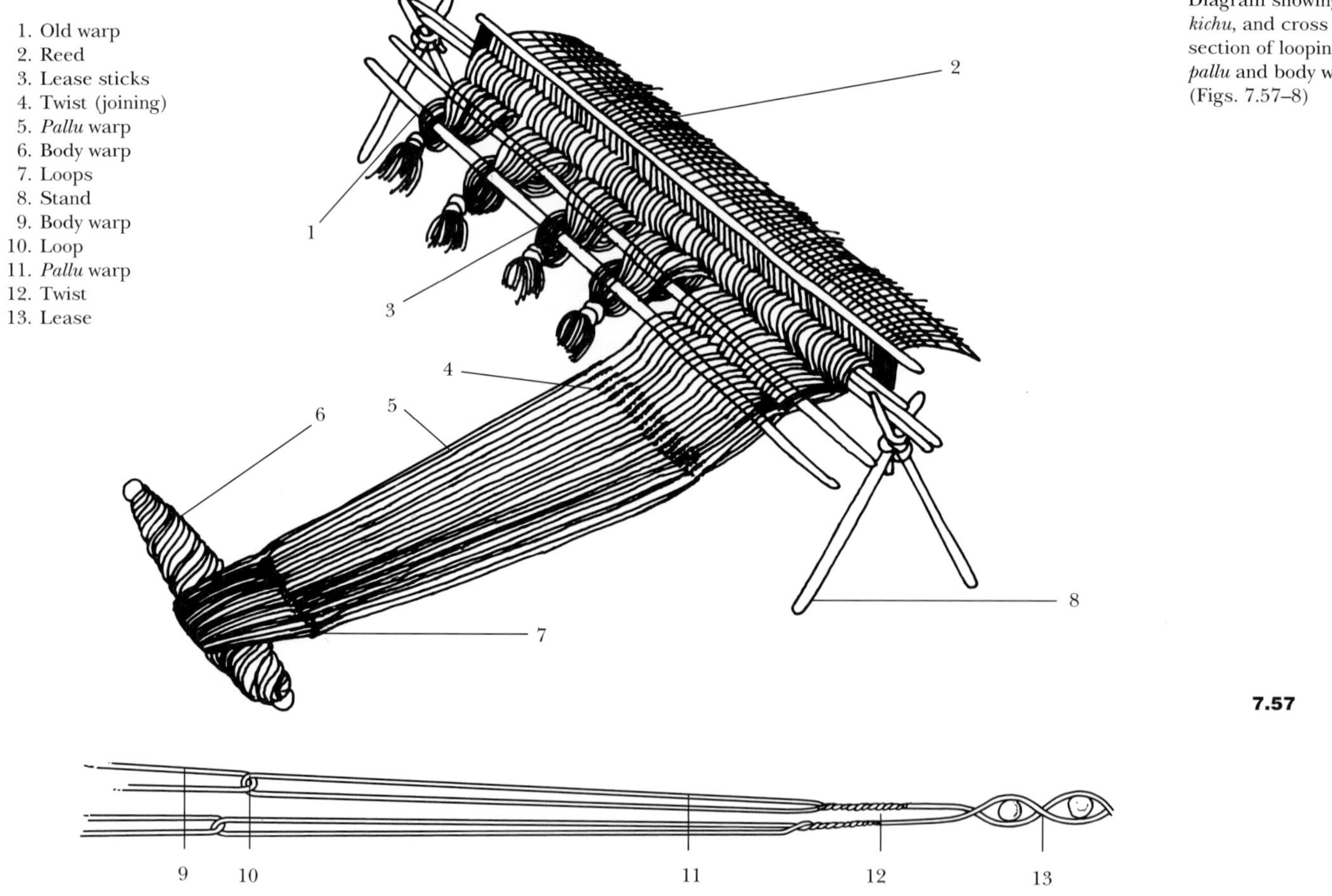

Diagram showing *kichu*, and cross section of looping of *pallu* and body warp. (Figs. 7.57–8)

7.57

7.58

Gadidadi, loom mechanism for lifting warp ends in weaving *Dharwad sari.* (Fig. 7.59)

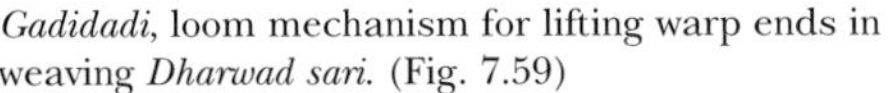

7.60

Gadidadi, border motif. (Fig. 7.60)

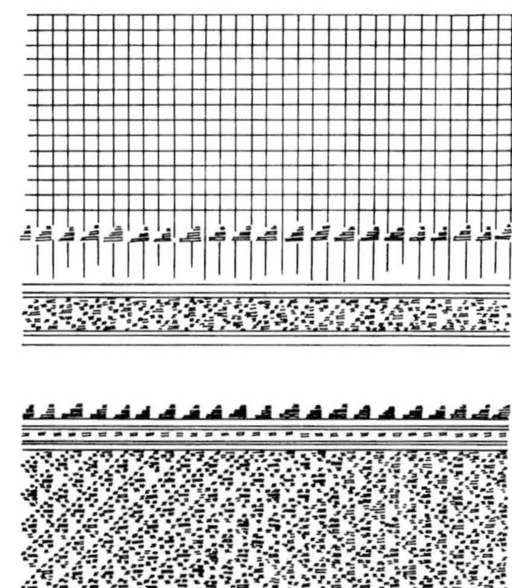

Chikki paras, border motif. (Fig. 7.61)

Types of Borders

Gadidadi: The mechanism used for lifting the extra warp threads is called *gada*; *dadi* is the border, hence the name *gadidadi* (see figs. 7.59, 7.60).

Paraspet: The word *paras* is derived from the local word 'pres' which is the term used for rhyming the last word in a line of verse. While weaving this border, the movement of the dobby is regular, and the sound produced is quite rhythmic, therefore this border came to be known as *paras*. *Peṭṭi* means boxes. The motifs formed with extra warp are in the shape of boxes.

There are two kinds of *paras* borders:

***Chikki paras*:** *Chikki* means small which has small boxes (fig. 7.61).

***Tori paras*:** *Tori* means bangle. The border resembles the zigzag of the bangle. Therefore the border is named *tori paras*.

Pallu

The *pallu* of all saris is always red. In some places the *pallu* length of the warp is tied and dyed; in others another warp is introduced by the looping method, before the warp is put on the loom. In the latter case, the *pallu* of the sari is heavier than the body ensuring a better fall as the warp ends are double in number. The *pallu* has bands of white/*bili* across the width of the sari between the two borders with small temple-like motifs at the edges. In some cases, these bands run across the whole width.

Traditional Motifs Used in the Border

***Tope teni*:** *Tope* is the red portion of the *pallu*. The motif formed by the three-shuttle technique where the white is introduced is called *teni* which means a bunch.

Paṭṭī kambi: Red and white bands which run throughout the width of the sari *pallu*. The bands of white are generally five in number. Numbers which are not divisible by two are used. Even numbers are considered unlucky. A maximum of seven stripes are used. This is done to save labour and reduce the cost of the sari.

These saris are mostly worn by the women of the agricultural class. However, the night after the harvest is a night of celebration, when silk saris are in demand.

Colour Significance

Certain colours have special significance. These include:

Red/*kempa*: Red is considered an auspicious colour and is always used for the border and *pallu*. It is essential for married women to wear this kind of sari with a red border and *pallu*. Widows wear either plain white or an all red sari. When a red, darker than the border red, is used for the body of the sari, it signifies a *sumaṅgalī*, a married woman in the same way as the *kumkum*, the vermilion used in the hair parting, or as the *bindī*, the red dot on the forehead. This sari is given at the time of marriage when the *maṅgalasūtra*, a special necklace, is worn.

Green/*hasiru*: Green is a sign of fertility. It is customary to give the girl a green and red sari at the time of puberty, along with rice. This is to signify that the girl is ready for marriage.

Indigo blue/*nīlī*: This type of sari is called *chandrekali*, and is presented to the bride from her nearest relative. It is given by a married lady, preferably the mother. Earlier, the term *chandrekali* was used to refer to an indigo blue sari on which tiny dots in white had been intricately embroidered in *kasuti*. It is very delicate embroidery associated with north-west Karnataka and adjoining territories. *Chandrekali* were usually gifted at the time of the *Saṅkrāṅti* festival to mark the winter solstice.

Ilkal Saris

Ilkal saris are woven in and around the villages of Kolhar, Ilkal, Kamatgi, and Nidigundi. The name originated from the village where the weaving of this type of sari was first started.

The loom used is the pit loom with barrel dobby mechanism attached for achieving the extra warp patterning in the border. The body of the *Ilkal* sari is woven in cotton along with art silk with a border and a *pallu* of art silk; or in mixed art silk and cotton with a pure silk border and *pallu*; or even a body of silk and cotton with a silk border and *pallu*. Saris with a pure silk body, border and *pallu*, are later introductions.

There are two types of borders for *Ilkal* saris: the *gomi* border and the *chikki paras* border. In *chikki paras*, there are two varieties: *kaddi chikki* which means small dots or stars and *kondi chikki* which means big dots or stars. The two borders are quite similar except that the triangles are smaller in the *kaddi chikki* and hence more can be used. For the big dots the repeat is of eight picks; the smaller type is complete in four picks. The extra warp is called *peti* and the border, *dadi*.

There are a large number of variations in the size and composition of the checks in the body of *Ilkal* saris. Accordingly, they have different names. The term for checks in Kannada is *tirki*. A description of these is given below:

1. Simple checks of two and four ends are the most common. These are called *ragavali* and the sari is called a *ragavali* sari.
2. Smaller checks are called *kaddi chikki*.
3. *Maṇdap* checks are formed with two warp ends and picks positioned at a certain distance to form larger checks.
4. More elaborate checks have been a recent addition. These have fancy names like *patang* checks or shirting checks.
5. *Dapla* checks are like the *maṇdap* but smaller.
6. Checks with three or more coloured ends are called *pañcaraṅgī*.
7. Broad stripes in the body are called *jabra paṭṭā*. This word originated from zebra stripes.
8. Finer lines are called *suzī paṭṭā*. *Suzī* means a needle.
9. Saris which have a plain bluish black body are called *chandrekali*. A red and orange combination is called *basanthī*. Other saris using the same colour in the warp and weft are named according to the colours, e.g. *bili* for white, *haldī* for orange, *kempo* for red.

Pallu of the Sari

The *pallu* is woven only in red and white. The *pallu* is woven in the single interlocking technique and is called *tope teni*. In Marathi, the *pallu* is called *padre* and in Kannada, *sharag*. The reed used is number 52s for the body and 80s for the border. The denting is two ends per dent throughout.

The *Ilkal* sari is woven in lengths of 6 or 9 yds. with a width of 48 in.

The price range of an *Ilkal* sari varies according to the materials used. The cheapest are worn by the poorer classes who use art silk and cotton in saris worn for special occasions. The *pallu* here is given more importance, because it is always worn over the head. People consider it prestigious to wear a sari of silken sheen with a *pallu* on special occasions (*Cotton Saris Bijapur*, NID, 1985).

Ilkal Sari Loom

The *Ilkal* sari loom is a pit loom with barrel dobby attachment (see fig. 7.62). Originally a throw shuttle pit loom, it has seen changes through time. Due to the demands on the weavers for faster production today, the loom has been converted to a fly shuttle pit loom. In the local language it is called *kumi-magga*, implying a pit formed below the ground level from where the weaver operates the pedals to lift the required shafts for shed formation. This pit (1) is approximately 2 ft. in depth, 2 to 2½ ft. in width, and 3 to 4 ft. in length. The weaver sits in a depression formed at ground level close to the centre of the cloth beam, so that his feet rests on the pedals. The beater (6) with the bamboo reed (5) is suspended from the wooden frame and is placed in between the cloth beam and the heddle shafts. There are two heddle shafts for weaving of the plain ground weave. The cloth beam has a take up peg (3) which the weaver operates manually for winding the woven fabric.

The barrel dobby (14), mounted on top of the loom frame is a locally made device for extra warp ornamentation. This dobby is cylindrical in shape and has permanent indentations according to the shedding requirement of the design. This type of dobby can be used only for a particular design, where the repeat size is small. Extra warp threads are taken through the heddles of small shafts placed behind the ground weave shafts on either side (9). These are weighed down with metal weights, bags or bolts, that is, anything that can provide the necessary weight to produce the bottom line of the shed (12). Each small extra warp shaft is harnessed to the levers of the dobby, which are activated by a pedal.

The warp beam is attached to another beam which is parallel to it. From the centre of the second beam is tied a rope (17). This rope is wound around the wooden peg behind the two warp beams holding the warp in tension and in position. This rope is then tied within an arm's length of the weaver, to a peg driven in the ground. The weaver uses this rope to release the warp towards him without his actually having to get up from the loom.

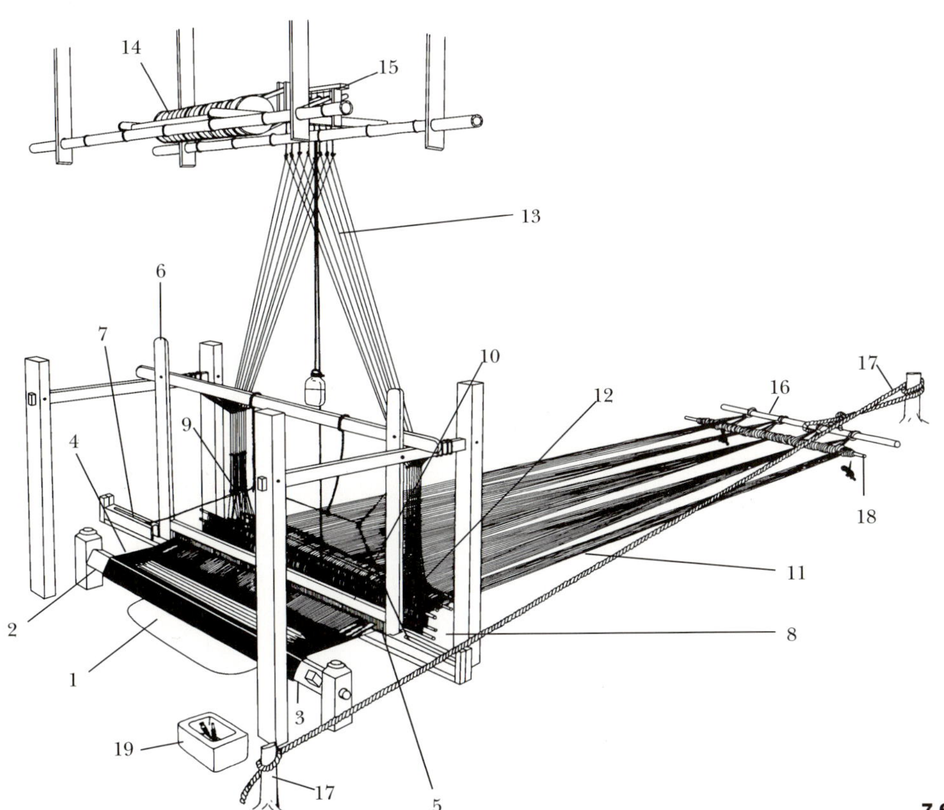

7.62

Ilkal pit loom with dobby attachment. (Fig. 7.62)

1. Pit
2. Cloth beam
3. Takeup peg
4. Woven sari
5. Reed
6. Sley
7. Shuttle box
8. Heddles
9. Extra warp harness
10. Lease sticks
11. Warp
12. Weight for harness
13. Twine tied to jacklever
14. Barrel dobby
15. Frame for dobby
16. Back rod
17. Rope for tensioning the warp
18. Warp for extra warp
19. Container for perns

Khans of Maharashtra and Karnataka

Khaṇḍa, now known as *khan*, means portion or piece in Marathi. One *khan* is a piece from the yardage of fabric woven specially for making sari blouses/*cholīs*. These blouse pieces are coordinated with or matched to complete the ensemble with the colour, pattern, composition and texture of the sari. According to Baba Kasurkar (Chairman, Sholapur District Small Industries Corporation), the craft of *khan* originated in Guledgudda, Karnataka, *c.* AD 1478. It flourished during Shivaji's time and spread all over Maharashtra. The migration following the famine of 1676–7 was also responsible for the spreading of this craft in Maharashtra (*Khans of Maindergi*, NID: 2). It is reported that in about AD 1890 the following categories in *khan* were being made:

1. Pure silk *khans* either plain or combined with gold threads, were produced in Belgaum and Kaladgi in Belgaum district, Hubli and Gadag in Dharwad district, Guledgudda and Ilkal in Bijapur district and in Nasik.
2. *Garbasūtī* (cotton warp and silk weft) plain or with *zarī kinārī*/border *khans* were produced in Gokak and Kaladgi in Belgaum district and also in Pune and in Ahmednagar.
3. *Khans* with cotton body and silk border were produced in Dharwad, in Ilkal and Guledgudda in Bijapur district, in Ravanda in Kolaba, in Pune, Sholapur, Bhiwandi in Thana and in Mumbai.
4. *Khans* with cotton body and *Garbasūtī* border were made in Rahuri and Sangamner in Ahmednagar district.

Around 1978 *khans* were woven in Maindergi, Kaladgi and Hiroli villages in Sholapur district, Maharashtra. In the Dharwad district, Karnataka, *khans* were also woven similar to those from Maindergi including checked ones.

Weaving Process

Khans are woven in narrow widths of 30 to 33 in. with borders. The main field is woven in a plain weave. It may be checked, plain or have all-over extra warp ornamentation. For this a lattice dobby, which can accommodate a larger number of steps, has to be used.

The process involved in the making of *khans* is similar to that of the *Ilkal* sari which is made on the pit loom. The main differences are in the type of dobby used and the manner in which the extra warp/patterning warp is arranged over the loom. In Maindergi, Padmini Tolat (*Khans of Maindergi*, NID: 2) observed that the village carpenter was the person who knew how to set up the pegs on the lattice according to the peg plan of the design by transferring this from the selected graph. He would do so by referring to his old graph book containing all the traditional designs.

After having set up the loom and the ground warp, the extra warp is raised above the ground near the warp beam at almost a 45° angle to the ground warp and hung from a bobbin suspended from the ceiling. A weight is suspended from this warp which keeps it in tension and allows its automatic and appropriate release.

The Loom

The loom used for weaving of the *khans* is similar in most respects to the Ilkal pit loom on which the *Ilkal* saris are woven (see fig. 7.63). The main difference is in the type of dobby that is used for weaving of the patterning with supplementary warp.

This dobby works on the same principle as the barrel dobby (fig. 7.43). However, as the lifting order for the *khan* designs is longer and more detailed, a chain with pegged lags or lattice is mounted on the cylindrical barrel. The chain is made up of lags equal to the number of picks in the repeat of the design. This is called a lattice. Each lag is pegged as per the peg plan of the design.

Just as the jacquard was developed in France as a solution for the production of the *kani paśmīnā* shawls which is an Indian product, it would appear that the lattice dobby is an English invention devised to facilitate the production in England of the *dhotī*, another Indian product. In 1868, the prototype of the lattice dobby, the Keighley Type, was patented by Hattersley and Smith. Holmes' description: 'The dobby machine is a shedding motion, used for conveniently working the healds up to 24 shafts or for *dhotī* borders upto 40. By the aid of pegs placed in a revolving lattice any heald can be selected and raised by the machine, a peg indicating a heddle up and a blank a heald down' (Holmes, 1914: 112).

Khan loom. (Fig. 7.63)

1. Pit
2. Warp
3. Peg
4. Tensioning rope
5. Shaft
6. Shuttle box
7. Heddles
8. Extra warp harness
9. Lattice
10. Warp for extra warp patterning
11. Weight
12. Frame for dobby

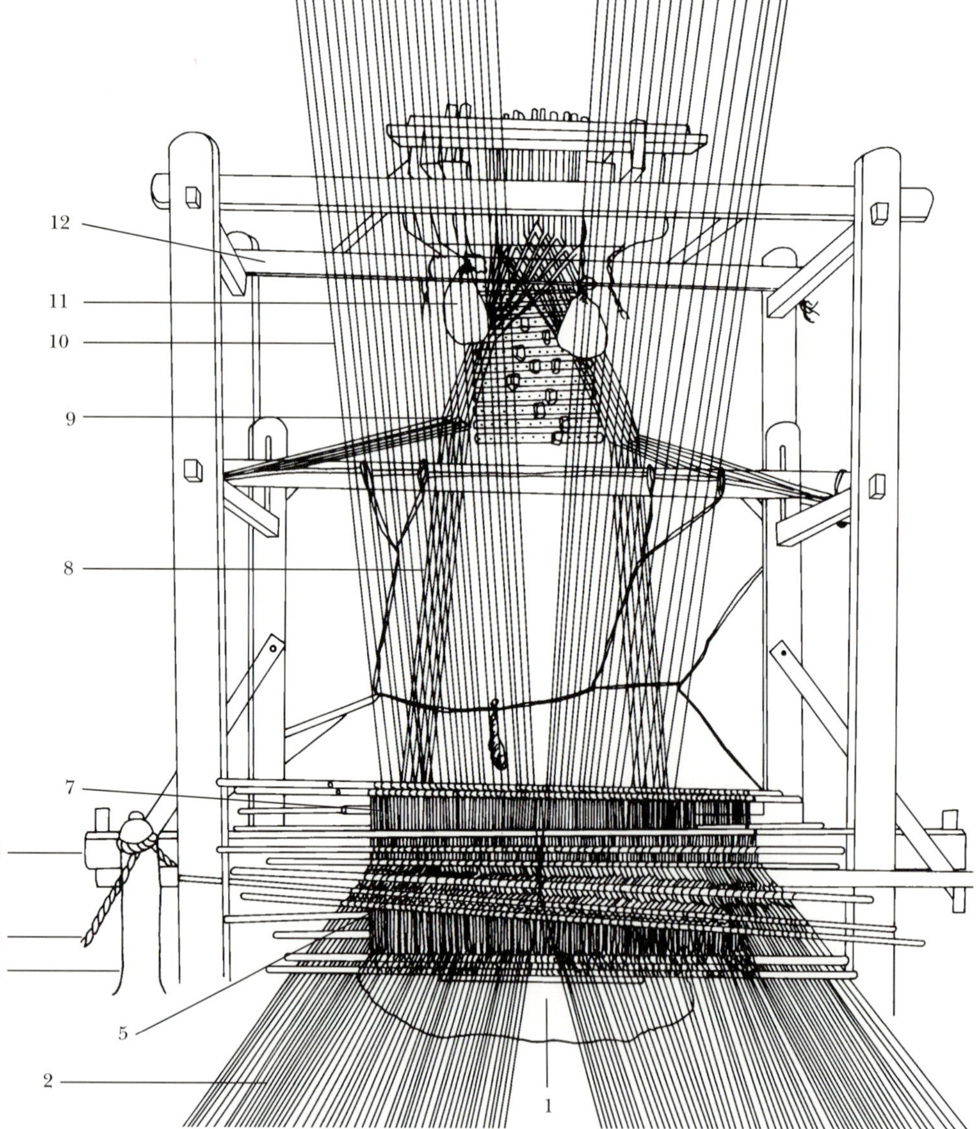

7.63

Upon the expiry of the patent, several concerns in England began to specialize in the manufacture of this contrivance, the most notable being Lupton & Place Ltd., Burnley (Holmes, 1914: 112).

The question arises as to whether the dobby was an Indian loom mechanism perfected in England and then re-exported to its place of origin, or, whether the origin itself was to be attributed to the English. The system of levers on which the barrel dobby is based (fig. 7.43) was not alien to the Indian tradition. Indeed, the barrel dobby with indentations seems to be an indigenously developed device arising out of the need for expediting and increasing the efficiency in production. The use of these dobbies appears to be restricted to the states of Maharashtra, Andhra Pradesh and Karnataka. It is of interest that the dobby was absent in areas with a strong tradition of ornamentation through pedal and shaft as in the *ghoḍa shāl* and *gaṭṭhuā* loom accessories as also in the areas with a strong orientation to *aḍai* and *jāla*. It would appear that where the dobby is found in Varanasi it might be taken as an intervention. On the basis of this cumulative evidence it would appear that there is a case for the assumption that the barrel dobby has been indigenously developed by Indian weavers.

Characteristics of the Fabrics

As the border warp is denser than the body warp, the reed used for it is denser. Some fabrics which are sixty to seventy-five years old, have been analysed revealing that the borders have 92–6 warp ends in one inch whereas the body has from 50–62 warp ends per inch.

The main distinguishing characteristics of the *khans* are the visual and textural qualities achieved with the patterning weave structures, the type of yarn, the colours and the patterns used. The details of the variations of colour and pattern are given here.

Variations in the Body

1. Plain body with warp and weft of the same colour.
2. Plain body with warp and weft of different colours achieving a shot effect due to interaction of colour and play of light called *dhūp-chhāon*/sun and shade.
3. Body with stripes formed with the plain weave and/or with supplementary warp.
4. Body woven in checks. This is done by the insertion of the desired coloured weft interlocking with the warp ends close to the border warp ends. Hence this yarn does not pass through from selvedge to selvedge. When required to be woven again the shuttle is inserted as before. Long floats are seen on the back face of the fabric showing that the weft was carried over after the required picks of the ground of another colour.
5. Body with small motifs patterned with supplementary warp along with stripes or checks. The vertical stripes may be woven using the supplementary warp or plain weave and the horizontal stripes forming the checks in plain weave.
6. Body with all over supplementary warp patterning.

Variations in the Border

1. Border with stripes in plain weave and supplementary warp.
2. Border with *gomi* and *paraspet* patterning using supplementary warp.
3. Border with stripes utilizing some of the patterning used on the body in supplementary warp.

The colours used for the warp of the main field of the fabric are red, purple, yellow, green, blue, pink, reddish brown, crimson, orange, black and white. The border warps are usually in dark colours and in contrast to the main field colours like maroon, deep crimson, dark blue, and black.

The weft is black, dark maroon, dark crimson or dark blue. Indigo blue-black predominate in warp colour.

Silk or rayon and cotton yarns are used in the making of these *khans*. Earlier gold and silver threads were used but these have now been replaced by lurex.

Simple geometrical forms such as squares, diagonal lines, triangles, and patterns inspired from nature and from ornaments locally worn, form the basis of the design repertoire of the *khans* (fig. 6.50).

In establishing core areas of different techniques the following parameters have to be kept in mind: first, presence of a number of variations in design along with their nomenclature in the given area; second, other things being equal, a continuity in design into present times with a clientele drawn from the sector unaffected by changes in fashion; third, use of traditional technology except in areas in which identifiable interventions have taken place. The south Kanara region satisfies all these criteria in relation to the design repertoire associated with the saris of the Dharwad–Hubli area which has spread throughout the Maharashtrian culture region.

In present times the surrounding areas of Hubli and Gadag are the main centres of weaving. Gadag and Dharwad are mentioned by E. Balfour as centres of production *c.* 1884 (Balfour, 1976: II, 831). This area had access to cotton and indigo and in an earlier period *Indigofera tinctoria* was cultivated here. The specifics relating to pattern of sari in the region is the presence of indigo shot colour or variegated checks in the ground.

The *molakalmuru* sari has three-shuttle weaving in the border but the technique of *ikat* may also be found (see figs. 7.64, 7.65). In the case of the *Udupi* sari, the *pallu* is tie dyed and since there are denser ends in the border than in the ground, the ground and border picks are adjusted through the method of dovetailing (fig. 7.66).

Today, *Ilkal* production is geared to the weaving of both saris and *khan*. Earlier silk *dhotar jode* and *uparna* were woven. The

7.64

7.65

7.66

Molakalmuru sari, Karnataka. Courtesy: Weavers Services Centre, Bangalore. (Fig. 7.64)

Molakalmuru sari, Karnataka. Courtesy: Weavers Services Centre, Bangalore. (Fig. 7.65)

Udupi saris. Courtesy: Weavers Services Centre, Bangalore. (Fig. 7.66)

dhotī was 4 yds. in length and the *uparna* 2 yds. The only ornamentation for the border was the *rudrākṣa*. Saris from Ilkal and *khan* from Guledgudda are now a part of gift exchange at marriages in Maharashtra. The bride's party can specify five types of *khan* and *Ilkal* saris which can be expected to form part of the gifts made by the family of the groom.

Narayanpet

Although today Narayanpet is located within the territories of Andhra Pradesh, in earlier times there appears to have been some Maratha affiliation. The town which had earlier served as the centre was called Lokayapalle. A neighbouring village called Dhanvada indicates that it may have been once the seat of the treasury. Marathi weavers appear to have migrated at the time of Rajaram, AD 1680–1700. Indeed the division between Andhra Pradesh and Maharashtra was none too clear in the medieval period. The link between the Telangana region and Maharashtra is demonstrated even today by the popularity of Pandharpur as a place of pilgrimage for the people of Telengana (oral communication: H.H., Raja Rameshwar Rao of Wanaparty).

Narayanpet saris, both of cotton, and mixed cotton and silk, are very similar to Karnataka products both in layout and weight. In fact, the 9-yds. sari made at Narayanpet was, prior to the 1950s, sold exclusively to buyers in Kolhapur, Miraj, Sangli and Poona, and here again, it is to be borne in mind that Sangli, Kolhapur and Sholapur formed part of the Nizām's dominions in earlier times. For reasons which are not clear but which may point to a dominant cotton tradition, Maharashtrians prefer lighter silks to the heavier ones popular in Tamil Nadu. The notable exceptions, however, are the *Paithani* and the *pītāmbara*. Merchants from Maharashtra came to Narayanpet and made spot purchases.

The market for *Narayanpet* saris in Andhra Pradesh opened up in the 1950s, while the all-India market began to be explored only around 1970. With the development of the power loom industry in Sholapur and Ichalkaranji, as many as 40 per cent of the handloom weavers migrated from Narayanpet to the new centres. This has led to some decline in Narayanpet.

In consonance with the Udupi and Kumbakonam saris, *pallu* tie dyeing has been introduced in recent times. Borders are ornamented by means of the barrel dobby. The pit loom prevails. The Punu type of *Paithani* was woven here in the past. Unlike Gadwal, tested *zarī* is used at Narayanpet in present times.

Gadwal

The history of the kingdom goes back to the year AD 1290 when Raja Pratap Rudra Deva of Warangal conferred this fiefdom in grant to Bukka Polavi Reddi. Thereafter the state had a checkered history with inevitable interjections from the Marathas (Mudiraj, 1934: 620 A, 620 X).

Gadwāl Sari

The *Gadwāl* sari derives its name from the town of Gadwal in the Mahbubnagar district of Andhra Pradesh where these saris originated and continue to be woven. In the Vishakhapatnam and Telangana regions of Andhra Pradesh, cotton saris with ground of natural unbleached yarn with red *zarī* borders were preferred as wedding saris. Brahmin women would dip their white saris into an impermanent turmeric solution.

The typical *Gadwāl* sari would be of 80 count cotton, 8 yds. in length with pure silk border and *pallu* in contrast colour (fig. 7.67). The width of the border would be about 2½ in. Extra warp patterning is done using *zarī* in the border and the extra weft patterning is done also using *zarī* in the *pallu* with the plain weave commonly used as the ground weave. The *zarī* is made of pure gold-plated silver wire which is wrapped around a silk core yarn. There was no earlier tradition of weaving the main field in silk. Checked designs were called *podi* (personal communication: Srilata Bhoopal, Hyderabad).

The art of weaving with the *jāla* though learnt from the north, was used in the *Gadwāl* sari, which developed its own identity amongst the south Indian category of saris. The two *Gadwāl* saris found on display in 1993 at the Salar Jang Museum, Hyderabad, provide a glimpse of what *Gadwāl* saris may have looked like prior to the design interventions of the 1950s. The patterning was strong and distinctive. The tope motif in the *pallu* was suggestive of the Karnataka–Dharwad tradition, the *karvathi kathi* motif in the border brought in elements from Maharashtra, while the miniscule temple motifs in three-shuttle weave in the border linked it with products of Tamil Nadu. All these elements were so well integrated that they suggested a mature tradition rather than a hybrid amalgam. In present times only the ground pattern seems to find an echo in the indigo-dyed saris woven at Uruvakonda in Anantapur district.

The presence of *jāla* rather than *adai* as a loom mechanism can be explained by the fact that Maharani Adi Lakshmi Devi, who died in 1953, sent weavers to the major centres of weaving in India, to learn new techniques. It was probably at this time that a new range of techniques was introduced extending from the double interlock of *Paithani* to the utilization of the *Banarāsī jāla*. Gadwal has no major repertoire of design with which it can now be identified. The small *bādām* motif in the *pallu* could, however, be related in shape to the local mango called *banganapalli.* Pure silk saris began to be woven in Gadwal in the late 1940s (see figs. 7.68). According to B.C. Mohanty (Mohanty, 1984: 179), the *jungu* system of lifting warp threads for brocading work was to be found in operation on some looms.

7.67

7.68

Gadwāl sari. Courtesy: Kumudini Devi of Gadwal, Hyderabad. Note hybrid element in design. (Fig. 7.67)

Gadwāl sari. Cf. Pune *Paithani.* Courtesy: Kumudini Devi of Gadwal, Hyderabad. Note hybrid element in design. (Fig. 7.68)

Weaving Processes

A throw shuttle pit loom with a *Banarāsī*-type *jāla* (see fig. 7.69), is commonly used for operating the warp threads for extra warp and extra weft patterning on the sari, whereas the ground weave is achieved with the manipulation of the heddle shafts through which the warp ends are drafted. These shafts are tied to pedals and are foot-operated.

The solid colour borders are achieved through the method of single inter-locking tapestry in three-shuttle weave as is done in *Kanchipuram.* Two shuttles with silk yarn are used for the border and one shuttle with cotton yarn is used for the body. As the beginning of the weaving of the *pallu* is reached it is woven in silk with one shuttle of the same colour as the border.

To effect a firm joining of the border to the body, very often a *kumbha* or *kotakomma,* also called a temple pattern, is adopted so that the silk weft of the border can be woven into the body. The three-shuttle interlocking style of weaving is locally called *kupudam* or *tippadamu* (Mohanty, 1984: 180). However, the saris may not always have the *kumbha* pattern as mentioned above, as they are also woven with the border wefts and the body weft in the single interlock technique to form solid coloured straight borders.

At the initial stages of draping the sari a section is tucked in at the waist. Skin contact with metallic yarn can be uncomfortable. Hence, in this section, extending between 2½ to 3¾ yds., extra warp patterning is sometimes done in cotton. For the rest of the border, the *zarī* extra warp ends for patterning are pieced together and the weaving is continued. Sometimes small motifs/*būṭīs* are interspersed in the main field and the border of the sari, using the localized extra weft method of patterning.

The method of piecing described above which is a common practice in other centres of weaving in India, is used here also. After the sari is woven up to the beginning of the *pallu,* each *pallu* warp end which is made of 2 ply, 20–2 denier silk is carefully twisted together with one end of the cotton body warp. The entire width of the *pallu* warp is connected to the body warp in this way, while maintaining the tension. This method of attaching the silken *pallu* warp to the sari body warp in cotton and the twisting of two different materials, silk and cotton, together is considered to be a speciality of the *Gadwāl* weavers making it a unique feature of the sari. It may also be regarded as a variation of the *petni* of the *Kanchipuram* tradition (figs. 5.52, 7.67).

The *pallu* is woven using coloured silk weft yarn. This matches the warp yarn of the borders and the *pallu* warp. Motifs that form a part of the *pallu* are woven with *zarī* in the localized extra weft patterning method. About 2 in. of the *pallu* warp ends are left unwoven on reaching the end of the warp. These unwoven *pallu* warp ends form the fringe. These are the uncut looped ends of the main field of the sari warp, which have been slipped off the warp beam. However, the border warp ends which also form a part of the total width of the sari are distinguished by the absence of loops in the fringe.

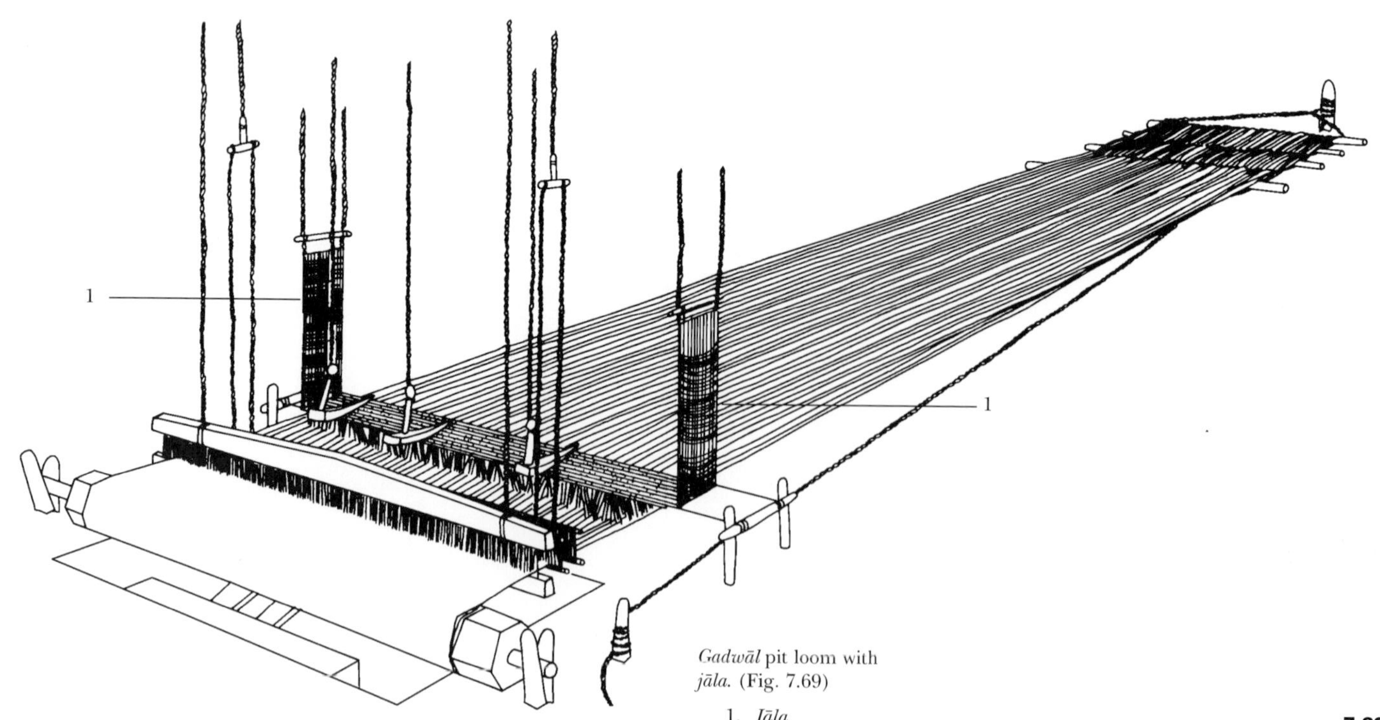

Gadwāl pit loom with *jāla.* (Fig. 7.69)

1. *Jāla*

7.69

Wanaparty and Kothakota

The nucleus of the state, later to be called Wanaparty, came into existence in the year AD 1510. In this year, Veera Krishna Reddy from Cuddapah, after establishing himself at Patapally, enlarged his sphere to include Sugur and Kothakota. He was recognized by Krishna Deva Raya of Vijayanagar, and the kingdom became known as the Samasthan of Sugur. The capital shifted from Sugur to Kothakota *c.* 1686. With the change in power, Asaf Jahi recognition was obtained and, in 1807, the capital was once again moved, this time to Wanaparty. From the point of view of weaving, the important consideration was access to *Gadwāl* weaving, the two kingdoms being separated only by the river Krishna.

According to Raja Rameshwar Rao of Wanaparty, Kothakota developed into a centre of silk weaving between *c.* 1930 and 1940. Weavers were brought in from Gadwal and saris in pure silk began to be woven. In 1958, largely at the initiative of Raja Rameshwar Rao, himself, a Pilot Weaving Centre was opened at Kothakota by the All India Handicrafts Board, New Delhi, and in 1960 the premises were handed over to a local cooperative society. During this period an intricate hybrid design repertoire was developed which showed a marked departure from traditional Andhra norms. Weavers were Maharashtrian as was also the market, and, inevitably, the *Paithani* repertoire was copied (see fig. 7.70). This explains the presence of the Pune *Paithani* type of sari in Andhra Pradesh. In present times the production at Kothakota appears to be virtually dictated by the demands of the marriage market. In view of the rich ornamentation of such saris, each warp is limited to one sari length of 6 yds. A weaver's shed would have at least two 6 yds. lengths stretched across from opposite ends of the room from one pit loom to the other (fig. 7.71). The *jāla* is provided by the master craftsman (fig. 7.72). Weaving is done by male weavers, women performing ancillary functions. If a *jāla* exceeds one hundred string loops, a helper is required. The latter can either stand at the side or sit across the width of the stretched warp on a bench (fig. 5.53).[6]

Under the *Nizāms*, the centre for the manufacture of medium weight silk saris for marriages had been Armoor. Around 1930, as we have seen in the case of Kothakota, the design inspiration had been derived from Varanasi. However, there was a large Maharashtrian settlement in Armoor and market demand also existed in Maharashtra. After 1930 shot colour silk with narrower borders began to be made in close consonance with the *Paithani* repertoire (see Das Gupta, 1983, for illustrative samples). In reviewing the weaving traditions of Gadwal, Wanaparty and Narayanpet it would appear that in Gadwal, until AD 1794, Maratha political authority was strong. Wanaparty was more open to Mughal influence (Mudiraj, 1934: 620 N, P, Q, S, W, 622, 624).

Weaving skills were nurtured at Gadwal and Wanaparty by royal enterprise. The *jāla* may have grafted more easily as local

Paithani motif and weave. Woven at Kothakota Courtesy: Artistic Saree Weavers Industrial Society Ltd., Kothakota. (Fig. 7.70)

Two looms stretched end to end along same width of room. Note second, warp below the loom. Weavers Complex, Kothakota. This practice is also followed in Tamil Nadu. (Fig. 7.71)

Jāla used in weaving *pallu*. Weavers Complex, Kothakota. (Fig. 7.72)

Fragment. Double interlock. Ahmednagar. Courtesy: Rani Saheb Kanchanmala Raje Bhonsle, Junior Bhonsle Palace, Nagpur. (Fig. 7.73)

Reverse face of fig. 7.73. (Fig. 7.74)

Pagaḍī woven at Ahmednagar. Courtesy: Deepak Shah, Pune. (Fig. 7.75)

weaving skills do not appear to have been of a very high order (figs. 5.53, 5.54). Early twentieth century designs, particularly at Kothakota which can be regarded as an offshoot of the Gadwal repertoire, show elements of Varanasi, Pune, *Paithani* and the very ornate Kanchipuram silks. Kothakota could develop as it did because it was not restrained by any traditional benchmarks. Narayanpet appears closer to the Dharwad/Maratha tradition. Gadwal takes fierce pride in using genuine *zarī* from Surat. Narayanpet and Kothakota do not show the same pride and there is evidence of a willingness to experiment with tested *zarī.* In Narayanpet the only areas requiring interlock is at the *tope* section of the sari. This is absent in the Gadwal tradition.

Amarchinta

The nucleus of the territorial unit which was later to bear the name, Samsthan of Amarchinta, appears to have come into existence in AD 1289. A tributary state to the Nizām in the district of Mahbubnagar, it comprised an area of 190 sq. miles. Wanaparty constituted its boundary to the east while the river Krishna separated it from Gadwal to the south. It was noted for its fine muslins (Mudiraj, 1934: 641–5). The speciality of Amarchinta lay in a cotton sari in which the ground was made up of 1/8 in. checks comprising eight ends and eight picks (personal communication: J.K. Reddiyya, Hyderabad). The border, 2 in. in width, and the *pallu,* 1 1/2 ft. in width, were ornamented in *zarī.* The *pallu* was further decorated by a running floral motif similar to the Poona *Paithani* (oral communication: Srilata Bhoopal, Hyderabad).

Asavali type *pallu* woven on *jāla* loom at Nardipur and Ridrol. Courtesy: Bharat Kala Bhavan, Varanasi. (Fig. 7.76)

Ahmednagar, Ahmedabad, Hyderabad, Tanjore

There are nineteenth century extant samples of *Paithani* ware attributed to a provenance of Ahmednagar, Ahmedabad and Tanjore. Ahmednagar, which was well within the Maratha culture sphere, appears to have manufactured *Paithani* type *paṭkās.* These have been documented (figs. 7.73, 7.74). Ahmednagar appears to be associated with a particular style in its articulation of the cone motif. This motif is apparent both in *paṭkās* as well as *pagaḍīs* (fig. 7.75). The running motif is also quite distinctive (figs. 7.73, 7.74).

Ahmedabad, a great centre for silk weaving and located within a short distance of the court of the Gaekwads of Baroda, has also been associated with the manufacture of *Paithani* ware (Birdwood, 1892: 28). Confusion has arisen because of a dual meaning assigned to the term *asavali.* In Paithan, *asavali* is the name given to one of the popular border motifs (cf. fig. 6.58). In the context of Ahmedabad brocades, it refers to a highly ornamented parakeet and floral motif worked in brocade (fig. 7.76, cf. fig. 5.42). The

derivation of the two terms appear quite unrelated. The term *asavali* could have its origin in *hansia* or border, plus *avāli*, series or line, which may be taken in the sense of a border with a flowing line. The Ahmedabad term *asavali* may have had some association with an earlier city at the site of presentday Ahmedabad, called *Asavali*. Alternatively, it may be derived from an earlier motif of *haṁsāvalī* or border with the mythological *haṁsa* bird. With regard to Hyderabad, J. Forbes Watson has a sample of a *Paithani* type turban piece manufactured at Hyderabad (Forbes Watson, 1873: no. 48; fig. 7.77). While it is possible that *Paithani* ware may have been woven in Hyderabad, the majority of this ware appear to have been commissioned at Paithan. These included *Ajantā* designs, which were developed by the firm, Ajanta Fabrics, founded by Ahmed Bahadur in Paithan (figs. 7.78, 7.79). This firm operated between 1950 and 1955 (oral communication: M.A. Munaf Ali Quli, Design Centre, Hyderabad). Since Tanjore had a reigning Maratha dynasty (cf. fig. 6.57), it would not be surprising to find *Paithani* weaving being practised here. However, if the tradition existed at this Centre, it appears to have been highly exotic and Forbes Watson has one example, no. 161, of a *Paithani* worked *pallu* said to have been woven in Tanjore (Forbes Watson, 1873: no. 161).

Turban piece, Hyderabad. Reproduced from Forbes Watson, 1873. (Fig. 7.77)

Ajantā designs. Paper patterns. Courtesy: Design Centre, Hyderabad. (Fig. 7.78)

Ajantā motif. Woven sample. Courtesy: Design Centre, Hyderabad. (Fig. 7.79)

Dharmavaram

The earlier tradition here appears to have been the weaving of coarse quality cotton and the *Manual of the Administration* (*Manual*: 272), makes no specific mention of weaving in relation to Dharmavaram. Around 1920 owing to a persistent shortage of cotton, silk weaving was initiated and up to about 1955 both silk and cotton were woven. After 1955 there was a concentration on silk. Until about 1957 only the 9-yds. sari was made, after which 6-yds. saris became increasingly popular. The main market was in neighbouring Karnataka rather than the more distant Andhra Pradesh.

The Thogathas were the dominant weaving community comprising 60 per cent of the work force. The other groups, also mainly Telugu, comprised 20 per cent Padmasali, 10 per cent Pattusali and 10 per cent Devanga (personal communication: M. Balakottaiah, Hyderabad). There were no Saurashtras. Cotton saris earlier made here had narrow borders woven through the dobby mechanism. When silk weaving began to be practised *c.* 1930, the *aḍai* loom was used. Initially the *Dharmavaram* product was lower in cost than the *Kanchipuram* one and the three-shuttle technique was not used. Ornamentation was restricted to the border, the *pallu* and ground being left without any intricate design. Later, *c.* 1960, *petni* joining for the *pallu* was introduced and the only decoration in it comprised parallel *zarī* lines.

Around 1950, some weavers who had visited Banaras introduced the *jāla*, and the *pallu* became increasingly complex. After 1954 the lattice rather than barrel dobby began to be used for the border and ground *būṭīs* in extra warp, while the *jāla* was used for extra weft and pallu decoration (fig. 7.80). The innovation of tie dyeing of the *pallu* on the loom was introduced around 1964 (oral communication: G. Kodayya, Dharmavaram).

Dharmavaram silk is lighter than Kanchipuram silk, the warp of the former being 18 to 20 deniers while that of Kanchipuram ranged from 22 to 24 deniers. The warp was filature silk while weft was *charkhā. Charkhā* silk was about 33 per cent cheaper than filature. The border could be woven in three shuttle. For the *pallu*, the unwoven warp ends would be loosened on the loom, and the portion to be dyed for the *pallu* would be doubled, leaving ends free for the next sari to be woven. The *pallu* portion would be fastened at one end with string and a rubberband, and the ends would be bleached and rinsed in fresh water. Synthetic dye would be prepared on the spot and the *pallu* portion dyed. The warp would then be realigned on the loom, and weaving would recommence as soon as the ends were dry (figs. 7.81, 7.82, 7.83, 7.84, 7.85).

Piecing the warp was done by women, setting up of the loom by men and, weaving was undertaken by both sexes. The basic material was supplied to the weavers, a loss of 15 gm. per sari being condoned. Five 6-yds. saris could be woven from one warp. Border and *pallu* designs are changed every one or two years. Warping was done in a separate unit, but *zarī* warping was done within the cooperative society. In preparing the warp, the length of two saris would be in one colour, the remaining three in another. If the warp was of the same colour, variations could be introduced by changing the weft colour thereby creating a shot effect, commonly seen in traditional ware.

Dharmavaram may have had a distinctive cotton sari but with the introduction of silk weaving the earlier tradition, *akin* to Kanchipuram *aḍai*, became hybridized. Use of *jāla* drew it closer to Gadwal and Wanaparty, but practice of three shuttle with temple effect elicit comparisons with Kanchipuram. Having no clear benchmark of its own in terms of design it would be of interest to see how this product evolves with time.

Elaborately designed *Dharmavaram* silk sari. Three shuttle border. Woven with double lattice dobby. *Jāla* has been used for ground *būṭī* and *pallu*. Workshop of O. Obilesu, Dharmavaram. (Fig. 7.80)

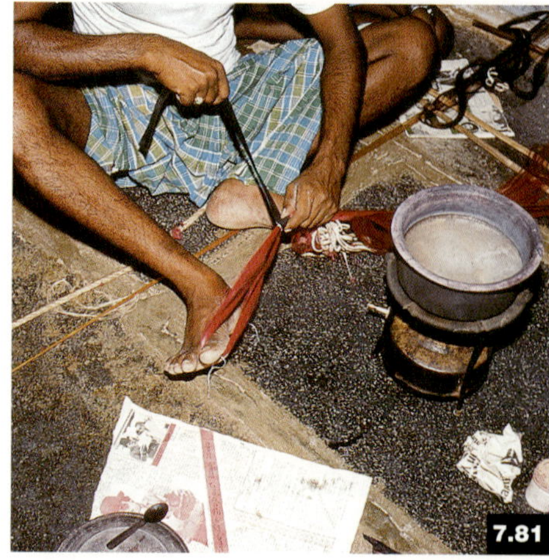

Stage 1. Retying of dyed ends on loom. Workshop of O. Obilesu, Dharmavaram. (Fig. 7.81)

Stage 2. Tie dyeing of *pallu* ends. Body portion of end blocked by tying of rubber tube. (Fig. 7.82)

Stage 3. Bleaching of *pallu* ends within loom area. (Fig. 7.83)

Stage 4. Tie dyeing of *pallu* ends after bleaching. Rubber bands are then removed. (Fig. 7.84)

Stage 5. The dyed ends are retied on loom. Workshop of O. Obilesu, Dharmavaram. (Fig. 7.85)

Notes and References

1. For the early history of *Chanderī* see Garde, 1948: 4–5, 8, 10, 11; Cunningham, 1972: 404–11.
2. *Imperial Gazetteer,* 1905: 348–9; Watt, 1903: 308; Luard, 1908: 212
3. The information pertaining to the construction of stitched double cloth has been compiled from the analysis done of the fabric at NID in 1993–4 by B.M. Anand and Krishna Amin-Patel. Discussions with B.M. Anand, Aditi Ranjan and Vijai Singh Katiyar, faculty members NID, also helped to clarify doubts and bring about a clearer picture of the weave structure. Giraben Sarabhai is thanked for making available the samples mentioned from the holdings of the Calico Museum of Textiles, Ahmedabad, as is the Calico Museum of Textiles for permitting Krishna Amin-Patel and B.M. Anand to study the samples.
4. For the composition of weavers' communities at Pune see Gadgil, 1952: 107, 115, 118–22, 200–3, 215; Edwardes, 1976: 52–6.
5. Dewar, 1901: 16, 29–30; Grant, 1870: 125–9, 131, 313; Nissim, 1909: 11; Dewar, 1904: 7–12; Harnetty, 1991: 457. The fine cotton and gold items of Burhanpur decorated with *zarī* were in demand in the markets of Turkey and Poland. Both these countries are known to have imported *paṭkās*. It is probable that double interlock *paṭkās* may have been involved in this trade to which a reference is being made. Blennerhassett, 1916: 42.
6. Documented at Weavers Complex, Kothakota at the workplace of D. Lakshmana, weaver.

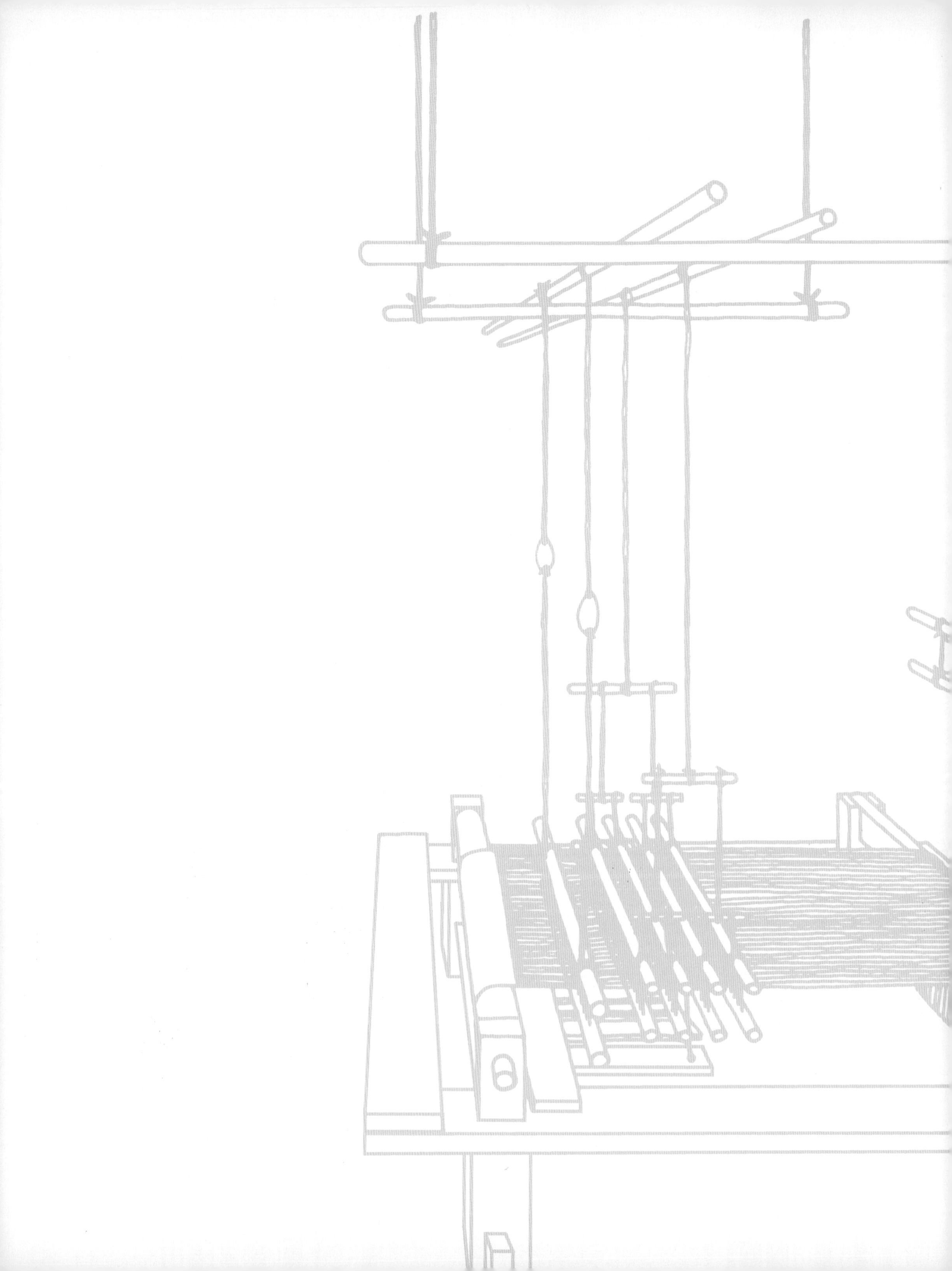

08 MAJOR WOOL TRADITIONS: NORTH INDIA

08 Major Wool Traditions: North India

Kani Paśmīnā: Kashmir

The technology involved in processes of manufacturing the *kani paśmīnā* of Kashmir and the *paṭṭu* of the Kinnaur and Kullu regions of Himachal Pradesh, no less than their attributes as unstitched apparel, draw these items within the landscape of this narrative. Whereas, in the case of the other loom products which have been surveyed in the text, the coverage of the social context has tended to be discrete, for these Himalayan states, the immediate social and ecological connotation is integral to an understanding of the technical procedures evolved.

The choice of the term, *kani paśmīnā*, rather than the more familiar appellation, Kashmir shawl, has been very consciously taken. *Paśmīnā*, Persian for woollen, refers to cloth woven from *paśm*. *Paśm*, in Persian, is a generic term for wool. As a descriptive term it is particularly well adapted to the Kashmir product. However, the meaning assigned to *paśm* in the context of *kani paśmīnā* is that of the underfur of certain Himalayan domesticated animals. *Kani* refers to the willow tree derived wooden spool used in lieu of bobbins in weaving coloured patterns in this product (fig. 8.1). Whereas the

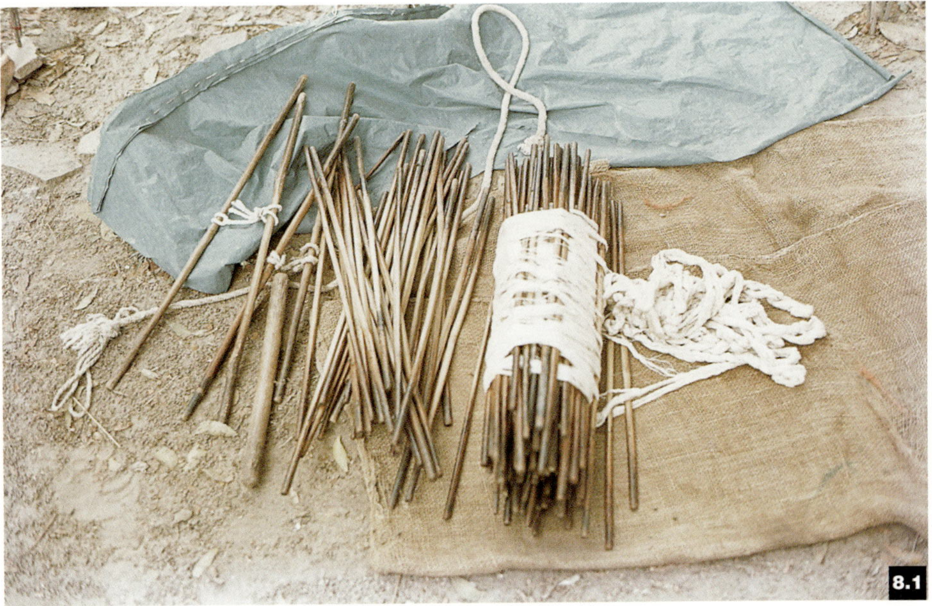

Kani/wooden spools. Kanihama. (Fig. 8.1)

term, Kashmir shawl, was very specific to the European market, the connotation was somewhat limited as it related exclusively to one item of clothing. *Kani paśmīnā* pertains to a technique of patterning in *paśm* and covers the entire product range woven in this manner. The market for these items has always been created by fastidious demand.

The *Paśm* Route

In some ways the trade route of *kani paśmīnā* during medieval times is analogous to that of the Silk Route of an earlier period. It was the Silk Route through which Chinese silk had made a major entry into commercial arteries. For both these items, it was the unique nature of the product which constituted their allure, and once the products began to be simulated to perfection elsewhere the demand in the respective countries of origin declined.

In the case of *kani paśmīnā*, the existing pattern of communications played an important role both from the point of view of securing the raw material as well as that of distributing the finished product. In an age when channels of communication depended on factors of physical geography, climate and on what human ingenuity could best exploit from human and animal energy resources, Kashmir was admirably placed to develop links with outside marts and secure access to the desired raw materials.

To the south, the Banihal Pass led to Jammu and the Marbal Pass provided access to Kishtawar. There were several routes to Punjab.[1] The Pir Panjal Pass led to Gujarat, while the river Jhelum could be followed from Baramula to Kohala. The 'Old Mughal Route' connected Srinagar via Budil with Akhnoor and Sialkot. Further to the west the Tosa Maidan Pass controlled communication from Srinagar to Poonch, and then to the Punjab plains, while the Haji Pir Pass led to Peshawar.[2] Kashmir was linked with Baltistan through Gurez. From Baltistan a route led to Astor and Gilgit and from Gilgit the traveller could push on to Badakshan, Bokhara and Samarkand (Hasan, 1959: 26). The most important, from the point of view of the wool trade, was the channel to Ladakh through the Zoji La Pass. From Ladakh ingress to Yarkand was provided through Sabu, Akhan and the Karakoram Pass. Yarkand was the point at which mountain routes from Ladakh, Tibet, China and Khotan joined those leading to Kashgar. Apart from Gilgit, Samarkand and Bokhara could also be reached through Kashgar.

There was a second, and more difficult route, leading from Ladakh to Chinese Turkestan via Baltistan and across the Karakoram (Hasan, 1959: 25, 244–5; Rizvi, 1985: 13–17). It was through Ladakh that Kashmir obtained entry to Tibet, China and Nepal. From Leh merchants proceeded to Kuti. From Kuti the way was open either through Shigatse to Lhasa, and thence to Sining and even Peking, or to Kathmandu and onward from there to Patna. Kashmiri merchants, called Baqal, played an important role in the trade between Bengal, Nepal and Tibet.[3] Such was the popularity of *kani*, and later, of *amlikar*/embroidered *paśmīnā* in Bengal, that when European markets dwindled towards the closing decades of the nineteenth century, it was the Bengal demand which helped to sustain the industry in Kashmir (Sufi, 1974, 569).

Wool: Wild and Domesticated Sources

The unique nature of Kashmir *paśmīnā* emanated from the attributes of the basic raw material utilized. The primary fibre was wool. While the quadruped ruminant mammals from which this was derived varied, the common factor was the manner in which each species adapted to its environment. In the arid mountains which constituted the natural habitat for animals growing *paśm*, the snow line existed at a higher level than equivalent eminences exposed to the monsoon. Snow could not accumulate on wind blown precipices and some plant life could sprout on any open ground (Schaller, 1980: 8–9). Among animals which grazed in the region during the winter months an undergrowth of down or inner fur was developed in response to the seasonal variations in temperature and prevailing icy winds. The additional growth commenced in autumn and was complete by winter by which time hair coverage had increased by 400 per cent. This inner fur assured the isolation and immobilization of a bed of air on the surface of the skin thereby providing warmth (Rougeot, 1984: 44). As the weather improved this additional growth was shed through the process of moulting.[4] Such underfur obtained from wild species was called *asalī tūṣ*, while that derived from the domesticated animal was called *paśm*.

Tibet played an extremely important role in the development of the *kani paśmīnā* industry in Kashmir. The Tibetan plateau is said to have been the first place where the wild *paśmīnā* goat, *Capra hircus*, was domesticated. Prior to 1950, Tibet wool was marketed through Ladakh. Although *paśm* was also procured from Ladakh, the latter region became a major producer of *paśm* after *c.* 1950 consequent to political events in Tibet (oral communication: P.N. Bhat, Indian Council of Agricultural Research, New Delhi). The Tibetan language distinguished between sheep and goat wool, the generic term for the former being *bāl*. Wool yielded by the southern variety of sheep called *yul bāl*, was a little coarse. The northern Chang breed, however, provided very soft wool. The trade name for Changthani wool in British India since the nineteenth century was *biangi* (Watt, VI, 1972: 609, 610). This is the term which is still used in Himachal Pradesh for the best quality of sheep wool earlier procured exclusively from Tibet. The wool was much prized for its silkiness and length of staple (Watt, V, 1972: 610). The Tibetan term for north is *byan*. The term, *byan-hbrog*, referred to herdsmen of the northern solitude of Tibet, the *Jan-thang* (Changthan) (Das, 1970: 883, *byan*, q.v). Neither variety of these

Yak. Kufri, Himachal Pradesh. (Fig. 8.2)

domesticated sheep grew *paśm,* although wool, *ok,* of the under chin area, was extremely soft.

The generic term for goat wool was *rapoo. Ra* was obtained from the upper covering of the animal and was of rough quality. The winter undergrowth, *paśm,* was called *lena* in Ladakh and western Tibet, and *khu-lu* in western Tibet.[5] Ordinary yak wool of coarse quality was called *tsipa* (fig. 8.2). However, yaks living at altitudes of between 8,000 to 16,000 ft. developed *khu-lu* during the winter months.[6] The spinning of ordinary wool required no special expertise. However, because of the fineness of its fibre the spinning of *lena* required very special handling (*G.K.L.*: 79). It would appear that the dynamics of the wool trade led to a natural specialization of labour. The spinners of the vale of Kashmir evolved a high degree of skill in the sorting and spinning of *lena* on which the entire industry of *kani paśmīnā* depended. The sorting of *paśm* was a major activity for the women of the valley during the winter months. The remaining Himalayan wool weavers preferred to use lower grade *paśm* which was easier to handle, as also wool yielded by sheep pastured at lower attitudes. *Paśm* of the first quality was mainly channelized to feed the *kani paśmīnā* producing units.

Asalī Tūṣ

Asalī tūṣ, called *shatoosh* in present times, although yielded by a variety of wild species, was uniformly silky, light and of superb warmth. The supply was not easily predictable as the animal had to be either hunted or snared, or the shorn fleece of the summer months collected from the undergrowth of inhospitable terrain. However, since the wool was matchless for warmth and texture its collection remained economically viable. *Asalī tūṣ* was obtained from three sources: a variety of wild sheep and goat, the Tibetan ibex and the Tibetan antelope. Precise zoological classification is made difficult by the fact that the Himalayan ecological zone is extraordinarily prolific in species of antelope, sheep and goats with hybridization occurring among some. Mixed characteristics have further confounded attempts at strict classification.

William Moorcroft referred to an animal, *bharal*, in 1816, which had a very beautiful brown undercoat beneath its brittle fur (Moorcroft, 1816: 457). Hooker refers to its Tibetan appellation, *gnow*. The animal was always to be viewed at very great elevations (Hooker, 1854: 132). The Hindi term was *bharal*. G. Schaller dismisses the name, blue sheep for the *bharal*, as he considers the *bharal* to be an aberrant goat with sheep like affinities (Schaller, 1989: 44). It was also confused with the deer in physical appearance.[7] It moves along the Tibetan highlands and neighbouring ranges, foraging at the timberline at 13,000 ft., or above the timberline up to an elevation of 18,000 ft., which constitutes the limit of vegetated space (Schaller, 1980: 5, 196). The mixed physical characteristics of this animal in the Tibetan region has earned it the appellation of *Pseudois nayaur nayaur* (Grzimek, 1972: 495; Schaller, 1989: 44).

The references to wild sheep (Moorcroft, 1816: 457) can be taken as the Punjab *urial*, *Ovis punjabiensis*, Lydekker, or *Nyan*, the Marco Polo's sheep, *O. ammon*, Polii. Another identification could be the Argali, *O. ammon*, Linnaeus. Great Tibetan sheep, is described as lightly built and very close to an antelope in appearance (Grzimek, 1972: 496, 502–3; Schaller, 1980: 4). Its habitat lay in the north Hunza district between the Khujerab and Kilik passes (Schaller, 1989: 74). The *O. punjabiensis*, the male being called *shapu* and the female *shamu* in Ladakh and Baltistan, is to be found around Peshawar, the Salt Range, north Tibet, and the area extending between Gilgit to Afghanistan (Watt, VI, 1972: 557–8; Arora, 1940: 236). Among the goat family the best wool was obtained from the Tibetan ibex, the *Capra ibex* Sibirica, *Pallas*. Vernacular names are *skin* or *sakin* for male and *dabmo* or *danmo* for female in Ladakh, *kail* in Kashmir and *tangrol* in Kullu. Cunningham cites Vigne to assert that every year between 100 to 200 of them were killed in Balti during the winter when they were forced to descend into the valleys. In Ladakh they were snared at night or shot at dawn when they ventured down to the streams to drink. According to Vigne the colour of *ibex tūṣ* was a dark, dull, brownish maroon, while Cunningham describes it as a light brown. The yield of one ibex was three times that of a goat.[8] The terrain preferred by the wild goat was that of cliffs and their vicinity while wild sheep preferred to occupy the plateaux above the cliffs and the undulating terrain along the bases (Schaller, 1980: 118).

Yet another source of *asalī tūṣ* was the Himalayan *tahr*, inhabiting the region between Kashmir and Bhutan. Not being a true *Capra* goat, it has been given the appellation, *Genus hemitragus* (Schaller, 1989: 150–1). The Himalayan *tahr* was called *Hemitragus jemlahicus,* H. Smith while the name for the Sikkim breed was *H. j. schaeferi*. G. Schaller is of the view that this division into a Sikkim species is unjustified and the term for the Himalayan *tahr* is sufficiently comprehensive to include the Sikkim *tahr* (Schaller, 1989: 42). The Himalayan *tahr* was also identified as *jharal*, Jerdon, *C. jharal*, and *C. quadrimammus*, Hodgson.[9] It is said to inhabit the most precipitous parts of the upper Himalayas at altitudes ranging between 7,000 to 12,000 ft. It is called *kras* and *jegla* in Kashmir, *jharal* in Nepal and *tehr* or *jehr* in the western Himalayan region. The vernacular term for the same animal found at lower elevations of 6,000 ft. was *khar-tehr*. The *paśm* derived from animals at the lower altitudes was not as fine as that available at higher elevations, nor was the staple length as long.[10] In August–September the oldest males are found close to the snowline and the females are viewed above the forest limit. As the snow falls they descend towards the valleys but never leave the precipices except to graze on the adjacent grassy slopes.

Another species of wild goat was *C. falconeri*, Wagner. Muslim Baltis called it *markhor* or snake eater. The Kabul variant was called *C. megaceros*, Hutton. A. Adams has found a similarity between the habits of this animal and those of the ibex (Adams, 1858: 525–6). In Ladakh the male was called *rapoche* and the female *rawachi*. *C. falconeri*, Hugel, and *C. megaceros*, Hutton, existed in a wild condition at higher altitudes. At lower elevations they were domesticated. The habitat included Central Asia, Afghanistan, Kashmir, Pakistan and north India. Although, according to some authorities, the *paśm* of the domesticated variety was much prized,[11] G. Schaller is quite specific that the maximum winter altitude for the *markhor* was 2.2 m. (Schaller, 1989: 70). This would, at best, have yielded poor quality *paśm*.

Among the better known sources for *asalī tūṣ* was the *Panthelops hodgsoni*, Abel, the Tibetan antelope, belonging to the gazelle family. The vernacular appellation is *chiru*. It was called *ktsos ortsros* in Ladakh and *sous* in the area bordering on Chinese Tartary. The Tibetan terms are *tsus*, *chiru* or *chuku* for the male and *chus* for the female. The word *tūṣ* may have been derived from *ktsos*, *ortsros*, *sousor tsus*.[12]

The underfleece of this animal is of a dark brown colour. Between the months of May and July the animal goes through the transitory state of switching over to its summer coat. At this time it could also play host to a fly which bores itself below the skin and deposits its eggs under the surface. When the eggs hatch, the pupae cause an irresistible urge to scratch. The animal rolls on the ground trying to obtain relief, losing some underfur in the process. If the animal was caught or killed, the fleece could be plucked out in handfuls. Should the animal be spared this, the problem of deletion of winter growth would be accomplished through natural moulting.

While there is some published literature on the different species yielding *asalī tūṣ*, very little is available on the domesticated animal yielding *paśm*. One source should have been domesticated *C. falconeri*, Hugel, and *C. megaceros*, Hutton, to which a reference has already been made. Gabriel Vial refers to *C. hircus* while Marie-Hél ene Guelton prefers the appellation, *C. hircus laniger* (G. Vial, 1987: 38; Guelton, 1987, 24). *The Oxford Dictionary*, under shawl goat, provides the term *C. lanigera*. According to Chinese sources, the animal, *chan-yang*, was introduced into China from Tibet towards the close of the Tang period, *c.* AD 904. The undercoat of this animal was extracted, hair by hair by means of the index finger and thumb. The wool was fine and supple and used for weaving the textile, *jong-ho*. Tibetans, or those of Tibetan descent, alone possessed the necessary expertise to weave this fibre. It is thus clear that one of the strains of the Tibetan *paśm* yielding animals had been domesticated well before the tenth century AD. The *paśm* derived from the portion between the underneck to the underbelly was considered to be of the first quality (Irwin, 1973: 4). The hair of the outer coat of the animal played no part in the fabrication of *kani paśmīnā*. This was diverted to other uses (Moorcroft, I: 311; Balfour, 1976: 599).

In 1867, Forbes Watson provided evidence of the heterogeneity of sources, apparently of domesticated species, of the wool used in *kani paśmīnā*. This included wool of the Kabul goat, Kerman sheep wool and camel hair (Mikosch, 1985 a: 10). To this Vigne, writing *c.* 1832–9, added the *paśm* of the shepherd's dog and yak (Vigne, 1981: 124). George Watt draws attention to the difficulty of identifying the domestic species in view of the many genetic strains introduced over time resulting from successive series of cross-breeding. However, a few animals are singled out for special mention in the context of *paśm*. The *changra*, the domesticated Himalayan goat and the common goat of Tibet, was recognized by Hodgson as a strongly marked derivative from the wild goat, *C. aegagrus*. The *chyapa* or *chapu* of the northern portion of the sub-Himalayan region had strong similarities, but also some dissimilarities, with the *changra*. The habitat was different and it was much smaller in size than the *changra* (Watt, VI, 1972: 562–4). This is the *paśm* yielding *chigu* of Spiti and Kinnaur. G. Watt also refers to two varieties of domesticated sheep as yielding very fine quality wool. The *hunia*, found in western and eastern Tibet, could, according to Hodgson, be compared with the *C. quadrimammus*, Hodgson or the *O. ammon*. The *hunia* was mainly used for carrying loads (*G.K.L.*: 86). The *silinga*, a variety of sheep found in eastern Tibet, also yielded superlative wool (Watt, VI, 1972: 572–3). The *silinga* may be the domesticated *tsilloo* to which J.D. Hooker makes a reference (Hooker, 1854: 85). Watt, however, makes no explicit reference to the *paśm* yielding propensities of either the *hunia* or *tsilloo*. As has been emphasized by Tsaring Wangyal (Tibet House, New Delhi), Tibetan sheep are not associated with the growth of *paśm*. Marie-Hélene Guelton has characterized the attributes of *paśm* as lack of elasticity, silk-like brilliance, melting softness, magnificent warmth and a remarkable affinity to dye (Guelton, 1987: 30–1). The lack of elasticity, advantageous in a draped garment like the shawl, arises from the fact that *paśm* fibre has no medullation.[13]

The domesticated goat was sheared with a knife once a year at the commencement of summer. If scissors had been used the outer coat/*pat*, would be separated along with the underfleece, making the task of separation extremely laborious. The outer hair was first cut away. The underfleece was then combed towards the head and cut off progressively. Even so, there was some admixture of hair, which had to be painstakingly sorted out by hand.[14] In present times the whole fleece is sheared with clippers. *C. hircus* enjoyed a wide dispersion in the mountainous regions contiguous to Kashmir. In earlier times the best quality came from Changthang and Rudok.[15] Rudok was the collection mart for flocks pastured in the Changthang area. The *paśm* was loaded at Rudok on to backs of sheep and sent to Leh where it was purchased by Kashmiri merchants.[16]

From Leh it was transported by shoulder load to Srinagar, the journey from Leh to Srinagar taking 18 days. The labour used was not free but *begār*, enforced service (Vigne, 1981: 126; Watt, VI, 1972: 641). In Ladakh, the shawl goat was reared in Nubra, Zanskar and Rupshu. Wool was also procured from Spiti but this was accorded a lower grading.[17] An important factor to be borne in mind with regard to *paśm* obtained from domesticated animals is that during the winter months migrant shepherds often enjoyed grazing rights to high altitude pastures thus ensuring a healthy growth of *paśm* for their livestock.

After AD 1800 a fresh source of supply began to be tapped. This was the *paśm* procured from goats reared in the Tien-shan mountains, the prime marts of collection being Turfan and Ush Turfan. Kashmiri traders, Baqals, brought the wool by caravan via the Kashgar, Yarkand, Leh route.[18] Irrespective of the colour of the outer coat, the inner coat of *paśm* tends towards shades of white, grey and dun.[19] Since the white variety had the greatest affinity to dye it commanded a better price, fetching 33 to 50 per cent more than the darker hues. In this context, the statement of Abul Fazl that at the time of Akbar, AD 1556–1605, the natural colours of *tūṣ* shawls comprised black, white and red is somewhat puzzling. White and red are understandable in the above context but black remains ambiguous.[20]

The Loom

As has been pointed out, A.F. Barker had noted *c.* 1933, that the *paśmīnā* loom was so designed that it placed the minimum strain on the warp. In the earlier prototypes, the walls themselves acted as supports along the horizontal plane while the other parts of the loom were suspended from the roof. Cutting across the horizontal plane, the warp was held taut by a vertical mechanism. The shed was shallow necessitating usage of proportionately lighter and smaller spools in weft patterning (Barker, 1933: 131). Usage of *kani* was the obvious solution. Since the weaving operation took place on the reverse side of the cloth, the warp ends received further cushioning.

The loom used for the weaving of the *kani paśmīnā* shawls and the *shatoosh* shawls of Kashmir as recorded in the year 1982 is a frame loom with four shafts and four pedals. On the *shatoosh* loom, a throw shuttle was used. Such a loom was seen in operation at the workshop of Gulam Mohammad Wani Kaniham, Srinagar, to weave *shatoosh* shawls in the 2/2 twill weave. The total weight of a *shatoosh* shawl was around 100 gm., the feel extremely soft and it could pass through a ring measuring approximately 1/2 in. in diameter (see fig. 8.3). *Shatoosh* shawls were also being woven in a type of backcloth twill using two colours, making the shawls reversible. As the fleece itself is extremely delicate, the weavers did not consider it feasible to weave the shawls in the twill tapestry weave.

Ring shawl woven in *asalī tūṣ*. Srinagar. (Fig. 8.3)

The frame loom on which the *kani paśmīnā* shawls are woven has a wooden framework that stands on the floor. It has a cloth beam/breast beam on which the woven fabric is rolled. In the interior of Kanihama village, Srinagar, one loom was located inside the weaver's house. As the room was very small, one side of the loom was supported by one of the walls. This side of the loom did not have the two vertical supporting beams (fig. 8.4). One vertical beam on the right hand side of the weaver was also absent. Two of the overhead beams were replaced by two wooden rods which were suspended from the ceiling. The lower back beam (3) of the loom works like a guide or a support over which the warp passes as a sheet of warp ends (9). These warp ends are carried down to the lower beam of the frame passing around it and are then brought straight up almost at right angles to the back beam, after which the warp beam, made up of two wooden/bamboo rods (4), is tied to a peg on the wall or the overhead beam of the loom with ropes balancing and tensioning the warp. This is done to avoid the *paśm* warp from tangling. If it were wound directly to the warp beam, the fibres of the yarn would get entangled with each other. Further overlapping of successive layers of ends would cause the yarn to flatten out and lose its original springy body. The tension of the warps may be tightened or released by cranking up the modern ratchets on the cloth beam (*Jamavar Shawls*, NID, 1983: 11). At the midway point of the tensioned warp are suspended the four heddle shafts (5). These shafts are counterbalanced by being hung on wooden heddle horses (6) that move up and down counter to each other, making the loom a counter balanced loom (fig. 8.5).

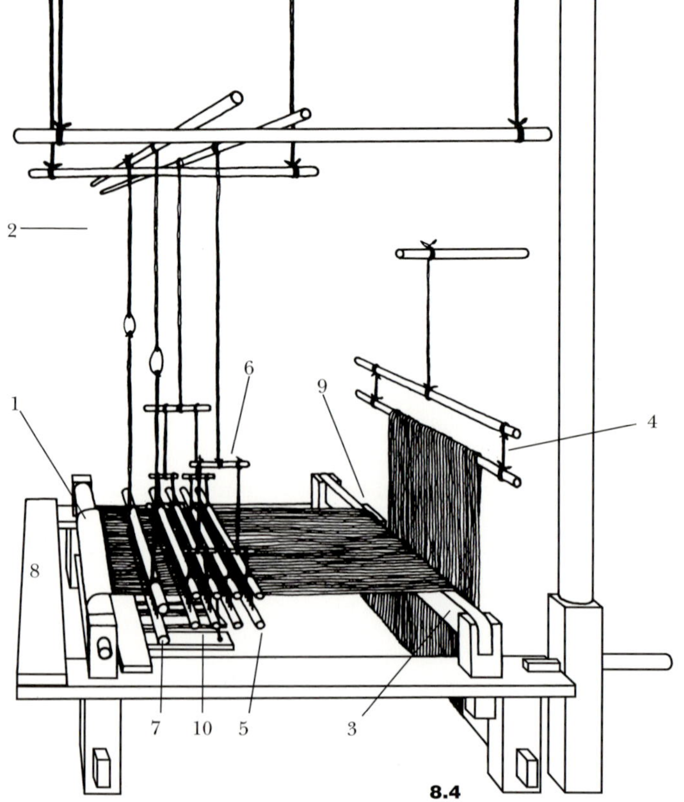

Kani paśmīnā loom. (Fig. 8.4)

1. Cloth beam
2. Wall
3. Lower back beam
4. Upper warp beam
5. Shaft
6. Heddle horses
7. Reed and beater
8. Weavers bench
9. Sheet of warp ends
10. Treadle

Loom for *kani paśmīnā*. Srinagar. (Fig. 8.5)

Woven *paśmīnā*. Note dead fibres which have been sorted out. Srinagar. (Fig. 8.6)

Kani paśmīnā. Courtesy: School of Design, Srinagar. (Fig. 8.7)

The first and third shafts are tied together as are the second and fourth shafts. Each shaft is tied to an individual heddle horse after which the pair of heddle horses is linked to one main heddle horse on either side of the shafts. These two main heddle horses are tied with rope to the overhead beam. Each shaft is made up of two rods, the distance between the two determining the length of the heddles. These shafts are each attached to a rod below, and are linked with the treadles (10) by cords according to the lifting order of the shafts. The heddles, known as *sāz*, are made of two interlinking cotton thread loops. The warp ends are passed through the space between the interlinking part of the two loops. In front of the shafts, the reed or *kangī* consists of bamboo splits laced together on two bamboo sticks at equal distance with cotton tape or *newār*. This comb is fitted into two slotted frames of wood, the outer edges of the top frame slope upwards at the centre. These frames are tied together with yarn and this beater is pivoted to the top of the loom overhead (7). A bamboo rod is used to clear the shed and is kept inserted while weaving in the open shed.

Raw Material

The fibres of *shatoosh*, and *paśm* fleece are extremely delicate. The dead fibres are sorted out by hand before spinning (fig. 8.6). No carding is done thereafter and the fibres are spun directly from the cleaned fibres with an ordinary *charkhā*. Anshoo Nath, in 1983, confirmed the continuity in practice as recorded earlier by Moorcroft (*Jamavar Shawls*, NID, 1983). To remove the impurities from the procured fleece, a solution of rice soaked in water is prepared. The water is drained off after 48 hours and the residue is left to ferment in a sealed container. This residue is ground into a fine powder and sifted. This is then sprinkled on the fibres, not only drawing out the moisture but also separating the hard fibres. A comb is used to separate these. Before spinning, the dead fibres are sorted out from the *paśmīnā* fleece by hand. Here also the process of carding is absent and the yarn is spun with the use of an ordinary *charkhā*.

The yarn is spun to form single lengths. This single ply yarn is doubled and twisted to form a two ply yarn. On completion of the process of spinning the yarn is converted to hanks and washed with *rīṭhā* the fruit of the *Sapindus trifoliatus*, Linn. The higher twist stronger yarn is used for the warp, whereas the yarn with a lighter twist is used for the weft. The weft count is usually higher than the warp count but the shawls are seldom if ever completely weft faced (Emery, 1980: 107).

Colours Dyes and Dyeing

The wool used in weaving *kani paśmīnā* was sorted into two categories: *asalī tūṣ* and a second quality called *phiri*. Owing to the delicacy of its fibre, the dyeing of *asalī tūṣ* posed a problem. Even if dyeing were to be attempted the fibre would probably only take on a darker colour of its pristine shade. Thinner categories of *paśm* yarn would be dyed in lighter colours. William Moorcroft noted *c.* 1821 that the contemporary colour palette comprised 64 shades, almost all being permanent. The dyeing agents were lac, cochineal, logwood, indigo and carthamus (*Moorcroft Papers*: 4, 9–11; *G.K.L.*: 72; fig. 8.7, cf. fig. 5.39). It is of interest that in the *Āin-i-Akbarī*, among colours graded according to value, *tūṣ* colour and *safid alcha* stood at the head.[21] The waters of the Dal Lake, Srinagar were said to be particularly well suited for the dyeing of *paśm* (Hasan, 1959: 21). If patterning was desired the warp could be of *paśm*, and, rarely, of *asalī tūṣ* while the weft could comprise *kani* of coloured *paśm*.[22]

Some examples and samples of old *kani paśmīnā* shawls seen at the S.P.S. Museum and the Design School, Srinagar, reveal the use of a combination of different natural fleece colours, as also the priority given to natural white and off-white as the ground colour with patterning in different colours. The coloured yarns used for weaving of the *kani paśmīnā* shawls were carefully and artistically matched. These colours were originally dyed in natural dyes. A few colours obtained from natural dyes are still in use at the Government Training and Weaving Centre, Kanihama, Srinagar.

Details of Natural Dyes

Yellow: Dried peels of the pomegranate fruit are kept in an earthen pot to which boiling water is added. The pot is then buried under the ground. After about 20 days, it is taken out, the crust skimmed off and the mixture sifted. The ratios of the peel and the water is 1:1 in weight. About 5 kg. of dye would dye almost 30 kg. of wool.

Pink: Beetroot is boiled in a *karhai*/round cooking vessel, and *phiṭkirī*/alum is added. About 10 ltr. of the juice when reduced to half the quantity would give 5 kg. of dye.

Today all the colours except the above (see *Jamavar Shawls*: 14), are dyed using synthetic dyes.

According to Gulam Mohammad Wani Kaniham, the colours in use now at Kanihama are listed below:

Local terminology	English
Burut zangari	Bottle green
Feel	Black
Chut	Off-white
Maroon	Maroon
Anārī (pomegranate)	Red
Ferozī	Turquoise blue
Malayee	Sky blue
Gulābī	Pink
Kafoorī	Yellow
Zard	Orange
Lāzvard	Violet
Zangarī	Green
Khudrang	Undyed greyish beige

Warping

Warping (fig. 8.8) is done for one shawl at a time, as the warp is not rolled on to a warp beam. The warp/ *tānī* is made around four pegs driven into the ground. The warp maker guides the yarn around these pegs with the help of a guiding stick which has a hook at the furthest ends of the stick. The yarn is passed through this hook. It may be fed straight from a bobbin which is kept on the ground. The warp maker may also hold a bamboo swift with the warp yarn on it in one hand while guiding the warp end around the pegs with the guiding stick in his other hand. He walks up and down, moving from one peg to the other, crossing the yarn over two pegs in figure of eight to form the lease. When this process was documented in 1983, 1500 warp ends were made for a 40 in. wide shawl with the length varying from 2 yds. 2 ft. to 3 yds. 1 ft. The finished warp is taken off the pegs and transferred to the loom. Bamboo sticks are inserted in the place of the lease thread.

Sizing

Paśmīnā warps are always sized. The starch, *kañjī*, is made from rice.

Kani

The *kani* or spool is made from locally available willow wood which is soft and easy to cut with an ordinary knife or blade. The ends of these spools are rounded. They are carefully polished with glass pieces that have been ground to powder. The ends of the spools are then smoothened over the fire. The yarn used for the weft is wound around the *kani*. Several such *kanis* with coloured weft yarn wound around them are used for weaving the *kani* shawls (figs. 8.1, 8.5, 8.16).

Drafting and Denting

Drafting and denting are done by the *baraiwala*. The drafting order is straight. Three ends are passed through each dent of the reed (*Jamavar Shawls*, NID: 16).

Tālīm

Despite the complexity and functional compartmentalization which marked the organization of the manufacture of *kani paśmīnā*, there was no wholesale market in dyed wool. As recorded by Moorcroft, for each item, wool would be piece dyed according to the amount estimated by the pattern drawers. Three persons were involved in this process. The initial pattern, the *siyatara*, would be outlined in black on paper by the *naqqāś turuhguru*. The *turuhguru*, or colour caller, beginning from the bottom of the pattern, would point out the colour spread of each repeat, calling out the number of ends to be interlaced by the relevant weft *kani* in that particular portion of the pick. While this was in progress the *tālīmguru* would prepare the *tālīm* encoding the directions of the *turuhguru*. Presumably this was

Warp making. Kanihama. (Fig. 8.8)

Narkayen kalam/ pen and *tālīm* used by carpet weavers. Srinagar. Kanihama. (Fig. 8.9a)

Sample of *kani* weaving with *tālīm*. Kanihama. (Fig. 8.9)

Colour shade card used in carpet weaving. Srinagar. (Fig. 8.10)

noted in a book as the *turuhguru* was also called the *kitāb wāla*. On completion of this task one copy of the *tālīm* was handed over for the use of the weavers (*Moorcroft Papers*: 13, 16; *G.K.L.*: 72). In some shawls in which the white background was thrown into relief by the two cross border sections in contrast colour, the warp ends in the two latter areas would be neatly tie-dyed (Mikosch, 1985 b: 26, fig. 8.10).

The *tālīm* is made of coded symbols, each symbol representing the colour and the number of ends that each individual coloured weft must pass through. For every shed opening, the movement of each coloured weft and its insertion (several spools with coloured weft yarn are used to complete one pick in an open shed) are translated into the language of coded symbols on paper, using the *narkayen kalam*/reed pen and ink. The whole drawn design is converted in this way (fig. 8.9). The weaver either follows the code or has another person to sing it out to him. A similar practice of using *tālīms* is also followed by the carpet weavers in Kashmir (figs. 8.9a, 8.10).

On crosschecking with different weaving centres in Kashmir in the 1980s, it was noted that while there was consistency in the use of symbols, the signs used for colours were different or that the same signs were used for different colours. It is possible that different workshops developed their own individual signs. Shri Raina of the School of Design, Srinagar, has also confirmed that the signs vary from one workshop to another.

According to Gulam Mohammad Wani Kaniham, Kanihama village, Srinagar, the meanings of *tālīm* codes are to be decoded as follows:

1	one red	O one	— red
1	one bottle green	O one	ω bottle green
1/2		9 half	
2	two *malayee*	two	∴ skyblue
2	two light green	two	light green
2 1/2		two and half	
3	three white	three	white
4	four maroon	four	maroon
5	five black	five	• black
6	six black	six	• black
7	seven green	seven	green
8	eight yellow	eight	// yellow
9	nine orange	nine	6 orange
10	ten *turquoise* blue	ten	*turquoise blue*
11		eleven	
20	twenty blue	twenty	∴ *skyblue*
21		twenty-one	

No sign is used for undyed greyish beige/*khudrang*.

Weaving Techniques

Kani paśmīnā is a weft faced fabric and in earlier times was woven both in twill as well as tabby. Although there are some deviations, the dominant structure is double interlock. The warp colour sets the tone for the main background shade, patterning being achieved by means of polychrome motifs woven in interlock twill tapestry. In weft faced plain tapestry the warp is usually concealed by the weft. In 2/2 twill tapestry, the predominant weave of *kani paśmīnā*, the warp is visible and becomes a component in the elements forming the pattern. In striped patterns, and the Qajar Court of Iran, *c.* 1793, set the trend in this style, the warp threads are aligned in coloured rows,[23] thus providing a series of empaneled colour backgrounds. Tapestry in tabby weave can be woven on the simplest of looms but twill tapestry demands the requisite number of pedals and shafts. A.F. Barker records the presence of pit looms for the weaving of *paśmīnā* in 1933 (Barker, 1933: 131).

Kani paśmīnā weave reverse face. Unwoven wefts carried over. (Fig. 8.11)

Fabric Structure and Process of Weaving

The *kani paśmīnā* shawls are woven in 2/2 twill with patterning that is identical on both the faces of the fabric. This is achieved through the use of discontinuous wefts to vary the weft colour and create distinct colour areas that are identical on the two faces of the fabric (Emery, 1980: 106). These (discontinuous) wefts are linked with their adjoining (discontinuous) wefts through the double-interlocking method. This is worked through the width of the warp at every change in shed to form the twill double interlocking tapestry weave, while a marked ridge is produced on the working face of the fabric (Emery, 1980: 106).

The reverse face of the *kani paśmīnā* shawl reveals some diagonal, horizontal and vertical jumps of unwoven wefts suggesting that these non interlocking wefts were carried or crossed over after many steps in the process of weaving had been achieved. These were then woven into the fabric when required (fig. 8.11).

John Irwin notes that in the 1860s, the reversible shawl was being produced. The pattern was identical on both sides of the cloth. This did not reflect any significant departure in technique, but was achieved by skilful trimming of the loose weft threads on the reverse side, and the outlining of all the main details in the pattern by needlework (Irwin, 1973: 17). Field investigations in 1982 confirmed that these work procedures were still being followed.[24]

As the twill weaves are characterized by progressive succession of floats in diagonal alignment (Emery, 1980: 92), colour patterning done with discontinuous wefts using the double interlocking twill tapestry weave, follows the characteristic 2/2 twill weave, lending the shapes a distinguishable angularity. This makes them distinct from the other kinds of tapestry woven fabrics (cf. fig. 8.7)

Evolution of *Kani Paśmīnā*

Having attempted to outline the geographical background and the techniques and processes involved in the manufacture of *kani paśmīnā*, the stage is set to follow the evolution of this trade item in a more cogent manner. Certain factors have to be borne in mind while tracing the development of *kani paśmīnā*. These encompass canons of design, colour predilections prevailing in pre-Islamic India, early references to *paśmīnā* and the attempt to analyse, in a more precise manner, the nature of inputs traditionally assigned to certain individuals in the development of this tradition. The antiquity or otherwise of *kani paśmīnā* would be important for the establishment of norms relating to the loom and its product in relation to India. This discourse cannot but be intimately interlinked with the history of other regional manifestations of the interlock and tapestry weave on the subcontinent.

The antiquity of the twill weave in India is yet to be established with precision. Extant twill brocades of Gujarat and Banaras date back only to the late eighteenth or early nineteenth centuries. However, Abul Fazl refers to a category of shawl called *safid alcha* or *tarahdār*, the word *tarahdār* meaning corded (Blochmann, 1965: 97, n. 1). F. Pelsaert, *c.* AD 1626, in describing the Kashmir shawl, which he calls *pāmarī*, suggests that these had a twilled effect (Moreland, Geyl, 1972: 36, n. 1). The usage of the word *pāmarī* by Pelsaert highlights the extraordinary continuity in Indian tradition. *Pāmarī* is used in the sense of a body covering by the Gujarati author Premanand, *c.* AD 1649–1750. It is traced to Sanskrit *prāvaraṇa*, by R.N. Mehta (Mehta, 1951: 177). In the *Bṛhadāraṇyaka Upaniṣad*, dated between 4000–1000 BC, *pravara* or *pravāra* is used in the sense of covering or woollen cloth (MacDonnel,

Keith, 1958: 40). Syrian wool weavers had been weaving tapestry since the second millennium BC. At the beginning of the Christian era, they developed a weft faced twill weave on a three shaft loom. This evolved into the four shaft loom and other refinements were added (Wulff, 1966: 173–5). However, it is not clear if these West Asian developments played any part in the evolution of the Kashmiri loom.

The generic term tapestry,[25] is given to a weave in which two basic principles are found at play—the hiding of the warp with a closely packed weft to secure solid planes of colour, and the weaving of independent wefts each confined to its own colour area within any given pick. The question arises as to what happens when two colour junctions intermingle. Any one of four methods may be followed thus providing variations in nomenclature accommodated within the range of tapestry. If the two weft picks interlock with each other it is called interlock; if the two weft picks interlock with each other as also around a common warp it is termed interlock around a common warp. If the interlocking takes place on alternate rows it is single interlock (fig. 8.12), and if it is practised on each row, double interlock (fig. 6.92). If two weft picks interlock around a common warp without simultaneously interlocking with each other, it falls within the category of dovetailed tapestry (fig. 8.13). If there is no interlocking at colour junctions this is slit tapestry or *kilim*[26] (figs. 8.14, 8.15). The *kilim* technique is in evidence in the border sections of *kambal*/blankets woven at Kishtwar and at the Intensive Handloom Development Project, Bandipur, 65 km. from Srinagar. The process of weaving in double interlock twill tapestry, so integral to the *kani paśmīnā* repertoire, was time consuming and when, in the mid-nineteenth century, motifs became highly enlarged and convoluted, to hasten the course of weaving, individual motifs began to be separately produced based upon the appropriate *tālīm*[27] (fig. 8.9). The motifs were then cut and joined to form part of the complex pattern. The manner in which the *tālīm* has been used has changed along with the organization of the producing unit and the scale of operations involved. Moorcroft, *c.* 1821, recorded the functioning of the industry during its most flourishing period (*Moorcroft Papers*: 16).

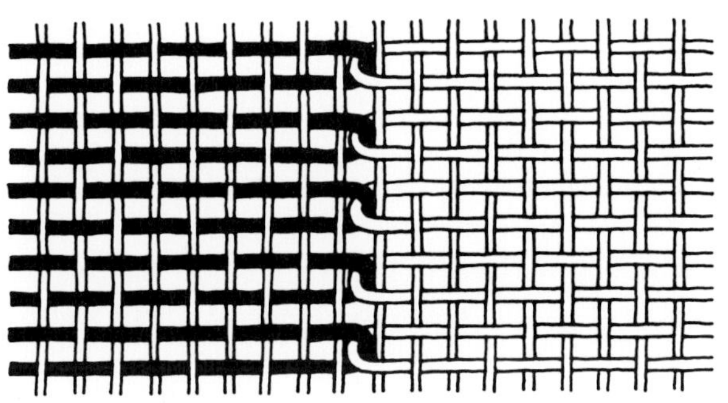

8.12

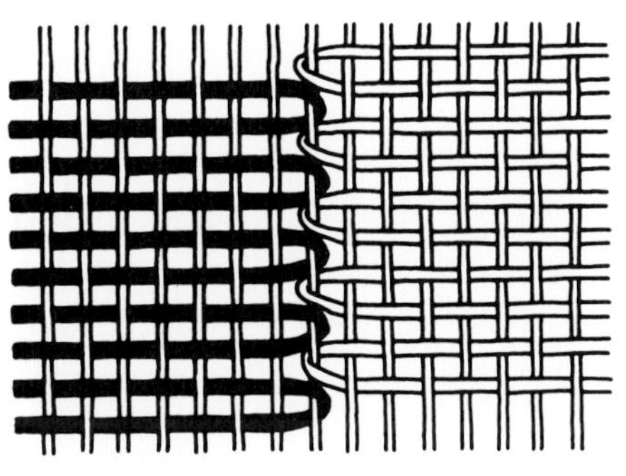

8.13

Single interlocking tapestry weave. (Fig. 8.12)

Single dovetailing tapestry weave. (Fig. 8.13)

Slit tapestry weave showing movement on vertical and diagonal planes. (Fig. 8.14)

Slit tapestry weave. Diagonal movement. (Fig. 8.15)

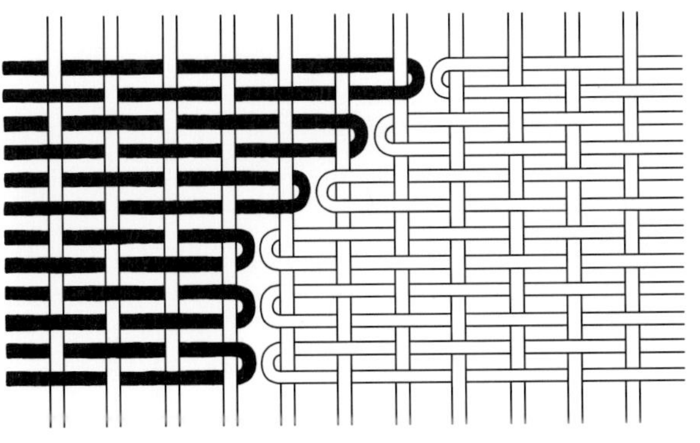

8.14

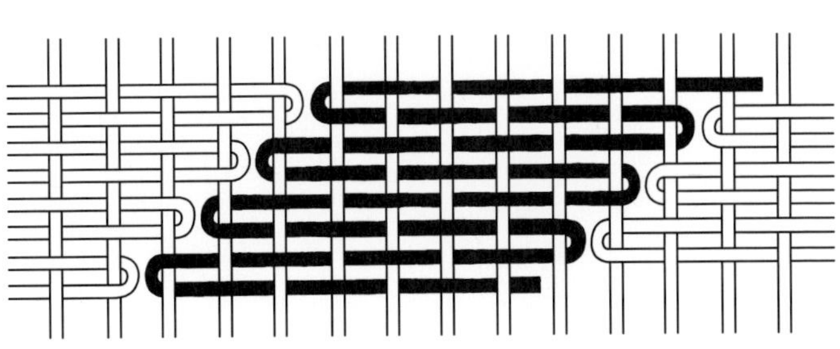

8.15

The shawl manufacturer was known as *karkhandar*. The *karkhandar* employed supervisors/*ustāds*, each *ustād* being responsible for between 25 to 30 weavers. The *karkhandar* distributed his looms in individual units/*dukān*. A *dukān* could have as many as 15 looms but the average stood at about 7. The number of weavers per loom varied between 2 to 3. The highest paid and most skilled weaver, who would sit on the left side, would manipulate 33 per cent more *kani* than his immediate subordinate. The *ustād* would call out the *tālīm*, the responsibility of maintaining the count devolving on the highest paid weaver on each loom (*Moorcroft Papers*: 16; *G.K.L.*: 69).

The method in Ludhiana, Punjab, appears to have differed from that prevailing in Srinagar, Kashmir. Victor Jacquemont, in his description of Ludhiana weaving *c.* 1828–32, records that there were three weavers per loom; the most experienced being seated in the middle (*Châle Cachemire*, 1982: 13–14). It is said that the weavers migrated from Kashmir *c.* 1800. The change in location may have brought about some changes in procedure including that of seating the most experienced weaver in the middle. In his analysis of a *kani paśmīnā* shawl made in Kashmir, G. Vial, who appears to accept the positioning of weavers as given by Jacquemont, states that the weaver on the left and the weaver in the middle were both more experienced than the weaver on the right (Vial, 1987: 42, 51–2). The statement of G. Vial could be reconciled with the documentation of Moorcroft if it were suggested that the most skilled weaver could be seated either on the outer left or middle position depending on the complexity of the pattern.

As described above, each line of the *tālīm* simultaneously embraced two rows of symbols. The top row represented the number of ends to be utilized for each colour step in the pattern while the code below indicated the precise colour to be used in the above procedure. Vertical bars indicate the end of the sequence in weaving. In present times the *tālīm* is recited in the workshop by one person in a singsong voice. Only the colour and relevant counts are called. Sometimes, the weavers respond in a chorus indicating that the count is being followed.[28]

In weaving, two ends constitute the basic unit of the warp, while one constitutes that of the pick (fig. 8.16). However, as the weft fibre is thicker, the weight of the weft is estimated to be 50 per cent higher than that of the warp. The weight of the yarn on each *kani*, also called *tili* or *tojil*, is approximately 4 grains (Moorcroft, II: 76; *G.K.L.*: 72–3). The extreme delicacy in operation is shown by the use of jewellers' weights in a weaving programme. The weaving is accomplished in the step from left to right and right to left, with the reverse side of the fabric facing upwards. The double interlocking thus takes place on this side above the floating warp and is, therefore, only visible in reverse (fig. 8.11). In single interlock, however, there is no ridge formation. Since the interlock is done on the double ends of the warp the *tālīm* count usually involves even numbers. On occasion, however, when a straight line is required as in the striped design, *mehramat*, or along the border edge, the *tālīm* may necessitate the enunciation of half counts. With the elaboration in pattern which took place in the course of the nineteenth century, it became necessary to provide tautness at the outer edges so that the shawl draped well. This was effected by the attachment of silken warp borders. The introduction of silk in the body warp is dated strictly to contemporary times.

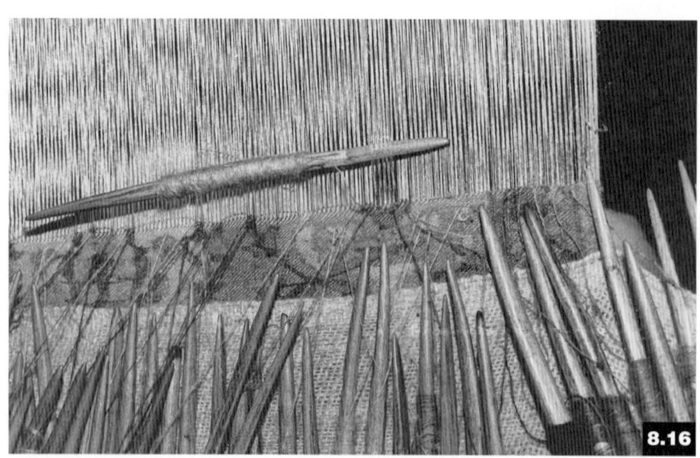

Interlocking in weaving of *kani paśmīnā*. Pashmina Weaving Centre, Srinagar. (Fig. 8.16)

It was noted *c.* 1844, that in items of the best quality three weavers could weave about $^1/_4$ in. per day. Even in contemporary times when designs have been greatly simplified, two to three weavers on the loom can weave between $1^1/_2$ to 1 in. per day. A museum quality *kani paśmīnā* could contain 38 ends and 42 picks per centimetre, while the number of *kani* utilized in each completed pick could range from 400 to 1500.[29]

On completion, the *kani paśmīnā* was washed by the washerman/*dhobī*. There were two stages in this operation. At the first stage the material was lightly washed after which faults were rectified. This was followed by a more thorough wash. The best water for this purpose was considered to be that found in the canal between the Dal Lake and the floodgates at Drogjun, and in the Tej Bal stream near Shalimar Garden (*G.K.L.*: 74; *Moorcroft Papers*: 20).

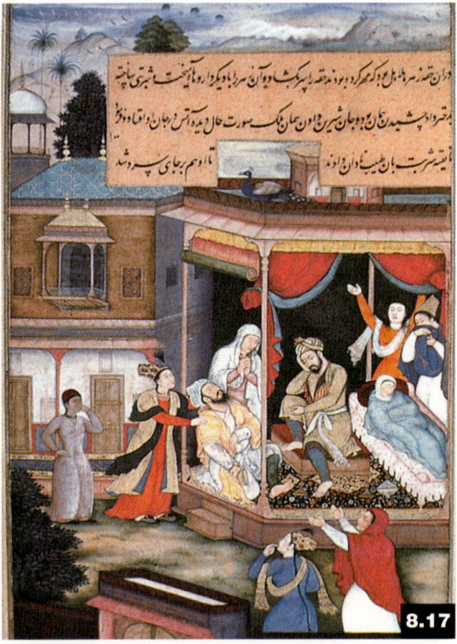

Miniature painting. Death scene by Manohar, sixteenth century. Note red shawl in middle panel. Courtesy: Bharat Kala Bhavan, Varanasi. (Fig. 8.17)

Illuminated manuscript, *Aṣṭasahasṛikā Praj naparamitā*, Newari *Samvat* 191/1071. Courtesy: Asiatic Society, *Kolkata*. Note costume. (Fig. 8.18)

Early Traditions in Design and Colour

While the *uparna*, male shoulder cloth, as a diaphanous garment, is in evidence in early sculpture, it is only in the paintings of the Mughal period dated to the reign of the Emperor Akbar, AD 1556–1605, that the shawl may be visually recognized (fig. 8.17). However, the item depicted is in monochrome and is devoid of any patterning. Starting from the period of the Emperor Jehangir, AD 1605–27, decorations begin to appear on the ground. V. Bérinstain states that in the initial stages, following precedents set by the Rajput School, motifs are geometrical (Bérinstain, 1987: 11–12). The latter repertoire concords well with the garment designs set out by King Somesvara of Dhar, AD 1127–38, in *Mānasollāsa*. Such designs translated well into *jāmdānī* weaving of which the earliest forms of patterning may be found in the cave paintings of Ajanta in Maharashtra.[30] The *jāmdānī* range (figs. 4.6, 8.18), geometrical rather than curvilinear in form, shares similarities with the early patterning of *kani paśmīnā* (Ames, 1984: 404, figs. 1–4). After *c.* 1610 the flowering scroll comes into increasing evidence and the earlier simplicity in layout is replaced by an increasing density in design.

The preference of the Hindu cultivated male towards white and natural colours (Varadarajan, 1984a: 70, 76, n. 7), the remark of the seventeenth-century French traveller, Thévenot, to the effect that among the range of colours available for the Kashmir shawl, Hindus preferred a background colour of dead leaf, is understandable (Irwin, 1973: 11). It has also to be remembered that until the advent of Empress Josephine, wife of Napoleon Bonaparte, *kani paśmīnā* had remained within the confines of male sartorial attire. If it be conceded, and a case in favour of this argument will be made, that the *kani paśmīnā* industry of Kashmir was one of long standing, specific contributions of different personalities, the history of this fabric may be viewed, not in terms of introduction of new techniques but within the range of a series of innovations introduced in terms of design and colour.

Early References

The *Ṛgveda* makes mention of both the sheep, *āvi*, as well as of the goat, *aja*. Among unstitched woollen garments are noted *pravara* or *pravāra*, and *śāmulya* which in the *Ṛgveda* is mentioned as a garment worn at night.[31] *Baudhāyana* opines that blankets of the hair of mountain goats should be cleaned with areca nuts (*Bühler*, 1882: 165: 1.5.8. 39–40). Gautama (600–300 BC), makes a reference to *kuṭapaḥ*, of which those in natural colour are recommended. The *Maskarī*, dated to the tenth–eleventh centuries AD, describes *kuṭapaḥ* as a woollen garment made from the hair of wild goat.[32] Further, a passage in the *Arthaśāstra*, dated between 300 BC and AD 100, describes duties to be respectively levied from articles made from goats, rams and the hair of *raṅku*/deer.[33] The *raṅku*, could perhaps be identified with the *Pseudois nayaur nayaur* discussed earlier. Thus, by the opening years of the Christian era, not only was sheep wool being differentiated from that of goat, but even in the latter category, a distinction was being drawn between the domesticated variety and its wild counterpart.

In attempting to trace the antiquity of *paśmīnā*, some attention needs to be devoted to the connection between projection of the mythical animal, the unicorn, and what is known about the habits of some of the *asalī tūṣ* yielding animals. Viewed at certain angles, even the domestic shawl goat appears to have only one horn in silhouette (Rizvi, 1990: 44). Ktesias, between 416 BC and *c.* 398 BC, in his description of India, referred to the animal, *monokeros* or unicorn. When pursued it would throw itself off a precipice. Since the shock was absorbed by its horns, the animal would escape unharmed (McCrindle, 1973: 94, 95, n. 65). In discussing the attributes of *C. aegagrus*, the wild goat included within the broader spectrum of *C. hircus,* Linnaeus, G. Watt delineates how, in falling, the shock is absorbed by the horns of the animal.[34] The parallel with the unicorn of Ktesias is too close to permit of easy dismissal. The case of the unicorn is further corroborated by Chinese sources. In the first and second centuries AD, the *tou-kio-cheou* also called *kio-touan*, the Chinese Unicorn, was said to have its habitat in the region earlier referred to loosely as Tartary, present-day Central Asia. In Tibet it was called *bse ru* from which was derived *serou*, identified with *chiru*, *P. hodgsoni*, Abel.[35] If this argument is accepted, the history of *asalī tūṣ* and *paśm* could be pushed back to approximately the fifth century BC.

From the point of view of continuity in techniques of production, the following excerpt from the *Arthaśāstra* is of considerable consequence: 'Woollen cloth is white, all red, part red, with threads laid in with a needle, variegated in weaving, with pieces joined together and with broken off threads slippery and wet as it were, fine and soft is best.'

In a note it is explained that the words, all white, red, part red, can also be taken as undyed, fully dyed and partly dyed (Kangle, 1972: 103, n. 97). However, these could also be taken as natural shades with the red verging towards russet. Further, the description of the best quality wool bears a remarkable concordance with *asalī tūṣ*. Moreover, the morphology of the weaving process delineated could well be transposed in terms of present-day weaving process on the *kani paśmīnā* loom if some concessions are made to the inevitable metaphor of linguistic transference.

On the basis of this evidence the assertion that shawl weaving was introduced to Kashmir from Central Asia during the reign of King Suhadeva, AD 1301–20, can be rejected (see *E.W.A.*, 1963: 970). Moreover, *paśmīnā* shawls are known to have been available in Delhi in the thirteenth century and were particularly popular in the reign of Alauddin Khilji, AD 1296–1320 (Pande, 1971: 30). Taking into account the extremely troubled period in Kashmir AD 1301–39, it is quite plausible to associate the revival of weaving with Mir Sayid Ali of Hamadan who visited Kashmir between AD 1378 and 1380. After his departure the ruling Sultan, Qutb-ud-din, AD 1373–89, is said to have continued to nurture the industry (Khazanchi, 1955: 18, 21; Koul, no date: 46). The next personage receiving mention is Sultan Zain-ul-Abedin of Kashmir, AD 1420–70. According to Srivara's *Rājataranginī* relating to the period AD 1459–86, resulting from the patronage of the Sultan, woollen cloth of the finest quality began to be made, and the patterns which were woven, 'filled even painters with surprise'.[36] The specific input of Zain-ul-Abedin appears to have been in the sphere of design rather than technique. The Mughal Emperor, Babur, AD 1526–30 is said to have imported weavers from Andijan who introduced the *kalgā*/ *bādām*/ *turañj*/paisley motif into *kani paśmīnā*.[37]

Vishnu image, Verinag. Courtesy: Archaeological Survey of India. New Delhi. Note motif in crown. (Fig. 8.19)

However, slit tapestry forms the dominant tapestry tradition of Central Asia. Moreover, the *kalgā*, in an incipient form, may be found among the triple crown motifs depicted in the sculpture of Kashmir (fig. 8.19). If not in weaving, the motif was in existence in sculpture in Jammu and Kashmir since at least the tenth–eleventh centuries AD. It is difficult, therefore, to define what exactly Babur may have contributed. Leaving aside the origin of the *kalgā*, this motif may have entered into *kani paśmīnā* at this time, but here again, there is the difficulty that extant samples suggest the steady appearance of this motif only to the turn of the eighteenth century

Kani paśmīnā. Courtesy: School of Design, Srinagar. (Fig. 8.20)

Kani paśmīnā, design derived from Persian deer horn pattern. Anavian collection, Tehran. (Fig. 8.21)

(figs. 8.20, 8.21). Naqhz Beg, in the service of Mirza Haidar Dughlat, is credited with the invention of a new design. It is not clear if this can be associated with the combination of *kani* with *bandhanī* as described by Moorcroft.[38]

The name of the Mughal Emperor, Akbar, AD 1556–1605, is linked with the development of *kani paśmīnā* in terms of new shades of colour, the manufacture of yardage material, *jāmavār,*[39] and new fashions in the draping of the shawl. Abul Fazl wished to emphasize Akbar's love for woollen garments to suggest a common trait with Sufi saints (Blochmann, 1965: 97, n. 3, 98). The question of a reversible shawl needs to be sorted out at this point. Akbar is said to have initiated the style of wearing the double shawl (Blochmann, 1965: 98). It appears more probable that Akbar may have adopted the Hindu practice of utilizing double lengths in shoulder cloth, rather than developing a fully reversible *kani paśmīnā.* During the Peshwa period in Maharashtra such double lengths in unstitched garment cloth were called *sāla-joḍī* and *dhotara-joḍī* (Chapekar, 1937: 57–9; Bhave, 1976: 264). Victor Jacquemont noted that in Ludhiana shawls were always sold in pairs, every attempt being made to match these as closely as possible (*Châle Cachemire,* 14).

Undoubtedly, some methods existed which provided an ostensibly reversible *kani paśmīnā,* the *do-rukha.* Gabriel Vial has noted that although the occurrence was rare, single interlock weave could provide a fully reversible length in *kani paśmīnā* (Vial, 1987: 40). There is also the instance of a Mughal painting, *c.* 1590, in which a courtier is depicted in an unpatterned shawl showing purple on one plane and a bright green on the other (Mikosch, 1985a: 10, Fig. 4). This could have been achieved through the back cloth technique. At the turn of the seventeenth century some *kani paśmīnā* shawls were woven with such delicate yarn that they could be regarded as being virtually reversible. Another variation in *do-rukha* comes into evidence *c.* 1860. In these, the interlock outlines would be worked over in embroidery thus obliterating any difference between the two sides (Mikosch, 1985 b: 25, 43). According to two contemporary craftsmen, Abdul Khaliq Gana, village Batapura, and Gulam Mohammad Wani, Srinagar, the technical aspect of handling *kanis* for weaving reversible *kani paśmīnā* would pose formidable problems. Moreover, if only certain areas were to have dual colouring the fabric would be proportionately thicker in those parts. It was only after the *amlikar*, embroidered shawl, had developed in the nineteenth century, that a fully reversible shawl became practicable (fig. 8.22). The interlock was camouflaged by delicate embroidered outlines. In larger areas where the ground colour showed, the twill effect was transposed through embroidery. The warp ends were split into half by the needle so that the colour remained only on one half of the surface. Such shawls were called *do-rukha* and *aksi.* Identical motifs could present differing colours on each face. The twill weave was so well simulated in embroidery that only a trained eye could identify it as embroidery rather than a weave (Dhamija, 1976: 6–7).

With the development of European demand there was a virtual boom in *kani paśmīnā.* Nadir Shah was said to have sent 15 elephant loads of Kashmir shawls with an embassy despatched to Constantinople in 1739. This constituted the first significant introduction of the item to Western Europe (Watt, VI, 1972: 645). Under the Afghans, AD 1752–1819, *kani paśmīnā* was in great demand in Iran, Afghanistan, Turkestan and Russia (Sufi, 1974: 563–4). Napoleon is said to have sent (*c.* 1798) a Kashmir shawl from Egypt to his consort, Josephine in France (Lawrence, 1967: 376). Kashmir shawls became a rage in France around 1800. Josephine Bonaparte modified it to serve as an accessory in female attire. Hitherto *kani paśmīnā* had catered to male sartorial needs alone. Convention now decreed that Kashmir shawls could be worn exclusively by married women. Between 1820–39, Kashmir shawls, both the original article made in Kashmir, as well as the European copy, began to form a part of trousseaux in France (*Châle Cachemire,* 17, 29, 33).

Changes that were now introduced included innovations in technique as in embroidered *amlikār* (fig. 8.22) as well as departures in design articulated through themes such as *Śikārgāh* hunting scenes. The initiation of outlining motifs in embroidery in *kani paśmīnā* to which a reference has been made, is not dated, but Pandit A. Koul, ascribes the innovation of the fully embroidered shawl to a resident of Srinagar,

Paśmīnā embroidery/ *amlikār*, Srinagar. (Fig. 8.22)

Saida Baba, under the administration of the Afghan Governor, Azad Khan, 1783–5 (Koul, no date: 46). John Irwin (Irwin, 1973: 3), dates the development to the year 1803, under the aegis of an Armenian, Khwaja Yusuf. According to J.L. Bhan (Centre of Central Asian Studies, University of Kashmir, Srinagar), *Śikārgāh* patterns are to be associated with Dogra rule, AD 1846–1947 (cf. Singh, Ahivasi, 1981: 11).

Having reached its highwater mark, a decline in European demand began to set in *c.* 1860–70 (Irwin, 1973: 18). As had been the case with chintz, the Indian painted and printed cottons which had resulted in European machine manufactured analogues, European experimentation in duplication of *kani paśmīnā* on the drawloom (fig. 8.23), led to the development and perfecting of the jacquard mechanism[40] (fig. 8.24). This led to further fracturing the demand for *kani paśmīnā*. There was a brief spurt before the candle was dimmed and W.R. Lawrence records that the finest shawls were being produced during the reign of Maharaja Ranbir Singh between 1865 and 1872 (Lawrence, 1967: 376). It was during this same period that H.E. Bates (Bates, 1873: 33), was led to comment on the miserable economic condition of the *kani paśmīnā* weavers of Kashmir.

8.23

8.24

Drawloom. Courtesy: La Maison des Canuts, Lyon. (Fig. 8.23)

Jacquard loom. Courtesy: La Maison des Canuts, Lyon. (Fig. 8.24)

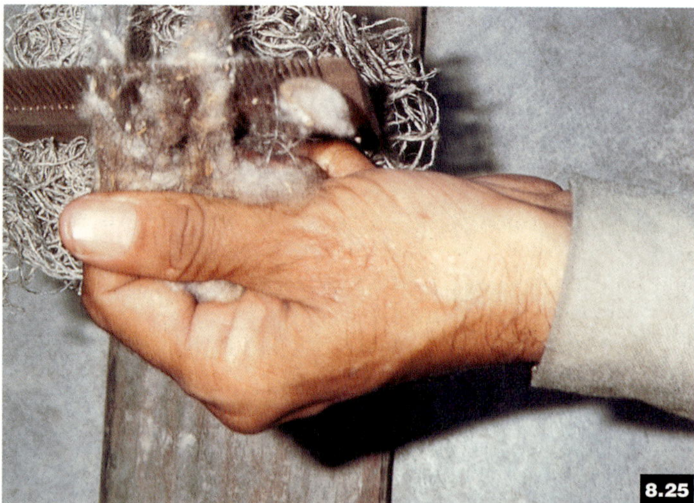

Carding of coarser quality *paśm* on teeth of comb held up right on wooden stand. Courtesy: Abdul Salam and Sons, Nurpur. (Fig. 8.25)

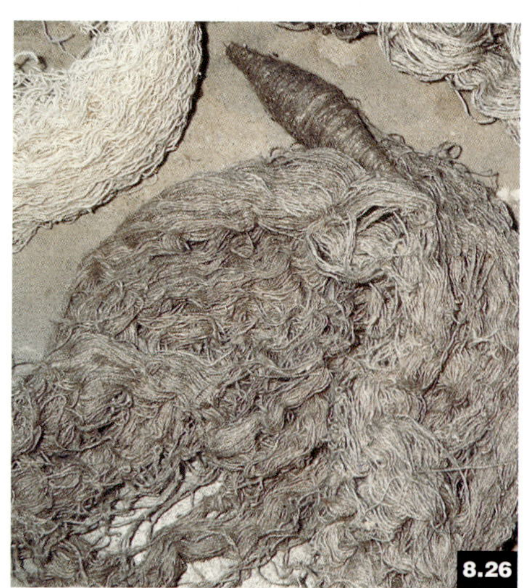

Natural colour *paśm*. Purchased at Kullu. Courtesy: Abdul Salam and Sons, Nurpur (Fig. 8.26)

Dispersal of Centres of *Kani Paśmīnā*

The popularity of *kani paśmīnā* resulted in the absorption of some of its motifs within the range of other kinds of textiles. *Jāmavār* patterns appeared in Banaras brocade and the *kalgā* motif made an entry within the format of Indian woven as well as painted and printed fabrics. *Kani paśmīnā* itself began to be woven in centres outside Kashmir, but invariably the weavers would be of Muslim Kashmiri origin. Ranjit Singh is said to have introduced this skill into Punjab in 1817 and the famines of 1833 and 1878 led to a dispersion of many weavers from Kashmir to these new centres.[41] Centres came up at Amritsar, Ludhiana, Jalalpur (Gujrat district), Dina Nagar (Gurdaspur district), Nurpur (figs. 8.25, 8.26) and Triloknagar in Kangra district.[42] Owing to royal patronage a centre came up at Bashohli. Production here appears to have been largely geared towards the French market, and with the steep decline in French demand after 1871, this centre collapsed, weavers migrating to Amritsar, Sujanpur and Pathankot.[43] The procurement of *paśm* and *tūṣ* was always deficient at these centres as the major supplies were still diverted to Kashmir. Because of this, admixtures were introduced, resulting in deterioration in quality. Among the wools used for this purpose were Kabuli goat *paśm*, used mainly at Lahore, Kirmani sheep wool used at Amritsar, and wool called *paṭ* obtained from Kokand via Bukhara.[44]

Paśmīnā in Kashmir Today

The superlative quality of *kani paśmīnā* has always been based on the purity of the basic raw material used. The highly labour intensive cleaning operations have ensured and maintained the desired quality. It is only in the post-Independence era that admixtures in wool have been introduced in Kashmir. This has been inevitable because of non-availability in the desired quantity of *paśm* and *tūṣ*. At present, the warp may be of *raffal*/ ordinary sheep wool, imported Australian merino or even of silk. Shawls with a silk warp are cheaper and do not really qualify to be called *paśmīnā*. Centres of such weaving are few. In Srinagar some weaving is done at the Pashmina Centre run by the Khadi and Village Industries Commission, at the Kashmir Government Arts Emporium Centre at Namehagal, and, on occasion, at the School of Design. At the private level, Cottage Industries Exposition (P) Ltd., Srinagar, may be mentioned. Under the aegis of the Intensive Handloom Development Projects, a Centre aimed at training and revival had been set up at Kanihama, a village 20 km. from Srinagar on the Srinagar–Gul Marg Road. There is a consensus that, irrespective of design input, unless steady supplies of *paśm* and *tūṣ* are assured, the industry can only limp towards a retrograde image of its former excellence. With the banning of *tūṣ* because of indiscriminate animal slaughter, the ring shawl has entered the shades of history.

Notes and References

1. The spelling and geographical terms used are generally those which enjoyed currency during the British period.
2. Hasan, 1959: 25, 244; Lawrence, 1967: 383; Rabbani, 1981: 23.
3. Hasan, 1959: 245; Bernier, 426 n.; Gaborieau, 1973: 15, 22.
4. Drew had drawn attention to the fact that among the domesticated and wild animals of Ladakh which developed this fur, yaks and dogs were also to be included. This undergrowth was combed out at the beginning of summer and sent to Leh where it was picked free of hair and either worked up locally or sent on to Kashmir. Drew, 1971: 299.
5. Moorcroft, I: 346–8; Moorcroft, II: 165; Sufi, 1974: 562. In the village of Lhari in Tibet, 101 leagues from Lhasa, virtually all the inmates were shepherds. Sheep, oxen and goats were reared. The fine silky hair of this wool was used in spinning *khu-lu* of the first quality as also for the Kashmir shawl. Hazlitt, 1928: 176, 319–20. Although the English observer failed to make the distinction, *khu-lu* must have been obtained from goat rather than sheep. Further, since sorting of *paśm* was best done in Kashmir, the Tibetan *khu-lu* must have been of inferior quality, perhaps comparable to the contemporary product of Himachal Pradesh.
6. Oral communication: Mr Zasak J. Tsaring, former Principal, Tibetan High School, Mussoorie; Mr Tsaring Wangpal, Tibet House, New Delhi.
7. In view of classical references in India to the hair of *ranku*, the deer, for which see above (infra p. 219), this allusion acquires additional interest.
8. Watt, II, 1972: 636; Adams, 1858: 523–4; Cunningham, 1970: 199–200; Vigne, 1981: 124.
9. Adams, 1858: 523; Atkinson, 1973: 33; Grzimek, 1972: 494.
10. Ibid., loc. cit.
11. Watt, II, 1972: 551, 637; Cunningham, 1970: 199–220; Honacki et al., 1982: 330.
12. Adams, 1858: 521–2; Ganhar, no date: 34–5; Honacki et al., 1982, 340; Hazlitt, 1928, 311–14.
13. Oral communication: Dr P.N. Bhat, Indian Council of Agricultural Research, New Delhi.
14. Baden Powell, 1872: 34; Balfour, VIII, 1976: 599; Watt, VI, 1972: 642.
15. Balfour, VIII: 599; Moorcroft, I: 360–1, 363. According to Balfour, (Balfour, VIII: 599), Rudok was a former constituency of Ladakh while Changthang comprised the southern mountainous region of Khotan. Changthang *paśm*, usually white in colour, was divided into two qualities, the first being labelled *khulchak*, and the second, *ralchak*. Baden Powell, 1872: 33. Since the term *ra* referred to the upper coat of hair of the goat, it is possible that *ralchak* was not as meticulously sorted as *khulchak*.
16. The importance of this trade to the economy of Kashmir in medieval times is illustrated in the treaty which terminated the Tibeto-Mughal-Ladakhi war, *c.* AD 1681–4. The Tibetan authorities undertook to supply the entire wool of the region to Ladakh. At the same time Ladakh was bound to divert this stock, along with its own produce, to Kashmir. This agreement remained in observance until the end of the eighteenth century. The first dent in the system was affected by W. Moorcroft. Datta, 1970: 18–19; Moorcroft, 1816: 451. With the development of *paśmīnā* weaving centres in Punjab in the nineteenth century, the degree of diversion of supplies away from Kashmir assumed greater dimensions but the continuing monopoly of Kashmir remained a constant factor.
17. The quality of *paśm* depended on climatic factors and the altitude at which the animal was to be found. See Balfour, VIII: 599.
18. Younghusband, 1970: 211; Irwin, 1973: 5; Koul, no date: 44.
19. Oral communication: Dr P.N. Bhat, Indian Council of Agricultural Research, New Delhi.
20. Blochmann, 1965: 97. Moorcroft uses black and dark coloured wool interchangeably. *Moorcroft Papers*: 2. Dr W.H. Siddiqui, Archaeological Survey of India, New Delhi, was kind enough to check the English translation against the Persian original (*Āin-i-Akbarī*, I, Part I, Neval Kishore edition: 116). The relevant passage is quite unambiguous in the usage of the word *siyāh* to indicate the shade of black in natural colour *tūṣ*. In this context it is of interest to note that *c.* 1900 the word *muśkī*/tawny was the local term for black. See Watt, 1903: 351. In contemporary times weavers use *siyāh* to indicate wool which has been dyed black. According to present-day specialists, *paśm* cannot be black in colour.
21. Blochmann, 1965: 97. In descriptions of procedures adopted prior to the actual weaving of *kani paśmīnā*, there is no stage suggestive of bleaching. G. Forster is the only traveller to state that the raw material was bleached. However, the observations of W. Moorcroft make it amply clear that the process ascribed to that of bleaching by Forster, was in fact only the method for de-oiling and cleaning the wool. Forster, 1798: 18; Moorcroft, II: 175–6.
22. Gabriel Vial has summarized several ways in which coloured *paśm* could be woven to form different kinds of pattern in *kani paśmīnā*. Vial, 1987: 45–9.
23. The Qajar dynasty reigned in Persia, AD 1786–1925. *c.* 1793–4, striped edgings, possibly woven in Kerman, come increasingly into evidence. Striped patterning was called *khatraaz* in Iran and *mehramat* in Kashmir. Scarce, 1988–9: 31–2; Scarce, 1987: 48, Fig. 5; Ames, 1986: 45, Fig. 5; Mikosch, 1985 b: 26, Fig. 4; Moorcroft, II: 193.
24. Documented during field visit by H. Perheentupa and K. Amin-Patel (NID).
25. See Emery, 1980: 78–81; Pope, 1939: 2178–9, n.3, 2178, n.1, 2179, Fig. 726, 2181.
26. See *Darī Fatehpur Sikri*, NID: 22, 25, 41, 50.
27. These were woven in panels against a neutral fabric across the width of the loom. For an illustration of this see Anavian, Anavian 1975: Plate 56b entitled, 'Border strips, Kashmir, 18th century'.
28. Informant: Maqbool, Cottage Industries Exposition (P) Ltd., Srinagar. For additional details relating to *kani paśmīnā* see Varadarajan, 1984: 238, n. 9.
29. Balfour, 1976, VIII, 601; Guelton, 1987, 23; *Moorcroft Papers*, 16.
30. Shrigondekar, 1939: 14, 89. According to Debala Mitra, paintings of Caves 9, 10 in Ajanta belong to the first and second centuries BC, and those found in Caves 1, 2, 16 and 17 are dated to the fourth and fifth centuries AD. The activity of painting murals at Ajanta continued up to the sixth and seventh centuries AD. Mitra, 1980: 8.
31. MacDonnel and Keith, 1958: I, 41. For *prāvāra* see p. 120.
32. Srinivasacharya, 1917: 13. *Yājñavalkya* (AD 100–300), also refers to *kuṭapaḥ* for which *Vijñāneśvara* (AD 1100), offers the same explanation as the Maskari cited earlier. See Vasu, Naraharayya,

1913: xxxvii, 33, 82. For other early references to wool see Singh, Ahivasi, 1981: 1–4.

33. Kangle, 1972: 142, 2.22.6, 147, 2.23.8. It needs to be emphasized that whereas there could have been a different pattern of diffusion of centres of manufacture in earlier times, owing to ecological and environmental factors, the sources of supply for *paśmīnā* and *tūṣ* must have remained the same.

34. Watt, II, 1972: 564; Grzimek, 1972: 475; Honacki et al., 1982: 330.

35. Hazlitt, 1928: 311–14. In the first decades of fifteenth century AD, the Chinese appear to have shifted the earlier association of unicorn from the *chiru* to the giraffe, recently introduced from Africa into the Celestial Kingdom. Willetts, 1964: 41.

36. Moti Chandra and V.S. Agrawala, 1959–62: 40. In verse 29 quoted in this reference, there is an allusion to a woollen fabric, *sophā*, earlier imported, but now locally made. Since *sophā* remains unidentified no inferences can be drawn. *Suf* is the generic term for wool in Arabic, and *sophā* may have referred to some category of Tiraz production.

37. Lawrence, 1967: 375. Lawrence refers to the motif as *jigha. Turañj* is derived from the Persian term for citron. One author, somewhat fancifully, traces the origin of this motif to the imputed footprint of the goddess, Parvati. *Rupam* 2, January 1926, Notes, anon., 'An Indian Choga from Zanzibar.' Although Abul Fazl (Blochmann, 1965: 97), refers to a category called *qalgha'i*, it is not at all clear what the appearance of this fabric might have been. There is a method of printing on silk which is called clamp resist. This technique is described by A. Bühler and E. Fischer. See Bühler, Fischer, 1977. V. Murphy and R. Crill refer to this technique as *culgee.* Murphy, Crill, 1991: 179–80. The clamp resist method of patterning is in evidence in some of the silk material used to frame Tibetan *tankas.* The term *qalgha'i* could perhaps be taken in the sense suggested in the text above as a motif rather than constituting a category to be included in woollen items.

38. Moorcroft, II: 191–2, *Sabzkar Safed* q.v. Under the Mughal Emperor Humayun (AD 1530–56), although Sultan Nazuk Shah (AD 1540–51), was the nominal ruler of Kashmir, the real force was Mirza Haidar Dughlat who exercised his authority on behalf of the Mughal Emperor. Sufi, 1974: 563.

39. The word *jāmavār* appears to have been a term related to a pattern rather than a technique. Alexander Hamilton, who was in India from 1668 to 1723, refers to the export of silk *jāmavārs* from Thatta. Steingass, 1973: 357: *jāmawār* and *jāmawārī*, q.v., provides the following meanings: fit or intended to be made into a garment; a flowered sheet or shawl. It is possible that the original meaning may have been that of floral patterning which may then have transited to a more restricted meaning in the context of the ornamental repertoire of *kani paśmīnā.*

40. Vial, 1987: 62–8; *Châle Cachemire,* 15–28; Adrosko, 1982: 89–117.

41. Kipling, 1887–8: 61; Khushi Mohammad, 1923: 181.

42. Baden Powell, 1872: 41. Plain shawls in colours such as ivory, white, scarlet, turquoise, blue and gray, made at Ludhiana, were called *Rampur Chaddar.* Birdwood, 1880: 281.

43. Ram and Raina, 1933: 225. For French demand in *kani paśmīnā* see *Châle Cachemire,* 17–21.

44. Baden Powell, 1872: 33–4; Kipling, 1887–8, 16.

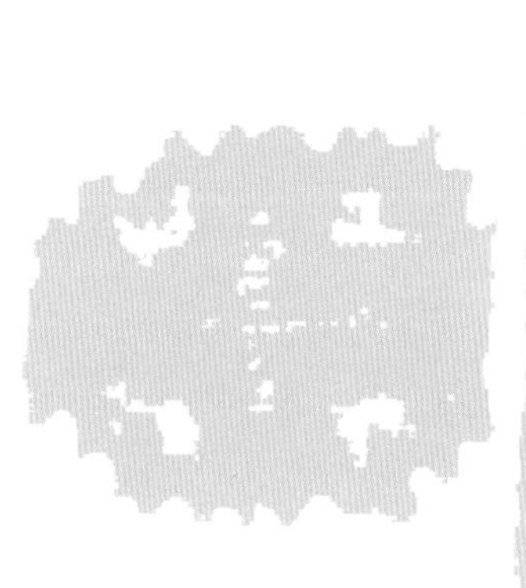

09

KULLU–KINNAUR, HIMACHAL PRADESH

09 Kullu–Kinnaur, Himachal Pradesh

Wool in the Kullu–Kinnaur Tradition

Kani paśmīnā of Kashmir represented an extremely complex commercial activity dependent on supplies of raw material secured from a widely dispersed geographical base; the Kullu–Kinnaur product, however, represented a seemingly simpler chain of operations.

Work procedure connected with wool and *paśm* has constituted one of the most important economic activities in the trans-Himalayan belt. This has caused a bifurcation in wool manufacture into two spheres – the rarified sphere of first quality *paśm* and *kani paśmīnā* catering to the elite, and that of thicker textured wool woven into *dohrū* and *paṭṭu* manufactured more at the subsistence level (see fig. 8.25). Despite the fact that the *paśm* trade constituted an important lifeline for both Ladakh and Tibet, domestic weaving in both areas was based predominantly on sheep wool and *paśm* of an inferior quality (cf. fig. 8.26).

Paṭṭu/dohrū, Kullu. Rear view. Courtesy: Nathi, Kullu. Note scarf worn as head covering. (Fig. 9.1)

The pivotal role of wool in the life of the tribal people in the higher reaches of Himachal Pradesh is reflected in the wool related belief systems operative at the folk level. Wool is believed to be innately pure being created by no less a deity than Brahma. While performing the *śrāddha* ceremony, a ritual connected with death, the householder had to wear wool and a woollen thread was placed over the *piṇḍa*, rice balls offered to the dead. Indeed, wool was the prescribed fabric for all ceremonial occasions. Amulets worn by villagers were tied with woollen thread, black being considered particularly potent in warding off evil spirits. A woollen cloth was placed above the bridal couple during the marriage ceremony, and, among the higher castes, a woollen fabric was draped around the funeral bier (*Rural Craft Survey*: 3–4).

Unlike *kani paśmīnā*, wool weaving at the subsistence level appears to have been executed in narrow width throughout these mountainous regions including Kashmir. Himachal Pradesh offers a fascinating variety and range in costume based essentially on the fabric of wool. The most important factor which has shaped usages in apparel in the upper reaches of Himachal Pradesh has been the climatic factor of extreme cold in the winter months and a cold to moderate climate in the remaining parts of the year. This has meant that in the range of unstitched garments, items of varying warmth have been used. It has also followed that items normally used for sleeping can also double as snow gear.

In female head covering a line of division is drawn between those who tie a variant of the *rūmāl*, kerchief/scarf (fig. 9.1) on the head and those who wear a cap. The *rūmāl*/scarf tradition is to be found stretching from the Punjab through Himachal

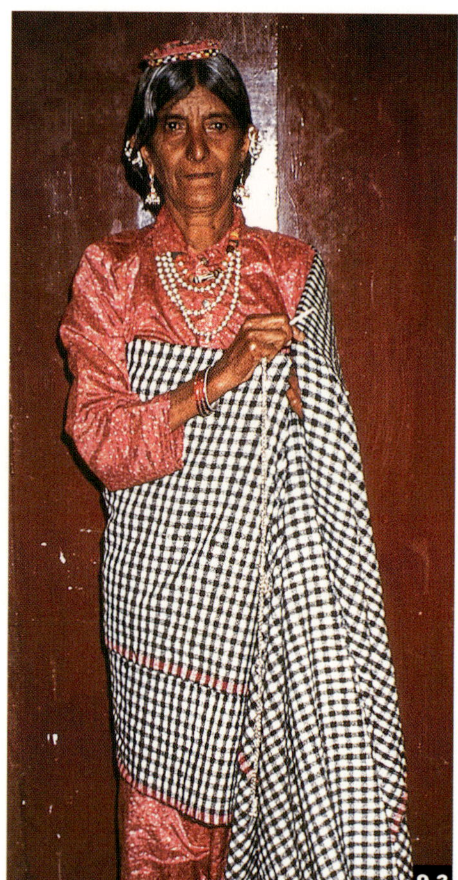

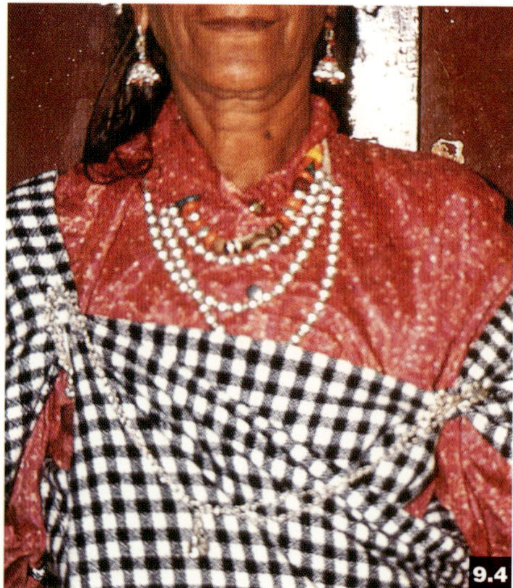

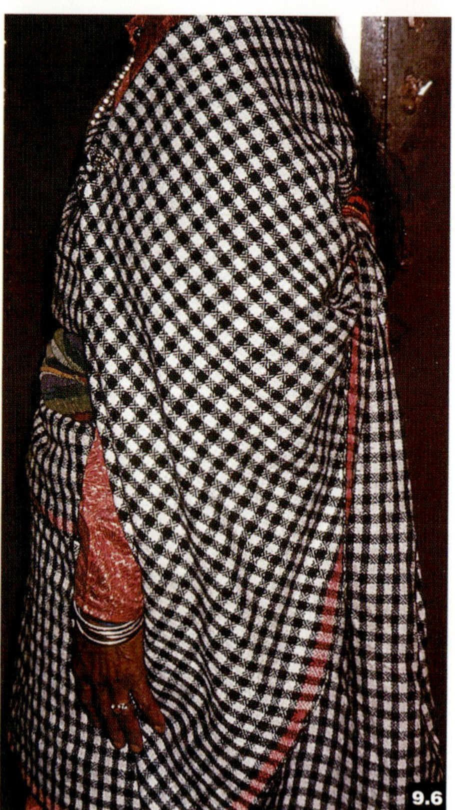

Pradesh to Kashmir. In the Simla region the scarf is called *dhatū*. The cap, known as *thepang*, is worn by both sexes in Kinnaur (fig. 9.2) and by males alone in Kullu. The upturned brim, earlier placed at the rear, could be pulled down over the ears in winter.

Lahul–Spiti, Ladakh and Tibet share the same mode in stitched costume. Kullu and Kinnaur form part of the wider South Asian tradition of a wholly unstitched garment, in this case, wool used as a single basic body drape by females. In Kangra and the Pangi sub-tehsil of Chamba, the unstitched and sewn usages merge. In Kangra, the Kinnaur–Kullu *dohrū/paṭṭu*, is draped over the *salwār-kamīz* of Punjab for winter warmth. In Pangi the basic costume is the *kamīz/choyel*, and a fully lined *dogri* of remarkable elegance called *chalana*. Above this a *dohrū/paṭṭu* is draped, the *dohrū* here being somewhat narrower in width than the Kinnaur prototype (figs. 9.3, 9.4, 9.5, 9.6). According to Chhallo Devi, village Gangeet, Chamba, this is called a *chādar*. Earlier this would have been an undyed garment probably of local manufacture, devoid of any ornamentation. Today Kullu–Kinnaur designs are favoured. According to the same informant there is yet another variation found in the costume of the women of Churah tehsil, Chamba. Here, the lower

Traditional Kinnaur female costume. Note *dohrū* and *thepang*/cap. Street scene. Labrang, Kinnaur. (Fig. 9.2).

Draping of *chādar*, Pangi. Stage 1. Courtesy: Chhallo Devi, village Gangeet, Pangi Valley. (Fig. 9.3)

Draping of *chādar*, Pangi. Stage 2. Courtesy: Chhallo Devi, village Gangeet, Pangi Valley. (Fig. 9.4)

Back view of *chādar*, Pangi. Stage 3. Note *jojee* in the hair plaited at the back. Courtesy: Chhallo Devi, village Gangeet, Pangi Valley. (Fig. 9.5)

Draping *chādar*, Pangi. Stage 3. Side view. Courtesy: Chhallo Devi, village Gangeet, Pangi Valley. (Fig. 9.6)

Rope made from waste goat hair. Kinnaur. (Fig. 9.7)

Chamba gaddi, costume. Male. Courtesy: State Museum, Shimla. Note black goat hair cord wound around waist. (Fig. 9.8)

Unsheared shawl goat. Tabo, Spiti. (Fig. 9.9)

Shawl goat after shearing. Tabo, Spiti. (Fig. 9.10)

garment is fashioned from a borderless plain or checked strip of cloth which is gathered in the front and kept in position with natural black goat hair cord (fig. 9.7) so integral to Gaddi costume (fig. 9.8). The head-dress both in Churah and Pangi is an embroidered strip of cloth (figs. 9.3, 9.5) called *jojee* in Pangi. This item is closer to the Ladakh repertoire rather than the scarf-cap usages of Kullu and Kinnaur. This evolution of mode in costume is the product of local ecology and environment. Specificity in the design repertoire can be traced to the matrix of religious practices. If a broad religious demarcation were to be made it could be stated that Hinduism reached up to Rampur–Bashahr. Between Rampur–Bashahr and Puh, Hinduism begins to fade into Buddhism while beyond Puh the dominant belief system is that of Tibetan lamaistic Buddhism.

The mountainous tribal regions of Himachal Pradesh enjoyed direct access to the Himalayan wool routes through which *paśm*, from domesticated *changra/chigu* goats, and *biangi*, from Changthang sheep, found their way to producer groups engaged in wool weaving. Chamba and Lahul–Spiti were contiguous to Ladakh, while Spiti and Kinnaur enjoyed direct egress to Tibet. Cutting across these regional boundaries, the fair at Gartok in western Tibet in September each year offered ample opportunities for trade transaction in wool (Singh, 1989: 25).

The area also has its own sources of wool. Sheep yielding *biangi*, are now reared in the Barabangal area in Kangra district. The domesticated *paśm* yielding goat, the *chigu*, has been raised in Spiti (figs. 9.9, 9.10) as well as in the Yangthan area, Puh sub-division, Kinnaur. Since these goats are non-migratory, the *paśm* is of inferior quality and has never been in demand in *kani* weaving. Sorting, which is locally done, is less than perfect and the material is used in regional products requiring a greater degree of warmth. Bhadarwah sheep, also known as *gaddi*, and Rampur–Bashahri sheep are the main sources of local wool. Sheep rearing is a traditional occupation among inhabitants of Lahul–Spiti, Kinnaur, Kangra, Mandi, Kullu and Simla. This vocation is popular in Pangi and Bharmour in Chamba. *Gaddi* sheep are herded by nomadic Gaddi shepherds who have an original home in Bharmour, district Chamba. Rampur–Bashahri sheep are found within the constituents of erstwhile Mahasu district[1] and the present-day district of Sirmaur. Natural colours comprise white, black, brown and grey.[2] Sheep are shorn two to three times in a year, during the months of February, June and October. The October shearing is considered to yield the best quality wool. Wool of the first shearing of lamb is called *imboo*. The length varies from 2 to 4 in. The common name of wool of coarser quality obtained from sheep of Rampur–Bashahr variety is known as *desar.* In Tissa, district Chamba, the term for *desar* is *palmi* while that used in Chamba is *thootu* (*Rural Craft Survey*: 18, 19; Singh, 1992: 298). In Kinnaur the output of one cropping is retained for personal use, the rest kept for commercial purpose (*Handicrafts of Kinnaur*: 44). Wool is spun on the *taklī,* spindle, rather than the *charkhā* as in Kashmir. The spindle provides for greater physical mobility and is much better suited to the needs of a migrant shepherd community. While weaving and dyeing are specialized activities, spinning cuts through all gender and social divisions.

In view of the contiguity of Spiti to Kinnaur, some notice has to be given to textile traditions of Spiti. While linguistically north Kinnaur merges into Spiti, costumes of Spiti call for a completely different range of perception in design. The importance of Spiti as a stage in the Tibetan wool route is demonstrated by the fact that the English annexed Spiti in 1846 in an attempt to acquire control over this route. In return for Tibetan wool, barley and wheat were sent from Spiti to Tibet. Society in Spiti was trifurcated between the *nono,* nobility, *chhazang,* the middle class and *pyipa,* the menial class which included weavers (*Third Revised Settlement*: V, 2). All woollens were milled. Throughout India processes of drying subsequent to milling appears to have followed the same procedure. Field experience in district Mahbubnagar, (fig. 9.11) Andhra Pradesh, corroborates the description given for procedures adopted in Himachal Pradesh. During the milling process the fabric is put in hot or cold water while one or two persons pound it with their feet. *Rīṭhā, Sapindus mukurossi* is then put in. The pounding continues. The cloth is then cleaned in running water without wringing. After this four persons hold it up, double it and tie up the four ends to the ceiling or on to a branch of a tree. The lower loop of the cloth is weighted down with stones to maintain tautness. The cloth dries within a day or two and is then ready for use (*Rural Craft Survey*: 40).

Two qualities of local Spiti weaves could be identified—the *therma* and the *purug. Therma* is made from *biangi* wool in a natural brown. Earlier it may have been made of *paśm* (Khachikian, 1967: 251). *Purug* is made from the wool of local sheep and is thicker in texture. Originally white, it is subsequently dyed. There is reference to a third quality, *shama* (*P.D.G.*, 1918: 287), but informants from Spiti say that this should be taken as *shulma,* the name of a garment worn by females. There is, however, a reference to a green dye obtained from the leaves of the *shoma* plant used by Tibetans in Darjeeling for dyeing wool (Mohanty et al., 1987: 166). Yellow dye was obtained from a wild plant, *potlo.* This could perhaps be the yellow *dholu* flower, a fine colouring agent, which springs up in the Himalayan foothills after the snow has melted. There is also reference to roots of a grass called *aikari* which yields a yellow dye (*Rural Crafts Survey*: 40; Mohanty et al., 268). The grass grows abundantly near marshy land. The roots of the plant are collected and dried in the sun for a few days, and then the desired quantity is ground to a powder, boiled in water and strained through cloth. Fresh water is added to this diffusion which is then reheated. Yarn is immersed in this liquid for half an hour, taken out, wrung and dried in the open. Blue was obtained from the plant *nydlo.* No further details are available with regard to this plant. It is possible that this may be a variety of *rum* (*Strobilanthes cusia*) used in the north-east. *Marpo,* madder, was procured from Kullu (*P.D.G.*, 1918: 275). In the absence of a full money economy, fairs, as in medieval Europe, played a crucial role in the chain of distribution. The most important fair in earlier times was the one held in the month of July in Patseo on the road to Leh. Traders from Kullu, Ladakh, nomads from Tibet and neighbouring uplands as also local persons from Lahul and Spiti participated in this fair. Transactions were mainly conducted in grain, wool and salt (*Third Revised Settlement*: I, IV).

Weighting down of Andhra *kambli.* Village Morikal, district Mahbubnagar. (Fig. 9.11)

Border designs. Kullu. Courtesy: Bhutti Cooperative Weavers Society, Kullu. (Fig. 9.12)

Shawl. Kinnaur. Detail, end piece. NID Coll., Ahmedabad. (Fig. 9.13)

The female unstitched body covering of Himachal Pradesh to which a reference has been made can be divided into two categories—items without ornamented horizontal and cross borders and those which were embellished with these (fig. 9.12). The plain items, earlier in natural colour, were of local origin of which some were woven by Gaddis. In Kullu and the upper reaches of Mandi, checked material which initially must have been woven from natural white and black wool, was very popular.[3] Ornamental body robes, undoubtedly, emanated from Kinnaur.

Historical Background

The history of Kinnaur during the medieval period is intimately linked with the fortunes of the kingdom of Rampur Bashahr. In 1685, Raja Kehri Singh of Bashahr signed a treaty which remained in force until 1962. This provided the basis for the wool and *paśm* trade with Tibet. Not only did this ensure Kinnaur participation in a trade which brought Tibetan *paśm* to markets such as Ludhiana and Amritsar, but a wool industry came into being in Rampur–Bashahr itself (Singh, 1992: 79–81; 1989: 17–21). However, apart from providing traders of Kinnaur with access to the highly coveted Tibetan wool trade, the rulers of Rampur–Bashahr provided no other input into the local wool industry. The range of woollen textiles in Kinnaur provide an indication of the repertoire available throughout this eco-region. Because of the number of dialects spoken, nomenclature of the same item vary from place to place.

In the scale of social hierarchy, the Negis constituted the upper social caste, the lower castes comprising the Chamang, Koli and Chanal. The Kolis, who enjoyed a slightly higher status, undertook the task of weaving apart from the other duties they had to perform. The patron provided the wool. The weaver dyed the wool according to the pattern and executed the weaving.[4]

There were several centres at which seasonal fairs took place to cater to the needs of the region. In August/September a fair was held at Kanam, where there was also a large Buddhist library. Among the fairs which took place at Rampur–Bashahr, the most important was the Loi or Lavi Fair. *Paṭṭu* and blankets from different parts of Kinnaur were brought here and those frequenting this fair bought their annual supplies of blankets at this time. A special part of the fair was reserved for those who were entitled to trade in Tibetan wool and *paśm* (*Simla Hill States*: 61–2).

Clothing and Patterning

Kinnaur enjoyed a very special role in the area of ornamentation of female body garments. Curvilinear designs within a geometrical grid appear to have acquired an overlay of symbolism resulting from the diffusion of the cultural impact of Lamaistic Buddhism radiating outwards from Tibet. Professor Lokesh Chandra concurs with the view that the repertoire should be labelled Kinnauri rather than Buddhist. The Chinese *svāstika* like design common in Kinnauri garment is called *yungrung* and *chhoktin*, corresponding to the Tibetan term *chorten*, and is taken to represent a Buddhist *stūpa* (fig. 9.13). The inspiration of other motifs appears to be derived from the range of eight auspicious symbols associated with Buddhism. The articulation of all these motifs appears, however, to be region specific. The items of clothing decorated with these seemingly auspicious symbols were not those associated with the performance of daily chores but were ceremonial garments reserved for auspicious occasions in the past. Under such circumstances the threshold between the sacred and the profane is extremely slim in modern times.

The terminology of items used in clothing reflects the dimension of multiplicity in usage. *Dohrū* was the general term used for the basic body cover garment used by females. Generally the dimensions were about 11 ft. 6 in. x 4 ft. 6 in. In Lower Kinnaur it was called *zang*. In Upper Kinnaur the term in use was *chugtu*. *Chugtu* were woven

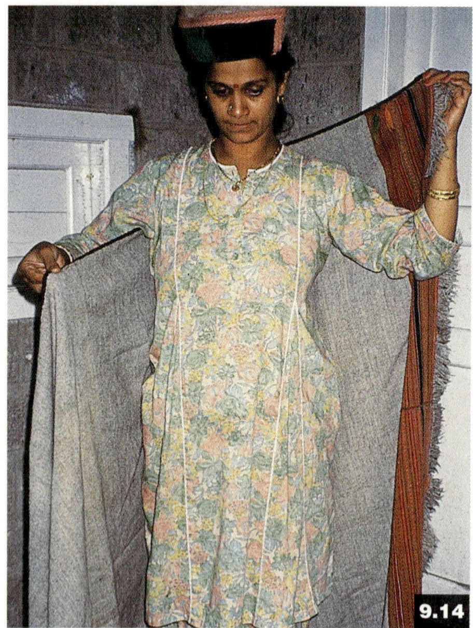

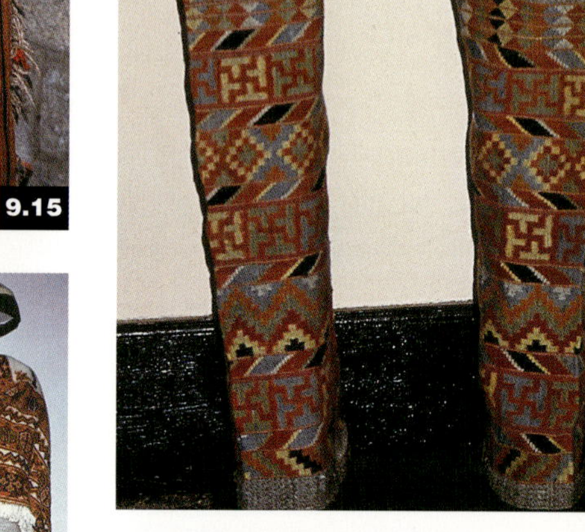

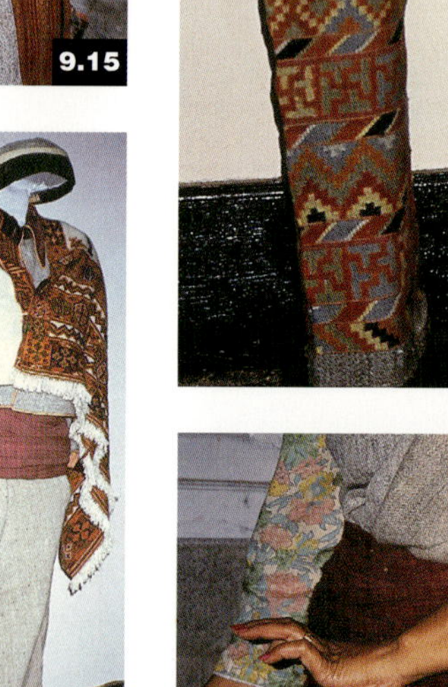

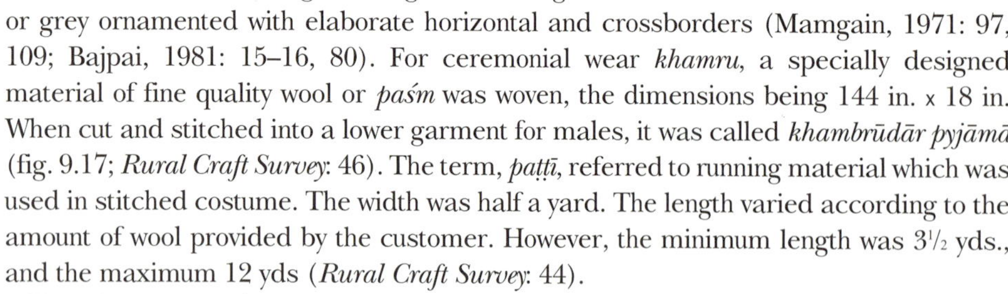

Wearing of *dohru*. Stage 1. Courtesy: Radhika Negi, New Mashobra. (Fig. 9.14)

Final stage draping *dohrū*. After the *gachchang* has been adjusted around the waist, the top segment of *dohrū* is slung back over it. Courtesy: Radhika Negi, New Mashobra. (Fig. 9.15)

Border design, *Topru-se-dohrū*. Work place of Chhanguram, New Mashobra. (Fig. 9.16)

Khambrūdar, Kinnauri male lower garment. (Fig. 9.17)

Traditional female costume. Courtesy: Shimla State Museum Shimla, Note *dohrū gachchang, chhanli, thepang*. (Fig. 9.18)

Dohrū in costume is associated with the *gachchang* wound around the waist. Courtesy: Radhika Negi, New Mashobra. (Fig. 9.19)

in Sangla in Kalpa block and at Rarang Labrang and Sanam in Puh block.[5] This item could also serve as a sheet or blanket. When used for the latter purpose the *dohrū* could be called *dohrī* or *paṭṭu*. While the term *paṭṭu* could be used synonymously with *dohrū*, the only meaning to be assigned to *dohrū* was that of the unstitched garment worn by females (figs. 9.14, 9.15).[6] Large *dohrī*, 12 yds. or more in length, and 6 yds. or more in width, made of coarser quality wool, could be woven to cover an entire family.

The *dohrū* could be made from first quality *biangi* wool. *Paśm* was considered too warm. Earlier, the ground colour came in a range of natural shades. It was made in several qualities. The simplest was called the *phatta orang*. It was without a border and could also be used as a blanket. The second category, which could have a simple border, was called *orang-se-dohrū*. The *topru-se dohrū* was taken as the highest quality (fig. 9.16). It was worn on festive occasions including weddings. The background was black or grey ornamented with elaborate horizontal and crossborders (Mamgain, 1971: 97, 109; Bajpai, 1981: 15–16, 80). For ceremonial wear *khamru*, a specially designed material of fine quality wool or *paśm* was woven, the dimensions being 144 in. x 18 in. When cut and stitched into a lower garment for males, it was called *khambrūdār pyjāmā* (fig. 9.17; *Rural Craft Survey*: 46). The term, *paṭṭī*, referred to running material which was used in stitched costume. The width was half a yard. The length varied according to the amount of wool provided by the customer. However, the minimum length was 3½ yds., and the maximum 12 yds (*Rural Craft Survey*: 44).

Female costume in Kinnaur comprised the *dohrū*, *gachchang*, *cholī*, *chhanli* and *thepang*/cap. The *gachchang* is a voluminous length of cloth, 2½ ft. x 22 ft., worn by females around the waist (figs. 9.18, 9.19). In a slightly shorter form it is also worn by males. It has more than one use—it keeps the midriff warm and acts as a rest in carrying back loads.[7] Earlier it was probably undyed *endi* from the north-east but the material worn in present times is dyed. The item is purchased at Kalimpong for local distribution. The *cholī*, stitched from *paṭṭī*, is a jacket worn on top of the *dohrū*. The *sanjab-se-cholī*, stitched with velvet cuffs, was earlier reserved for special occasions.

The *chhanli*, called *yangluks* in Upper Kinnaur, is a unisex shoulder wrap and can be woven of *paśm*. The female garment measures 5¾ *hath* x 1¾ *hath*/11 ft. 6 in. x 4 ft. 7 in. (fig. 9.18). The item worn by men is slightly larger. Unlike women who always use the *chhanli* as an item of clothing, men can sling it over the shoulder in summer and also use it as a carry cloth.[8] The term *loi* is used in Simla district to denote the *chhanli* worn by males.[9] The word is probably introduced from Punjab where a woollen *chādar* worn by males is called by this name. According to Kashyap, Himachal State Handicrafts and Handloom Corporation, Shimla, the dimensions of *loi* are 4 ft. 1 in. x 7 ft. 4 in.

Loom Types Used in Himachal Pradesh

The loom types traditionally used in Himachal Pradesh are (a) the throw shuttle pit loom, and (b) the horizontal ground loom (figs. 9.20, 9.21, 9.39).

These looms were placed inside the weavers' homes, or outside the house in any open space available such as the verandah. The loom is stretched horizontally on the ground and is normally supported by four posts, fixed deep in the ground in a rectangular form. The posts may be six in number in some cases. These are about 1½ ft. above the ground. The beater and balance of heddles are attached to two or more wooden sticks suspended horizontally with a cord near the ceiling (*Rural Craft Survey*: 8). The pit loom is about 1 ft. 5 in. width, 2 ft. 7 in. in length and approximately 2 ft. 6 in. in height (*Woollen Products H.P.*, NID: 14).

The looms have four shafts with thread heddles each operated by a separate treadle. The shafts are tied to heddle horses or pulleys and attached to one of the overhead rods. Bamboo reeds are used.

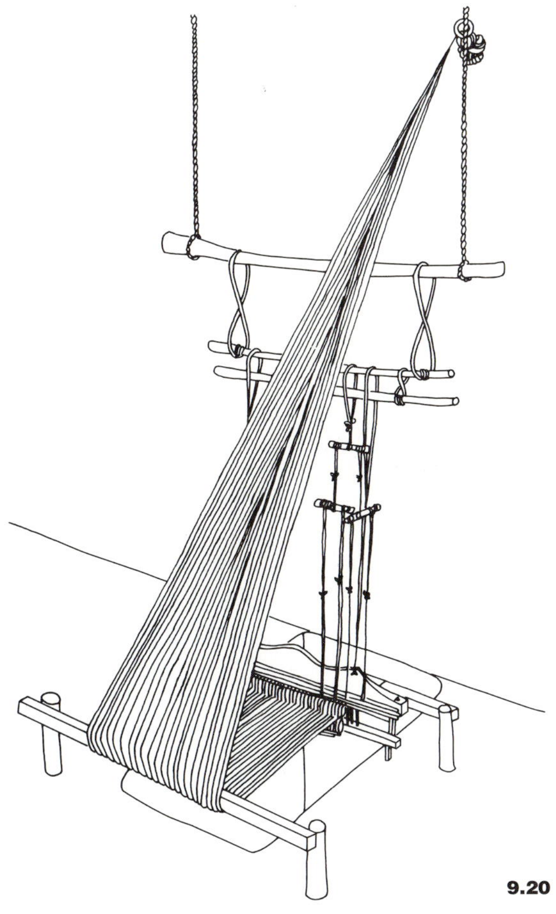

9.20

Pit loom, Kullu. (Fig. 9.20)

9.21

Kullu shawl weaving on pit loom. Village Baran, Kullu. (Fig. 9.21)

9.22

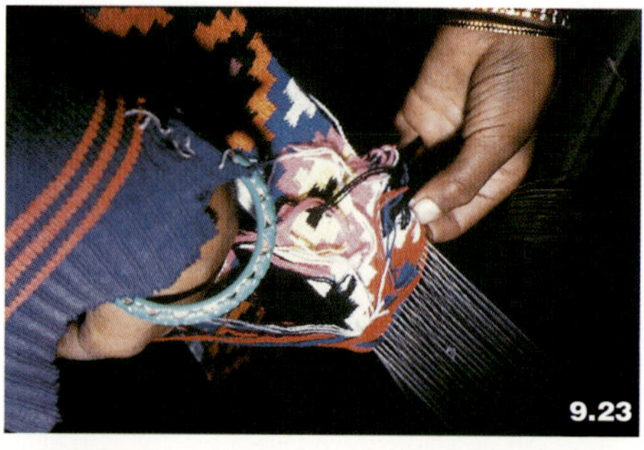
9.23

9.24

9.25

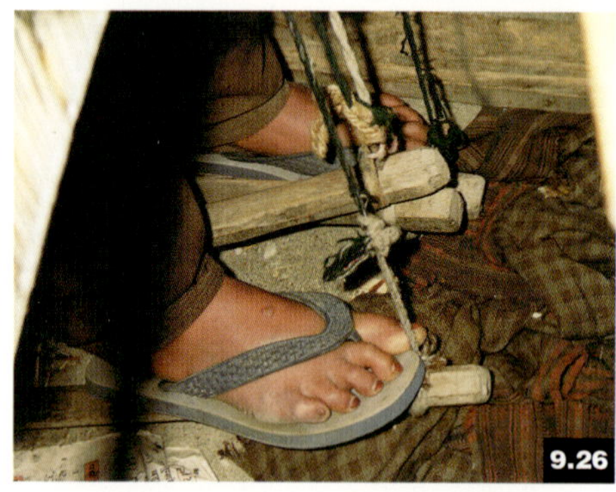
9.26

9.27

Frame border loom. Courtesy: Bhutti Weavers Cooperative Society, Kullu. (Fig. 9.22)

Frame border loom. Detail. Courtesy: Bhutti Weavers Cooperative Society, Kullu. (Fig. 9.23)

Loom used for *Kinnaur* shawl. Kinnaur. (Fig. 9.24)

Loom set up. Kinnaur. Detail. (Fig. 9.25)

Manipulation of pedal loom. Kinnaur. (Fig. 9.26)

Hollow bamboo tube which serves as shuttle. Traditional Kinnaur loom. New Mashobra, Shimla. (Fig. 9.27)

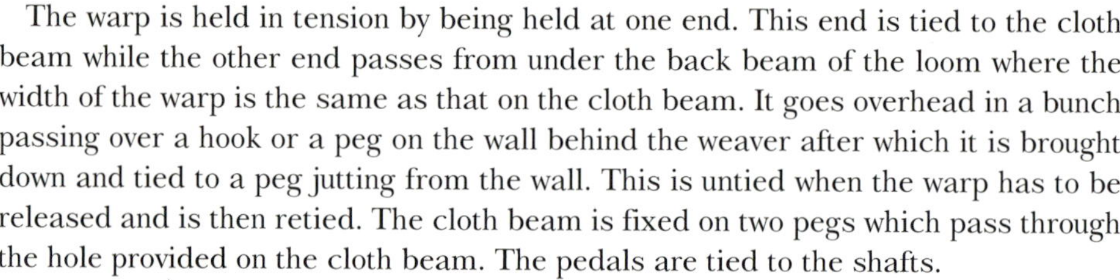

The warp is held in tension by being held at one end. This end is tied to the cloth beam while the other end passes from under the back beam of the loom where the width of the warp is the same as that on the cloth beam. It goes overhead in a bunch passing over a hook or a peg on the wall behind the weaver after which it is brought down and tied to a peg jutting from the wall. This is untied when the warp has to be released and is then retied. The cloth beam is fixed on two pegs which pass through the hole provided on the cloth beam. The pedals are tied to the shafts.

Traditional fabrics, mainly for unstitched garments and for household use were produced on the pit loom. It has been observed that the fabrics woven on throw shuttle pit looms operated by one weaver tended to be narrow. This was mainly due to the fact that the weaver operating the loom had to throw the shuttle from one end of the shed with one hand to be received at the opposite end by the other hand. As the fabrics woven on these looms are narrow, two identical pieces are stitched together to achieve a larger width resulting in a join running down its centre (cf. fig. 3.36). While these fabrics were suited for traditional use by the local people they were not viable for the commercial urban markets. At a later date, frame looms with wider widths were introduced for this purpose.

The border loom is a smaller version of the above loom, its width is small, and sometimes more than one border is woven on these looms. This loom is called the border *khaddi* and is used in Kullu (figs. 9.22, 9.23).

Most looms are now converted to frame looms and hence while many of the parts are the same, the weaver is seated on a wooden plank near the cloth beam. The earlier frame looms were simple outdoor looms propped up against a wall as in Kinnaur (see figs. 9.24, 9.25, 9.26).

Accessories

The shuttle is made of a hollow bamboo tube tapering and closed on one side. It is called *nalu* or *nali* in the local Kinnauri language. Instead of a bobbin, the weft pick is lightly coiled around a wooden stick which is then inserted into the hollow of the bamboo tube shuttle. The stick is then removed leaving the yarn inside. When the shuttle is thrown in the shed, the weft yarn flows out easily (fig. 9.27, cf. fig. 2.14).

Raw Material

The handspun yarn was originally supplied to the weaver by the people who wanted the yarn to be converted to fabric for their own use. Locally available fleece is used by the people in the state. It is spun at home in several counts with the use of a spindle. This consists of a 10 in. long piece of bamboo and a wooden whorl. The bamboo piece, almost as thick as a lead pencil, serves as a spindle and tapers at the upper end. The lower pointed end helps in its swift twisting. The wooden whorl, is fixed about 2 in. above the lower end. This serves as a weight and helps to sustain the rotation of the spindle. This tool is used for spinning particularly in the higher regions of Himachal Pradesh. The term for it in Kinnaur is *pang* or *panghach* and in other places it is known as *taklī* or *taokali* (*Rural Craft Survey*: 9, figs. 9.29, 9.30). The *charkhā* is also used for spinning yarn.

The raw wool or fleece obtained from sheep undergoes several processes before it is ready to be used as yarn for weaving. The first stage requires the sorting of fine wool from coarse wool. This is then washed with soap to remove the grease and other impurities. After the fibres are dried these are teased by hand or by pulling small quantities of fleece through a wooden comb. The small wooden comb is fitted on a stand with the teeth of the comb facing upwards. The fibres are then arranged to form a *rolag*. It is from these that the spinner draws the fibres while spinning the yarn. Single ply yarn is then doubled and twisted. The twisted yarn is used for the warp whereas the weft yarn is single ply.

The fleece, *paśm*, is also obtained from the *chigu* goat. The sorting of the dead hair from the fine hair is done by hand by the same process as is followed in Kashmir. In order to remove the grease about 250 gm. of whitish clay *gollu* or *makol* is mixed thoroughly in it. This makes teasing easier (*Rural Craft Survey*: 9). The wooden comb may also be used to tease the fibres (fig. 8.25). The process of making the *rolag* and spinning is the same as in sheep wool spinning. However a much finer yarn is obtained from *paśm* fibres. The yarn counts spun are dependent on the quality of fibres and the end product to be woven. These range between 16s to 60s count. The *paśm* used for weaving the *kharcha* is coarse and thick and is approximately between 4s and 16s count.

Preparation for spinning of wool for weaving of *kharcha*. Kinnaur. (Fig. 9.28)

Spinning of wool for weaving of *kharcha*. Kinnaur. (Fig. 9.29)

Spinning of wool for weaving of *kharcha*. Kinnaur. Note the implement is resting against the ground. (Fig. 9.30)

Fabric Characteristics

The traditional fabrics used as unstitched garments are made from wool and are woven in the 2/2 twill. Sometimes instead of a straight weaving order it is reversed to create a 2/2 horizontal herringbone twill. Often pointed drafting is done to achieve symmetrical pointed 2/2 vertical herringbone twill and diamond twill. These are patterned by the use of colour in checks, stripes and patterns executed in tapestry weave.

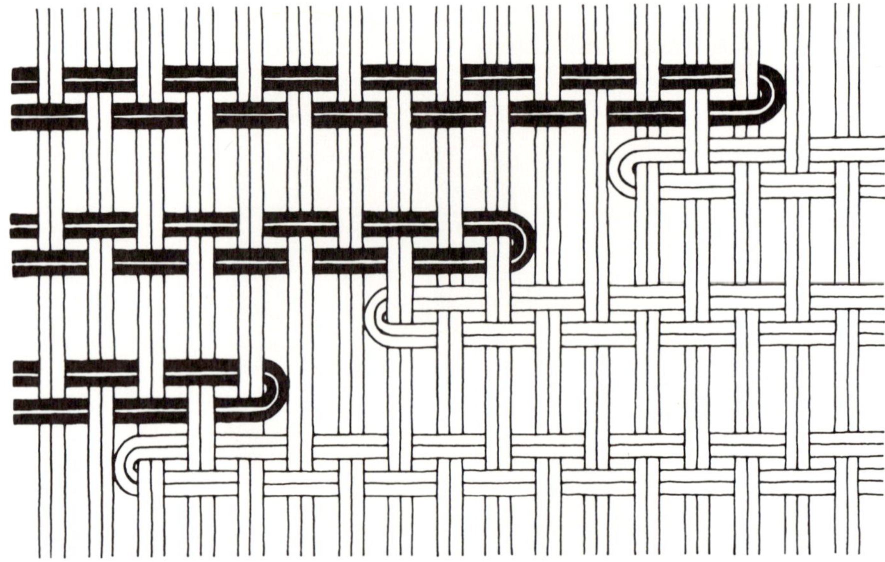

9.31

Interpenetrating type of dovetailing in 2/2 plain or basket tapestry weave. (Fig. 9.31)

Paṭṭu

While the daily wear *paṭṭu* is comparatively plain with a red border woven along its edges in the single interlock or dovetailed in 2/2 twill tapestry weave, the structure of the main fabric is in a 2/2 twill. The *paṭṭus* used for special or festive occasions are more elaborately patterned. The patterning is done using geometrical patterns woven in bright colours in the 2/2 dovetailing and 2/2 plain/basket tapestry weave. These techniques of weaving in Kullu approximate to those of Kinnaur. However in some cases where the weaver wishes to create diagonal margins the patterning weft of one colour may move two or three warp threads in one step in one direction and return in the next pick (fig. 9.31). At the same time the adjoining patterning weft would move towards it in one step on the same 2 or 3 warp threads and return in the next pick. This has been defined as an interpenetrating type of dovetailing by Irene Emery (Emery, 1980: 80). In this weave wefts from adjacent areas alternatively pass into the other area. They turn on adjacent warps instead of turning alternatively on a common warp.

In the Mandi district *paṭṭus* are woven in a variety of checks mainly in black and white using the natural colours of the fleece. The *paṭṭu* is worn by women. In Kullu the woman drapes the *paṭṭu* around her body in such a way that it covers the lower and the upper parts of the body. The two corners of one end of the fabric are brought from the back of the shoulder to the front and secured on each side by a pin/*bumani*. The two pins may be linked with a chain (cf. fig. 9.4)

Dohrū

The fabric is woven and worn in and around Kinnaur district. Its use is similar to that of *paṭṭu* and when worn has some resemblance to that of a sari. However, this garment has pleats at the centre back rather than the centre front when worn.

The two ends have cross-borders woven with colourful stripes. The ground is woven in natural wool colour mainly grey and shades of grey.

Dohrūs are worn by women in upper hilly areas of district Mahasu and in Kinnaur. They can be worn at night or they may be used as covers in place of quilts. They are prepared in different sizes. Some *dohrūs* are of such large dimensions, 12 yds. or more, that four or five members of a family can easily sleep together under it. Such *dohrūs* are used two-fold or four-fold and are warm enough to keep the cold away in winters. The width, which is about 2 yds. or a little more, is made by stitching up two pieces of equal lengths. The normal width is 1 yd. and the length may be 3 to 6 yds. (*Rural Craft Survey*: 44).

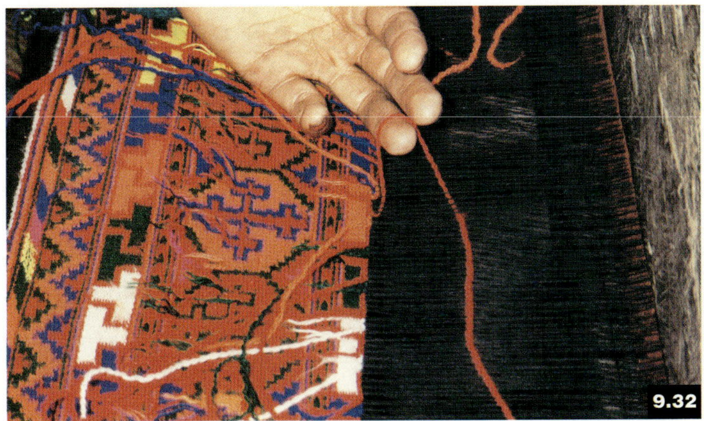

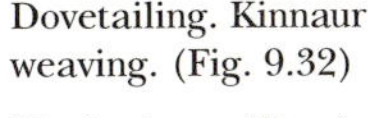

Dovetailing. Kinnaur weaving. (Fig. 9.32)

Single dovetailing in 2/2 plain or basket tapestry weave. (Fig. 9.33)

Gudma. Woven in Kinnaur with Tibetan *biyangi* wool, fulled and raised. Courtesy: Khadi Bhandar, Shimla. (Fig. 9.34)

The *dohrūs* used for ceremonial and festive occasions are elaborately and colourfully patterned. Geometrical shapes mainly based on a square grid that have been inspired by the Buddhist *stūpa*, the *chorten*, are used along with *palpe*, constituted by two interlinking rectangles. The design repertoire is extremely wide with each design having a name. The colours used for the patterning are the basic colours red, blue, yellow, orange, green and white. Black and bright pink are also used (figs. 9.32, 9.13).

For all vertical movements in the pattern and specially when requiring fine detailing, the 1/1 dovetailing (fig. 9.33) in a 2/2 plain or basket tapestry weave is used. For diagonal movements the slit tapestry weave in a 2/2 plain weave is used. In some cases the slits are clearly visible. A number of times one also finds the 2/2 or 3/3 dovetailing tapestry in the basket weave. The patterning with these techniques is also done along the two edges of the *dohrū*.

Turning to headgear, in earlier times the *thepang* was unlined and devoid of any decorative edging. The velvet adornment (fig. 9.14) appears to be a later embellishment.[10] Apart from these relatively lightweight items, for the severe winter cold, heavier ones are made to spread on the ground or cover the body. It has also to be remembered that the Kinnaur shepherd never uses a tent, preferring to sleep in the open or take shelter in caves. Some wool items have been developed which are reputed to be waterproof. This brings us to the range of floor coverings, the *gudma, kharcha* and *thobi.* Needless to say, these items also have multiple uses.

Gudma is a floor covering measuring 1½ yds. x 5 yds. made of *biangi* wool in natural white or black (fig. 9.34). Sometimes the borders can be defined through red stripes. The villages of Sunam and Kanam in erstwhile Chini tehsil are noted for *gudma* weaving.[11] Blanket weaving owes much to the efforts of the Moravian Mission, which, soon after establishing itself in Puh in 1854, set up an outpost at Chini. Father T. Schreve is known to have introduced several innovations into the local wool industry.[12] The introduction of the instrument for carding and the fuller's teasel, *Dipsacus fullonium*, can almost certainly be ascribed to the intervention of this Mission. Regional nomenclature, *tigal* in local parlance and *burash* in Punjabi, are themselves suggestive of foreign derivation. By raising the surface with the teasel, air pockets were created and the warmth of the fabric thereby enhanced.

Carding of wool for *kharcha* weaving. Kinnaur. Stage 1. (Fig. 9.35)

Carding of wool for *kharcha* weaving. Kinnaur. Stage 2. (Fig. 9.36)

Carding of wool for *kharcha* weaving. Kinnaur. Stage 3. (Fig. 9.37)

Carding of wool for *kharcha* weaving. Kinnaur. Stage 4. (Fig. 9.38)

Weaver weaving *kharcha*. Kinnaur. Note position in seating at ground level. (Fig. 9.39)

Kharcha loom. Creation of shed. Kinnaur. (Fig. 9.40)

The *kharcha* is known as *phug dohri* and *chugdan* in Upper Kinnaur.[13] It is a type of rug which is also used as a mattress by the village people. It is made of inferior quality *paśm* in natural colours, mainly black and white. *Paśm* from the annual shearing of the *chigu* goat is utilized, between 5 to 10 kg. of wool being required for one *kharcha.* The hair, *shālī*, is separated from *paśm* and sold to merchants for making *namda* and ropes fashioned by Gaddi shepherds[14] (fig. 9.7). The *paśm* is then carded (see figs. 9.35, 9.36, 9.37, 9.38) and spun into a coarse yarn called *paśm ke bāl* (fig. 9.28; see Dixon, 1979: 26–9). For the warp, the distaff is manipulated on the thigh for providing a higher twist (fig. 9.29).

A special selvedge is not planned. After weaving, the *kharcha* is given to the customer who mills the material by pounding it with the feet in water for 5–6 hours. After this it is dried and is then ready for use (*Rural Craft Survey*: 26, 33, 47). Two lengths are sewn together. This not only provides a larger width but the material can also be transformed to be utilized as a hooded garment in case of need. *Kharcha* is said to be waterproof (see fig. 9.42). Coats and snow boots can be made of this material as well as small bags to carry food grains loaded onto goats and sheep. It can also serve as a rug or mattress. There is a major concentration of *kharcha* weavers in the village of Labhrang in Puh block.[15] In Pangi, Tissa and Brahmaur, district Chamba, the *thobi*, an item resembling the *kharcha*, is woven. The *thobi* may be made either with *chigu* or yak *paśm.* It is woven in 6 or 9 in. width strips which are then stitched together to the requisite size (*Handicrafts of Chamba*: 60).

Horizontal Ground Loom for *Kharcha* Weaving

The horizontal ground loom with one heddle rod with half heddles made around every alternate warp end is used for weaving the *kharcha.* This loom has no reed (fig. 9.40).

Four wooden pegs are driven into the open ground, the distance between them being determined by the width and length of the *kharcha*. The heddle rod is suspended from a stick which is propped up on two forked branches driven vertically into the ground close to the weaver, thereby raising the heddles and one set of alternative warp ends. The heddles fall into the fixed heddle category. The cloth beam is propped up on stones and is placed behind the two wooden pegs and in front of the weaver. The warp beam is placed behind the two wooden pegs fixed at the far end of the loom. The warp is continuous and is held in tension as it is stretched between the warp beam and the cloth beam. As one shed is already formed by the fixed heddles the weaver inserts a wooden sword which is propped up on its edge to widen the shed opening for easy insertion of the weft yarn (figs. 9.40, 9.41). The weft is wound around a long wooden stick which serves as a shuttle (fig. 9.41). This is inserted into the open shed. The weft is beaten in by the wooden sword after turning it flat on the side of its tapered edge. The second shed is formed by the flat wooden shed rod which is placed behind the fixed heddles. This shed rod is brought forward closer to the heddles after which the sword or batten is inserted as in the first pick, the weft is inserted and beaten in. This process is repeated as long as there is enough space for formation of a reasonable shed for the passage of the shuttle. The woven fabric is then shifted from the cloth beam. This is moved down under the heddle rod and the top layer of the warp, as the warp is continuous. Once the weaving is completed the loops forming the heddles, and the two beams are removed. The weaver sits on the ground near the cloth beam. A type of plain weave is used for the weaving of these *kharchas*.

The Kullu *paṭṭu* or *dohrū* is woven in a narrow width, and, when joined, it has a dimension of 4 ft. 11 in. x 6 ft. 6 in. to 9 ft. 10 in. A.F.P. Harcourt, *c.* 1869,

9.41

9.42

Weaving of *kharcha*. Kinnaur. Note shuttle which is passed through. (Fig. 9.41)

Kharcha, doubled and sewn on top to serve as garment. Courtesy: Radhika Negi, New Mashobra, Shimla. (Fig. 9.42)

Draping of *paṭṭu/dohrū*. Courtesy: Premlata Thakur, Bhutti Weavers Society, Kullu. (Fig. 9.43)

Paṭṭu/dohrū worn for festive ocassion. Street scene. Note difference in fall as compared to Kinnaur style. (Fig. 9.44)

describes two kinds of *paṭṭu*. There are those with a plain colour ground in light grey, deep maroon and brown, embellished with borders rising in several multicoloured tiers; the other, the checked variety, is borderless (Harcourt, 1959: 43–4). As late as 1918 no undergarments were worn even in winter (*P.D.G.*: 79–80). There is an oral tradition that patterned borders and cross borders were introduced *c.* 1680 by a group of migrant weavers from Bashahr.[16] The articulation of Kullu ornamentation is dramatically more striking than Kinnaur, but lacks the delicacy of the latter. This can be attributed to the greater commercialization in Kullu because of its easier accessibility. In present times an upper garment, *chola* is worn in addition to the *salwār*. The *paṭṭu* is used as an overgarment outside the home, girls donning this at puberty (figs. 9.43, 9.44, 9.1). The *paṭṭu* with borders are called *phūltār paṭṭu* or *bushahr dohrī* now.

In earlier times the most important local fair was the Dussehra Fair at Sultanpur which attracted Baltis, Tibetans, Lahulas, the *zamindārs* of Kullu and merchants from the plains. The volume of sales in Kullu woollens was high (*Third Revised Settlement*: 3, 13). In present times Kullu wool weaving has received maximum attention. Kinnaur and Spiti, earlier sitting astride the major trade routes in wool, are now considered less accessible. Political events in China and Tibet have caused grave disruptions in the wool routes and in Kullu; acrylic has made an entry in the absence of ready availability of dyed wool of stable colour. While the technique of weaving has remained unchanged, the new product, introduced *c.* 1942, the shawl, 3 ft. 3 in. x 6 ft. 6 in., enjoying an all-India market, continues to project traditional designs but these have been mediated through the efforts of trained designers and modern marketing organizations.

The back strap loom is not in evidence in the region. Simple frame and pit looms appear to have constituted a parallel tradition in the past, both having evolved in conformity with local ecology and lifestyles. The ground material in Kullu *paṭṭu* is woven in 2/2 twill while the design portion is invariably in a weft faced plain weave. The coloured yarn used for the design is not wound around a *tilli* but is left hanging loose on the loom. As is the case of *pañja darī*, dovetailing occurs only in areas of vertical alignment in colour elements. In Kinnaur, delicacy in design is facilitated by the manipulation of single ends. Kullu ornamentation, bolder in surface impact, is based on units of two ends. The maximum width of the *paṭṭu* on the loom is 26 in., two such widths being joined in the final garment to constitute a width of 52 in. If the border is defined by a plain edging in contrast colour, two shuttles are used on the loom. If the design on the border is worked as a panel involving 14 to 20 warp ends, it is called *kanni* while the term *kushti* is used for broader panels with 24 to 30 warp ends. Alternatively, borders may be woven separately and subsequently attached (*Tweed for Kullu*, NID: 21–3).

The Kullu–Kinnaur tradition in wool weaving offers an interesting contrast both to *kani paśmīnā* of Kashmir as well as to other items in the handicraft category produced on commercial demand. Apart from differences in technique, there are several indicators as to the non-commercial nature of the first in the years prior to entry of the shawl, an item created by the drive for diversification, so important in the post-Independence period. Unit multiplication to serve market demand is facilitated in commercial weaving of items such as the sari by the process of piecing (fig. 5.4). The *paṭṭu* warp has not lent itself to lengthening as the raw material has always been provided by the user. Each item, therefore, has necessitated the setting up of the loom afresh. More often than not weaving has been a part-time occupation catering to specific local demand. Unlike *kani paśmīnā*, there has been little emphasis on division of labour and functional differentiation within the marketing agencies is noticeably absent.

Notes and References

1. The district of Mahasu was created on 15 April 1948 after the integration of Simla Hill States with the Punjab Hill States. It comprised the tehsils of Arki, Solan, Kasumpti, Theog, Kumharsain, Rampur, Rohru, Chini, Jubbal, Chopal and Suni. Consequent to subsequent reorganization the tehsils were re-apportioned and the district ceased to exist. M.G. Singh, 1992: 239–44.
2. Oral communication: B.M.S. Rajwar, Himachal Pradesh Wool Federation, Simla; *P.D.G.*, 1918: 287.
3. Oral communication: T.G. Negi, Mashobra, Shimla; *Rural Craft Survey*: 42.
4. Oral communication: Bansi Lal Sharma, Shimla.
5. Oral communication: Partab Singh Negi, Kinnaur.
6. Oral communication: V.C. Ohri, Simla; Rural Craft Survey: 44.
7. Oral communication: T.G. Negi, Mashobra.
8. Oral communication: T.G. Negi, Mashobra.
9. Oral communication: T.G. Negi, Mashobra.
10. Oral communication: T.G. Negi, Mashobra.
11. *Handicrafts of Kinnaur District*: 43, 60, 81; *Shimla Hill States*: 60; *Rural Craft Survey*, 43, 472; oral communication: T.G. Negi, Mashobra.
12. *P.D.G.*, 205, 206.
13. Oral communication: Partab Singh Negi, Kinnaur.
14. Oral communication: Gulam Ali, *paśmīnā* weaver, Nurpur.
15. *Handicrafts of Kinnaur District*: 90; oral communication: T.G. Negi, Mashobra.
16. Oral communication: Prem Lata Thakur, Bhutti Colony, Shamsi district, Kullu.

10

SILK: THE INDIAN IDENTIFICATION

10 Silk: The Indian Identification

Indian Silkworms

Silk producing lepidopterous insects are of many species possessing marked structural differences (Maxwell-Lefroy, Howlatt, 1909: 485–7). They belong to the order, Lepidoptera, members of two families, *Bombycidae* and *Saturniidae.* While all *Saturniidae* spin silk the same cannot be said of all *Bombycidae* (Wardle, 1881: 2). The reeling of silk from cocoons formed by these insects has proved economically and commercially viable only in the case of a limited number of species. In India, these comprise worms of the *Bombyx* group and *Saturnia* worms such as *Philosamia ricini* producing *eri, Antherea assamensis,* producing *mūgā* and *A. mylitta,* producing *tasar.* In their wild state these worms can feed on a variety of host plants but even so output is optimized by specific feed inputs. While the commercial exploitation of silk is associated with the north-eastern section of India, wild silk must have enjoyed a wider dispersion. Even currently 20 per cent of global production is estimated to be that of wild silk (Bouvier, 1984: 20).

The imagery quoted used by the twelfth century wandering poet-saint of Karnataka, Akka Mahadevi, born in Shimoga, shows an intimate awareness of silkworm and cocoon: 'Like that of the silkworm that spins a cocoon around its body and finds its end for the sake of others I have been encircled by my own desires and wishes in this world without finding place to come out. Oh Lord! Relieve me of my bandages which have encircled me and dissolve me in you' (Nakhate, 1981).

Morphology of *Bombyx mori*

The life cycle of the *Bombyx mori* proceeds through the stage of copulation, oviposition/egglaying, hatching into larvae/caterpillars, five stages known as instars, of growth achieved through four successive moults and finally the spinning of the cocoon (see figs. 10.1, 10.2, 10.3, 10.4, 10.5,

Egg laying. Note dead moth. Stage 1. Courtesy: Sericulture Institute, Wai. (Fig. 10.1)

Worms after hatching. Stage 2. Courtesy: Sericulture Institute, Wai. (Fig. 10.2)

Mulberry leaves chopped very fine for initial feeding. Stage 3. Courtesy: Sericulture Institute, Wai. (Fig. 10.3)

Worms removed from paper to tray with a feather. Stage 4. Sericulture Institute, Wai. (Fig. 10.4)

Thatte, bamboo trays coated with cowdung. *B. mori* worms feed on this during rearing in Karnataka. Courtesy: Rajesh Tandon, Department of Botany, University of Delhi. (Fig. 10.5)

10.6, 10.7 and 10.8). Some larvae go through three or five moults but these are of no economic significance. At the fifth instar the silk glands are sufficiently well developed to account for 40 per cent of the body space (*Silkworm Rearing*: 29–30). On completion of the cocoon the larva is transformed into a pupa and the pupa (see fig. 10.9) into a moth. The adult moth, the imago, slits its way through the pupal skin and pierces through the cocoon by softening the shell by an alkaline salivary secretion (fig. 10.7). K.N. Mehrotra, of the Indian Agricultural Research Institute, New Delhi, in an oral communication, has drawn notice to several aspects of the life cycle of *B. mori* at this stage of its existence. Both male as well as female moths have rudimentary mouths. The male moth is able to fly. In cocoons which have been laid by a common female progenitor, male moths emerge first to be followed by the females. Female moths, on emergence, let out a pheromone to attract the males. The male moths thus attracted, will fly to the nearest female. Unlike the male moth, the female moth does not fly. Since copulation takes place soon after emergence from the cocoon for the female, there is no need for it to fly (see fig. 10.7).

The spinning period lasts between 48–72 hours (see figs. 10.10, 10.11), the pupal stage between 8 to 14 days (see fig. 10.9) while the life cycle of the adult moth is completed within 3 to 10 days (*Silkworm Rearing*: 30; see figs. 10.12, 10.1, 10.2, 10.3, 10.4, 10.5, 10.6, 10.7, 10.10, 10.11; cf. figs. 10.8, 10.9). The life cycle of *B. mori* is repeated with minor variations in that of other species of silk worms. The *eri*, *mūgā* and *tasar* worms, like *B. mori*, evolve from larvae to pupae through four moults and five instars. In warm tropical climates the chrysalis stage inside the cocoon lasts about 15 days. The female moth lays between 600 to 800 eggs. The insect only feeds at the larval stage and the imago spends its life in copulation and reproduction.

10.6

10.7

10.8

10.11

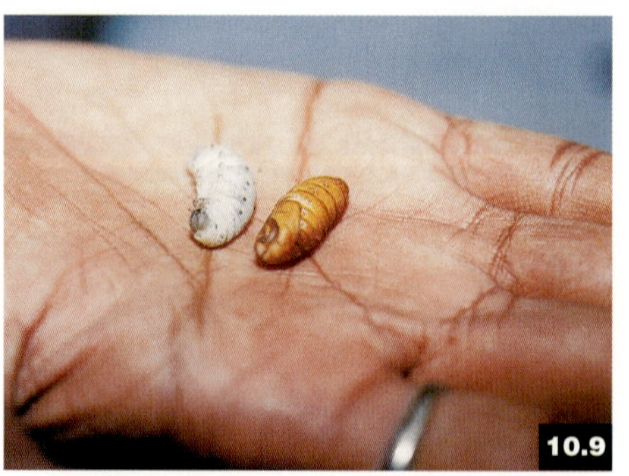

10.9

10.10

10.12

As worms grow, they feed on larger size leaves. Worms rearing upwards are going through major stages in growth, instar after, which they will move into the next phase in their life cycle. Stage 5. Courtesy: Sericulture Institute, Wai. (Fig. 10.6)

Moths emerging from seed cocoons. Stage 6a. Courtesy: Seed Multiplication Centre, Sericulture Institute, Mahabaleshwar. (Fig. 10.7)

Multivoltine cocoons, after boil to kill the worm and prior to the reeling. Stage 6b. Courtesy: Sericulture Institute, Wai. (Fig. 10.8)

Pupae. Reddish pupa indicate a mature stage. Male pupa are larger than the female ones. Courtesy: Seed Multiplication Centre, Sericulture Institute, Mahabaleshwar. (Fig. 10.9)

Cocoons of silk worms placed in bamboo frame/ *chandrike*, during spinning stage. Courtesy: Prof. Mohan Ram, Department of Botany, University of Delhi. (Fig. 10.10)

Chandrike, after completion of cocoon formation. Courtesy: Rajesh Tandon, Department of Botany, University of Delhi. (Fig. 10.11)

A. mylitta, *tasar* moth. Male. Kosa Centre, Development Corporation of Vidarbha, Armori, dist. Chandrapur. (Fig. 10.12)

The initial colour of the egg when freshly laid is yellow. In adapting to cold climates, the embryo stops developing half way, the diapause stage, and goes into hibernation until the return of warm weather. When this happens the eggs become grey in colour. If the cycle proceeds uninterrupted there is no change in the colour of the egg and hatching takes place within ten days of laying. The life term of the moth is between 5 to 10 days.[1] The occurrence of diapause and hatching determine the number of generations a race can produce each year. Among *Bombyx* and other silkworm races, some hatch once a year, others twice a year, and yet others, more than twice a year. Such worms are respectively called univoltine, bivoltine and multivoltine. Multivoltine eggs do not go through the stage of diapause. The life cycle of the worm is determined by climate and ecology. Uni and bivoltine worms follow the rhythm of the seasons in temperate regions, while the life cycle of the worms is determined by the torrid temperature obtaining in the tropics. The quality of uni and bivoltine silk, measured in terms of length reeled from the individual cocoon and its thickness, is superior to that obtained from the multivoltine variety (Vijay, 1985: 219–20). S.N. Chaudhury, in a personal communication to this author from Shillong, has noted that the bivoltine is the original, and largest race, from which the univoltine evolved in colder climates and the multivoltine in warmer latitudes.

Cocoon Formation and Silk Reeling

The body of the *Bombyx* worm comprises fourteen segments (*Silkworm Rearing*: 13–28). The silk glands, found in a pair, are situated between the fourth and eighth segments and occupy two-thirds of the body weight and volume. The ducts from each of these glands unite in the two spinnerets below the mouth of the silkworn and the entire structure is taken as the silk gland. Each gland is made up of three parts, the posterior, the middle and the anterior. The posterior ends are closed while the fore ends of the anterior parts converge to join together into a common excretory duct opening into the spinnaret. The posterior part secretes the silk substance, fibroin, which is then stored in the middle section. Here a gummy substance is secreted, sericin, which coats the fibroin. The anterior part moulds the sericin coated fibroin and conveys it to the spinnaret. The Fillippi glands, situated on either side of these ducts secretes a substance which prevents its hardening. The fibroins, called brins, are extruded through the two spinnarets and harden into the silk filament or bave only on contact with air (*Silk Reeling*: 41). Silk filament is made up of 75–80 per cent fibroin and a coating of 20–25 per cent sericin, or gum, both being protein substances. Sericin constitutes a gelatinous envelope or varnish for the fibre. The *mūgā* and *tasar* varieties of the worm also secrete a cement which causes a drab colour to develop, making bleaching a more laborious process. Fibroin is insoluble but sericin may be removed by boiling. The filament of *B. mori* is easily dyed owing to the homogeneous structure of its fibre which absorbs tinctorial matter uniformly. Moreover, this fibre has chemical affinity with many kinds of dyes, tans and mordants.[2]

The *Bombyx* cocoon is formed by the worm laying the fibre through a figure of eight movement (Riley, 1882: 16). The outer layer is made first. The thread is continuous in the case of *B. mori*, *mūgā* and *tasar*. In *eri*, however, it is discontinuous (Bahadur, 1915: 1). The colour of fibroin is white but that of sericin varies. Sericin is pigmented by carotinoids or flavones which pass from the blood to the silk gland. The female chrysalis, because of the presence of eggs, is heavier than the male, but the silk secretion of the male is higher (Sarkar, 1977: 37, 50).

Silk of the *Bombyx* variety, triangular in cross section, has a capacity to reflect light and the layers of protein impart it with a pearly sheen. *Tasar* silk has a flatter structure which is the reason for its dull appearance. *Mūgā* is rounder and more lustrous. These differences may be related to the varied fluidity of the sericin which is secreted.[3] The *Bombyx* and *Mūgā* imagos cut their way through one end of the cocoon thus making the filaments discontinuous. The *tasar* moth secretes an enzyme, cocoonase, which softens the cocoon shell facilitating emergence. The *eri* moth spins an open ended cocoon through which the imago can easily fly to full term (Wardle, 1881: 21, 26; Jolly et al., 1975: 12).

In commercial production cocoons are divided into two categories: seed cocoons used for propagation, and cocoons used in silk production. Since multivoltine worms do not go through the stage of diapause and hibernation, the transportation of such seed cocoons has necessarily been circumscribed. Seed cocoons of the uni and bivoltine variety, however, do not suffer from the same restriction. They can be transported over greater distances provided there is no radical change in temperature (Vijay, 1985: 220). This has meant that such cocoons could be transported and reeled in places other than those of their origin. In order to ensure uniform lustre, the silk used in commercial production has to be reeled rather than spun. This necessitates the use of uncut cocoons. Even in the case of the *tasar* worm, the emergence of the imago leads to the tangling of the thread forming its cocoon, and in this category also, the uncut cocoon is preferred. In order to obviate the emergence of the imago, the chrysalis within the cocoon is killed by exposure to heat. Heat stifling of the chrysalis is undertaken not by the rearer but by the reeler. In Bengal the preferred method, which is, perhaps, a pointer to its antiquity, was that of exposure to the sun. Boiling is, however, the most widespread method (*Reports and Documents*: Appendix C, 20; see fig. 10.13). Cocoons of the Bengal multivoltine worms could be stored between six to eight weeks prior to reeling. The *mūgā* cocoon was much more perishable and cocoons had to be reeled either by the rearers themselves or at a place in close contiguity to the place of rearing (*Reports and Documents*: 20; Nanavaty, 1965: 209–10).

Boiling of multivoltine cocoons. Detaching of outer stringy fibre. Courtesy: Sericulture Institute, Wai. (Fig. 10.13)

Throwing. Shakeel Silk Twisting Industry, Kollegal, Karnataka. (Fig. 10.14)

The drawing out of the silk filament also necessitates a prior stage of boiling. During this process 75 per cent of the sericin is removed. Care is taken to retain 25 per cent gum to prevent tangling in course of reeling. If the boiling is insufficient the filament tends to snap in course of drawing, while excessive boiling leads to an increase in the amount of waste silk (Sarkar, 1977: 51). The thread drawn from the cocoon is of uneven consistency, being finer at the beginning and the end. It is possible that terms used by Europeans during the Company period to define different categories of silk such as *cabessa*, head, *bariga*, belly, *pee*, foot, may have been linked to this attribute of the filament.[4] In order to provide uniformity in consistency, additional filaments were drawn in course of reeling, the process being called throwing (see fig. 10.14). The fewer the number of cocoons utilized, the finer would be the thread produced. The quality of silk from each cocoon is assessed in terms of denier, one denier being taken as about 492 yds. in length and 0.05 gm. in weight.

Because of the structure of the cocoon the process of reeling also generates a residue of waste fibre. In multivoltine *Bombyx cocoons* about 45 per cent of the product is floss. This material is spun. Spun fibre of the mulberry fed worm, whether from pierced cocoon or waste fibre, is called *maṭkā*. In the case of the *mūgā* worm 25 to 50 per cent may be spun. With regard to *tasar* only about 10 per cent of the filament can be reeled. Spun *mūgā* is called *juthorī*, while for *tasar* the terms used included *kete*, *ghicha* and *jhurī*.[5] The transformation in the nature of the filament is reflected in the finished weave.

Ahiṃsā and Silk

In killing the worm, a common practice in the preparation of silk yarn, the question may be raised regarding the attitude towards using silk by Indian religious groups such as the Buddhists and Jains known for their adherence to the principle of *ahiṃsā*/ non-violence. The Buddhist position *c.* AD 671 was well summarized by I-Tsing. In a chapter entitled, 'Necessary Food and Clothing,' he stated:

> As to fine and rough silk, these are allowed by the Buddha. What is the use of laying down rules for the strict prohibition of silk? The prohibition was laid down by someone; though intended for lessening complication, such a rule increases it. The four *Nikāyas* (i.e. *Āgamas*) of the *Vinaya* of all five parts of India use (a silk garment). Why should we reject the silk that is easy to be obtained and seek fine linen that is difficult to be procured? Is not this the greatest hindrance to religion? Such a rule may be classed with the forcible prohibitions that have never been laid down (by the Buddha).

If, the argument continues, silk is to be abjured on grounds of the killing of life, it would, indeed, be difficult to sustain life as humans have to destroy knowingly or unknowingly in order to nurture their own being. However, what can legitimately be regarded as reprehensible is for a monk to beg for cocoons containing silkworms while to witness the killing of worms was not permissible even for the laity. Nevertheless, gifts of silk could always be accepted (Takakasu, 1982: 58–9).

The ambivalence on this issue receives support from a legendary account. It was believed that when silk was introduced into Khotan following what may have been an apocryphal marriage between a scion of the ruling dynasty of Khotan and a Chinese princess, the latter had decreed that cocoons were no longer to be boiled but that silk was, henceforth, to be spun from pierced cocoons (Boulnois, 1986: 180–1).

In the case of Jainism, in the early period, wool was the favoured fibre. However, the only prohibition placed on the religious while begging for raiment was that of accepting expensive gifts (Jacobi, 1895: II, 5.1.157). By the twelfth and thirteenth centuries, the model for community life of the gods was assumed to be that which obtained in Rajput courts. This orientation received considerable impetus with the growth of the northern Indian Vaiṣṇavite cult, centred around the deity, Srinath. This sect was also called the *Puṣṭi Mārg*.[6] This opulent way of life was imbibed both by the Jain laity as well as by their *munis*. Silk began to be used to great effect in decorative details connected with the performance of rituals. Use of flowers, which earlier would have been abjured as constituting a violation of the sanctity of Vanaspati, the vegetative universe, now began to emerge in a myriad different ways.

During the sixteenth century, perhaps as a reaction to this growing extravagance, the Sthānakavāsī sect came into being. This was marked by an absence of image worship and there was a total eclipse of silk in ritual.[7]

It would, therefore, appear that while there may have been fringe groups in existence among Buddhists and Jains which were totally opposed to the use of silk from uncut cocoons, they could not be taken as representative of majority opinion. It is, therefore, clear that there were no religious proscriptions to fetter the Indian silk industry in the past.

Genera of Indian Silk Producing Worms

Mulberry feeding worms

The Indian *Bombyx mori* thrives best on the mulberry (*Morus alba*) leaf.[8] Thomas Wardle, *c.* 1879, listed domesticated and wild species found in India as under:

1. Domesticated Silk Yielding Worms
   a. *B. mori*, uni and bivoltine;
   b. *B. textor* (Hutton), or the univoltine *Boro polu* of Bengal, an introduction from Italy *c.* AD 1716.
   c. *B. sinensis* (Hutton), a Chinese multivoltine.
   d. *B. croesi* (Hutton), also called *Nistari* and *Madrāsī*. It was, by origin, a south-western Chinese worm. Dr James Anderson, Physician General in Madras, succeeded in acclimatizing this worm in Madras in 1770. Around 1780 it was despatched to Bengal. A multivoltine worm, the best yields occurred in March and in the months between July to October.
   e. *B. fortunatus* (Hutton), the *desī* or *chhoto polu* of Bengal, a multivoltine spinning a small cocoon of golden yellow colour. The best yields occurred during the months of October and November.
2. Wild Silk Yielding Worms
   a. Theophila huttoni (Westwood), north-west Himalayas, feeding on indigenous mulberry trees.
   b. *T. sherwelli* (Moore), the wild silkworm of the south-east Himalayas.
   c. *T. bengalensis* (Hutton), the wild silkworm of Bengal and Chota Nagpur, feeding on Monkey Jack or *Artocarpus lakoocha*.
   d. *T. religiosae* (Helfer), the *joree* of Assam and the *deomuga* of Cachar. It feeds on the banyan (*Ficus benghalensis*) and *pīpal* (*Ficus religiosa*). The cocoon yields a large quantity of strong, coarse and dark-coloured silk. This thread is said to be have been used for fishing lines in Bengal (*Silk in Assam*: 28).
   e. *T. mandarina* (Moore), a wild silkworm with a probable home in China.
   f. *Ocinara moorei* (Hutton), a multivoltine found in Mussoorie and north-west Himalayas, spinning a small yellow cocoon.
   g. *O. diaphana* (Moore), found in Meghalaya.
   h. *Trilocha varians* (Walker), found in north and south India.

Origin of *B. mori*

It may be asked as to which geographical region the mulberry silkworm may have originated from, and from where further diffusion may have taken place. This aspect is quite distinct from that of commercial exploitation of mulberry silk. After initial diffusion biological changes may have taken place pointing to the possibility of several ancient centres for production of *B. mori* derived silk.

Hutton was of the view that all domesticated forms of mulberry feeding insects came to India from China. He held that they originated in the northern colder tracts of China and were, therefore, purely exotic in the tropical districts of Bengal. N.G. Mookerji maintained that all domesticated mulberry insects originated somewhere on the slopes of the Himalayas. Indeed, he fixes the home of *Bombycidae* somewhere in the region around Mount Everest. There was diffusion from here to the sunnier Chinese regions to the north and those of the subcontinent to the south. K.N. Mehrotra, in an oral communication, has averred that the north-east region of India has been a rich cradle of natural biodiversity both in the past as well as in present times. He notes this in terms of the honeybee, *A. assamensis* and *A. mylitta*, and in food grains such as rice as well as in many forms of flora. This region may also have been the original home of the *B. mori*. Japanese authors have stated that the bivoltine worm is the original race from which the univoltine evolved in colder climates and the multivoltines in warmer climates. The home of the original species should have been located in a temperate clime and both the Indian Himalayan tracts as well as the south China zone could qualify for this. It should be noted that silkworms raised in the north and east China are bivoltine. In south China, particularly the Kwantung province, there are polyvoltine races yielding six to seven crops annually (Nanavaty, 1965: 49).

G. Watt mentions that the occurrence of a truly wild species of *B. mori* in a country, accompanied by a species of *morus*, might be accepted as affording the best claim to being the original birthplace of sericulture. Based on this the north-eastern region has a better claim than Bengal. The climate is cooler, and in earlier times, a large number of *Saturniid* and *Bombyciid* species were to be found, particularly in the hilly tracts. The same writer presented a case for the hypothesis that the home of the *B. mori* may have been Manipur where mulberry was also to be found in a wild state (Watt, 1903: 562). This appears unlikely. Silk rearing of *B. mori* in Manipur was probably introduced from outside as it was exclusively practised by the Lois. This was a group who were said to have migrated from their initial home in the west. All those expelled from their original caste in Manipur were absorbed into this group. The silk weavers were Kubos, so called because they had originally migrated from the Kubo Valley in Burma.[9]

S.N. Chaudhury, in a personal communication, has also summarized the question from the point of view of insect genetics. The two nearest to the species *B. mori*, the domesticated silkworm, are both of the wild variety. These are *B. (T). mandarina* and *T. religiosae.* One variety of *B. (T). mandarina* is to be found in north China, Manchuria and the Himalayan region, while another exists in north Japan. *T. religiosae* is mainly found distributed in the Himalayan foothills of eastern India.

Chromosome studies of *Bombyx* species show that *B. mori* has the haploid number of chromosomes n=28; the continental variety of *B. mandarina* has n=27. It was earlier concluded that one of the chromosomes of *B. mandarina* might have split into two chromosomes during the process of evolution as one of the 27 chromosomes of *B. mandarina* pairs with two chromosomes of *B. mori* at the time of maturation of F1 hybrids. In view of the recent discovery of a 28 chromosome continental variety of *B. mandarina*, a theory has now been propounded that the continental variety may have been the ancestor from which the island variety of *B. mandarina* has been derived, presumably by the association of two chromosomes out of n=28, and not by fragmentation as was earlier suggested. It had been proposed that *B. mori* may have evolved from *B. (T). religiosae*, a wild indigenous variety widely distributed in the Himalayas. *B. (T). religiosae*, however, has the haploid chromosome number n=31. It has, therefore, been excluded as a possible progenitor of *B. mori* since *B. (T) mandarina* is to be found in the Himalayan region apart from its presence elsewhere, noted above, the case for identification of its place of origin remains open. It now becomes necessary to emphasize that a distinction needs to be made between the origin of the *B. mori* and that of the practice of sericulture.

G.K.C. Liu has drawn attention to the fact that as early as the reign of the first emperor of the Hsia dynasty, which extended from 2205 to 1766 BC in China, there are already allusions to the assorted types of silk being paid as tribute by the different states of the Huangho basin (Liu, 1952: 133–4). There are no parallels in India for analogous developments in the area of indigenous sericulture. Since the predominant fibre in India has been cotton, it is possible that the development of sericulture did not receive adequate attention.

Saturniidae

The *Saturniidae* group in India comprises *Philosamia ricini*, producing *eri*,[10] *Antheraea assamensis*, producing *mūgā* and *A. mylitta*, yielding *tasar.*

Eri

Eri is derived from *eraṇḍa*, Sanskrit for castor plant. Assamese derivatives are *eraṇḍī*, *enḍī* and *eri.* The multi-voltine worm, *P. ricinis* is domesticated and is easier to rear than the mulberry worm being more resistent to disease and extremes of climate. *Eri* culture in Assam extends from the plains up to an altitude of 5000 ft. with the temperatures ranging from 50° F to 98° F, and humidity levels varying between 80 and 100 per cent. The three main nutrients for the worm include firstly, castor (*Ricinus communis*)/*eraṇḍī* (Hindi), secondly, *keseru* (Assamese), (*Heteropanax fragrans*), thirdly, *bor pat* (*Ailanthus grandis*) and *kanukpa* (Nepali, Bangla), (*Evodia fraxinifolia*). Results are favourable if *keseru* is fed from the hatching to the fourth stage and castor is reserved for the last stage. Results are better if castor is fed at the second stage and *kanukpa* is fed at other stages. Castor and *keseru* can be given as feed at all stages and seasons. However, worms which are exclusively fed on the *keseru* spin smaller and more compact cocoons as compared to those fed on castor. Among other food plants are included cassava (*Manihot utilissima*), papaya (*Carica papaya*), *champak* (*Michalia champaca*), *udumbara* (Sanskrit), *gūlar* (Hindi), *Ficus glomerata.* However, castor is the most compatible feed material.[11] There is also the wild variety, the bivoltine, *P. cynthia*, also called *Atticus ailanthus*, which is plentiful in Assam. It feeds on the leaves of *Ailanthus glandulosa.* The silk produced is highly elastic.[12]

According to S.N. Chaudhury, Shillong (personal communication), the host plant, castor, is considered to have originated in the Himalayan foothills. However, there are also those who believe that the plant may have evolved independently both in India as well as in Ethiopia. With regard to *A. ailanthus*, the term, *Ailanthus*, is traced to the Sulawesi group of islands in Indonesia. However, the worm is plentiful in Assam. *P. ricini* has the chromosome number 14n while that of *P. cynthia* is 13n. The former has developed from the latter by the splitting of chromosomes in the course of evolution (*Report*

1963: 3). S.N. Chaudhury (personal communication) leans towards the view that both these worms originated in Assam.

The *eri* worm spins an open ended cocoon and the departure of the imago causes no additional displacement of the yarn. Pupae, if consumed as a food item in the tribal diet, are manually recovered; otherwise the cocoon is boiled after the imago has flown out. Boiling is necessary to loosen the thread which is then spun.[13] Plantain leaves, leaf ash, or ash from wheat stalks, paddy straw, or the *mūṅg* plant (*Vigna radiata*), *matikalai* (*P. aconitifolia*), or pieces of green papaya, can be used for purposes of degumming in the course of boiling (*Notes on Eri Silk*: 7–8).

The silk obtained from *Cricula trifenestrata*, the *haum-pottanee* of the Assamese, like *eri*, is spun. T. Hugon noted *c.* 1837 that this silk was not used because it caused an itching sensation on the skin. It had been recently introduced to Cachar district and had already made an entry into Burma. It fed on the tea shrub, the black varnish tree and the wild mango tree. In Moulmein, the worm is fed on the cashewnut tree, *Anacardium occidentale.* J. Geoghenan observed *c.* 1872, that this silk was used for the adulteration of *eri* fibre. However, the itching sensation soon caused the wearer to recognize the nature of adulterant used. The cocoon was lustrous and yellow in colour.[14]

S.N. Chaudhury draws attention to the advantages of mixing ramie (*Boehmeria nivea*), with *eri*. Work practices point to the possibility that this constituted a traditional usage. *B. nivea*, a perennial shrub, was to be found in a wild state in the north-east. At the level of domestic industry, the processes for degumming and spinning approximated those for *eri*. When blended in weaving with *eri*, it prevented stretch and shrinkage while imparting elasticity to the finished product.

Eri silk filament can be white or brick red in colour. The red *eri* is called *bhunoku* and is manufactured for the Bhutan market. Since the fibre is spun rather than reeled it has little lustre. According to T. Hugon, *c.* 1837, *Laccifer lacca*, lac dye, *Rubia cordifolia*, *manjit*, madder, and *Indigofera tinctoria*, indigo, were used as dyes.[15] The dyeing of *eri* must have been laborious as the contemporary silk specialist M.S. Jolly, notes that the fibre has little affinity to dye. Although lacking in lustre, *eri* was strong and the life of a garment was listed as being thirty years. There was a good internal market for tribal weaving of this fibre in the nineteenth century. There was also a demand for this in Tibet and Bhutan. As a *borkāpor*, worn as a wrapper in cold weather, there was a market in Bengal. For the latter market a second item was also made. This was a body wrapper for the Bengali male called *enḍī* in which the warp was *mūgā* and the weft *eri*. There were several items, mainly in the unstitched garment category, in which various proportions of *eri*, *mūgā* and Assam *B. mori* silk could be used. After the Industrial Revolution, cotton began to increasingly replace *eri*.[16]

A. assamensis, the *Mūgā* Worm

A. assamensis is to be found in a wild state in the eastern Himalayas, Nagaland, Assam, south Tripura and north Burma. Its presence is also recorded in the valleys of Kumaon and Kangra (Chaudhury, 1981: 4, 7). Affiliated wild species on the subcontinent are *A. helferi*, found in Sikkim as also *A. compta*, but these have not been commercially exploited (Chaudhury, 1981: 7–8; Wardle, 1880: 6).

The *mūgā* worm is multivoltine, breeding four to five generations annually. There is seasonal variation in the cocoons and a few yields are subject to commercial exploitation. For reeling purposes, the best brood is between the months of October and November, while for selection of seed cocoons, the season, December to February, is preferred (Choudhury, 1981: 34–5). The worm goes through four stages of moulting. In the early stages of feeding young trees are selected. In general the leaves of trees having a growth period of three to five years, yield leaves which are best suited to *mūgā* rearing (see fig. 10.15). One tree can support between 1,000 to 1,500 worms (*Silk in Assam*: 15; Maxwell-Lefroy, Ansorge, I, 126). *A. assamensis* is semi-domesticated. Cocoon formation, pairing, egg laying and hatching take place indoors. In egg laying, all those laid after the first three days are rejected, as the emergent worms are less robust (*Silk in Assam*: 13–14; *Report 1963*: 4). After hatching the larvae are released on the selected tree. After the foliage is finished the grubs crawl down (see fig. 10.16). They are sorted according to growth and placed on bamboo sieves on the next tree (Jolly et al., 1975: 69). On completion of cocoon formation the pupae of cocoons selected for reeling are stifled by exposure to heat.

10.15

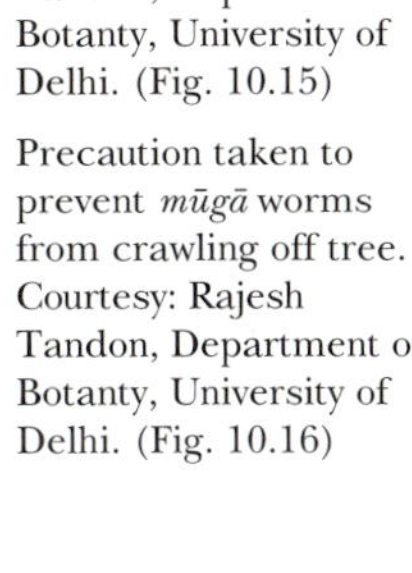
Host tree of *mūgā* worm. Gauhati. Courtesy: Rajesh Tandon, Department of Botanty, University of Delhi. (Fig. 10.15)

Precaution taken to prevent *mūgā* worms from crawling off tree. Courtesy: Rajesh Tandon, Department of Botanty, University of Delhi. (Fig. 10.16)

10.16

The medium can be the domestic fire followed by exposure to the sun. The best seed cocoons today are obtained from east Goalpara and south west Kamrup (Chaudhury, 1981: 13, 16).

A. assamensis is a delicate worm to rear. The transportation of cocoons has to be undertaken with care–rain, vibrations and jerks are to be avoided, and to provide protection against heat, cocoons are transported during the night. The pre-hatching and moulting stages are also hazardous for the worm. The process of emergence of the imago is difficult, as both the flossy cocoon and the tough pupal skin have to be pierced. Moreover, a large number of the larvae are killed by birds and insects during the feeding stages (Chaudhury, 1981: 2, 19, 37).

Only the middle section of the cocoon can be reeled, and as much as 25 to 50 per cent of the cocoon is spun as waste silk. Spun *mūgā* is called *juthori*, and when *mūgā* waste is combined with *eri* in weaving, the fabric is called *ereea* (Chaudhury, 1981: 66, 85; *Silk in Assam*: 18). As many as seven to twenty cocoons may be thrown in course of reeling. As is the case with *tasar*, twisting is done on the thigh by means of the palm and underpart of forearm. Women undertake both reeling and twisting (*Silk in Assam*: 18). The natural colour of *mūgā* is much prized in regional costume and has also been used in embroidery. However, indigenous methods for bleaching and dyeing have been perfected and colours which have been achieved in natural dye include red, yellow, blue, green and black.[17]

A. assamensis feeds on a variety of trees. In upper Assam the main host plant is *Som, Machilus bombycina*. This tree, like the *Sualu, Litsea polyantha* used in lower Assam, is suitable for purposes of rearing after five years of growth. *L. polyantha*, grows in elevations reaching up to 3,280 ft. The *Som* is absent on the plains but may be found in elevations up to a maximum of 1,640 ft. The tree lends itself to breeding twice a year, and those bearing leaves with a sweet taste are preferred for rearing. Silk from the *Som* fed worm yields a more delicate silk (Chaudhury, 1981: 12, 24–5; *Silk in Assam*: 13).

Among secondary host plants *Mejankeri* (*Litsea cubeba*) and *Chapa* (*Magnolia pterocarpa*), have yielded fine quality silk. The *Mejankeri* plant as used in silk rearing, appears to be a product of *jhūm*, slash and burn cultivation. The shrub was to be found in abandoned clearings in the jungle as recorded *c*. 1899, and the leaves were suitable for rearing only at the earlier stages of the growth of the tree. In the first year when the plant was still at the shrub stage, the leaves yielded the best quality silk. The silk became coarser in the second year, and in the third, the texture was the same as ordinary *mūgā*. The *Mejankeri* tree was not cultivated, and rearers had to move from clearing to clearing. There was the further hazard of exposure to birds and insects because of close proximity to the jungle. Only vigorous worms could survive on this plant. The silk, creamy in colour, was much prized fetching five times the price of *mūgā* derived from *Machilus odoratissima*. The area around Jorhat was famous for this product but *c*. 1895 it was reported that it was very difficult to procure.[18]

The silk which was produced from the host plant, *Chapa*, was legendary. The Lahing and Holongpur subdivisions of Jorhat district were famous centres of production of this silk. However, E. Stack noted *c*. 1884 that it could no longer be procured. The *Chapa* tree was also called *panchapa* and *titachapa*. Under the Ahom kings, *c*. AD 1288–1824, white *chapa* silk was in great demand. The production of *chapa* silk was laborious and the effective rate of rearing was very low. Only worms hatched on the first day, considered to be the most robust, were reared on this leaf. The fibre content of the leaf was high and it was difficult to digest. *Chapa* silk was reputed to be stronger and more pleasing than *paṭ*, silk of *B. mori*.[19] At times of leaf scarcity, leaves of *bumroti* in Assam, *Symplocos grandiflora, ber, Ziziphus jujuba* and *bajramani, Xanthoxylum alatum*, could be utilized. *Z. jujuba* was also utilized in *tasar* and in lac culture. For the more mature worm additional supplies of food could be drawn from *dighloti, Litsea salicifolia* and *patichanda* (Assamese), *Cinnamonum obtusifolium* (*Silk in Assam*: 13; Chaudhury, 1981: 19). Under the Ahoms there was also a special silk called *khesa* or raw *mūgā*. According to legend, after the final stage of moulting and the last instar, the worms were prevented from spinning a cocoon for three days. This was said to result in the accumulation of the silk within the body of the worm. The head was then plucked off and the bodies were thrown into a vessel with warm water. The end of the fibre was located and the filament was reeled off. This silk was supposedly highly prized and constituted royal privilege (*Silk in Assam*: 19). This story may have considerable mystique but in view of the morphology of the silk glands and nature of the bave, the process described is patently impractical.

Origin of *Mūgā* and *Tasar*

After detailed investigations M.S. Jolly has proposed that north-eastern India was the centre of origin for the *mūgā* and *tasar* worm and that it was from here that diffusion and divergence took place (fig. 10.17). *A. assamensis* may be considered as the ancestral stock from which all other species have originated. *A. mylitta*, the *tasar* worm, derived from *A. assamensis*, merges into a trail which branches off into several directions. Along the sub-Himalayan belt the temperate species of the worm, *A. roylei* is to be found. In China, *A. pernyi*, and in Japan, *A. yamamai* represent the same species. M.S. Jolly further hypothesizes that *A. roylei* and *A. pernyi* must have had a common base in north-eastern India.[20]

Tasar is produced by a variety of worms in India. These include *A. mylitta* in eastern, central and south India (see fig. 10.12); *A. frithii* in Assam; *A. sivalika* in north-western India and *A. roylei* in the sub-Himalayan belt (Sarkar, 1977: 55). The *tasar* belt in India is divided into two, the tropical and temperate. A distinct belt of humid and dense forest sprawling over the central and southern portion of the Decccan plateau constitutes the home of the tropical *tasar* worm. There are virtually nineteen eco-races of *A. mylitta* scattered over this belt (fig. 10.18). Some of the tribal groups such as the Bhils of Bihar and Gonds of Madhya Pradesh, involved in *tasar* production, are also counted among the earliest inhabitants of the subcontinent. The fact that tribals have dominated in the area of *tasar* production is an indicator of the antiquity of the tradition (figs. 10.19, 10.20). The knowledge possessed by tribal rearers is demonstrated by the fact that the highest quality seed cocoons are associated with the wild species feeding on *sāl.* The best category seed cocoons are obtained from Singhbhum and the erstwhile Mayurbhanj state. Tribal groups have their own nomenclature for local worms. Because of the cementing fluid so painstakingly spread by the worm in the course of spinning its cocoon, reeling of this filament has posed serious problems (figs. 10.21, 10.22). However, according to Wardle, towards the final decades of the nineteenth century, *koshtas*, the weavers, of Chanda, Maharashtra, and Sambalpur, Orissa, boiled *tasar* cocoons in a special lye made from a local plant, so that not only was the chrysalis killed but the mucilage was dissolved, thus facilitating subsequent reeling (Wardle, 1880: 16–20). This process is reflected in current methods of heating cocoons in Bhandara and Chanda districts, Maharashtra (*Kosa Weaving* 16). The temperate *tasar* tract extends from Jammu and Kashmir in the west to Manipur in the east including Himachal Pradesh, the Kumaon and Garhwal divisions of Uttaranchal, Meghalaya and Nagaland. The races are uni, bi and

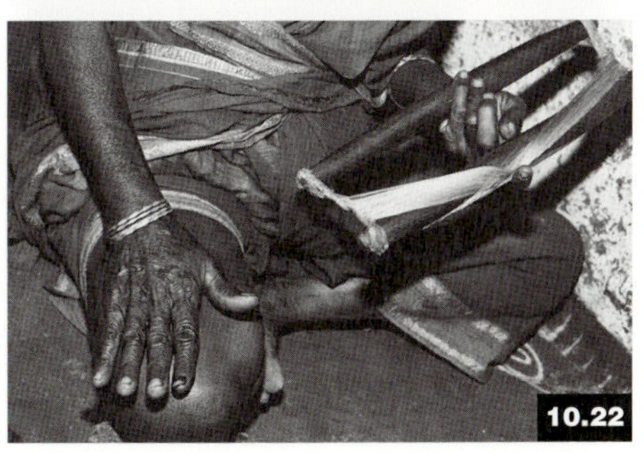

Reeling of *tasar* filament on pot. Weavers Services Centre, Imphal. (Fig. 10.17)

Penduncles of *tasar* worm. Brown/ *mukhi*, yellow/ *dhaba*. Courtesy: Tasar Development Centre, Armori. (Fig. 10.18)

Tasar cocoons placed in pot, topped with straw, for boiling. Photographed at Hakimpeta village, district Mahbubnagar, Andhra Pradesh. (Fig. 10.19)

Lower pot surmounted by a second inverted pot during boiling of *tasar* cocoons. Lower pot has water. Upper pot has straw packed cocoons. Boiled for 1 to 2 hours. Photographed at Hakimpeta village, district Mahbubnagar, Andhra Pradesh. (Fig. 10.20)

Tasar reeling. Village Hakimpeta, district Mahbubnagar, Andhra Pradesh. (Fig. 10.21)

Reeling of *tasar* with one hand, winding with other. Detail of fig. 10.21 (Fig. 10.22)

trivoltine, being reared from the months June to January.[21]

A. mylitta feeds predominantly on three food plants, *sāl, Shorea robusta, arjuna, Terminalia arjuna* (fig. 10.23) and *asan* (*T. alata:* syn. *T. tomentosa*). Among subsidiary food plants which number nearly thirteen, the more important are, *ber* (*Ziziphus jujuba*), *sidha, Lagerstroemia parviflora,* Roxb., *kumbi, Careya arborea,* Roxb., and *jāmun* (*Syzygium cumini*).[22]

With the exception of worms feeding on *sāl* which are wild, the other host plants nourish worms which are semi-domesticated. The activities of hatching, feeding of larva, and spinning of cocoons are conducted in the open. Emergence of the moth, coupling and laying of eggs, takes place indoors. Like the *mūgā* worm, *tasar* larvae are sensitive to sound, and any sudden noise may interrupt feeding.[23] Unlike silk from *B. mori* and *A. assamensis, tasar* silk is more resistant to dye. Even if dyed, the material would lack the brilliance of *B. mori* (Wardle, 1880: 23–5).

Apart from *A. mylitta,* there is another *tasar* yielding worm, *A. frithii* or *A. paphia.* It is to be found only in Assam where it feeds on *katkaria* (*Vangueria madagascariensis*), or a shrub, *phutuka* (*Melastoma malabrathicum*) available throughout the year. This silk was called *kutkuri* in Assam. Although it was glossy and strong, it was ranked lower than *mūgā* and could not compete with the cheaper and better quality *tasar* of Bengal (*Silk in Assam,* 25–7; *Report 1963*: 4).

Tasar worms which feed on oak are *A. roylei, A. pernyi* and *A. yamamai. A roylei* is an oak feeding worm found in the sub-Himalayan belt. It has not been commercially exploitable in the past as the worm spins a double shelled cocoon of which the outer shell has proved unreelable (Jolly et al., 1975: 49). The Chinese oak fed *tasar* worm, the *tsartsan,* or *A. pernyi,* is uni and bivoltine. Exploitation of this worm can be traced back to the Han (206 BC–AD 220), and Wei (AD 220–65), dynasties. Chinese oak *tasar* has been produced in the provinces of Liaoning, Shantung, Honan, Kweichow and Sichuan. The univoltine Japanese variety, *A. yamamai,* is said to have its origins in China. Its commercial exploitation in the Ariake area of Nagano prefecture, Japan, is dated to the nineteenth century.[24] A hybrid worm, *A. proylei,* has been reared in the north-east by crossing the indigenous *A. roylei* with the Chinese worm, *A. pernyi.* Commercial exploitation started in *c.* 1966 and is still continuing.[25]

B. mori—The Chinese Tradition

The natural history of the silk worm: wild, semi-domesticated and domesticated, suggests that the properties of the filament were known since the time that the spindle, and whorl came into evidence at archaeological sites (fig. 1.1). As has been already emphasized, there is, however, a great deal of difference between subsistence usage and commercial exploitation. It is only at the latter level that sericulture assumes identifiable dimensions.

Tasar worm on *T. arjuna.* Courtesy: Professor C.R. Babu, Department of Botany, University of Delhi. (Fig. 10.23)

Sericulture can best be practised when the worm is fully domesticated. It is this factor which has drawn preponderant attention to the *B. mori.* Although G. Watt has made out a case for the hypothesis that the original home of *B. mori* was Manipur in north-east India (Watt: 1905), the question remains open. However, the argument that the original home from which commercial diffusion of *B. mori* took place was China remains uncontested. The earliest evidence of silk and silk fabrics is provided by the archaeological discovery of the Liangzhu civilization which existed *c.* 3000 BC in the Jiangsu and Zhejiang provinces on the east coast of China.[26] Professor Tan Chung (oral communication) concludes, that the home-land of sericulture was probably the Yangtze River valley. It was from here that sericulture was gradually introduced into the northern regions of China which were, at the time, more highly developed. The dyeing of silk in the Shantung province area is dated to 2200 BC. Silk remnants preserved on bronzes found in excavations in Anyang, a capital of the Yin period, *c.* 1324–1065 BC, show that the experimental stages in silk production had been surmounted by this date, long reeled yarn being in evidence (Vivi Sylwan, 1949: 15, 19, 22).

By the time the Han dynasty came to power, three regional types of silk were in evidence, that of the basin of the Hwang Ho, the valley of the Yangtze River and the Sichuan region (Liu, 1952: 138, 140).

According to Yang Xianyi, the Greek and Roman appellation, Seres, for China, was derived from the kingdom of Shu, in presentday Sichuan province. The Chinese ideograph for *Shu* is, in fact, a silkworm, and it is known that there has been a gradual shift from north to south China in the area of silk production.[27] It is in this context that the different myths relating to the origin of the silk industry in China are to be assessed.

Central Asian temple banner, ninth to tenth century AD. Courtesy: State Museum of Prussian Culture, Berlin. (Fig. 10.24)

Popular legend, at one level, ascribes the origin of sericulture to Empress Si-ling-chi/Lei-tsu, wife of the great culture hero, Huang-ti/Yellow Emperor. The Emperor Fu Shih, whose dates are unknown, is also credited with the same achievement (Liu, 1952: 133–6). However, since the development of sericulture is a process in which many individuals play a role, stress on one personage rather than another may be taken as apocryphal rather than factual. Professor Ji Xianlin has stressed that the whole process of domestication of the silk worm must have been a very long drawn out one in which many individuals would have played a part.[28] The Si-ling-chi legend must have been a composite creation rather than an approximation to some actual culture heroine of known chronology. In view of the richness of archaeological data available, the veracity or otherwise of these eponymous cult figures no longer appear relevant to the history of Chinese silk.

Although the date for the commercial exploitation of *B. mori* in China is open to doubt, evidence drawn from several sources attest to the ceremonial, if not ritual status of silk. As early as the ninth and tenth centuries BC the silkworm was depicted in jade (*Chine*, 1988: 286, no. 189). Not only were some taxes paid in terms of silk but it also enjoyed a role in the local currency system. There was a range of terms in Chinese to denote various aspects of the life of the silk worm. The terms *tze, tsan, quie* and *loo* denoted egg, larva, pupa and moth respectively, while *touy, tsan-sa, woong* and *tsan-kuang* denoted shed skin, excrement, cocoon and floss. The first hatching was called *mio* (Liu, 1952: 165). Seed cocoons were also selected with meticulous care (Liu, 1952: 165–7). The words *si* and *ssu* were Chinese terms for silk, while *tso* referred to a coarse red coloured silk (*Karlgren Dictionary*: q.v). E. Loubo-Lesnichenko points out that during the Han period, while *po* and *seng* were synonymous terms for silk, *po* often denoted non-dyed silk, while *seng* referred to silk of better quality. According to the later Han poet, Tsai Young, *seng* was a heavier, costlier non-dyed silk. Monochrome patterned textiles were called *ki* and *ling*, terms which are still in use. *Kin*, the particular character being dated to the eighth century BC, denoted cloths woven with polychrome yarn. During the second and third centuries AD, the term *tsi-tcheng* was used with reference to cloths patterned with metallic elements (Jera-Bezard, 1977: 54–5; Pariset: 1843, 15).

The view that China attempted to maintain a tight monopoly over the spread of sericulture does not appear to be substantiated by factual evidence. The argument presented by E. Pariset in 1843 has been carried further by M.G. Raschke. E. Pariset has pointed out that between 2367 and 1125 BC, the production of silk was small and the main use was for standards and parasols (Pariset, 1843: 13, 15, 21–3; fig. 10.24). Raschke dates the rise of steppe nomadism to the eighth century BC, which served to isolate China from the civilizations of the Middle East in antiquity. However, Chinese objects such as silks, mirrors and lacquer ware were buried in the graves of the nomad warrior aristocracy. These goods were obtained through raids, warfare, tribute and gift exchange. Silk became a status symbol and was lavishly used in burials. Raschke suggests that silk may have acquired some symbolic value which would justify its ostentatious visibility in burial sites (Raschke, 1978: 606–11, 695, n. 83). Silk would thus have been distributed through nomad networks through several different forms but always in small quantities. Commercial transactions perhaps cannot be precluded. It was this pattern of exchange through which Chinese silk found its way into the sixth century BC Hallstadt D Iron Age graves in the upper Rhine and Danube. J.P. Wild suggests that the point of entry may have been the port settlement of Massalia/Marseilles, founded in *c.* 600 BC (Wild, 1983: 32; Raschke, 1978: 611).

While scholars may debate the chronology of the entry of silk as a major item in commercial exchange in Chinese external trade, within China it appears to have had a link with royalty and courtly privilege since early times. Such an association would also point to some form of restricted production. Under the reigning power of the Chou, 1122–256 BC, royalty itself was expected to participate in sericulture. During the Han period the production of silk was a state monopoly. Apart from its role in gift exchange, silk was also used in payment for services rendered to the reigning house. Use of silken robes, again, was a sartorial privilege emanating from royalty. Indeed, the origin of the word, *pu-i*, to denote the commoner, is dated to the Han period. *Pu* was the generic term for non-silken fabrics. The term *pu-i*, referred to one who could only wear *pu* in clothing, which soon became synonymous with the common man.[29]

By the turn of the Christian era the commercial exploitation of silk appears to have sustained some impetus. The *Periplus of the Erythraean Sea*, dated to the late first and early second century AD, states: 'There is a great inland city called Thinae from which raw silk, silk yarn and silk cloth are brought on foot through Bactria to Barygaza and are also exported to Damirica by way of the river Ganges.'

Damirica has been identified as the country of the Tamils (Schoff, 1974: 48, 205; Mathew, 1975: 156). This increasing importance of silk is reflected in the first active state intervention in the area of sericulture when the Wei dynasty, AD 386–557, directed that each farmer had to plant a stated number of mulberry trees each year. Under the Sung (AD 960–1127) and Yuan (AD 1280–1368) dynasties, the policy towards sericulture became much more systematized.[30]

The special status of silk in court ceremonial and the initial distribution of Chinese silk through channels of gift exchange rather than commercial supply would also explain the early diffusion of sericulture. The view that knowledge of sericulture was a closely guarded Chinese secret cannot be supported even at the level of natural history. The existence of host trees and spread of wild ancestors of *B. mori* could scarcely be subject to royal fiat. Indian tribal expertise in the area of *tasar* would reinforce the view that the origin of *B. mori* silk reeling should have been anonymous to which an eponymous origin was imputed at a later period. However, China undoubtedly enjoyed special ecological advantages for this form of sericulture which provided the country with unrivalled opportunities for the early exploitation of mulberry silk. Moreover, unlike other moths, *B. mori* is incapable of flight. This is believed to be the consequence of continuous domestication (*Silkworm Rearing*: 17). Hence, while the characteristics of the wild mulberry silkworm are suggestive of natural diffusion, the morphology of the domesticated worm would lend itself to restrictive practices if the host country so desired. However, it would appear that since China was so well endowed to develop *B. mori* sericulture, it was not necessary to take any overt steps to maintain its monopoly at the early stages. Success in sericulture demanded much practical experience and knowledge evolved over time and it was Chinese superiority in these areas which provided the country with its unparalleled advantages.

The Dispersion of Sericulture

B. mori appears to have reached Korea *c.* 1200 BC. Japan became conversant with this fibre about the third century BC, while Annam became familiar with it a little later in the second and third centuries AD (*Silkworm Rearing*: 1). The north Burmese worm, *B. arracanensis*, of Chinese origin, is, although undated, of considerable antiquity. In Assam this worm, called *nyapolu*, is raised by late immigrants from Burma such as the Khamti, Singpho and Phakial.[31]

Khotan had earlier been under Chinese control but some time between AD 105 and 125, along with Yarkand and Kashgar, it had slipped out of this control (Boulnois, 1986: 90). *c.* AD 419, resulting from a marriage alliance between China and Khotan, to which a reference has already been made, sericulture was introduced into Khotan. This knowledge spread from Khotan to Bokhara, Khiva and Samarkand. Persia was introduced to silk manufacture in the seventh century AD, but the Persians had monopolized the silk route well before they had begun to rear the worm (Geoghenan, 1872: iii). Between the seventh and ninth centuries AD, Zandaniji, present day Zandane, near Bokhara in Tajikstan, Central Asia, became a great centre for silk manufacture. Several samples from this centre found their way into European church repositories (Jera-Bezard, 1977: 65; Bongard-Levin, Vigasin, 1984: 235).

The term *zendado* was in use in Europe during the medieval period. It referred to a medium quality silk taffeta used in Europe for banners, tents and covers. However, the place of origin had, by this time, been forgotten (Heyd, 1886: 701).

Identification of *Byssus* and Coan Silk

Before turning to European and Indian traditions of sericulture and early European perceptions of silk, we take up the question of the identification of the origin of the fabric, *byssus*. *Byssus* is not synonymous with Coan silk to which Aristotle, 384–322 BC, has made reference (Pariset, 1843: 62–3). In Egypt the term *byssus* implied high quality linen, a royal monopoly, and in the third century BC a centre was known to be producing this in connection with the worship of Ammon in Thebes (Vogler, 1982: 162; Hall, 1986: 9). In the Bible the term *shesh* was used as a synonym for the high quality linen, *byssus* (Pariset, 1843: 26, 61, n. 2). As is to be encountered in the case of the Sanskrit derived Indian term *paṭṭa* which moved from its original meaning of cloth to that of silk, the term *byssus* shifted from linen to the material obtained from the *pima squamosa*, a species of the marine mollusc, *Lamelli branchiate*. It was by means of its *byssus* threads of protein origin that the mollusc was able to attach itself to the rocks at the bottom of its shallow watery habitat. During the Hellenic period, fourth to first century BC, it was discovered that these threads could be dried, spun and woven into fabric. The mollusc was found at several places in the Mediterranean but came to be concentrated at Tarentum in Italy which was the main centre of production. Syrian merchants distributed *byssus* in China where it was much prized. Ignorant of its origin the Chinese assumed that the yarn was obtained from the down of a 'water sheep'.[32] In his analysis of an eighth century silk piece in the collection of the Mus ee Historique de Tissu, Lyon (see fig. 5.36), G. Vial has come to the conclusion that although the item had been classified as *byssus*, it was, in fact, raw silk.[33]

The genesis of Coan silk, to which Aristotle made a reference, anteceded that of *byssus*. According to Aristotle, the first person to weave this silk was Pamphyle of Cos, daughter of Latous (Pariset, 1843: 62–3). The worm has been identified as either *Pachypasa*, or *Laosiocampa otus* or *S. pyri*. It was a native of the Mediterranean region and was to be found in the areas of the Adriatic Sea, southern Italy, Greece, Romania, Turkey, Armenia and Israel. The host plants comprised pine, ash, cypress, juniper and oak.[34] During the period of Pliny, first century AD, three kinds of silk were available, from China or Seres, Assyria and Cos. Men wore Coan silk while women, that of Assyria and Seres. According to Aristotle Coan silk was spun, and Pliny testifies to the finer quality of Assyrian yarn. Pariset assumes that the latter was reeled (Pariset, 1843: 256).

The Roman Tradition of Silk

Before turning to Assyrian and Seres silk, we will look at the Roman silk weaving and dyeing tradition. Around 150 BC the Hioung-nou insurgency was brought under control by the Hans who proceeded to establish colonies in Central Asia (Pariset, 1843: 89). Increased access to silk led to significant developments in Syrian weaving. The earliest examples of twill damask weave in silk[35] are those of Syrian provenance dated to between the first and fourth centuries AD (de Jonghe, 1983: 48–9). J.P. Wild feels that silk weaving in the Roman Orient should have started first in Syria (Wild, 1989: 213). The base filament would have been Chinese silk. Apart from imports of yarn, woven silk items were unravelled and the yarn reused. This may explain why Romans chose to spin silk on occasion.

Around AD 250 Roman weavers in the eastern provinces, probably Syria, developed the multiple heddle rod loom for weaving damask silk, and the Edict of Diocletian, found at Aerzani, Phrygia, in AD 301, included the price to be paid for a silk damask loom.[36] During the first century AD centres of weaving were also located at Antioch and Jerusalem, while the main centres for dyeing were to be found at Sidon and Tyre (Varron, 1938: 352). The silk of Seres which proved more compatible to purple dye, soon displaced the silk of Cos (Boulnois, 1986: 58). Silk yarn which was dyed and was ready for weaving was called *Sericum*, a term which later became a generic for silk. *Oloserica* had both warp and weft in silk. The term *tramoserica* or *dramioserica* referred to linen warp and silk weft while *subserica* or *stamoserica* covered the category in which the warp was silk while the weft could be of linen, wool or cotton. The term *metaxa* or *mataxa*, raw silk, first appears in the fourth century AD in a law of Arcadius and Honorius. Subsequent to this Byzantine writers make frequent mention of it (Pariset, 1843: 195–6, 208–10).

After the establishment of the Eastern Roman Empire at Byzantium or Constantinople in AD 330, a silk industry was developed, initially in imperial workshops. From the fourth century AD onwards silk began to be used in shrouds by wealthy Christians, and by the fifth century the church began to use silk during celebration of feasts, for altar and tomb covers and even in liturgical vestments worn by priests.

In the meantime the silk industry was also being nurtured in Persia under the Sassanids, AD 226–651. The Egyptian and Syrian silk industries had simultaneously to face the Sassanian and Byzantine inroads into their spheres of activity. In AD 360, the Sassanid monarch, Shapur II, entered into hostilities with Byzantium and succeeded in capturing some of the best Syrian weavers whom he transplanted to Susa. He thus laid the foundation of the Persian silk industry.

The Byzantine emperor, Justinian, AD 527–65, attempted to invigorate the industry by introducing sericulture into Byzantium. Around AD 552, eggs of *B. mori* were smuggled into the Byzantine Empire. According to Procopius this was done through the agency of Indian monks who had lived a long time in Serindia, Chinese Turkestan. Another point of view is that the worms were smuggled in from Cambodia while A. Muthesius suggests that the source could perhaps have been Syria which had a pre-sixth century tradition of indigenous silk.[37] The Arabs were later to inherit the Perso-Byzantine tradition of silk weaving.[38]

Assyrian Silk

There is considerable ambiguity with regard to the identity of Assyrian silk. To the Roman mind there was little difference between Assyria and Syria, and Syria itself was believed to be close to India (André, Filliozat, 1986: 404, n. 440). Roman writings contained several references to India. Among the items originating in Arabia, Parthia, and mainly in India, on which taxes had to be paid, Marcinus, in the *Digesta*, dated post AD 217, lists raw silk, cloth of silk, or mixed silk and silk yarn among the articles covered. Servius, fourth to fifth century AD, stated in his *Commentary on Virgil* that in Ethiopia, among the Indians and in Seres there are on trees a worm called *Bombyx* which spins, as spiders do, fine thread from which silk is derived. The *Cosmographia Anonyma* attributed to Ethicus, sixth century AD, states that from *India Major* there comes silk. Although the term *India Major* was extremely vague covering the area eastwards of Kerala up to an indefinite point, as a segment it was demarcated from China; and within the region the most significant area for silk was north-eastern India.[39] Assyrian silk was yellow and appeared to be reeled. E. Pariset assigns it a Himalayan provenance either on the Indian subcontinent or Sichuan in China (Pariset, 1843: 77–9, 256). A case will be made for the assumption that this Assyrian silk was either the *B. fortunatus* of eastern India or a variety of *tasar*, the wild silk of Indian origin.

One way of solving the problem would be to attempt to trace the nomenclature for silk and its imputed place of origin, China, in different linguistic zones. The term for China in India and Malaysia was Cin.[40] This is the nomenclature in the *Arthaśāstra*. In the *Mahābhārata*, the term *Mahācīna* is also used. According to Yang Xianyi, the Indian term, *Cīna*, may be a corruption of Chiang/Jiang, an early group residing to the south-west of presentday China. The states of Chin, Qin and Shu were formed by the descendants of the Chiang/Jiang.[41] During the ninth century BC, Ch'in or Ts'in was the westernmost state in the conglomeration of states constituting what later came to be known as China. The expansionist movement of these states can be dated *c.* third century BC. The term *Cin* may have been brought to India through the peoples of Central Asia in touch with this western state.[42] The Arabs absorbed this term as *Sin*.

Ptolemy referred to the inhabitants of south China as *Sinde* with a capital city, Thinae, probably located at or near modern Canton. Cosmas Indico pleustes, of the first half of the sixth century, called China *Tsinitsa*. The term *Tsinitsa*, as a term for south China needs to be demarcated from the term Seres, used in Europe for north China. Greek authors bestowed the name, Serica, to the place of origin of the silk worm. During the medieval period the term *Kattaii* came to be used being derived from a Persian corruption of the name of the reigning dynasty, the Kitan, which remained in power in north China from AD 1118 to 1235 (Giles, 1978: 37).

The Chinese called silk *see* or *szu*; the Mongols referred to it as *sirkek*; the English term, silk, is derived from the Russian *chelk*, itself derived from the Mongol, *sirkek*; the Korean word, *sir*, was identical to the Greek term for silk and the Armenian term for silkworm was *sheram* (Klaproth, 1827: 53–7; Pariset, 1843: 95–6).

The Greeks and Romans coined the word Seres or Serica, the name given by them to China derived from the prime commodity, silk, with which the country was associated. The Chinese themselves used a different series of appellations. Professor Tan Chung discusses the shades of meaning to be associated with the terms, *Tianzi*, Son of Heaven and *Tianxia*, territorial limits of what was taken to constitute China, and *Chung-kuo*, the Middle kingdom which was beginning to supersede *Tianxia c.* 162 BC (Tan Chung, 1986: 60–6). The term, *Chung-kuo* has been reflected in the names used by the neighbouring countries. The Manchu name for China was *Doulimba-i Gouroun*; the Mongol, *Doumda-iin oulous*; the Tonkingese, *Djoua kwok*, the Japanese, *Tsiou-kokue*; the Burmese, *Alai prai dai*—all carrying the meaning of Middle Kingdom. Other terms included, *Thian tchhao*, the Celestial Empire, *Thian hia*, under the sky, *Chin tan*, Eastern Dawn, found in Buddhist works, and *Morou Kossi*, the Japanese translation of Eastern Dawn. In erstwhile Cochin-China, the term for the Chinese people was *Ngo* and China was called *Noo oek ngo*. The Tibetans called China *Ioulbou*. They referred to the Chinese as *Djanag* or *Ghianag*, white inhabitants, in contrast to *Dja Gar* and *Ghia Gar*, embodying an imagery of black, and reserved for the inhabitants of the Indian subcontinent (Pariset, 1843: 95–8; Klaproth, 1827: 58–61).

It is, therefore, clear that the Roman perception of China was different from that of China's immediate neighbours. Japan, a close neighbour, which received both silk and Buddhism, largely through the aegis of China, adopted terms which denoted both a geographical entity and a linkage through religion. The Indian term reflected exposure at a particular period in time. The significant feature is that not only did India associate China with a particular reigning dynasty but that it had a range of names for silk of which only one category, *cīnāṁśuka* or *cīnapaṭṭa*, reflected Chinese origin. The fact that the *tasar* worm was dispersed from its place of origin in north-east India to South-East Asia, China and Japan is perhaps reflected in the Japanese legend of the Princess with the Golden Hair.

The Princess is said to have lived in Madhyadeśa, India, *l'empire du milieu de l'Inde*. Having survived a series of harrowing incidents engineered by her wicked stepmother, she finally emerged in Japan, where she died, transforming herself, in the process, into a silkworm which spun yellow silk. Pariset reads the legend at several levels: as reflecting the life history in allegorical terms of the silkworm spinning yellow silk and the geographical dispersion of this worm through China to Japan. However, Pariset identifies the filament as *B. mori*. The Chinese term for the species is *Kin-cul-tsan*, the child with the cover of gold. Some indication of dating is suggested on the basis that the sign for *kin*, meaning golden, is dated to *c.* eighth century BC (Pariset, 1843: 77–9). The Chinese term, *Kin-cul-tsan* can relate to a yellow cocoon, not uncommon in multi-voltine *Bombyx* species. These cocoons cannot be transported over long distances and the reeled silk, if coloured, still has a highly neutral shade. Assyrian silk was described as yellow. As such there would be closer parallels to *tasar* rather than to *B. mori*. While *tasar* is reeled there is a larger residue which is spun. The silk is less conducive to dye than *B. mori* and, because of its structure, is less lustrous. On both counts the explanation as to why the Romans preferred the silk of Seres become evident.

It is also of some interest to review the mention of silk in Chinese sources outside the conventional confines of China. M. Kordosis quotes a text, *Wei-lio*, dated by him prior to AD 363. This appears to be synonymous with the work, *History of the Northern and Eastern Wei Dynasties* (AD 386–534 and AD 534–50), written by Wei Shou in the sixth century AD (Fairbank et al., 1976: 125). The products of the region, *Ta-Ch'in* or *Ta-ts'in* are described. According to M. Kordosis (Kordosis, 1992: 194–205, 216) whose argument receives some indirect support from G. Fussman (Fussman, 1993: 123–30), the term, *Ta-ch'in* covered the geographical area of the Roman-Byzantine state west of the river Euphrates as also north-west India including Central Asia. The products included domesticated mulberry silk, 'down of water sheep' (*byssus*) and silk of wild worms. At the same time, 'they were always anxious to get Chinese silk for severing it in order to make *hu-ling* ...' (Kordosis, 1992: 247–8). The last statement clearly referred to Roman methods in weaving Chinese silk. The reference to *byssus* presents no problem. Since the reference to *B. mori* is found in a text dated to the sixth century AD, the provenance could be either Khotan or the Byzantine Empire, the latter country having embarked on the initial stages of sericulture at this time.

Silk Tradition in Kashmir and Karnataka

Because of its favourable ecological and climatic conditions it would be natural to suppose that Kashmir would form an extension of the Central Asian tradition. There is a reference in the *Sabhā Parva* in the *Mahābhārata* to a gift of clothing, 'woven with the threads spun by insects' by the Bāhlīkas to Yudhiṣṭhira during the performance of the *Rājasūya* sacrifice. However, D.C. Sircar has identified Bāhlīka with Balkh rather than Kashmir (Haldar, Roy: no date, 115; Sircar, 1971: 32, n. 6). According to *Kalhaṇa*, during the reign of King Śaṅkara-varman, AD 883–902, the capital city, Patan, was a famous silk weaving centre. Zain-ul-Abedin, AD 1420–70, and, at a later date, the Mughals, are known to have encouraged sericulture in Kashmir.[43]

Kashmir has a univoltine worm, and tree rather than bush mulberry. In 1873, C.E. Bates noted that there were as many as seven varieties of mulberry in Kashmir (Bates, 1873: 63). At eight centres of Bhairava (Śiva) worship in Kashmir, the god is worshipped as a mulberry tree (Ganju, no date: 190). Thomas Wardle noted in 1890 that *B. mori* were plentiful and wild cocoons were collected (Parey, 1977–8: 138). This latter activity was farmed out by the state until AD 1894 (Lawrence, 1967: 367). Despite these advantages, as noted by William Moorcroft in the second decade of the nineteenth century, silk production in Kashmir was not sufficient even to meet domestic needs (Moorcroft: II, 155). The industry was not self-sustaining and all cocoons were procured from Gilgit, Baltistan and Bokhara (Wardle, 1904: 15; Sufi, 1974: 574). T. Wardle noted that prior to 1869 when Maharaja Ranabir Singh revived silk production on a significant scale, the silk industry in Kashmir had subsisted in an 'unorganized crude state in which it had probably existed for centuries'. However, even this effort failed as in 1878 disease struck and virtually the entire cocoon stocks were wiped out (Wardle, 1904: 15; Lawrence, 1967: 367). The reason for the weakness of the Kashmir sericulture industry may perhaps be ascribed to the fact that the *paśmīnā* industry was so demanding and highly specialized that there was little scope for the development of an equivalent second major tradition. Indeed, it was noted by S. Geogheghan, *c.* 1880, that shawl weavers also constituted the silk rearers, *kirmkash*. Since the worm was univoltine, and tree mulberry only yielded one leaf season annually, duality in this vocation was possible. However, there was no question that the latter was a secondary occupation. *Kani paśmīnā* constituted a much more integrated economic activity. Visual evidence suggests that whereas in Assam *mūgā* is an integral part of regional costume, in Kashmir, in the past, silk was conspicuous by its absence, the common wear being wool, available in various degrees of warmth in accordance with seasonal requirements.

Turning to Karnataka, in view of the importance of the Karnataka silk industry in present times, it is commonly assumed that the tradition is an ancient one. B.A. Saletore makes a reference to a Tuluva king, King Dharma, who is said to have utilized silk as part of his sartorial requirements made at Kavuru, Boluru and Iravaduru. The king is unidentified, but since he is said to have later become the *bhūta* (spirit) of Todakinar, the account can be taken as legendary. However, since a European chair forms part of the setting, this legend can be anchored to the colonial period (Saletore, 1936: 484–5, 537). It is possible that the centres mentioned may have been places of silk manufacture, but silk rearing does not follow as a necessary corollary. The major commercial production of *B. mori* in Karnataka commenced *c.* AD 1795 when *B. meridionalis* introduced into Mysore from China (Nanavaty, 1965: 129).

The North-eastern Link

Early Indian references point to the eastern and north-eastern sectors of the subcontinent as the prime silk producing centres. In Assam, the Boḍos, a Tibeto-Burman group, have traditionally been associated with silk weaving. According to Suniti Kumar Chatterjee, the original home of the Boḍos may have been the Sikiang region in China from where they moved towards India. Zhu Changli discusses in detail the southern overland route linking Yunnan and Sichuan, with north Burma and Assam. The main road originated at Chengdu, passed through Xichang, crossed the Jinsha River and reached Dali, Yunnan. From here it penetrated into Burma via Yougchang. At Zhuge Liang it separated into two branches, the southern and the northern road. The southern route found its way to Manipur via the Shan

States, the Chin and the Naga Hills. The northern route could be traced from Tengchong, Yunan, via Myitkyina to Mogaung, Mungkuan and Hukawang Valley in Burma, after which it proceeded through the Naga Hills and Patkai passes to reach Assam. Stages in the Brahmaputra Valley included Sadiya, Pragjyotiṣpura, modern Guwahati, Haruppeśvara, modern Tezpur and Kopili, in Nowgong district. The kingdom of Kāmarūpa enjoyed a special advantage in having this access to the north Burma–Yunnan passageway (Zhu Changli, 1993: 31). A. Datta, on the basis of jadeite raw materials found in the Sadiya frontier zone, traces linkages with Yunnan as early as the neolithic period in Assam dated to the first millennium BC. Zhu Changli corroborates this.[44]

The existence of this route since very early times helps to explain the development of silk weaving in Assam which must have been a two way process, at least in the early stages. The Boḍos were referred to in Sanskrit literature as *kirāta* in the centuries prior to the Christian era. Although they had moved into Bengal, they were associated with an original home in the Himalayan slopes, and more particularly with Assam. These associations pointed to their routes of migration. About the second century BC, the Boḍos were the intermediaries in the trade which passed through China to the north-east and Bengal. Among the items which entered into this trade was Chinese silk, the *cīnāṁśuka* and *cīnapaṭṭa* of Indian texts.[45]

During the early period, textual references differentiated between varied categories of silk on the basis of colour rather than the parameters of wild and cultivated. The earliest archaeological evidence of silk in India is a twine of this fibre used for stringing beads dated *c.* 1500–1000 BC found at excavations at Nevasa, district Ahmednagar. The silk is white in colour and spun from cut cocoons (Gulati, 1961: 55, 57). The earliest sample in India of woven silk is the fragment found in a relic casket excavated at Devni Mori, district Sabarkantha, Gujarat, dated *c.* AD 375 to the reign of the Kṣatrapa king, Rudrasena (Mehta, Choudhury, 1966: 120; figs. 10.25, 10.26, 10.27). Both samples cited await detailed technical analysis.

Written sources are far more prolific than archaeological finds. The former can, moreover, be linked with living ethnological evidence to provide a reasonably reliable picture. The generic term for silk in Sanskrit was *kauśeya.* The earliest literary references in Brahminical religious literature were to *kṣauma*, linen, rather than to *kauśeya.* These included references in the *Maitrāyaṇīsaṁhitā*, 3.6.7, and the *Taittirīyasaṁhitā*, 6.1.1-3 (Varadarajan, 1984: 62). The *Śrauta* texts, dated to between the fifth and third centuries BC also refer exclusively to *kṣauma.*[46] However, Patañjali, second century BC, has translated the word *kośād* used by Pāṇini, 4.3.41, fifth century BC, as *kṛmau*, silkworm, and *koś* as cocoon (Abhyankar, 1952: 234).

With regard to mulberry, T. Ganapathy Sastri has translated the word *tūla* found in *Arthaśāstra* 14.2.6, 8, synonymously with

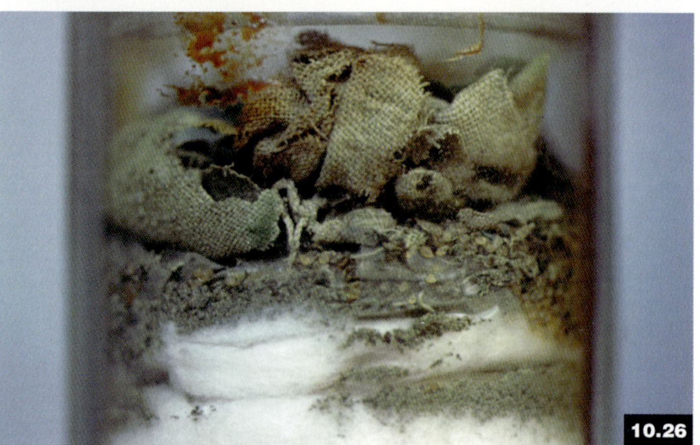

Silk bags *c.* AD 375, excavated at Devni Mori, district Sabarkantha. Courtesy: Department of Ancient Indian History and Archaeology, M.S. University, Vadodara. (Fig. 10.25)

Detail of fig 10.25. (Fig. 10.26)

Stone casket excavated at Devni Mori. Pali and Sanskrit inscriptions in Brahmi script. The copper box was found inside the casket which contained relics of Buddha. Courtesy: Department of Ancient Indian History and Archaeology, M.S. University, Vadodara. (Fig. 10.27)

brahmakāṣṭa. Monier-Williams has identified *kramukha* with mulberry, *Morus indica.* The *Śodhala-Nighaṇṭu* identifies *tūḍa* (*tūlaphala*) with *kramukha* and *brahmakāṣṭa.*[47] However, the overt link between *kṛmi,* the worm, and Sanskrit synonyms for mulberry, *Morus indica,* are lacking. R.P. Kangle has translated *kṛmijātānām* in *Arthaśāstra,* 2.22.6, as silk cloth (Kangle, 1972, 145, n. 6). According to the *Bhāvaprakāśa, tūṭa, tūḍa, tūla, kramūkha* and *brahmadārū,* are all to be taken synonymously. Among north Indian vernacular languages the word for mulberry is: Hindi, *śahatūta, tut*; Bengali, *tunt*; Marathi, *tutai*; Gujarati, *shetur* (Misra, Vaisya, no date: 581). Caraka, dated to some time between AD 80 and 180, mentions the fruit, *tūḍāmāmla.* The commentary of Cakrapāṇīdata, dated to the tenth century AD, states that this fruit *tūḍam,* is obtained from *Uttarāpatha* (Shastri, 1929: 295, 1.27.131). According to Professor R.S. Singh of Banaras, in the Hindu Kush area, the word *tul,* relates to mountain pass, while *tūla* is a generic term for floss derived from assorted plants. In Kashmir the folk term for mulberry, *morus alba,* is *tūla.* In the Kullu region of Himachal Pradesh, the term *pun,* derived from *pūgā,* is indicative of mulberry. *Pūgā,* originally synonymous with *kramūkha,* later shifted in meaning to areca nut.

Despite all these references to the mulberry, material from Ayurvedic texts seems to demonstrate that the fruit alone was availed for its medicinal properties. There is no mention of the leaves being used for the rearing of silkworms. However, this factor by itself cannot be taken as conclusive evidence as to whether or not *B. mori* was being reared. Ethnological evidence in the north-east would support a contrary view, particularly in view of archaeological evidence pointing to links since neolithic times between Assam and the Yunan–Sichuan region in China.

Among terms for textiles which could be identified as silk are *kauśeya, dhauta kauśeya,* and *patrorṇa.* Words which enjoy a certain ambiguity are *neta, aṁśuka* and *paṭṭa.* The term *kaliṅgam* stands in a category apart. According to the Jain canon, *Anuyogadvārasūtra,* dated to *c.* second century AD, five kinds of cloth are produced by the saliva of insects: *paṭṭa, malaya, aṁsuga, cīnāṁsuya,* and *kimīrāga. Paṭṭa* and *kimīrāga* appear to be silk of a golden colour. This could be *mūgā* or a variety of *tasar* (Agrawala, 1953: 78). The *Bṛhatkalpa Bhāṣya* mentions a silk, *suvanna,* which, according to the commentator, can be taken as a golden coloured thread produced by a certain variety of silkworm (Jain, 1947: 129). *Kanagga* cloth, is another variety referred to in the *Niśītha Cūrṇi,* being woven from the golden coloured yarn of certain insects (Sen, 1975: 153).

According to the *Amarakoṣa* (fifth century AD) II.6, III, *kauśeya* is obtained from the egg (cocoon) of *kṛmi. Patrorṇa* or *dhauta kauśeya,* is very costly and precious—*mahādhanam* (II.6, 113). The term *dhauta* may be taken in the sense of being washed or bleached (Sastri, 1970: II.6, III.113). The *Pañcatantra* epitomizes the qualities of silk, thus: 'People do not go by birth alone but by virtue or qualification, *kauśeya* is born of a worm but this is highly respected.'[48]

Medini, thirteenth century AD, commenting on *netra* in *Amarakoṣa,* 3.1.180, associates this with a root of a tree, suggestive thus of *bast* fibre. Amarasiṁha refers to *aṁśuka* synonymously with *netra.* It would appear that the term *aṁśuka* related both to silk as well as some other kind of fibre. The term *nata,* possibly derived from *netra,* was associated with silk or mixed fabrics in sixteenth and seventeenth century *Varṇak* literature of Gujarat (Mehta, 1951: 152). J.C. Jain has surmised that the term *aṇḍaga,* used in the Jain work, *Anuyogadvārasūtra,* could perhaps be taken in the sense of *aṇḍī* (*endi*) or *eriṇḍi, eri* silk (Jain, 1947: 130, n. 91). The *Aṅgavijja,* a fourth century AD work, refers to *cīnāṁasuga,* as a gossamer fine Chinese silk, and *cīnapaṭṭa* as Chinese silk of ordinary quality.[49] The term *paṭṭa* in reference to silk began to be freely used only in the fourth century AD (Gulati, 1951: 3).

Kṣauma, dukūla and *tārpeya,* in their earlier connotations, related to the category of bast fibre rather than silk. *Kṣauma,* as has been stated above, referred to linen. The official Sanskrit to Tibetan Glossary, the *Mahāvyutpatti,* transcribes *dukūla* as *dukūla'ras.* The Chinese translator gives the term 'cloth of *ko',* *ko* being a climbing pea, *Pueraria thunbergiana,* Bentl.[50] However, over a course of time and especially after AD 1305, *dukūla* began to be used in the sense of fine silk (Thaker, 1970: 39). *Tārpeya,* in the *Atharvaveda* (18-4-31) and later texts, denotes a garment made of some material the nature of which is uncertain. Commentators of Kātyāyana, *Śrauta sūtra* and the *Śatapatha Brāhmaṇa* suggest that a linen garment, or one thrice soaked in ghee, or one made of the *tṛpa* or of the *tṛparṇa* plant is meant. According to Sāyaṇa, AD 1300–86, who commented on the *Atharvaveda, tārpya* was a cloth woven from a kind of grass, *tṛpa,* which was softened by butter or ghee to form the yarn. The meaning of the word under *abhiṣecanīya* given by Goldstücker in his Dictionary is silken garment. S. Eggeling accepts this meaning. The question, however, remains open.[51] With regard to the Tamil term, *kaliṅgam,* S. Krishnaswamy Aiyangar identifies it as cotton from Kalinga, Orissa. C. Minakshi states that this is a term which indicates clothing, but, more particularly, silken clothing. Paul Wheatley identifies it as silk from Takkola, Kidaram or Kalagam in Malaysia.[52] Ma Huan, *c.* 1416, noted the rearing of the mulberry worm in north Sumatra. However, the art of reeling was unknown and silk was used as a wadding material. According to the *History of the Sung Dynasty* (AD 920–1279) Book 489, sericulture was known and practised in Java (Groenveldt, 1887: 15, 20). Ma Huan is a late source and this could be taken as a reference to *sonket* of Thailand and Malaysia. Alternatively it could apply to the *tasar* silk industry of Orissa.

Yet another category is encompassed in the term, *vāḍaga* cloth, found in the *Bhagvatsūtra,* an early Jain text. The commentator translates this as *tasar,* a word enjoying currency in the regional language of the time (Jain, 1947: 156; Gopal, 1964–5: 39–40). The earliest archaeological evidence of *A. mylitta, tasar,* is recorded in a Coptic tapestry stylistically dated to the sixth century AD. *Tasar* is found in the warp, the extra weft and in the two border portions (Vial, 1985: 50). The word *tasar* in the sense of coarse silk is found in *Karpūramañjarī,* a Prakrit work written by Rājaśekhara, AD 880–920 (Konow, 1963: 232). Turner (Turner, 1966, no. 5744, 27) links the word *tasar,* as related to fibre, to Prakrit *tasara,* masculine, 'a kind of thread.' The vernacular term, *tasar,* denoting a coarse silk derived from *A. paphia/A. frithii,* appears to be a derivative from the Prakrit, *tasara,* rather than Sanskrit *tasara,* or Pali, *tasara,* meaning a shuttle of a loom. Wardle, however, suggests that the transition in meaning from shuttle to fibre is to be traced to loom usages in the weaving of mixed fabrics. He cited the instance of a cloth woven in the erstwhile Central Provinces in which the warp was cotton and the weft, shot through the shed by means of the shuttle, was that of *tasar* silk (Wardle, 1880: 17).

Another explanation is offered by S. Mukhopadhyay, Indira Gandhi National Centre for the Arts, New Delhi in a personal communication which is based on the dating of Prakrit literature. The word *tasar* is derived from the eastern Prakrit dialects, Māgadhī and Ardhamāgadhī. Mithila and Bengal constituted the heartland of these dialects, but their range extended to Kāmrūpa in the east and to the Bengal–Orissa border in the south. These dialects were also associated with the articulation of the Jain religion. It is significant that in their geographical span the dialects encompassed the earliest centres of sericulture in India. According to Mukhopadhyay, the Prakrit meaning of *tasara,* as thread, cannot be postulated to a date earlier than the second century AD. In eastern India, the heyday of Ardhamāgadhī is confined to the sixth and seventh centuries AD, while Māgadhī enjoyed currency until the ninth century AD being the chosen vehicles of the Jain religion, the dialects found an entry into western India when the Jain faith established itself there. Until their codification by western Indian grammarians such as Hemachandra Sūri, AD 1089–1173, both dialects had essentially existed at a spoken level. This period of literary codification coincided with the increasing popularity of silk in ritual among Jains of western India. Gujarat also enjoyed a rich tradition in silk weaving. It would not, therefore, be unusual, to find a lexicographical preoccupation with words suggestive of categories of silk.

The east Prakrit term, *tasar,* may well have existed only at the level of oral expression until this codification of dialects had taken place in distant Gujarat. These factors could, perhaps, explain the late appearance of the word, *tasar* in Bengal literary expression. It was only *c.* AD 1500 that folk literature in Bengal began to make a distinction between *kauśeya* and *tasar.* It is curious that despite the existence of the term *tasar* in vernacular, the English, in referring to wild silk, preferred to use the term *herba* or *erba* between the sixteenth and eighteenth centuries (Das Gupta, 1935: 281; Temple, 1900: 339). Against this, the Chinese term, *tsartsan,* for oak fed *A. pernyi* (Liu, 1952: 189), appears to be derived from the Prakrit, *tasar.*

The *Arthaśāstra* throws a great deal of light on the early history of silk in India. For purposes of taxation three categories of cloth are mentioned: *kṣauma, dukūla* and *kṛmitāna,* the last being translated as silk yarn.[53]

As for *patrorṇā,* it comes from the Magadhas, the Puṇḍras and Suvarṇakuḍya. The *nāga* tree, the *likuca,* the *bakula* and the banyan tree are the sources. That from the *nāga* tree is yellow. That from the *likuca* is wheat coloured. That from the *bakula* is white. The remaining one is the colour of butter. Of these, that from Suvarṇakuḍya is best. By that are explained the silk and silk cloth (*kauśeya* and *cinapaṭṭana* in the Sanskrit text), from the land of Cina (Kangle, 1972: 104–5, 2.11. 107–14).

Magadha is taken as a unit comprising the Nalanda and Gaya districts of Bihar. Magadha may also refer to a region located in *Prācya* which comprised the areas lying eastwards of eastern Uttar Pradesh. According to the seventeenth century work, *Śakti Sangama Tantra,* Magadhadeśa comprised the tract between Varanasi and Monghyr. The Nandas are said to have ruled over Prācya comprising Bihar, Bengal, eastern Uttar Pradesh and the eastern fringe of Orissa.[54] The *Aitareya Brāhmaṇa* refers to the Puṇḍra as a group of outcastes living on the borders of Aryan occupied territory. They settled in Bengal before the third century BC. Moti Chandra refers to an identification by H. Shastri which includes Malda, portions of Purnea east of the Kosi including a part of Dinajpur and Rajshahi as Puṇḍra country. On the basis of inscriptional evidence, S.B. Chaudhury identifies Pauṇḍravardhana with Mahasthan, seven miles north of Bogra in Bangla Desh.[55] R.P. Kangle, quoting H. Shastri, identifies Suvarṇakuḍya with Karṇasuvarṇa which included Murshidabad and Rajmahal in Bengal. K. Barua prefers an identification with the town, Sonkudha, in Kamrup, Assam.[56] Irrespective of the question of the validity of specific identification, all three centres lie within the eastern Indian belt of *tasar* and mulberry silk production areas. It is also of interest that *c.* 1903, N.G. Mukarji (Mukarji, 1903: 8) recorded the Pundas, as constituting the hereditary silkworm rearing caste in Bengal.

With regard to the trees mentioned, the *nāga* tree could be *keseru, Heteropanax fragrans,* and host to the *eri* worm, or *champaka* (*Michelia champaca*) host to the *mūgā* worm. Since *patrorṇa* was a high quality silk the latter identification appears more appropriate. The *likuca* tree is *Artocarpus lakoocha,* host to *B. (T). bengalensis,* and *T. affins,* wild silkworms of lower Bengal and Chota Nagpur respectively. It

Nestling of *tasar* cocoon within leaf frame *T. arjuna.* Courtesy: C.R. Babu, Department of Botany, University of Delhi. (Fig. 10.28)

is significant that although wild, this is the first worm of the *Bombycidae* with which a link may be established in early texts.[57] The *bakula* is *Mimusops elengi.* This served as the host plant to a wild moth of the *Bombycidae, Trilocha varians* or *Ocinara varians* (*No. 23, Silk*: 6). Indeed, one identification of *patrorṇa*, seeks to link it to the term *patra*, leaf (fig. 10.28). *Trilocha varians*, found in abundance in the plains prior to the onset of the rainy season, spun its cocoon snugly on the leaf or bark of the host tree.[58] As has been stated above *Ficus indica*, the banyan tree provides nutriment to *T. religiosae*, Helfer.

The varieties of silk mentioned in the *Arthaśāstra* have a range of colour but all fall within the category of wild silk. *A. assamensis* and *A. mezankooria*, Moore, yielding *mūgā* would, respectively, provide white silk when fed on *champaka, Michelia champaca* and *mejankeri, Litsaea citrata, L. cubeba*, Perb. It should be remembered in this context that indigenous methods of bleaching were in practice to which attention has already been drawn.

The material from the *Yuktikalpataru*, dated to *c.* eleventh century AD can provide some clues to chronology (Gopal, 1964–5: 95–6). In this work it is stated that worms providing silk cocoons can be divided into four castes. The Brāhmaṇa variety, of which the threads are exceptionally white, are small in size and are to be found in coastal districts and forests near the south-east sea. If this is taken as a reference to cultivated *Bombyx* silk it could point to a chronology for the presence of *B. fortunatus*, the Bengal *desī* worm, considered to be native to this country. *B. fortunatus*, also called *chhoto polu* could yield either a white or a yellow cocoon. The yellow colour was derived from yellow pigment present in the body of the worm.[59] According to D.W.H. Speed there were five bunds ranging from 40 to 110 days. Of this the best yield was that obtained in the months of October and November.[60] Since there is a reference here to coastal districts, presumably of low elevation, a reference to silk of *Trilocha varians* cannot be ruled out. The Kṣatriya type of worm is not described but the Vaiśya type, with yellowish white threads, was to be found in forests and marshy land near the western sea. There was no known commercially viable silkworm rearing in western India at this time, but if one were to grade worms the Vaiśya variety could be taken as *tasar*. The Śūdra worms produced heavy threads of many types and could be found in forest and inland places in many areas. This could be taken as *eri* and inferior types of *tasar*.

The question that could be raised is whether the first category could be associated with Assam *paṭa*, the fine white silk of the *Bombyx* variety reserved for royalty by the Ahom kings of Assam. The rearing and reeling of this silk was confined to Katani Jugis. The latter, a weaving community, had been driven out of Bengal by Ballāl Sena, AD 1159–79, and his successors. It was possibly at this time that *B. fortunatus* was introduced to Assam. The silk industry in Assam received a great deal of encouragement under the Ahoms, particularly Swargadeo Sutāpha, AD 1268–1364.[61] Hence, although the *Arthaśāstra* associates Suvarṇakuḍya, Sonkudiha in Kamarupa, Assam, with the best kind of *patrorṇa*, in view of the dating of the *Yuktikalpataru* and the relatively larger size of the *mūgā* cocoon, the first category cited from the *Yuktikalpataru* appears to relate exclusively to Bengal.

The discussion of silk in eastern India would be incomplete without some reference to the commodity referred to as *paṭṭa vastra* used on ritual occasions in Bengal and Orissa. As has already been demonstrated, silk has enjoyed special favour as the textile to be worn by both sexes on occasions of Hindu ritual.[62] Mulberry silk has always been an expensive item and *paṭṭa vastra* appears to have been a cheaper equivalent. In Bengal, the poorer women wore saris made of Sunn hemp (*Crotalaria juncea*) called *kauma, khemi, khani* or *khaniya.* There is also a reference to a coarse sari called *megila*, woven from the fibre of white jute (*Corchorus capsularis*).[63] *Paṭṭa vastra*, which appears to have approximated silk more closely, may have been woven from the fibre of *ulatkambal* (*Abroma augusta*) the fibre of which was much valued and which could substitute for silk (Wardle, 1886: 119). The fine quality nettle cloth referred to in the *Rāmāyaṇa* was probably made of the fibres of *Urtica heterophylla*, the *horoo surat* of Assam. In Bhutan this fibre was called *herpah.*[64]

During his visit to India, *c.* AD 629, Yuan Chuang had declared that in India, a number of fibres were utilized in garments among which he included *k'au-she-ye, kauśeya*, which was the product of the wild silkworm. *Koś* silk carries the same connotation in Bihar today. It could, therefore, be tentatively concluded that while the development of *eri* and *tasar* could be related to movements of the Boḍo groups, the development of *B. fortunatus* silk, associated with Bengal, may be located at some point in time between *c.* seventh century AD when Yuan Chuang visited India and the eleventh century when the *Yuktikalpataru* was composed. This conclusion is not based on written evidence alone. Sequential developments have become clearer only because of the richness of traditional data which underpins a support system, linking historical data with a living tradition in which the chronological past is found embedded in a vibrant ethnological present.

Notes and References

1. Varadarajan, 1986: 189; Varadarajan, 1988: 561; Maxwell-Lefroy, Ansorge: 145, 148; Bouvier, 1984: 21; Sarkar, 1977: 36–8.

2. Wardle, 1891: 9, 10; Maxwell, Howlatt, 1909: 487. Twisting of silk yarn while conferring strength reduces its lustre. For this reason in India twisted yarn has usually been confined to the warp. The presence of sericin makes it possible to ensure the adherence of several threads together without the necessity of twisting. This ensures superior lustre and such untwisted filament is used for the yarn in both warp and weft. When silk is to be dyed, however, degumming is essential to ensure depth in colour.

3. Wardle, 1880: 21; Watt, V, 1972: 55, 176; Hyde, 1984: 7.

4. Om Prakash, 1985: 55. However, since these words are of Portuguese origin an alternative explanation could be sought. The terms *cabecca* (head), *barriga* (belly), *pes* (feet), were used by the Portuguese in the camphor trade to denote the best, medium and inferior qualities of the product. Soares, 1988: 8, 40, 284. There may have been an unthinking transfer of the same terms in the silk trade to denote qualitative differences.

5. Sarkar, 1977: 72; Nanavaty, 1965: 201–2; Choudhury, 1981: 85.

6. For further details of this Vaiṣṇavite cult see Thoothi, 1935: 83; Talwar, Krishna, 1979: 1–15, 128–134.

7. Oral communication: Dalsukhbhai Malwania, Ahmedabad and H.C Bhayani, Ahmedabad.

8. For details of mulberry propagation and cultivation in India and the rearing of *Bombyx mori* see *Silkworm Rearing*; *Mulberry Cultivation*; Sarkar, 1977: 16–26, 32–41; Nanavaty, 1965: 56, 109–19; Dey, 1914: 8–14. In China a number of plants were used as alternative feed plants to mulberry. *Artemisia* was given at the first instar; *Lactuca, Zizyphus* and *Cudrania* were given to caterpillars after the third moult. Rice, bean and dry mulberry leaves in powder form were offered after the last moult. Liu, 1952: 181.

9. Brown, 1874: 90; Dun, 1975: 12; Hardiman, 1901: 3–4.

10. See *Altas* and *Eria* group in Wardle, 1880: 5–6. Also refer to Sarkar, 1977: 59–61; Nanavaty, 1965: 204–6.

11. *Short Note 1965*: 1–2; *Notes on Eri Silk*, 6–7.

12. *Report 1963*: 3; Liu, 1952: 190; Nanavaty, 1965: 196, 204.

13. Removal of the pupae along with the last skin shed is deemed to be advantageous for spinning. However, nineteenth century English observers, state that the pupa is removed after boiling. *Silk in Assam*: 10. The most valued cocoons were those in which the chrysalis had been taken out immediately after completion of spinning. Chaudhury, 1981: 38, 45, 65.

14. Geoghenan, 1872: 176; Wardle, 1880: 7; Hugon, 1837: 21, 33.

15. See Allen, 1899: 17. For dyeing of *eri* with lac, *Laccifer lacca*, madder, *Rubia cordifolia* see Duncan, 1896: 20, 45. Although the dye indigo, *Indigofera tinctoria* is known in Assam it is probable that the more common *rum*, *Strobilanthes cusia*, was used for the dyeing of *eri*. Duncan, 1896: 28–9, 48–50.

16. *Silk in Assam*, 10–12; Jolly et al., 1975: 88–9; Nanavaty, 1965: 206; Allen, 1892: 5–10, 18–24.

17. Chaudhury, 1981: 98–100. Tavernier noted *c.* 1666 that in Assam a lye was made with the ashes of fig leaves which bleached the silk as white as snow. In Kasimbazar, the ashes of plantain, *Musa paradisiaca* were used to good effect. Crooke, 1977: 3, n. 2, 220–2. The Assam methods for bleaching and dyeing noted by S.N. Chaudhury, could, therefore, be traced well back into the medieval period, if not earlier. See Allen, 1899: 13–14; Chaudhury, 1981: 98–9.

18. *Silk in Assam*: 21; Allen, 1899: 15–16; Chaudhury, 1981: 19, 22, 25.

19. *Silk in Assam*: 21; Allen, 1899: 15–16; Chaudhury, 1981: 19, 22, 26.

20. Jolly, 1980: 17–18. S.N. Chaudhury also holds a similar point of view. *A. chengtuana*, of Szechwan, China, is the most closely affiliated worm to *A. assamensis*. An allied species, *A. youngei*, is found in Java. The chromosome numbers are respectively: *A. assamensis*, n=15; the Indian tasar worm, A. mylitti, n=31; A. roylei, n=30; the Chinese *tasar*, *A. pernyi*, n=49; Japanese *A. yamamai*, n=31; *A. frithii*, also called *A. paphos* and found exclusively in Assam, n=31; *A. polyphamus*, present in USA, n=30. Chaudhury, 1981: 7; *Report 1963*: 4.

21. Jolly et al., 1975: 7–8; Jolly, no date: 4–6; Nanavaty, 1965: 199, 203; Sarkar, 1977: 57.

22. *Report 1963*: 4; Jolly et al., 1975: 7–8; Sarkar, 1977: 56.

23. Sarkar, 1977: 58. For aspects of spinning the cocoon, pairing, egg laying, hatching, techniques of seed cocoon production and utilization of *tasar* waste see Sarkar, 1977: 56–9; Nanavaty, 1965: 200–2; Hugon, 1837: 21, 32, 43; Wardle, 1880: 16–28; *Kosa Weaving*, 15–16. About seven cocoons are utilized in reeling, the combined filaments being brought together and rolled between the hand and left thigh of the spinner. Women always carry out this operation. Wardle, 1880: 20.

24. Liu, 1952: 189; Jolly et al., 1975: 49; Nanavaty, 1965: 196–9.

25. Jolly, 1974: 3–9; Jolly et al., 1975, 51–63; Chaudhury, 1981: 101, 103.

26. Courtesy: Professor Tan Chung, New Delhi, quoting *Studies in New China*: 155.

27. Courtesy: Professor Tan Chung, New Delhi, citing Yang Xianyi, 1983: 172.

28. Courtesy: Professor Tan Chung, New Delhi, quoting Ji Xianlin, 1982: 52.

29. Personal communication: Professor Tan Chung, New Delhi.

30. Liu, 1952: 143–6. It was under the Sungs that two important treatises on sericulture were produced by government officials. These were, *Tsan Sa*, or *Treatise on Silkworm*, and *Keng Chi Tu* (*Agriculture and Sericulture in Picture*), written by Ch'in Kuan, AD 1049–1100, and Lou Shou, *c.* AD 1131–62 respectively. Under the Yuan dynasty, AD 1280–1368, a special ministry for agriculture and sericulture was created. In AD 1285 the *Nung Sang Chi Yao* (*Essentials of Agriculture and Sericulture*), was issued by special imperial edict (Liu, 1952: 146). It is of interest that this active propagation of sericulture in China coincided with the trading activities of the Radanite Jews, centred at Ahvaz, in the post ninth century AD period. The ambit of their enterprise extended to China. Varadarajan, 1987: 99; cf. Pariset, 1843: 23.

31. Silk Museum (Musum Label), Yokohama; Geoghegan, 1872: ii-iii, 105, 108; Hardiman, 1904: 25. Personal communication: Dr S.N. Choudhury, Shillong.

32. Kordosis, 1992: 172–4; Pariset, 1843: 26, 61, n. 2, 208; Needham, 1976: 155; *Chau, Ju Kua*, 141; Nanavaty, 1965: 194.

33. G. Vial tentatively ascribes the 'Orient' as its provenance. It was

a fragment from a dalmatic attributed to Saint Lambert, Bishop of Maastricht, AD 708. The weave is figured gauze. China was the first country to weave figured gauze and archaeological evidence supports the existence of this tradition from the Han to the Tang periods, *c.* second century BC to the tenth century AD, Vial, 1983: 51, 53, 57–8.

34. Granger-Taylor, 1987: 29, n. 65; Nanavaty, 1965: 193.

35. The earliest example of damask weave in linen occurs at a much later date. This practice has been traced, *c.* AD 1453, to the region comprising the urban centres of Bruges, Tournai, Brussels and Gand in the Low Countries. Prinet, 1983: 45–6.

36. Granger-Taylor, 1987: 25–6; Wild, 1987: 459, 465, 470.

37. Willetts, 1958: 207; Muthesius, 1989: 136–7; Wild, 1989: 213.

38. With the Moorish conquest of Sicily in AD 827, silk weaving was established there. In AD 1080 the Normans replaced the Moors. Roger II, AD 1101–54, after his victorious campaign in Greece, AD 1147, brought back weavers who were settled in Sicily. Emperor Frederick II, Hohenstaufen, AD 1194–1250, introduced silk to lower and central Italy. The centre of Tuscan silk weaving was initially Lucca but Florence soon became a rival. Silk weaving was introduced to France from Italy *c.* AD 1080 but it was only in the sixteenth century that Florence was to be displaced by Lyon. The revocation of the Edict of Nantes in 1598 in France led to the migration of Huguenots to England. The art of silk weaving was brought by these refugees to their new home. Moorish presence in Spain led to the initiation of the industry there. Varron, 1938: 353; Reininger, 1939: 965–6; Birrell, 1959: 28. This geographic dispersion constituted the essential contours of the silk map of Europe at the time of the Renaissance, Reformation and the Discoveries.

39. André and Filliozat, 1986: 144, 250, 312; Logan, 1951: 202.

40. Giles, 1978: 46; Sircar, 1971: 103–4; Kordosis, 1992: 204.

41. Courtesy: Professor Tan Chung, New Delhi, quoting Yang Xianyi, 1983: 172.

42. Oral communication: Professor Tan Chung, New Delhi.

43. Ganju, no date: 190; Moti Chandra and Agrawala, 1959–62: 40; Sufi, 1974: 574; Briggs, 1910: 445.

44. Zhu Changli, 1993: 31; Datta, 1992: 36; Mukherjee, 1992: 13, 25.

45. Chatterjee, 1968: 2; Chatterjee, 1970: 16–18, 22–3, 30, 32; Chatterjee, 1974: 26, 35–7; Nath, 1948: 13–14; Choudhury, 1966: 340.

46. Oral communication: B.V. Bapat, Poona.

47. Ganapati Sastri, 1925: 222; Monier-Williams, 1976: 32, *kramūkha* q.v.; Sharma, 1978: 64, *tūḍa* (*tūlaphala*) q.v. Sodhala is dated to the twelfth century but the manuscript used in this text is ascribed to AD 1412.

48. *Pañcatantra*, I, 103. The author of the original text probably belonged to the period between AD 200–400. The earliest available version of the *Pañcatantra* cycle of stories, the *Tantrākhyāyikā*, was written *c.* AD 300

49. Muni Sri Punyavijayaji, 1957: 40. The *Divyāvadāna*, dated to *c.* fourth century AD, mentions silk of three varieties, *pāṭṭāṃśuka, kauśeya* and *dhautapaṭṭa*. Cowell, 1951: 136.

50. Sastri, 1970: II, 6, p. 113; Monier-Williams, 1976: 483, *dukūla*, q.v.; Lévi, 1925: 32. Prior to the ninth century AD translations of Buddhist texts were inconsistent as Tibetan terminology for Sanskrit words had yet to be standardized. King T. Ralpbatsan, *c.* AD 866–901 ordered Kga Sog to compile a glossary of Sanskrit–Tibetan terms. The Tibetan name for the work was *lo pan mang pos mzad pa'i bye brag rtogs byed chen mo (The Great Work Made by Many Lotsawas and Pandits that Brings Comprehension of Particulars)*. It was known as *Mahāvyutpatti* in Sanskrit. http://www.tibet.dki/pktc/tdiction.htm

51. MacDonnel and Keith, 1958: 308; Eggeling, 1963: 85; *Jaiminī*: X, IV, 13.

52. Krishnaswamy Aiyangar, 1941: 398; Minakshi, 1977: 193; Wheatley, 1966: 183, 209.

53. Kangle: 1972, 145, 2.22.6. In n. 6, p. 145, Kangle refers to *kṛmijāta* as silk cloth. For the Sanskrit text refer Kangle, 1960: p. 74. In referring to treatment of artisans (ibid., 75, 2.23.8), a similar fibre distinction is observed with the exception that the term *kṛmitana* rather than patrorṇa or *kauśeya* is used for silk. With regard to wages, those weaving patrorṇa were to receive higher emoluments than weavers of *kauśeya* (Kangle, 1972: 255, 4.1.10). P.V. Kane dates the *Arthaśāstra* to between 300 BC to AD 100 (Kane, 1941: XI). Professor B.N. Mukherjee, Calcutta, in an oral communication, says that this work may have been composed prior to the first century AD. According to M. Winternitz the *Arthaśāstra* cannot be dated earlier than the third century AD (Winternitz, 1985: 631).

54. Sircar, 1971: 31, n. 4; 38, n. 1; 46, 76, 80, 212.

55. Sircar, 1965–6, 253; Moti Chandra, 1945: 110; Choudhury, 1955: 176, 180; Chakrabarti, 1992: 22–3; Morrison, 1970: 14–15, 30; Majumdar, 1940: 352.

56. Haraprasad Shastri linked the name with the soil which 'is red like gold.' See Kangle, 1972: 104, n. 102. K. Barua points out that the Sonkudha area still produces *mūgā*, important centres being Sualkuchi and Bhomlahat, both lying within 15 miles of Sonkudha. Neogi 1973: 222–3, 232, and S. Lévi, 1925: 30–6, attempt to localize the site of *Suvarṇakuḍya* to a spot south-east of India in the direction of Java. However, since the silk tradition of north-east India is not in doubt, S. Lévi's identification in this context appears somewhat specious.

57. Wardle, 1881: 4; Geoghegan, 1872: 163; Watt, V, 1972: 3.

58. Swinhoe, 1885: 302; Maxwell-Lefroy and Howlatt, 1909: 478, 483–4; Fletcher, 1914: 407.

59. Oral communication: D.C. Sarkar, Calcutta.

60. Watt, V., 1972: 12; Dey, 1914: 27.

61. Hugon, 1837: 22; *Assam I*: 270; *Silk in Assam*: 2–3; Das Gupta, 1969–70: 19.

62. Kane, 1941: 672; Mukarji, 1903: 66; Mookerji, 1895: 2.

63. Das Gupta, 1935, 278-80; Forbes Royle, 1855, 242-3; Varadarajan, 1984, 54, n. 13.

64. D.K. Myers classifies the nettle from which cloth was woven in east Nepal as *Girardinia palmata* (Fonsk) Gaudich. Myers, 1987: 50. This was also known as Nilgiri Nettle. K.A. Chowdhury records that two species of *Boehmeria*, ramie, are indigenous to India: *B. macrophylla* and *B. platyphylla*, growing in the Himalayan foot hills near Dehra Dun. The earliest samples of *B. platyphylla* have been identified at the site of Atranijikhera dated to between 1200–600 BC. Chowdhury, 1977: 43–4, 69. For weaving of nettle cloth also see Elwin, 1959: 68–9.

SRINAGAR
Kohala
Bandipur
Peshawar
Hazi Pir Pass
Gulmarg
Leh
JAMMU AND KASHMIR
Zaskar Range
Poonch
Taxila
Pir Panjal Pass
Budil
Banihal Pass
Kishtwar
Jhelum
Akhnoor
Tisa
Patseo
CHAMBA
LAHUL
Gujrat
Sialkot
Pathankot
HIMACHAL PRADESH
KANGRA
Dina Nagar
PAKISTAN
KULLU
Puh
Amritsar
MANDI
Kanam
Bashohli
Pangi
Kumharsain
Rampur
Sultanpur
SHIMLA
Rohru
Nurpur
Arki
Jubbal
Kasumpali
Theog
Ludhiana
Solan
Chaupal
CHANDIGARH
SIRMAUR
Mussoorie
PUNJAB
DEHRADUN
Sunam
GARWA
I N D I
Sikri
HARYANA
Tanda
Great Indian Desert
(Thar Desert)
DELHI
DELHI
UTTAR P
RAJASTHAN
Mathura
Agra
Aravali Range
Fatehpur Sikri
JAIPUR

MAPS

INDIA AND SURROUNDING COUNTRIES

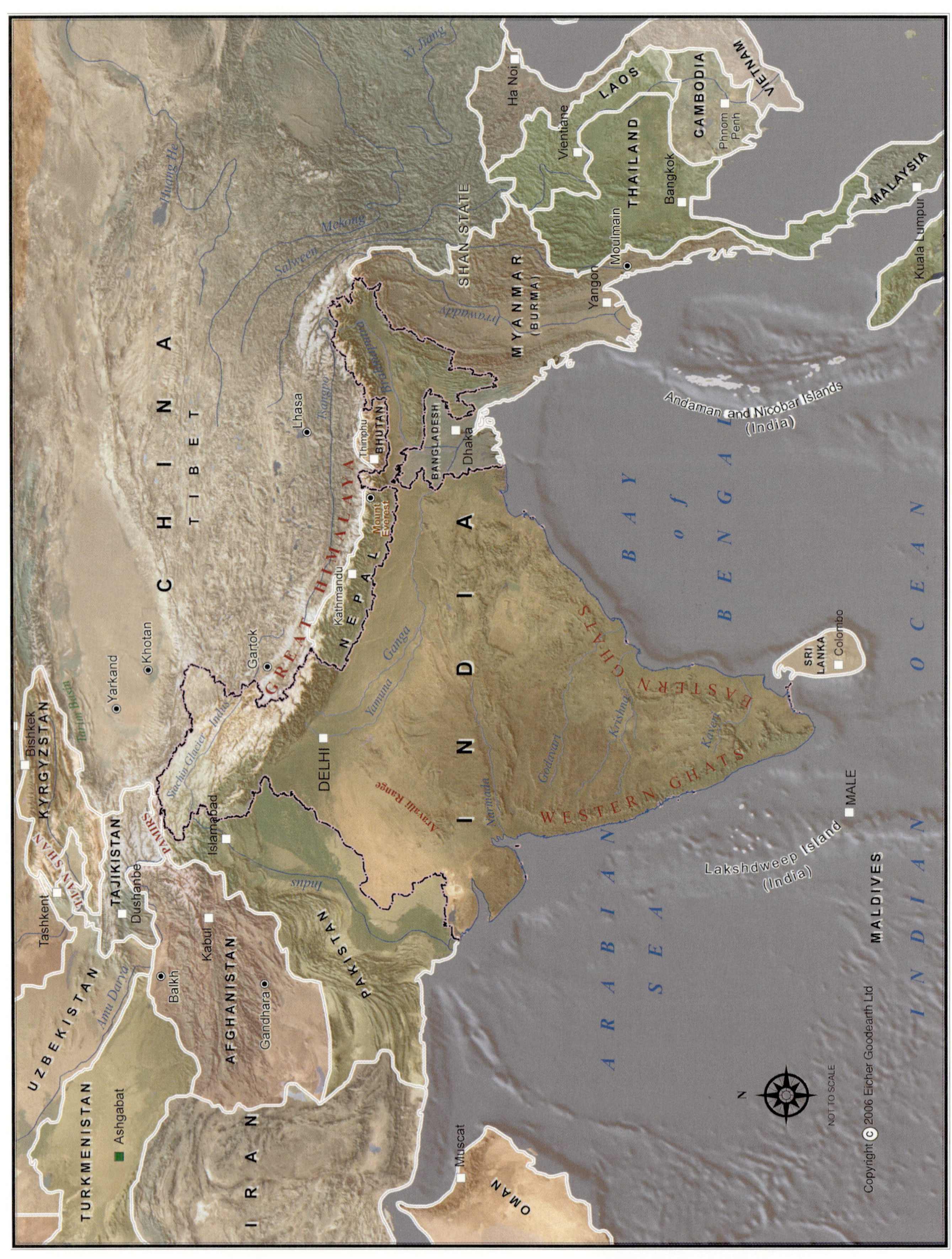

NORTH-EASTERN INDIA

NORTHERN INDIA

NORTH-WESTERN INDIA

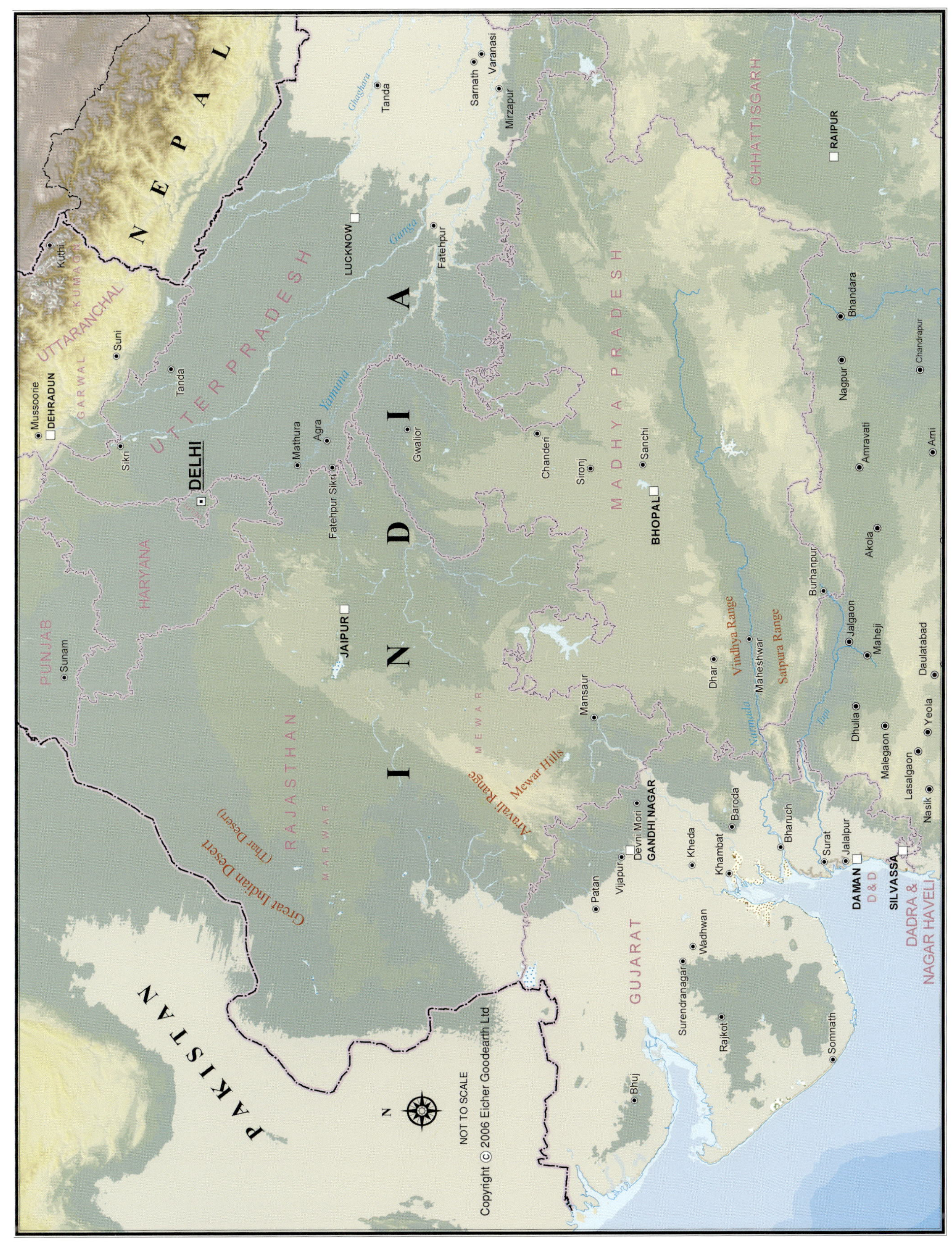

SOUTHERN INDIA

Abbreviations

GENERAL

A.E.D.T.A. — Association pour l'Etude et la Documentation des Textiles d'Asie, Paris.

Assam I — Census of India, 1891, *Assam, I,* Report, Shillong, 1892.

Aurangabad — *Gazetteer of Aurangabad,* published under orders of His Highness the Nizam's Government, Bombay, 1884.

Aurangabad 1977 — Maharashtra State Gazetteers, *Aurangabad District,* revised edition, Bombay, 1977.

Bernier — A. Constable, ed., *Travels in the Mugal Empire, A.D. 1656-1688, by Franscois Bernier, 1891,* rpt. New Delhi, 1972.

B.L.C.I.E.T.A. — *Bulletin de Liaison du Centre International d'Etude des Textiles Anciens,* Lyon.

B.N., N.A. — Bibliothèque Nationale, Nouvelles Acquisitions, Paris.

Châle Cachemire — *La Mode du Châle Cachemire en France,* Exhibition Catalogue, 19 May–31 October 1982, Musée de la Mode et du Costume, Paris.

Chau Ju Kua — Friedrich Hirth, W.W. Rockhill, *Chau Ju Kua: His Work on the Chinese and Arab Trade in the Twelfth and Thirteenth Centuries Entitled Chufan-ch'i,* New York, reprint, 1966.

Chine — *Chine Cielet Terre, 5000 Ans d'Inventions et de Découvertes,* Bruxelles, 1988.

Encyclopaedia of Textiles — *Encyclopaedia of Textiles,* Eaglewood Cliffs, New Jersey, 1960.

E.W.A. — *Encyclopaedia of World Art,* VII, New York, 1963 Toronto, London, 1963.

G.K.L. — *Gazetteer of Kashmir and Ladakh,* New Delhi, 1974, reprint, 1890.

Handicrafts of Chamba — *Survey Report on Handicrafts of Chamba (Himachal Pradesh),* Simla, 1969.

Handicrafts of Kinnaur — *Survey Report on Handicrafts of Kinnaur District, H.P.,* issued by the Director of Industries, Himachal Pradesh, Simla, 15 August 1965.

Imperial Gazetteer — *Imperial Gazetteer,* I, Simla, 1905.

Jaiminī — *Jaiminī's Pūrvamīmāṁsā-Sūtra with the Bhāṣya of Śabara and the Tantravārtikā and Tup-ṭīkā of Kumārilabhaṭṭa,* Anandāśrama Series, Parts I-VI, Poona, 1930–53.

Karlgren Dictionary — B. Karlgren, *Analytic Dictionary of Chinese and Sino-Japanese,* place and date of publication not cited.

Karnataka 1968 — *Karnataka Through the Ages.* Government of Mysore, 1968.

Khandesh — Gazetteer of the Bombay Presidency, XII, *Khandesh,* Bombay, 1880.

Kokad — A.D. Kokad, 'Paithani', *Stree,* published by Kirloskar, Poona. Translated by B.D. Arjun-Wadkar, Nagpur, 9 November 1975.

Kosa Weaving — Census of India, 1961, X, Maharashtra, Part VII-A(3), Handicrafts of Maharashtra, *Kosa Silk Weaving at Ganeshpur,* Delhi, 1961.

Linschoten — A.C. Burnell, P.A. Tiele, eds., *The Voyage of John Hughen Jan Linschoten to the East Indies,* Cambridge, 1855.

M.S.G. — Maharashtra State Gazetteers, *History,* Part I, Ancient Period, Bombay, 1967.

Manual — *Manual of the Administration of the Madras Presidency,* III, Madras, 1893.

Marathi Theatre — *The Marathi Theatre, 1843 to 1960,* Marathi Natya Parishad, Bombay, 1961.

Maxwell-Lefroy, Ansorge — H. Maxwell-Lefroy, E.C. Ansorge, *Report on an Enquiry into the Silk Industry in India,* I, Calcutta, 1917.

Moorcroft — H.H. Wilson, prepared for press from original, *Journals and Correspondence, Travels in the Himalayan Provinces of Hindustan and Punjab; in Ladakh and Kashmir; in Peshawar, Kabul, Kunduz and Bokhara:* by Mr. William Moorcroft and Mr. George Trebeck from 1819 to 1825, I, II, New Delhi, 1971, reprint.

Moorcroft Papers — India Office Library, London, *Moorcroft Papers,* Mss. Eur. E. 113.

M.S.K. — *Maharashtra Shabda Kosh,* 5, Poona, 1936.

Mulberry Cultivation — G. Rangaswamy, M.N. Narasimhanna, K. Kasi-viswanathan, C.R. Sastry, M.S. Jolly, Sericulture Manual, I, *Mulberry Cultivation,* New Delhi, Bombay, Calcutta, 1992.

N.I.D. — National Institute of Design, Ahmedabad, India.

Notes on Eri Silk — *Notes on Eri Silk,* compiled and published by Department of Sericulture and Weaving Assam, Shillong, no date.

No. 23 Silk — Imperial Institute Series, Handbooks of Commercial Products, Indian Section, *No. 23, Silk,* Calcutta, 1895.

Palmyra, 1987 — *Meisterwerke aus Palmyra,* Berlin, 1987.

P.D.G. 1918 — Punjab District Gazetteers, XXXA, *Kangra District,* Parts II, III, IV, Lahore, 1918.

Report 1963 — *Report on the Activities of the Regional Sericultural Research Station,* Titabar, Assam, Shillong.

Reports and Documents — *Reports and Documents connected with the proceedings of the East India Company in regard to the culture and manufacture of cotton-wool, raw silk and indigo,* London, 1836.

Rural Craft Survey — Census of India, 1961, XX, VII-A, No. 2, Himachal

	Pradesh, *Rural Craft Survey, The Art of Weaving,* field investigations and draft by Lakshmi Chand Sharma, ed. *Ram Chandra Pal Singh.*
Selected Handicrafts of Assam	E.H. Pokytein, *Selected Handicrafts of Assam,* Census of India, III, Assam, Part III-A, 1961.
Short Note 1965	*A Short Note on Production of Eri Seeds in Government Eri Seeds Grainage,* Nongpoh, Directorate of Sericulture and Weaving, Assam, Shillong.
Silk in Assam	E. Stack, Notes on some Industries of Assam from 1884 to 1895, *Silk in Assam,* place and date of publication not cited.
Silk Reeling	S. Krishnaswamy, M. Rao, S.K. Suryanarayan, T.S. Sundaramurthy, Sericulture Manual 3, *Silk Reeling,* New Delhi, Bombay, Calcutta, 1991.
Silkworm Rearing	S. Krishnaswamy, M.N. Narasimhanna, S.K. Suryanarayan, S. Kumararaj, Sericulture Manual 2, *Silkworm Rearing,* New Delhi, Bombay, Calcutta, 1992.
Simla Hill States	*Simla Hill States,* no. 2: Bashahr State Gazetteer, Part A, 1910.
Studies in New China	*Xin Zhongguode Kaogu faxian he yanjiu*/Archaeological discoveries and studies in New China, Institute of Archaeological Studies of the Chinese Academy of Social Sciences, Beijing, 1984.
Symposium 1984	*Le Prob elme des Fibres dans les Textiles Anciens,* A.E.D.T.A., Paris, 4–6 July 1984.
Third Revised Settlement	*Final Report of the Third Revised Settlement 1910-1913, of the Parol, Lag Maharaja and Lag Sari, Rubi, Lahul and Apiti Waziris of the Kulu Subdivision of the Kangra District,* by J. Coldstream, ICS.
Varṇakramī	*Mahārāshtra Til Khedyāñchi va Shaharāñchi Varṇakramī,* Census Department of Maharashtra, Bombay, 1965.
Watt, 1905	G. Watt, 'Comments on paper presented by T.C. Hodson, Manipur and its Tribes', *Journal of the Society of Arts,* 2733, LIII, Proceedings of Society.
Watt, II, 1972	G. Watt, *A Dictionary of Economic Products of India,* II, Part II, New Delhi, 1972, reprint of 1893.
Watt, V, 1972	G. Watt, *A Dictionary of the Economic Products of India,* V, Part III, New Delhi, 1972, reprint of 1893.
Watt, VI, 1972	G. Watt, *A Dictionary of Economic Products of India,* VI, Part II, New Delhi, 1972, reprint of 1893.
Wealth of India III	*The Wealth of India: A Dictionary of Indian Raw Materials and Industrial Products, Raw Materials,* III, New Delhi, 1952.

ABBREVIATIONS: NID DOCUMENTATION AND PUBLICATIONS

Faculty

Chakraborti, D.K. and Katiyar, V.S.	*An Ergonomic Enquiry into the Paṭolā Loom,* NID, 1993, unpublished.
Perheentupa, Helena	*How the Banaras Jāla and South Indian Technique of Adai Differ,* NID, 1982, unpublished.
Shirali, A.	*Textile and Bamboo Crafts of the Northeastern Region,* NID, Ahmedabad, 1983.

Students

Banaras Brocades	Reena Singh, *Banaras Brocades,* Craft Study and Documentation, NID, 1978, unpublished.
Ahmedabad Brocades	Uma de Jhala, *Brocades of Ahmedabad,* Craft Study and Documentation, NID, 1977.
Camel Belts	K. Nagarajan, *Camel Belts,* Craft Study and Documentation, NID, 1992.
Cotton Saris of Bijapur	Vibha Narula, Shraddha Desai, Bharati Jagdale Hasmukh Prajapati, *Cotton Saris of Bijapur and Dharwar,* Craft Study and Documentation, NID, 1985.
Dari Fatehpur Sikri	Sanchita Das Gupta, Smita Mehta, *Dari Weaving of Fatehpur Sikri,* Craft Study and Documentation, NID, 1980, unpublished.
Dhablas of Kutch	Krishna Amin-Patel, *Dhablas of Kutch,* Diploma Project, NID, 1975, unpublished.
Gyaser	Preeti Gopinath, Aditi Butail, Priya Kainikara, *Gyaser,* Craft Study and Documentation, NID, 1993, unpublished.
Jala Banaras	Meena Parikh, Roshan Michael, *Jala Weaving of Banaras,* Craft Study and Documentation, NID, 1986, unpublished.
Jamavar Shawls	Anshoo Nath, *Jamavar Shawls of Kashmir,* Craft Study and Documentation; NID, 1983, unpublished.
Khans of Maindergi	Padmini Tolat, *Khans of Maindergi,* Craft Study and Documentation, NID, 1978, unpublished.
Moirangphee	Vinay Manapura, *Moirangphee of Imphal,* Craft Study and Documentation, NID, 1984, unpublished.
Navalgund Durries	Chandrasekhar Bheda, *Navalgund Durries of Karnataka,* Craft Study and Documentation, NID, 1986, unpublished.
Pattu Weaving	Sunil Munshi, P.T. Girish, Anu Tandon, *Pattu Weaving of Western Rajasthan,* Craft Study and Documentation, NID, 1987, unpublished.
Products Tamil Nadu	Abhilasha Gupta, Chetna Bhatt, Jayshree Poddar, S.M. Kulkarni, *Traditional Products of Tamil Nadu,* Craft Study and Documentation, NID, 1982, unpublished.
Rhythm of Fingers	Jyotilakshmi Sethu, *Rhythm of Fingers,* Craft Study and Documentation, NID, 1991. Supplemented by Priya Kainikara, Aditi Butail, Field Study Notes, NID, 1994, unpublished.
Tangalias	Ashwini Bhuta, *Tangalias of Gujarat,* Diploma Project, NID, 1977, unpublished.
Woollen Products of H.P.	Abhilasha Gupta, *Re-designing and Diversification Woollen Products of Himachal Pradesh,* NID, Diploma Product, vol. 1, 1983–4, unpublished.
Woven Products Kachch	Vaishali Varshney, *Development of Woven Products for Exports in Kachch,* Diploma Project, NID, 1992, unpublished.

COMMISSIONED STUDIES

Pañja Darīs	Girija Kaimal, *Pañja Darīs,* NID, 1993, unpublished.
Tweed for Kullu	C. Bhatt, *Design Development of Tweed for Kullu,* NID, 1984, unpublished.

Bibliography

Abhyankar, K.V. ed. — *Vyakaraṇa Mahābhāṣya by Patañjali*, trans. into Marathi by V.S. Abhyankar, IV, Pune: 1952.

Adams, A. — Remarks on the Habits and Haunts of some of the Mammalia found in Various Parts of India and the Western Himalayan Mountains', in *Proceedings of the Zoological Society of London*, Part XXVI, 1858, pp. 512–31.

Adelaar, K.A. — 'Les Langues Austronèsinnes at la place du Malagasy dans leur Ensemble', in *Archipel*, 1989, pp. 25–52.

Adrosko, R.J. — 'The Invention of the Jacquard Mechanism', in *B.L.C.I.E.T.A.*, I, II, Lyon: 1982, pp. 89-117.

Agarwal, R.C. — 'A Study of Textiles and Documents from Kharoshthi Documents', in *Bharati Vidya*, XIV, 1953.

Agrawala, V.S. — *Harṣacarita–Ek Sāṅskritika Adhyayana* (Hindi), Patna: 1953.

— *The Deeds of Harsha*, Varanasi: 1969.

Albers, A. — *On Weaving*, Middle Town CT: Wesleyan University Press: 1965.

Allen, B.C. — *Monograph on the Silk Cloths of Assam*, Shillong: 1899.

Altekar, A.S. — 'A History of Important Ancient Towns and Cities of Gujarat and Kathiawad from Earliest Times to About 1330 AD', *Indian Antiquary*, III–IV, 1924–5.

Ames, F. — 'Lost Horizons, The Stylistic Developments of the Kashmir Shawl', *Hali*, vol. 4: 1984, pp. 403–7.

— *The Kashmir Shawl*, Suffolk: Woodbridge: 1986.

Anavian, R. and Anavian, G. — *Royal Persian and Kashmir Brocades*, Kyoto: 1975.

André, J. and Filliozat, J. — *L'Inde vue de Rome*, Paris: 1986.

Apte, V.S. — *The Practical Sanskrit English Dictionary*, III, Poona: 1959.

Archana — *The Language of Symbols, A Project on South Indian Ritual Decorations of a Semi Permanent Nature*, Chennai, Crafts Council of India, no date.

Ardika, I.W. and Bellwood, P. — 'Sembiran: The beginning of Indian contact with Bali', *Antiquity*, 1991, pp. 221–32.

Arora, R.C. — *In the Land of Kashmir, Ladakh and Gilgit*, Agra: 1940.

Atkinson, E.T. — *The Himalayan Gazetteer*, II, Part I, Allahabad, 1882; rpt. Delhi: 1973.

Baden Powell, B.H. — *Handbook of the Manufactures and Arts of the Punjab*, II, Lahore: 1872.

Bahadur, B.C. Rai — *The Silk Industry of Assam*, Shillong: 1915.

Bajpai, S.C. — *Kinnaur in the Himalayas*, New Delhi: 1981.

Balfour, E. — *Encyclopaedia Asiatica*, II, VIII, 1884; New Delhi, rpt 1976.

Barker, A.F. — 'Textile Industries of Kashmir,' in *The Indian Textile Journal*, XIII, January, no. 508, Bombay: 1933.

Basak, R. — *The Prākṛt Gāthā Saptaśatī*, Calcutta: 1971.

Basham, A.L. — 'The Mandasor Inscription of the Silk Weavers', in B.L. Smith, ed., *Essays on Gupta Culture*, Delhi, Banaras, Patna: 1983.

Bates, H.E. — *A Gazetteer of Kashmir*, Calcutta: 1873.

Becker, J. — 'Silk Weaving Techniques of Han China: The Monochrome Patterned Weaves', *B.L.C.I.E.T.A.*, 1981, pp. 1, 21–43, 53.

Bellwood, P. — 'A Hypothesis for Austronesian Origins', *Asian Perspectives*, 26, Part I, 1984–5, pp. 107–17.

— *Pre-history of the Indo-Malayan Archipelago*, London: 1985.

— 'The Pre-history of Island Southeast Asia: A Multi-disciplinary Review of Recent Research', *Journal of World Prehistory*, 1987, nos. 1, 2, pp. 171–224.

— 'Cultural and Biological Differentiation in Peninsular Malaysia; the Last 10,000 Years', paper prepared for *Second International Conference on Malay Civilization*, Kuala Lampur, 16–20 August 1989, pp. 1–45, unpublished.

Bellwood, P. and Koon, P. — 'Lapita Colonists Leave Boats Unburned! The Question of Lapita links with Islands in Southeast Asia', *Antiquity*, 63, no. 240, 1989, pp. 613–22.

Berinstain, V. — 'L'Histoire du Ch ale Cachemire Travers les Miniatures de l'Inde', in *Quelques Aspects du Châle Cachemire*, A.E.D.T.A., Paris, 1987, mimeo, pp. 4–22.

Bhave, V.K. — *Peśavekālīna Mahārāṣṭra*, Poona: 1935, rpt. New Delhi: 1976.

Bier, C., ed. — *Woven from the Soul, Spun from the Heart*, Washington: 1987.

Birdwood, G. — 'The Collections of Indian Art in Marlborough House and Sandringham Hall', in *The Journal of Indian Art*, IV, 1892.

— *The Industrial Arts of India*, London: 1880.

Birrell, V. — *The Textile Arts*, New York: 1959.

Biswas, A. — *Indian Costumes*, New Delhi: 1985.

Blancard, P. — *Manuel du Commerce des Indes Orientales et de la Chine*, Paris: 1806.

Blennerhassett, A. — 'The Cotton Fabrics of the Central Provinces', in *The Journal of Indian Art and Industry*, XVII, no. 132, 1916.

Blinks, A. — 'An Unusual Shedding Device from Thailand',

in I. Emery and P. Fiske, eds., 1977, pp. 76–7.

Blochmann, H. The *Āin-i-Akbarī*, 1871; rpt. Delhi: 1965.

Blust, R. 'Austronesian Culture History: Some Linguistic Inferences and Their Relations to the Archaeological Record', *World Archaeology*, vol. I, no. 8, 1976, pp. 19–43.

Bolland, R. 'An Old Pakistan Loom', *Tropical Man*, I, 1968, pp. 109–15.

Bongard-Levin, G. and Vigasin, A. *Image of India, the Study of Ancient Indian Civilization in the U.S.S.R.*, Moscow: 1984.

Boulnois, L. *La Route de la Soie*, 2nd edn., Paris: 1986; 1st edn., 1963.

Bouvier, H. 'La production de soie par les vers, evolution de la soie dans le monde au cours des derniers siecles et situation actuelle', in *Symposium 1984*, 1984, 20–5.

Brennig, J.J. 'Textile Trade of 17th Century Northern Coromandel', Ph.D. thesis, University of Wisconsin: 1975.

Briggs, John *History of the Rise of the Mahomedan Power in India Till the Year AD 1612*, IV, Calcutta: 1910.

Broudy, E. *The Book of Looms*, London: 1979.

Broughton, T.D. *Letters Written in a Mahratta Camp During the Year 1809*, London: 1813.

Brown, Rachel *The Weaving, Spinning & Dyeing Book*, London: 1979.

Brown, R. *Statistical Account of the Native State of Manipur*, Calcutta: 1874.

Bühler, A. and Fischer, E. *Clamp Resist Dyeing of Fabric*, Ahmedabad: 1977.

— *The Patola of Gujarat*, I, Basle: 1979.

Bühler, G. *The Sacred Laws of the Āryas*, Part II, Oxford: 1882.

Burnham, Dorothy K. *Textile Terminology Warp & Weft*, London, 1980.

Carroll, D.L. *Looms and Textiles of the Copts: First Millennium Textiles in the Carl Austin Rietz Collection of the Californian Academy of Sciences*, Seattle: 1988.

Chakrabarti, D.K. *Ancient Bangladesh*, New Delhi: 1992.

Chapekar, L.N. *Pésvāīcyā Sāvalīnta*, Pune: 1937.

Chatterjee, S.K. *Bolts and Aryans in their Indo-European Background*, Simla: 1968.

— *The Place of Assam in the History and Civilization of India*, Guwahati: 1970.

— *Kirāta-Jana-Kṛti: the Indo-Mongoloids, Their Contribution to the History and Culture of India*, Calcutta: 1974.

Chattopadhyaya, Kamaladevi *Indian Carpets and Floor Coverings*, All India Handicraft Board, 1984.

Chaudhury, S.B. *Ethnic Settlements in Ancient India*, Calcutta: 1955.

Chaudhury, S.N. *Mūgā Silk Industry*, Guwahati: 1981.

Chisti, R.K. and Sanyal, A. *Madhya Pradesh, Saris of India*, New Delhi: 1989.

Choudhury, P.C. *The History of Civilization of the People of Assam to the Twelfth Century* AD, Guwahati: 1966.

Chowdhury, K.A. *Ancient Agriculture and Forestry in North India*, Bombay: 1977.

Collingwood, P. *The Techniques of Rug Weaving*, London: 1968.

— *The Techniques of Tablet Weaving*, London: 1982

— *Textile and Weaving Structures, A Source Book for Makers and Designers*, London: 1987.

Conway, S. *Thai Textiles*, London: 1992.

Cowell, E.B., ed. *Divyāvadāna*, Cambridge: 1951.

Crockett, C. 'Card Weaving', in Emery and P. Fiske, eds., 1977, pp. 27–9.

Crooke, W., ed. *Travels in India by Jean-Baptiste Tavernier*, rpt. Delhi: 1977.

Cunningham, A. *The Stupa of Bharhut*, London: 1879.

— *Ladak, Physical, Statistical and Historical, with Notices of the Surrounding Countries*, New Delhi: 1970.

— Archaeological Survey of India, *Reports*, vol. II: *Four Reports made during the years 1862, 1863, 1864, 1865*; rpt. Delhi: 1972.

Dalal, C.D., ed. '*Pṛthvīcandra caritra* by Mānikyasundarasūrī', in Prācīna Gurjara Kāvya Saṅgraha, Gaekwad Oriental Series, XIII, Baroda: 1924.

Das, S.C. *A Tibetan English Dictionary*, Delhi, Patna, Banaras: 1970.

Das Gupta, R. 'Medieval Assamese Textiles as Depicted in Paintings', in *Art and Life*, Bulletin no.19, Art Study Centre, Banaras: 1969–70, pp. 8-22.

Das Gupta, Sanchita *Silk Handloom Weaving in Andhra Pradesh, Technology and Design Appreciation*, Foundation to Aid Industrial Recovery, New Delhi: 1983 (no pagination).

Das Gupta, T.C. *Aspects of Bengali Society*, Calcutta: 1935.

Dastur, J.F. *Useful Plants of India and Pakistan*, Bombay: 1977.

Datta, A. *Neolithic Culture in West Bengal*, Delhi: 1992.

Datta, C.L. 'Significance of the Shawl Trade in Western Himalayan Politics', in *Bengal Past and Present*, 1970.

Dave, I.R. *The Saurashtras in South India*, Rajkot: 1976.

Dave, K.B. 'Some Light on the Salvis of Patan', in *Maha Gujarat* (a Gujarati Weekly), 14 May 1964a.

— 'From where did the Salvis Hail?', in *Maha Gujarat* (a Gujarati Weekly), 23 July 1964b.

de Bone, M.G. '*Patolu* and Its Techniques', in *Textile Museum Journal*, vol. IV, no. 3, 1976, pp. 49–62.

de Jonghe, D. and Tavernier, H. 'Un métier a tisser primitif d'Extrême Orient', in *B.L.C.I.E.T.A.*, vol. 49, no. 1, 1979, pp. 39–50.

— 'Le métier "Sonket" de la côte est de Malaisie', in *B.L.C.I.E.T.A.*, 53, 1, 1981, 70–7.

de Jonghe, D. 'Quatre types de damas sergé 3 lie I du 1er siècle au 14ème siècle apres Jésus Christ', in *Rapport No. 2*, A.E.D.T.A., Paris, 1983, 48–62.

— 'Resumé de l'intervention de Monsieur Daniel de Jonghe, Les Moyens de façonage et leurs caracteristiques', in *Symposium 1985* (3, 4, 5 Juillet), *Le métier à tisser aux baguettes: Influences orientales dans les costumes polonais et ongrois*, A.E.D.T.A., mimeo, Paris: 1985.

Deglurkar, G.B. 'Cultural History of Marathwada', Ph.D. thesis, Faculty of Mental, Moral, Social Sciences, University of Poona: 1971.

Dermigny, L. *Cargaisons Indiènnes, Solier et cie 1781–1793*, II, Paris: 1959.

Desai, S.V. *Social Life in Maharashtra under the Peshwas*, Bombay: 1980.

Dewar, F. *Monograph on the Silk Fabrics of the Central Provinces*, Nagpur: 1901.

— 'Silk Fabrics of the Central Provinces', *Journal of Indian Art*, 1904.

Dey, M.N. 'Instructions for Rearing Mulberry Silk Worms', in *Bulletin No.39*, Agricultural Research Institute, Pusa, New Delhi: 1914.

Dhamija, Jasleen *Catalogue for Exhibition of Royal Iranian and Kashmir Termeh from the Collection of Rahim and George Anavian of Khaneh Aftab*, mimeo, Tehran: 1976.

Dixon, M. *The Wool Book*, London, New York, Sydney, Toronto: 1979.

Dongerkery, K.S. *The Indian Sari*, New Delhi: All India Handicrafts Board: 1909.

Drew, F. *The Northern Barrier of India*, A Popular Account, rpt. Jammu: 1971.

Dubey, K.C. and Jain, H.C. Census of India, 1961, VIII, Madhya Pradesh, Part VII A, Handicraft Survey Monographs, no. 2, in *Handloom Sari Industry of Maheshwar*, Delhi: 1965.

Dun, E.W. *Gazetteer of Manipur*, Calcutta: 1886; rpt. New Delhi: 1975.

Duncan, W.A.M. *Monograph on Dyes and Dying in Assam*, Shillong: 1896.

Eberhard, W. *A Dictionary of Chinese Symbols*, London, New York: 1986.

Edwardes, S.M. 'Silk Fabrics of the Bombay Presidency', in *Journal of Indian Art*, X, 1904.

— A Monograph on the Silk Fabrics of Bombay Presidency. *Art and Industry Through the Ages, Monograph Series on Bombay*, New Delhi: 1976.

Eggeling, J. *The Śatapatha-Brāhamaṇa*, III, 1894; rpt. Delhi, Banaras, Patna: 1963.

Elwin, V. *The Art of the North-East Frontier of India*, Shillong: 1959.

Emery, I. and Fiske, P., eds. *Looms and their Products*, Round Table on Museum Textiles, Proceedings, Washington: 18-20 April 1977.

Emery, I. *The Primary Structure of Fabrics: An Illustrated Classification*, Washington D.C.: 1980.

Endrei, G.W. L' apparition en Europe du metier à marches, *B.L.C.I.E.T.A.*, 8, 22–27 July 1958.

Fabri, C. *A History of Indian Dress*, Calcutta: 1968.

Fairbank, J.K. Reischauer, E.O. and Craig, A.M. *East Asia, Tradition and Transformation*, Boston, Tokyo: 1976.

Fleet, J.F. *Corpus Inscriptionum Indicarum*, III, Banaras: 1970.

— 'Sanskrit and old Kanarese inscriptions, no. 162, Mandasor inscription of Kumaragupta and Bandhuvarman', *Indian Antiquary*, XV (no date), pp. 195–200.

— *Tagara*; Ter, *Journal of the Royal Asiatic Society*, 1901.

Fletcher, T.B. *Some South Indian Insects*, Madras: 1914.

Forbes Royle, J. *The Fibrous Plants of India*, London: 1855.

Forbes Watson, J. 'The Textile Manufactures and Costumes of the People of India', *Edinburgh Review*, CXXXVI, 1867.

— *Collection of Specimens and Illustrations of the Textile Manufactures of India*, London: 1873.

Forster, G. *Journey from Bengal to England through the Northern Part of India, Kashmire, Afghanistan and Persia into Russia by the Caspian Sea*, II, London: 1798.

Foster, W., ed. *A New Account of the East Indies* by Alexander Hamilton, I, London: 1930.

Fox, R.B. 'The Prehistory of the Philippines', *Hemisphere*, October 1968, pp. 10–16.

Fraser-Lu, S. *Handwoven Textiles of South-East Asia*, Singapore: 1988.

Fussman, G. 'L'Indo-Grec Ménandre ou Paul Demièville Revisité', in *Journal Asiatique*, 1993, pp. 61–138.

Gaborieau, M. *Recit d'un Voyageur Mussulman au Tibet*, Paris: 1973.

Gadgil, D.R. *Poona, A Socio-Economic Survey*, Part I, Poona: 1945.

— *Poona, A Socio-Economic Survey*, Part II, Poona: 1952.

Ganapati Sastri, T. *The Arthaśāstra of Kauṭalya*, Trivandrum Sanskrit Series, no. LXXXIII, III, Trivandrum: 1925.

Ganhar, G.N. *The Wildlife of Ladakh*, Srinagar: no date.

Ganju, M. *Textile Industry in Kashmir*, Delhi: no date.

Garde, M.B. *A Guide to Chanderī*, Gwalior: 1948.

Geoghegan, J. *Some Account of Silk in India*, Calcutta: 1872.

Gervers, V., ed. *Studies in Textile History: In Memory of Harold B. Burnham*, Toronto: 1977.

Ghurye, G.S. *Indian Costume*, Bombay: 1966.

Giles, H.A. *A Glossary of Reference on Subjects Connected with the Far East*, 1900; rpt., 3rd edn., London, The Hague: 1978.

Gittinger, M. 'Introduction to the Body-Tension Looms and Simple Frame Looms of Southeast Asia', in I. Emery, P. Fiske, 1977, pp. 54–68.

Gluck, J., Gluck S.H. and C.J., Penton *A Survey of Persian Handicraft*, Tehran, New York, London, Ashiya: 1977.

Godlewski, J. *The Kashmir Shawl*, Yale University Art Gallery, 12 February–6 April, Catalogue of the Exhibition, New Haven, CT, 1975.

Goetz, H. 'Some Early Indo Muslim Textiles', in *Oriental Art*, vol. I, no. 1, 1955, pp. 22–7.

Gogoi, L. *The Tai Khamtis*, Chowkham, Lohit District: 1971.

— 'The Textile Industry of Early Medieval India, circa 700–1200 AD', in *Journal of the Asiatic Society*, nos. 39–40, Bombay: 1964–5.

Granger-Taylor, H. 'Two silk textiles from Rome and some thoughts on Roman silk weaving', in *B.L.C.I.E.T.A.*, 65, 1987, pp. 13–30.

Grant, C. *The Gazetteer of the Central Provinces of India*, Nagpur: 1870.

Groenveldt, W.P. 'Notes on the Malay Archipelago and Malacca compiled from Chinese Sources', in *Straits Branch of the Royal Asiatic Society*, 1887.

Griffiths, J. *The Paintings of the Buddhist Cave Temples of Ajanta, Khandesh, India*, vol. I, London: 1896.

— *The Paintings of the Buddhist Cave Temples of Ajanta, Khandesh, India*, vol. II, London: 1897.

Grzimek, H.C.B. *Grzimek's Animal Life Encyclopaedia*, 13, New York: 1972.

Guelton, M.H. 'Le duvet de chèvre dans les ch ales cachemires', in *Quelques Aspects du Châle Cachemire*, A.E.D.T.A., Paris, mimeo, 1987, pp. 23–37.

Gulati, A.N. *The Patolu of Gujarat*, Bombay: 1951.

— 'A note on the early history of Silk', in J. Clutton-Brock, V. Mittre, A.N. Gulati, *Technical Reports on Archaeological Remains*, Poona: 1961.

Gupta, A. *Minorities on India's West Coast: History and Society*, Delhi: 1991.

Gupte, B.A. 'Thana Silks', in *Journal of Indian Art*, vol. 1, no. 5, 1886.

Guy, J. 'Trade Ceramics in Southeast Asia: Popular or Princely Patronage', in J. Masselos, ed., *Popular Art in Asia: The People as Patrons*, Visual Arts, Working Papers, no. 1, The University of Sydney Centre for Asian Studies (no date), pp. 22–33.

Haldar, H. and Roy, P.C., eds., trans. *The Mahābhārata*, 2nd edn. Calcutta (no date).

Hall, K.R. *Trade and Statecraft in the Age of the Colas*, New Delhi: 1980.

Hall, Rosalind *Egyptian Textiles*, Aylesbury: 1986.

Harcourt, A.F.P. *The Himalayan Districts of Kooloo, Lahoul and Spiti*, rpt. Delhi: 1950.

Hardiman, J.P. *Silk in Burma*, Rangoon: 1901.

— 'Silk in Burma', *The Journal of Indian Art*, X, 1904, nos. 81–8.

Harnetty, P. 'Deindustrialisation Revisited: The Handloom Weavers of the Central Provinces of India, circa 1800–1947', in *Modern Asian Studies 25*, 3, 1991, pp. 455–510.

Hasan, M. *Kashmir under the Sultans*, Calcutta: 1959.

Hazlitt, W. *Travels in Tartary Tibet and China 1844–48*, II, trans., Huc and Gabet, London: 1928.

Hecht, A. *The Art of the Loom*, London: 1989.

Heyd, W. *Histoire du commerce du Levant au moyen age*, II, Leipzig: 1886.

Hodivala, S.H. *Studies in Parsi History*, Bombay: 1920.

Hodivala, S.K. *Parsis of Ancient India*, Bombay: 1920.

Hoffman, M. 'Looms and their Products, An Introduction', in I. Emery, P. Fiske, eds., 1977.

Holder, E. *Monograph on Dyes and Dyeing in the Madras Presidency*, Madras: 1896.

Holmes, J. *Manuscript Notes on Weaving for First Year, 1914.* Further details not cited, available in Knowledge Management and Information Centre, NID, Ahmedabad: 1914.

Honacki, J. Kinman, K.E. and Koeppel, J.W. *Mammal, Species of the World: A Taxonomic and Geographic Reference*, Lawrence, Kansas: 1982.

Hooker, J.D. *Himalayan Journals*, II, London: 1854.

Hornell, J. 'The Origins and Ethnolgical Significance of Indian Boat Designs', in *Memoirs of the Asiatic Society of Bengal*, VII (1918–23), Calcutta: 1930.

Hugon, T. 'Remarks on the Silk Worms of Assam', *Journal of the Asiatic Society of Bengal*, VI, 1837.

Huzayyin, S.A. *Arabia and the Far East; Their Commercial and Cultural Relations in Graeco-Roman and Irano-Arabian Times*, Cairo: 1942.

Hyde, N. 'The Queen of Textiles', in *National Geographic*, vol. 165, no. 1, 1984.

Irwin, J. *The Kashmir Shawl*, London: 1973.

Jacobi, H., trans. *Gaina Sūtras*, Part II, Oxford: 1895.

Jacobs, J. *Hill Peoples of Northeast India: The Nagas, Society Culture and the Colonial Encounter*, New York: 1990.

Jain, J.C. *Life in Ancient India as Depicted in Jain Canons*, Bombay: 1947.

Janse, O.R.T. 'Some Notes on the Sa-huynh Complex', in *Asian Perspectives*, III, 2, 1961, pp. 109–11.

Jayakar, P. 'Naksha Bandha of Banaras', in *Journal of Indian Textile History*, 7, 1967, pp. 21–44.

Jenkins, R. *Report on the territories of the Raja of Nagpur submitted to the Supreme Government of India*, 1827; rpt. Nagpur: 1923.

Jera-Bezard, M. ed. *L'Asie Centrale, Histoire et Civilization*, Paris: 1977.

Ji Xianlin 'Zhongguo cansi shuru Yindu wentide chubu yanjui/A tentative study of the problem of Chinese silk imported into India', in *Zhong-Yin wenhua guanxishi lunwenji/A collection of articles on the history of Sino-Indian Cultural Relations*, Beijing. 1982.

Jolly, M.S. *A Decade of Tasar Research*, Ranchi: 1974.

Jolly, M.S., Chowdhury, S.N. and Sen, S.K. *Non-mulberry Sericulture in India*, Bombay: 1975.

Jolly, M.S. 'Distribution and Differentiation in Antheraea Species (*Saturniidae: Lepidoptera*)', paper presented at *XVIth International Congress of Entomology*, Kyoto, 3–9 August 1980, pp. 1–20.

— *Tasar Culture for Tribal Uplift*, Central Silk Board, Bombay: No date.

Julien, S. 'Sur les pays et les peuples étrangers tirées des géographies et les annales Chinoises, Thien Tchou, l'Inde', in *Journal Asiatique*, 1847.

Kane, P.V. *History of the Dharmaśāstra*, II, Part I, Poona: 1941.

— *History of the Dharmaśāstra*, IV, Poona: 1953.

Kangle, R.P. *The Kauṭilya Arthaśāstra*, I, Bombay: 1960.

— *The Kauṭilya Arthaśāstra*, Part II, 2nd edn., Bombay: 1972.

Kaplan, A. *Tzitzith, A Thread of Light*, New York: 1984.

Khachikian, Levon 'Le registre d'un marchand arménien en Perse, en Inde, au Tibet (1682–1693)', in *Annales Économies Sociétés, Civilisations*, 2, 1967.

Khan, Bahadur C. Khushi Mohammad, *Kashmir, Part I, Report*, Census of India, XXII, Lahore: 1923.

Khazanchi, T.N. 'A Chronological Chart of the History of Kashmir', in *Marg*, VIII, 2, 1955.

Kipling, J.L. 'The Industries of the Punjab', in *The Journal of Indian Art*, II, no. 23, 1887–8.

Klaproth, M. 'Sur les noms de la Chine', *Journal Asiatique*, 10, 1827.

Konow, Sten *Rāja Cekhara's Karpūra Mañjarī*, 1901; rpt. Delhi, 1963.

Kordosis, M. 'China and the Greek World, I Hellenistic - Roman - Early Byzantine Period (2nd century BC–6th century AD)', in *Historicogeographica Meletemata*, no. 2, Thessalonica: 1992.

Koul, Pandit Anand *Geography of the Jammu and Kashmir State*, New Delhi, Rohtak, Jammu (no date).

Krishna, Rai Anand and Vijay Krishna *Banaras Brocades*, New Delhi: 1966.

Krishna Sastri, H. ed., trans. *South Indian Inscriptions*, III, Part III, Madras, Madras Museum Plates of Uttam Chola, 1920.

Krishnaswamy Aiyangar, S. *Ancient India and South Indian History and Culture*, I, Poona: 1941.

Kulkarni, A.R. *Maharashtra in the Age of Shivaji*, Poona: 1969.

Kwang-chih Chang Grace, G.W. and Solheim, W.G. 'Movement of the Malayo-Grace, Polynesians: 1500 BC to AD 500', in *Current Anthropology*, 5, 5, 1964, pp. 359–406.

Lamb, V., Lamb, A. 'The Classification and Distribution of Horizontal and Treadle Looms in Sub-Saharan Africa', in *Textile History*, 11, 1980, pp. 22–62.

Lawrence, W.R. *The Valley of Kashmir*, Srinagar: 1895; rpt. Oxford: 1967.

Lévi, S. 'Ptolemée, le Niddesa et la Brhatkatha', *Etudes Asiatiques publiées a i'occasion du vingt cinquième anniversaire de l'École Française d'Extrême Orient*, II, Paris: 1925.

Ling Roth H. *Studies in Primitive Looms*, 3rd edn. Bankfield Museum, Halifax: 1950.

Liu, G.K.C. 'The Silkworm and Chinese Culture', in *Osiris*, 10, 1952, pp. 129–93.

Logan, W. *Malabar*, I, Madras: 1951.

Luard, C.E. *Gwalior State Gazetteer*, I, Calcutta: 1908.

MacDonnell, A.A. and Keith, A.B. *Vedic Index of Names and Subjects*, 1, II, Delhi, Banaras, Patna: 1958.

Mahajan, T.T. 'Industry, Trade and Commerce in Western Maharashtra during the Peshwa Period', Ph.D. thesis, 1980, University of Poona.

Majmudar, M.R. *Gujarat and Its Art Heritage*, Bombay: 1968.

Majumdar, G.P. 'Dress and Other Personal Requisites in Ancient India', in *Indian Culture*, October: 1934.

Majumdar, R.C. 'Physical Features of Ancient Bengal', in B.C. Law, ed., *Ācarya-Puspānjali Volume*, Calcutta: 1940.

Mamgain, M.D. *Kinnaur*, Simla: 1971.

Marawade, A. *Paithanī of Paithana* (Marathi), Aurangabad: 1979.

Markham, C.R., ed. *Narratives of the Mission of George Bogle to Tibet and the Journey of Thomas Manning to Lhasa*, 1st edn. 1876; rpt. New Delhi, 1971.

Mate, M.S. 'Miniature Painting', in Saryu Doshi, ed., *Shivaji and Facets of Maratha Culture*, Bombay: 1982.

Mathew, G. 'The dating and Significance of the Periplus of the Erythraean Sea', in H.N. Chittick, R.I. Rotberg, eds., *East Africa and the Orient*, London, 1975, pp. 149–63.

Maxwell-Lefroy, H. and Howlatt, F.M. *Indian Insect Life*, Part II, Calcutta, Simla: 1909.

Maxwell-Lefroy, H. Ansorge, E.C. *Report on an Enquiry into the Silk Industry in India*, and III, Calcutta: 1917.

McCrindle, J.W. *The Commerce and Navigation of the Erythraean Sea*, Amsterdam, rpt., 1879, 1882, Calcutta, London.

Mehta, R.N. 'Clothes and Textiles as Described in the Mamerun by Premanand', in *Budhhiprakash*, 98, 5, May 1951, pp. 146–52; 98, 6 June, 1951, pp. 174–7.

— 'Bandhas of Orissa', *Journal of Indian Textile History*, VI, 1961, pp. 62–74.

Mehta, R.N. and Chaudhuri, S.N. *Excavation at Devni Mori: A Report of the Excavation Conducted from 1960 to 1963*, Baroda: 1966.

Menon, Padmanabha, Menon, K.P. and Krishna, T.K.D., eds. *History of Kerala*, 2, New Delhi: 1983.

Merrit, Ruhlen, ed. *A Guide to the World's Languages*, I, Classification, Stanford, California: 1987.

Mikosch, E. 'The Scent of Flowers, Kashmir Shawls in the Collection of the Textile Museum', in *The Textile Museum Journal*, 1985a, pp. 7–22.

— 'Catalogue of Kashmir Shawls in the Textile Museum', in *The Textile Museum Journal*, 1985b, pp. 23–54.

Mills, J.P. *The Lhota Nagas*, London: 1922.

Minakshi, C. *Administration and Social Life under the Pallavas*, Madras: 1977.

Misra, Pandit Sri B. and Vaisya, Rupalalji, eds. *Bhāvaprakāśa of Sri Bhāvamiśra*, Part I, Kashi Sanskrit Series, no. 130, Banaras (no date).

Mitra, D. *Ajanta*, 8th edn., New Delhi: 1980.

Mohanty, B.C. *Brocaded Fabrics of India*, I, II, Ahmedabad: 1984.

Mohanty, B.C. Chandramouli, K.V. and Naik, H.D. *Natural Dyeing Process in India*, Ahmedabad: 1987.

Monier-Williams, M. *Sanskrit-English Dictionary*, 1899; rpt. Delhi: 1976.

Monnas L. and Granger-Taylor, H. *Ancient and Medieval Textiles, Studies in Honour of Donald King*, London: 1989.

Mookerji, Baboo G. 'Silk Industries of Murshidabad', in *Journal of Indian Art*, V, nos. 38–45, 1895.

Moorcroft, W. 'A Journey to Lake Manasarovara in Un-des, a Province of Little Tibet', in *Asiatick Researches*, XII, Calcutta: 1816.

Moreland, W.H. and Geyl, P., trans. *Jahangir's India, the Remonstrantie of and Francisco Pelsaert*, Cambridge, 1925; rpt. New Delhi, 1972.

Morrison, B.M. *Political Centres and Cultural Regions in Early Bengal*, Tucson: 1970.

Morwanchikar, R.S. 'Paithan Through the Ages', Ph.D. thesis, Marathwada University, Aurangabad: 1978.

— 'Textiles', in Saryu Doshi, ed., *Maharashtra*, Bombay, 1985, pp. 155–62.

Moti Chandra *Geographical and Economic Studies in the Mahābhārata: Upāyana Parva*, Lucknow: 1945.

— *Costumes, Textiles, Cosmetics and Coiffure in Ancient and Medieval India*, 1st edn. 1963, rpt. 1973.

Moti Chandra and Agrawala, V.S. 'A Note on Some Cultural References', in *Śrivara Paṇḍita's Rājataraṅgiṇī, Prince of Wales Bulletin*, no. 7, 1959–62.

Mudiraj, Krishna Swamy R. *Pictorial Hyderabad*, II, Hyderabad: 1934.

Mukarji, N.G. *A Monograph on the Silk Fabrics of Bengal*, Calcutta: 1903.

Mukherjee, B.N. *External Trade of Early North-Eastern India*, New Delhi: 1992.

Murphy, V. and Crill, R. *Tie Dyed Textiles of India, Tradition and Trade*, Ahmedabad: 1991.

Muthesius, A. 'From Seed to Samit: Aspects of Byzantine Silk Production', in L. Monnas, H. Granger-Taylor, 1989, pp. 135–49.

Myers, D.K. 'Costume and Ceremonial Textiles of Bhutan', in *The Textile Museum Journal*, 1987, pp. 25–53.

Nadkarni, D. *Balgandharva and the Marathi Theatre*, Bombay: 1988.

Nakhate, G.G. 'Sericulture in Karnataka 800 Years Ago', in *Reshmi Krishi*, trans. S. Shadakasheriah, Kannada, Department of Sericulture, Bangalore: 1981.

Nanavaty, M.M. *Silk, from Grub to Glamour*, Bombay: 1965.

Narayanan, K. *Camel Belts*, Ahmedabad: 1992.

Nath, R. *The Background of Assamese Culture*, Shillong: 1948; 2nd edn. Guwahati: 1978.

Needham, J. 'Abstract of Material Presented to the International Maritime History Commission at Bierut', in M. Mollat, *Sociétés et Compagnies de Commerce en Orient et dans l'Océan Indien*, Paris: 1976.

Neogi, M., ed. K. Barua's, *Studies in the Early History of Assam*, Guwahati: 1973.

Nilkanta Sastri, K.A. *The Colas*, Madras: 1955.

— *Foreign Notices of South India from Megesthenes to Ma Huan*, Madras: 1972.

Nissim, J. *A Monograph on Wire and Tinsel in Bombay Presidency*, Bombay: 1909.

Om Prakash *The Dutch East India Company and the Economy of Bengal, 1630–1730*, Princeton: 1985.

Padhye, K.A. 'A Few Notes as Regards the Custom of Wearing Garments Peculiar to the Deccani females', *The Journal of the Anthropological Society of Bombay*, vol. XV, no. 5, 1933–4.

Pande, M.C. 'Kashmir Shawls in the State Museum, Lucknow', in *Bulletin of Museums and Archaeology in U.P.*, vol. 7, June 1971, pp. 29–39.

Pandey, B.P. *Banaras Brocade, Structure and Functioning*, Banaras: 1981.

Parey, M.I. 'The Silk Industry in Kashmir, 1858–1947', M.A. thesis, University of Kashmir, Srinagar, 1977–8.

Pariset, E. *Histoire de la Soie*, Lyon: 1843.

Picard-Schmitter, M.Th. 'Deux métiers horizontaux utilisés en Egypte vers 2000 avant J.C., in *B.L.C.I.E.T.A.*, 26, 1967, 13–39.

Picton John, and John Mack *African Textiles*, 2nd edn., London: 1989.

Pirazzoli T'ser-Stevens, M. 'The Bronze Drums of Shizhai - Shan, Their Social and Ritual Significance', in R.B. Smith and W. Watson, *Early South-East Asia*, New York, Kuala Lumpur: 1979.

Pope, A., ed. *A Survey of Persian Art*, vol. III, Text, London, New York: 1939.

Prinet, M., 'Le damas de lin: origine historique', *Rapport No. 2*, A.E.D.T.A., Paris, 1983, 44–6.

Prinsep, W. 'Memorandum upon the Specimens of Silk and Silkworm from Assam', in *Journal of the Asiatic Society of Bengal*, VI, 1837.

Priya Devi *The Master Weavers*, Bombay: 1982.

Punyavijayaji, Muni Sri, ed. *Aṅgavijya, Science and Divination through Physical Signs and Symbols*, Banaras: 1957.

Rabbani, G.M. *Ancient Kashmir*, Srinagar: 1981.

Ram, A. and Raina, H.N. Census of India, 1931, XXIV, *Jammu and Kashmir State*, Part 1, Report, Jammu: 1933.

Ramaswamy, G. 'Traditional Adai Techniques of South Indian Silk Weaving', Weavers Service Centre, Madras, unpublished (no date).

Ramaswamy, V. 'Notes on the Textile Technology in Medieval India with Special Reference to the South', *The Indian Economic and Social History Review*, XVII, 2, 1980, 227–41.

Raschke, M.G. 'New Studies in Roman Commerce with the East', in H. Temporini, and W. Haase, eds., *Aufstieg und Nierdergang der Romischer Welt*, IX, Part II, Berlin, 1978, pp. 604–1076.

Rau, W. *Weben und Flechten in Vedischen Indien*, Akademie der Wissenschaften und der Literatur, Abhandlungen der Geistes-und Sozialwissenschaftlichen Klasse, Jahrgang 1970–NR 11, Akademie der Wissenschaften und der Literatur, Mainz, Wiesbaden: 1970.

Ray, H.P. 'Seafaring in the Bay of Bengal in the Early Centuries AD', in *Studies in History*, 6, 1, n.s., 1990, pp. 1–14.

Reddy, R., Soma 'History of Religious Institutions in Andhra Desa from 1300 to 1600 AD', Ph.D thesis, Osmania University, Hyderabad: 1980.

Reininger, W., 'The Florentine Textile Industry', *Ciba Review*, 27, 1939.

Rhys Davids, T.W. *The Questions of King Milinda*, Part II, rpt. Delhi, Banaras, Patna: 1965.

Riboud, K. 'Techniques and Problems Encountered in Certain Han and T'ang Specimens', in I. Emery and P. Fiske, eds., 1977, 153–69.

Riley, C.V. *The Silkworm, Being a Brief Manual of Instructions for the Production of Silk*, Washington: 1882.

Rizvi, J. 'Peasant Traders of Ladakh: A Study in Oral History', *India International Quarterly*, 12, 1, March 1985, 13–27.

— 'Woven Textiles', in J. Jaitley, ed., *Crafts of Jammu, Kashmir and Ladakh*, Ahmedabad: 1990, pp. 36–57.

Rondot, N. 'Notice sur le métier a tisser le jong et le ho', in *Journal Asiatique*, IX, 1847, pp. 332–50.

Rougeot, J. 'Evolution de la Toison des Moutons en Relation avec la Technologie Textile', in *Symposium*, 1984, pp. 42–8.

Saletore, B.A. *Ancient Karnataka*, I, Poona: 1936.

Sandesara, B.J. 'Some Light Thrown on the History of the Salavis, the Patolan Weavers' (Gujarati), in *Buddhiprakāśa*, 100, 2, February 1953.

Sandesara, B.J. and Mehta, R.N. *Varṇaka-Samuccaya*, Part II, Cultural Study and Indices, Vadodara: 1959.

Sankalia, H.D. 'Lata, Its Historical and Cultural Significance', in *Journal of the Gujarat Research Society*, XXII, no. 4/88, October 1960.

Sardesai, G.S., ed. *Peshwa Daftar*, 32, Bombay: 1933.

Sarkar, D.C. 'Sericulture and Silk Industry in India', in *Indian Museum Bulletin*, XII, 1977, pp. 7–75.

Sastri, Pandit H. *Nāma-liṅgānusāsana or Amarkoṣa of Amarasiṃha*, Kashi Sanskrit Series 198, Banaras: 1970.

Scarce, J.M. 'Vesture and Dress: Fashion, Function and Impact', in C. Bier, 1987, pp. 33–56.

— 'The Persian Shawl Industry', in *The Textile Museum Journal*, 27 & 28, 1988–9, pp. 23–39.

Schaller, G.B. *Stones of Silence, Journeys in the Himalaya*, New Delhi: 1980.

— *Mountain Monarchs, Wild Sheep and Goats of the Himalaya*, Chicago, London: 1989.

Schoff, W.H. *The Periplus of the Erythraean Sea*, rpt. Delhi: 1974.

Schuette, M. 'The Geographical Distribution of Tablet Weaving', in *Ciba Review*, 117, November 1956, pp. 9–18.

Sebeok, T.A., ed. *Current Trends in Linguistics*, no. 8, The Hague, Paris: 1971.

Sen, Madhu *Cultural Study of the Niśitha Cūrnī*, Banaras: 1975.

Sergeant, R.B. *Islamic Textiles, Material for a History up to the Mongol Conquest*, Beirut: 1972.

Sharma, P.V., ed. *Śodhala-Nighaṇṭu (Namasaṅgraha and Guṇasaṅgraha) of Vaidyācārya Sadhala*, Baroda: 1978.

Shastri, Narendranath, ed. *Carakasaṁhitā*, I, Lahore: 1929.

Sherwani, H.K. and Joshi, P.M. *History of Medieval Deccan* (1295–1724), II, Hyderabad 1974.

Shirali, A. *Textile and Bamboo Crafts of the Northeastern Region*, Northeastern Council, Shillong: 1983.

Shrigondekar, G.K., ed. *Mānasollāsa*, II, Baroda: 1939.

Singh, B.P. *Life in Ancient Banaras*, Delhi: 1985.

Singh, C. and Ahivasi, D. *Woollen Textiles and Costumes from Bharat Kala Bhavan*, Banaras: 1981.

Singh, J. Banks *Gods and Government*, Stuttgart: 1989.

Singh, K.S. *Anthropological Atlas—Ecology and Cultural Traits, Languages and Linguistic Traits, Demographic and Biological Traits*, People of India, National Series, XI, Delhi, Bombay, Calcutta, Madras: 1993.

Singh, Mian Govind *Himachal Pradesh, History, Culture and Economy*, 3rd edn., Shimla: 1992.

Singh, R. *Banaras—Sacred City of India*, London: 1987.

Sinha, N.C. *Tales the Thankas Tell: An Introduction to Tibetian Scroll Portraits*, Gangtok: 1989.

Sircar, D.C. 'Cosmography and Geography in Early Indian Literature', in *Indian Studies Past and Present*, 7, 1965–6.

— *Studies in the Geography of Ancient and Medieval India*, Delhi: 1971.

Soares, A.X. *Portuguese Vocables in Asiatic Languages, from the Portuguese Original of Monsignor Sabastiao Rodolfo Dalgado*, Baroda; rpt. Madras: 1988.

Solheim, W.G., II 'Further Relationships of the Sa - Huynh - Kalanay Pottery Tradition', in *Asian Perspectives*, VIII, 1965, pp. 19–211.

Soman, G.R.S., ed. *Eknathi Bhagwat*, Satara: 1973.

Soundara Rajan, K.V. *The Art of South India*, Tamil Nadu and Kerala, Delhi: 1978.

— *Glimpses of Indian Culture, Architecture, Art and Religion*, II, Delhi: 1981.

Speiser, N. 'The Techniques and the Methods of Braiding, a Technological Research in Progress', in *B.L.C.I.E.T.A.*, 1977, pp. 42–53.

Srinivasacharyar, L. *Gautam Dharma Sūtra with Maskarī Bhāṣya*, Mysore: 1917.

Stacey, A. *Visiting India*, London: 1986.

Staudigel, O. 'Tablet Weaving and the Technique of the Ramases Girdle', in *B.L.C.I.E.T.A*, nos. 41–2, pp. 71–98.

Steingass, F. *A Comprehensive Persian–English Dictionary*, rpt. New Delhi: 1973.

Sufi, G.M.D. *Kashmir, Being a History of Kashmir*, II, rpt. New Delhi: 1974.

Sukul, K.N. *Varanasi Down the Ages*, Patna: 1974.

Swinhoe, C. 'On the Lepidoptera of Bombay and the Deccan', in *Proceedings of the Zoological Society*, London, 1885, pp. 287–307.

Takakasu, J. *A Record of the Buddhist Religion, as Practiced in India and the Malay Archipelago by I Tsing*, Oxford: 1896; rpt. Delhi: 1982.

Talwar K. and Krishna, Kalyan *Indian Pigment Paintings on Cloth*, Ahmedabad: 1979.

Tan Chung *Triton and Dragon*, Delhi: 1986.

Temple, R.C. 'Extracts from the Log of a Voyage along the Coast of India in 1746', in *Indian Antiquary*, XXIX, 1990.

Thaker, J.P., ed. *Laghu-Prabandha-Sangraha*, Baroda: 1970.

Thoothi, N.A. The Vaishnavas of Gujarat, London: 1935.

Thurston, E. *Castes and Tribes of Southern India*, I, rpt, London, New York: 1965.

Tulpule, S.G. *An Old Marathi Reader*, Poona: 1960.

— *Classical Marathi Literature, from the Beginning to 1818*, Wiesbaden: 1979.

Turner, R.L. *A Comparative Dictionary of the Indo-Aryan Languages*, New York, Toronto: 1966.

Varadarajan, L. 'Indian Textile Technology in the Pre-Industrial Period', in *Indica*, 21, 1984, pp. 61–9.

— 'Designs in Cotton: Horizons of the Past and Present', in *India International Quarterly*, 11, 4, 1984a, pp. 69–77.

— 'Kani Paśmīnā of Kashmir', in Varadarajan, L., A. Ray, H. Sanyal and S.C. Ray, eds., *Indian Studies, Essays Presented in Memory of Prof. Niharranjan Ray*, Delhi, 1984b, pp. 231–40.

— 'Silk—Identifications within the Indian Tradition', in *Indica*, 23, 1986, pp. 189–98.

— 'Commodity Structure and Indian Participation in the Trade of the Southern Seas, circa Ninth to Thirteenth Centuries', in Satish Chandra, ed., *The Indian Ocean Explorations in History, Commerce and Politics*, New Delhi, London, 1987a, pp. 90–108.

— 'Arboreal Motifs in Kalamkari', in M.L. Nigam, ed., *Decorative Arts of India*, Hyderabad, 1987b, pp. 141–9.

— 'Silk in Northeastern India: The Indigenous Tradition', in *Modern Asian Studies*, 22, 3, pp. 561–70.

— 'An Ethnographic Approach to Medieval Textile Technology', in *Journal of the Japan-Netherlands Institute*, III, 1991, pp. 207–22.

— 'Kabbala and the Chintz Tradition', in *Mare Liberum*, 9, 1995, pp. 423–31.

Varron, A. 'The Origins and Rise of Silk', in *Ciba Review*, II, July 1938.

Vasu, S.C. and Naraharayya, S.N. *The Sacred Laws of the Aryas as taught in the School of Yajñavalkya and explained by Vijñāneśvara in his well known commentary named Mitākṣara*, III, Allahabad: 1913.

Venkatarama Sastriar, J.S. *A Short History of Saurashtra*, 2nd edn., Madurai: 1971.

Verma, O.P. *The Yadavas and their Times*, Nagpur: 1970.

Vial, G. 'A propos d'une soierie fasçonée dite de Byssus', in *B.L.C.E.T.A.*, 1983, pp. 50–60.

— 'Deux exemplaires de soie non-murier (Tussah) dans les tissus anciens', in *B.L.C.I.E.T.A*, 1985, pp. 50–5.

— 'Le technique du ch ale Indien', *Quelques Aspects du Châle Cachemire, A.E.D.T.A.*, Paris, mimeo, 1987, pp. 38–47.

Vidyarthi, L.P. *Art and Culture of Northeast India*, New Delhi: 1986.

Vigne, G.T. *Travels in Kashmir*, vol. II, London, 1842; rpt. New Delhi: 1981.

Vijay, S.R. 'Bivoltine in the Tropics', in *Sericologia*, 25 (2), 1985.

Vivi Sylwan *Investigations on Silk from Edsen-Col and Lop Nor*, Stockholm: 1949.

Vogler, H. 'The Craft of Dyeing in Ancient Egypt', in *Textile History*, 13, 2, 1982, pp. 159–63.

Vollmer, J.E. 'Archaeological Evidence for Looms from Yunnan', in I. Emery, P. Fiske, 1977, pp. 78–89.

Waddle, L.A. *Buddhism and Lamaism of Tibet*, rpt. New Delhi: 1979.

Wakeman, F., Jr. 'Voyages', in *American Historical Review*, 1933, pp. 1–15.

Wardle, T., *The Wild Silks of India, Principally Tusser*, London: 1880.

— *Handbook of the Collection Illustrative of the Wild Silks of India*, London: 1881.

— *Royal Commission and Government of India, Silk Court*, London: 1886.

— *On the History and Growing Utilization of Tussar Silk*, London: 1891.

— *Kashmir: Its New Silk Industry*, London: 1904.

Watson, A.M. 'The Rise and Spread of Old World Cotton', in V. Gervers, ed., 1977, pp. 355–68.

— *Indian Art at Delhi, 1903*, being the official Catalogue of the Delhi Exhibition, 1902–1903, Calcutta: 1903.

Weir, S. *Spinning and Weaving in Palestine*, London: 1970.

Wheatley, P. *The Golden Khersonese*, Kuala Lampur: 1966.

Wild, J.P. 'Early Finds of Chinese Silk in North-West Europe', in *Rapport No. 2*, A.E.D.T.A., Paris, mimeo, 1983, pp. 32–6.

— 'Roman Silk and the Horizontal Loom in the Maas-Rhine Region', in L. Monnas, H. Granger-Taylor, 1989, pp. 209–21.

— 'The Roman Horizontal Loom', *American Journal of Archaeology*, 91, 1987, pp. 459–71.

Willetts, W. *Chinese Art*, I, London: 1958.

— 'The Maritime Adventures of Grand Eunuch Ho', in *Journal of Southeast Asian Studies*, 5, 2, 1964, pp. 25–42.

Williams, C.A.S. *Outlines of Chinese Symbolism and Art Motives*, Shanghai: 1932.

Winternitz, M. *History of Indian Literature*, III, Delhi: 1985.

Wulff, H.E. *The Traditional Crafts of Persia*, Cambridge, MA: 1966.

Yang Xianyi *Yiyu ou shi/Titbits, Besides Translation*, Beijing: 1983.

Younghusband, F. *Kashmir*, rpt. New Delhi: 1970.

Yusuf Ali, A. *A Monograph on Silk Fabrics*, rpt. Ahmedabad: 1974.

Zhu, Changli 'The Southern Overland Silk Route: Eco-cultural Exchanges between China, India and Burma', in *Indica*, 30, 1993, pp. 23–45.

Acknowledgements

Andhra Pradesh

Gadwal

Institution: Gadwal Handloom Weavers Cooperative Society.

Individuals: Akala Surendar, Pulla Reddy, Sankarrao Ghotre, S. Ramulu.

Dharmavaram

Institutions: Panduranga Silk Handloom Cooperative Production and Sale Society Ltd., Silk Weavers Cooperative Society.

Individuals: D. Obilesu, B.A. Venugopal, G. Kodappa.

Warangal

P. Yadagiri, L. Rajalingam.

Mahbubnagar

Tasar Khadi Silk Weavers Cooperative Society Ltd., The Tasar Khadi Silk Production and Sales Cooperative Society Ltd., Pothreddypally dist.

Kothakota

Kothakota Artistic Saree Weavers Industrial Cooperative Society Ltd., Kothakota.

Hyderabad

Late J.K. Reddy, late N. Ramesan, late M. Balakotaih, late Raja Rameshwara Rao of Wanaparti, Abdul Hai, P.M. Reddy, A.V.V.K. Prasad, M.A. Nayeem, M. Kumudini Devi of Gadwal,.

Narayanpet

Vasudeva Rao Kalyani, V. Rama Aiyer.

Bengal

Kolkata

Institutions: Resham Silpa Samabay Mahasangha, Weavers Service Centre, Sericulture Directorate, Regional Office of Central Silk Board.

Individuals: Late Nihar Ranjan Ray, Ruby Pal Chowdhary, D.C. Sarkar, B.N. Mukherjee, Archana Roy, Ashok Majumdar, K.K. Ganguly, A.K. Basu, Tapan Kumar Mukherjee, Prabhas Sen, A.K. Bhoumick, Santosh Basu, Shiten Chakravarty, Geeta Mallick.

Gujarat

Mehsana

Vithal Jesinghbhai Patel, Nagji Jesingbhai, Village Paraina.

Bhavnagar

Late Khodidas Parmar.

Ahmedabad

Late Z.A. Desai, Gira Sarabhai, Mrinalini Sarabhai, Toofan Rafai.

Vadodara

S.G. Kantawala.

Patan

Chotulal M. Salvi.

Himachal Pradesh

Kullu

The Bhutti Weavers Production cum Sale Cooperative Industrial Society Ltd.

Shimla

Renu Dhar, Deepak Sanan, Sarojini Thakur, S.N. Varma, Vinod Lall, P.S. Kashyap, B.M.S. Rajwar, Bansi Lal Sharma, Sudarshan Vashisht, V.C. Ohri, T.G. Negi, Chhangoo Ram.

Chamba

Institution: Abdul Salam and Sons, Nurpur, district Chamba.

Individuals: Challo Devi, Village Ghangeet, Gulam Ali.

Manali

Khuhi Ram Thakur.

Haryana

Panipat

D.N. Vij.

Karnataka

Bangalore

Institution: Weavers Service Centre, Sericulture Department.

Individuals: Late V. Chandramouli, G.R. Gurumurthy, Kasturibai, Jayachandra, H. Sanjiva Shetty.

Ilkal

Workshop of Phaniraj Konappa Guledgud.

Mysore

Manjeet Jolly, Shastri, S. Shadakasheriah.

Navalgunda

Birappa Kuruba, Joshi.

Kashmir

Srinagar

Institutions: S.P. Museum, School of Design, Tosh Khana.

Individuals: Maqbool, Mushtaq Ahmed, G.N. Naik, M.A. Khan, J.L. Bhan, K. Kaul, M.G. Sapru, M. Shafi Mir, Gulam Mohammad, Iqbal Singh Dewan, P.N. Kachru, Triloki Kaul, Niranjan Nath, Bashir Ahmed.

Maharashtra

Nagpur

Saboo Textiles; Vidarbha Record Office.

Paithan

Bala Sahib Patel, S.G. Patil.

Aurangabad

Late T.V. Pathy, late K.V.S. Sampath Kumar, R.S. Morwanchikar, Thambre, Bhalchandra Nemade.

Bombay

Institutions: Weavers Service Centre; Bhau Daji Lad Museum; Central Silk Board, Maharashtra Small Scale Industries Cooperation Ltd.

Individuals: Late Setu Madhav Rao, late Karl Khandalawala, R.B. Patankar, Ashok Ranade, Gautam Vaghela, A.N. Datta, Raji Bhatavadekar, Kumud Damle, S. Pednekar, R.S. Deboo, K.T. Deshmukh, S. Gorekar, Kalpana Desai, H.S. Rama Rao, B.M. Parker, S.H. Dange, D.G. Godse, Devangana Desai.

Kolhapur

Late G.M. Pawar, V.A. Sangave, Sashikala Pandit, R.L. Tawde, Dilip Joshi, V. Deshpande, Inamdar of Malikwada.

Yeola

P.V. Khandar, G.T. Patil, Bhagwan Balasa Pahwan, Abdul Gani, Abdul Khalliq.

Nagpur

Late Ajay Mitra Shastri, V.V. Mirashi, V.B. Kolte, G.B. Deglurkar, P.M. Muley, Raja Uday Singh, S. Chitnavis, D.M. Soukusare, S.G. Chatti, P.G. Kapale, Champak Rangachari.

Pune

Institutions: Balgandharva Natya Mandir; Bharatiya Natya Samshodhan Mandala; Jadhav Natya Sansar.

Individuals: Late G.H. Khare, late R.N. Dandekar, late H.D. Sankalia, Kamalini Bai Peshwa, Kamalabai Gokhale, Ramabai Patwardhan, Rani of Jamkhandi, V. Deshpande, R.L. Tawde, Sashikala Pandit, M.A. Sangave, Walimbe, A.R. Kulkarni, Kamte, S.G. Chatti, M.S. Mate, Mahasabde, A.M. Ghatage, P.V. Bapat, Yuvarani of Miraj, Rani of Kurundwad, Leelavati Devi Patwardhan of Jamkhandi, Balasahib Purandare, Venkappa Karli, A.M. Ghatage, M. Sankalia, M.S. Mate, Bhikulal Kantilal Shah, Chamanlal, Deepak Shah, Kelkar, Venkappa Karli, Pradip Sinha.

Wai

Late Laxman Shastri Doshi, Professor R.G. Jadhav.

Satara

G.R. Soman, Rajmata Sumitra Devi Raje Bhonsle.

Sangli

Kusum Karmakar, Daji Saheb Karmerkar, Maharani of Sangli.

Aundh

Lalitagouri Pant Pratinidhi, M.D. Kulkarni.

New Delhi

Instituitions: Division of Entomology, Indian Agricultural Research Institute; Department of Botany, University of Delhi; Indira Gandhi National Centre for the Arts, Janpath; Weavers Services Centre, Bharatinagar.

Individuals: Late Kamaladevi Chattopadhyaya, late C. Sivaramamurti, late Pupul Jayakar, N. Ramakrishnan, K.N. Mehrotra, H.Y. Mohan Ram, M.C. Joshi, Sampath Narayanan, Tan Chung, S. Mukhopadhyay, Amarnath Khanna, Lokesh Chandra, Bharatlal, B.B. Bhasin, Krishna Lall, S. Baxi, K. Vatsyayan, Jasleen Dhamija, D.N. Saraf, Pepita Seth, Bunny Page, K.S. Singh.

Rajasthan

Jaipur

Institution: Maharaja Sewai Mansingh II Museum, City Palace.

Individuals: R.C. Agarwal, R. Nath, Chandramani Singh.

Jaisalmer

Amba Ram Sanook, Kastur Ram.

Tamil Nadu

Chennai

Late K.V. Soundara Rajan, Reddappa Naidu, S. Ramani, Venugopal, T.S. Natesan, R. Nagaswamy, Gita Ram, Shanta Guhan.

Salem

Shankar Kumar.

Kanchipuram

N. Veerappan.

Uttar Pradesh

Varanasi

Late J. Supakar, late Kalu Hafiz, late Jafar Ali, Badrudin, Anjan Chakravarti, Kalyan Krishna, Anand Krishna, R.S. Singh.

Bangladesh

Dacca

S. Ghuznavi.

France

Paris

Late Valerie Berinstain, late Alice Thorner, Sophie Desrosiers, Monique Lévi-Strauss, L. Boulnois, Françoise Cousin, R.de Montfalcon, Françoise and Donal Hayes, Marie-Hel ene Guelton.

Mulhouse

Jacqueline Jacqué.

Holland

Leiden

Rens Heringa.

Special Thanks to:

Late R.N. Mehta, late Umakant Shah, late K.T.M. Hegde, Vadodara; Late Acharya Dalsukhbhai Malwania, late H. Bhayani, Vikas Satwalekar, Aditi Ranjan, M.P. Ranjan, Helena Perheentupa, I.S. Mathur, Hansmukh Shah, Sudarshan Khanna, Shilpa Das, Ahmedabad; Late Krishna Riboud, France.

Uzma Mohsin, Visual Vibe; Swati Mitra, Madhumita Rao, Eicher; Ramesh Jain, B.N. Varma, Siddharth Chowdhury, Bansi Ram, Manohar Publishers & Distributors, New Delhi.

Lotika Varadarajan

I take this opportunity to thank all those who have made this book possible. Helena Perheentupa, my guru has been the prime mover. Her gentle, firm and persistent prodding gave me the strength to both involve myself in this project and more importantly, complete it. I am grateful to Giraben Sarabhai for her encouragement and incisive analytical comments. The project was initiated at the National Institute of Design in 1980 when Ashoke Chatterjee was the Executive Director. Vikas Satwalekar, former Executive Director has been a mentor who has been extremely supportive of the excercise and I must thank him for the unwavering trust he has placed in me.

The NID community which I joined as faculty member in 1981 has been the mainstay of this exercise. For his support as teacher, friend and advisor and his helping out in the oddest situations I thank S. Sethuraman. Sethu has been nothing short of a saviour. Aditi Ranjan and M.P. Ranjan have been constant supports. Aditi Ranjan has not only looked at the technical aspects of this work but also helped me believe in my work and myself. Ranjan provided invaluable inputs on the classification and organization of visuals. Suranjana Satwalekar has been both great support as a friend and as Chairman, Design Services and Research (DSR), facilitating administrative matters pertaining to the work. I.S. Mathur earlier Chairman, DSR not only provided the necessary administrative backing, but edited the first round of photographed materials. R.K. Banerjee went through my very first drafts to ensure that I produced a readable text. The late Suresh Sharma undertook the initial groundwork in patiently locating most of the places for our maps. Tridip Suhrud provided valuable advice on the overall structure of the book. S. Bodiwala helped in transferring the initial manuscript from IBM to Macintosh and back to IBM software.

I cannot thank Nilam Iyer Ghosh enough for being available whenever I needed her or Binita Desai for her friendship and active support of the writing. I must also thank Mithoon, Mookesh, Bhanuba, Gangaba, Sarlaben and Rameshbhai for their endless patience while I took off on the many field trips that this project demanded. Need I say that it has been wonderful collaborating with Lotika Varadarajan without whom the book would not have been this particular book? My education in history and research would have remained incomplete but for her.

Late Shankar Singh Rajwadi, called 'Masterji' by all, has pushed and bullied me along with Rajendra Patel and Krishnan Kutty who have all supported the arduous activity that is part of the very long gestation of this book.

For their help in various ways I thank the following people: B.M. Anand, Vijai Singh Katiyar, D. Chakrabarti, M.F. Jadliwala, Anjan Chakravarty, Veena Rajpal, Michael Preston, Sir Peter Collingwood, Subrata Bhowmick, Girija Sharan, Siddhartha Ghosh, Kurma Rao, Errol Pires, Ushma Tolikar, S. Varadarajan, Meera Mehta, Madhurima Patni-Bhowmick, Punam Zutshi, Romanie Jaitly, Nina Sabnani, Jan Baker, Nirmala Khadpekar, Shakirbhai, the late Devki Ahivasi of Banaras, late Supakar and the late Ghanshyam Pandya.

I must also acknowledge the help of Swati Mitra of the Delhi Cell of NID and of N.V.R.K. Murthy, also in the Delhi Cell. Nitin Chitnis of Bombay who photographed Paithani textiles from the collection of Dilip Shah, Poona.

Suresh V. Modia apart from refining some illustrations organized with meticulousness, all visual materials related to the project. Chanda Naidu typed the endless versions of the initial manuscript and Sarita R. Gundeti who typed in many versions of corrections.

I have been very fortunate in having access to the archives and resources of the following institutions: The Calico Museum of Textiles and the Sarabhai Foundation in Ahmedabad, Central Cottage Industries Emporium in Bombay, Maharashtra Small Scale Industries in Bombay and Paithan. Weavers Service Centre in Bombay, Kanchipuram, Madras, Indore, Ahmedabad, Hyderabad and Calcutta, the Raja Kelkar Museum in Poona and the Bharat Kala Bhavan in Banaras.

The various departments and services of NID have been unstinting in their support. The staff of the Computer Studio, the Printing Studio, the Textile & Apparel Design Studio and the Delhi Cell have contributed greatly. But I must thank very specially the staff of the Resource Centre, the Photography Studio, Design Services and Research Department and the Design Services Studio.

My special thanks to: Dr Darlie O. Koshy, Director, National Institute of Design, Development Commissioner (Handlooms) Ministry of Textiles, Government of India, Handloom Artisans of India, Dr Lotika Varadarajan, Sudarshan Khanna, Uzma Mohsin, Shilpa Das, Ramesh Jain, Siddharth Chowdhury, Bansi Ram and Manohar Publishers & Distributors, New Delhi.

Krishna Amin-Patel

Index

D

E